k-SAMPLE ($k \geq 3$)	DESIGNED EXPERIMENTS	ASSOCIATION
One-way analysis of variance and Fisher least significant difference (FSD) method, chap. 11	Completely randomized design, sec. 17.2	Simple linear regression, chaps. 14, 15
(FSD method based on trimmed means, sec. 12.2)	Randomized block design, sec. 17.2	Multiple regression, chap. 16
(FSD method based on ranks, sec. 12.4)	Replicated two-factor design, sec. 17.3	
Chi-square test for homogeneity, sec. 13.3		Chi-square contingency test, sec. 13.3

Introduction to Contemporary Statistical Methods   Second Edition

Introduction to Contemporary Statistical Methods   Second Edition

Lambert H. Koopmans
University of New Mexico

 Duxbury Press   Boston

PWS PUBLISHERS

Prindle, Weber & Schmidt •🌸• Duxbury Press •♠• PWS Engineering •⋀• Breton Publishers •⚙
20 Park Plaza • Boston, Massachusetts 02116

Library of Congress Cataloging-in-Publication Data

Koopmans, Lambert Herman, 1930–
  Introduction to contemporary statistical methods.

  Bibliography: p. 615
  Includes index.
  1. Mathematical statistics.   I. Title.
QA276.K65   1987   519.5   86-25150
ISBN 0-87150-073-6

Printed in the United States of America

87 88 89 90 91 — 10 9 8 7 6 5 4 3 2

Sponsoring Editor: Michael Payne
Production Coordinator: Susan Graham
Production: Greg Hubit Bookworks
Interior Design: Susan Graham
Cover Design: Susan Graham
Cover Photo or Illustration: © Michael Lasuchin
Interior Illustration: Carl Brown
Typesetting: Bi-Comp Inc.
Cover Printing: Phoenix Color Corp.
Printing and Binding: The Maple-Vail Book Manufacturing Group

Preface

This book represents both an enlargement and revision of my text, *An Introduction to Contemporary Statistics*. While the basic approach to exploratory data analysis and statistical inference remains the same as in its predecessor, this edition contains many changes. Probability is given more complete coverage in an attempt to set the presentation more uniformly at the level of statistics courses for students of business and for students of the natural and social sciences. Also, I have added new chapters on multiple regression and the two-way analysis of variance. Most chapters have been extensively rewritten, and the earlier chapters on exploratory data analysis have been "tightened up" by eliminating nonessential topics and, where possible, by using simpler methods.

Robust methods for the measurement variable comparison problem are now collected together into a single chapter (chapter 12), making it possible to treat them in a more unified way. This arrangement also makes the presentation of the classical inference ideas in other chapters more self-contained.

In the brief period since the publication of the first edition, the statistical computing scene has changed considerably. The rapid infusion of microcomputers into the business and scientific worlds has made interactive statistical computing a reality on a wide scale. In an effort to prepare students for the use of these machines in their careers, many colleges and universities are acquiring microcomputers in sufficient quantities to make possible their use in the teaching of statistics. Because of the computational effort required to make some of the exploratory graphs and to carry out the more complex inference procedures presented in this book, the courses for which it may be used are prime candidates for computerization. Programs are now available for doing essentially all the exploratory and inferential analyses presented in the book on a wide variety of computers, including microcomputers. An excellent example of such a program is MINITAB. The discussions of methods are pointed toward the use of computers; however, directions are still given in sufficient detail to carry out (almost) all calculations with a scientific hand calculator.

There is enough material in the text for a two-semester or three-quarter course. However, as is the case at the University of New Mexico, beginning statistics courses are often only one semester or two quarters long. Several options are available for subsetting the material in the text to accommodate such courses. I view a "basic" one-semester course to contain the material in chapters 1 through 10. Chapter 10 now contains all of the t tests commonly used for comparison problems, including the paired difference test.

Topics can be selected from the other chapters to fill out the remaining time. These can be treated to varying depths by making further selections of material within chapters. Chapter 13 covers the statistical inference for the categorical variable problems treated descriptively in chapters 3 and 4, and I usually cover this material at least through the chi-square test for contingency. Chapters 10 and 11 cover inference for the measurement variable comparison problems from chapter 3, and I also tend to include chapter 11 when I teach the course. I seldom find time for more than the descriptive regression methods in chapter 14 in a one-semester course. However, for instructors interested in covering multiple regression and experimental design, it would be possible to branch directly from chapter 10 to chapters 14 through 17. The two-sample t test and the one-way analysis of variance are treated as special cases of multiple regression problems in chapter 16, and the one-way analysis of variance reappears in an experimental design context in chapter 17.

Robust alternative methods for comparison problems are given in chapter 12 and, for regression, in the starred sections of chapter 15. Since I generally emphasize the diagnostic features of the exploratory methods, and because the data come primarily from real problems, students are sensitive to the presence of outliers in data by the time we begin the study of statistical methods. Consequently, I cover methods based on both the trimmed mean (because of the immediate availability of confidence intervals) and on ranks for at least the two-sample problem. The two-sample t test based on ranks is essentially equivalent to the Wilcoxon two-sample test, and the text presentation emphasizes its robust properties, thus setting it in what I believe to be the proper context for applications.

Although not my personal preference, it is possible to teach a rather conventional course on classical methods using this text by omitting the section on trimmed means from chapter 2 and not covering chapter 12 or the starred sections of chapter 15. My reason for opposing this strategy is that the robust methods add very little to either the computational or conceptual burden already present in the classical methods; yet they add considerably to the scope of applicability of the methods. Thus, a small investment in time will provide students with tools that can be more safely and satisfactorily used on real data.

During the course of writing this revision, I have had help from several sources. My colleague, Ed Bedrick, read an early version of the manuscript and made suggestions that led to a number of improvements. Chris Stidley, a graduate student in Statistics at UNM, has done an excellent job of error checking. Her meticulous work is greatly appreciated.

I have been fortunate in receiving a great deal of support from several people at PWS. Excellent editorial assistance has been given by Michael Payne, Karen Garrison, Susan Graham, and Helen Walden. Special thanks are due to Betty O'Bryant. Greg Hubit Bookworks of Larkspur, California did a magnificent job of the final production.

Important suggestions and feedback are provided by reviewers and I have benefitted greatly by the advice I received from the following individuals: Jeffery Birch, VPI and State University; Roger Carlson, University of Missouri—Kansas City; Kathryn Chaloner, University of Minnesota—Twin Cities; Rudy Freund, Texas A & M University; Brian Joiner, Joiner & Associates; Diane Lambert, Carnegie-Mellon University; Sue Leurgans, Ohio State University; David Lund, University of Wisconsin; Frank B. Martin, University of Minnesota—Twin Cities; Thomas Moberg, Grinnell College; James Stapleton, Michigan State University; and Judy Vopava, Grinnell College.

While not actively involved in this writing project, as she was in earlier ones, my wife Sharon provided vital moral support and even forgives me for the many hours spent with my computer that should have been spent with her. For this, among other things, I am grateful.

Contents

◢ PART III
Statistical Inference Methodology

Introduction to Contemporary Statistical Methods   Second Edition

PART I

Exploratory
Statistics

In the beginning of statistics, some three or four centuries ago, little besides data description was possible. Modern methods of statistical inference could not evolve without certain mathematical tools that were developed relatively recently. Statistical inference is almost exclusively a product of this century. As statistical inference was developed, the numerical and graphical methods classified as descriptive statistics were pushed into the background and are now regarded by many as primarily of historical interest. Bar graphs, histograms, frequency polygons, and other graphical displays of frequency distributions have been used in various forms for many years in both scholarly and popular publications; along with certain numerical descriptions, they are the principal staples of descriptive statistics. These graphs are often constructed late in the analysis to depict what has been found by other methods.

Some applied statisticians were greatly disturbed to see graphical methods pushed into this secondary role. Not only were the answers being sought often clearly visible in the graphs; problems that could seriously affect standard inferential methods could also be detected by using them. Under the leadership of J. W. Tukey, who devised many innovative numerical-graphical methods, descriptive statistics is undergoing a much deserved renaissance. It is now known by the more appropriate name of exploratory data analysis.

The main emphasis in these first chapters will be the introduction of statistical problems through the exploratory analysis of data. Important classical descriptive tools such as the bar graph and histogram will be covered as well, but they will be either augmented or replaced by more effective exploratory displays. We will learn to "see" the solutions to statistical problems graphically. Exploratory analysis will help us detect problems that will adversely affect the standard inference procedures, making it possible to take appropriate corrective action. Statistical inference will then follow as a natural capstone for measuring and confirming the strength of the solutions found in the preliminary exploratory analyses. The two fields of exploratory data analysis and statistical inference form a pair of coequal and interdependent subjects, which together make up one of the most powerful tools of modern science.

Variables and Their Frequency Distributions

Introduction

People have a need to understand the world and to devise strategies for getting along, and ahead, in it. They need information for many types of activities and pursuits. For example, the physician interested in using a new medication for arthritis needs to know whether the drug is effective and whether it has any unacceptable side effects. The businessperson wanting to market a new product needs information about the product's potential appeal to consumers. The government has a voracious appetite for information; it gathers statistics on our health, the state of our employment, the cost of our food, and practically everything else about our lives.

Because of the ever-accelerating pace of nearly all human activities, it is increasingly important, and at the same time increasingly difficult, to get information that is both timely and accurate. Decisions based on faulty or outdated information can have very unfortunate consequences. In an attempt to keep information up to date, computers are now used for the storage and rapid retrieval of large volumes of data. However, computers alone cannot guarantee that information is both timely and accurate. Methods pioneered in the science of statistics for gathering, processing, and interpreting data must play an important role in the information acquisition process if these criteria are to be met.

Statistics and Variability

vary

Statistics is concerned with information about phenomena that **vary.** For example, the responses of prospective patients to the above-mentioned arthritis remedy would vary from person to person. Some individuals would probably respond well, and others would not show any improvement. Some might even respond adversely to the drug. Individual differences among the patients would lead to a variety of reactions. It is the unpredictable variation

population

of drug responses from one individual to another in the **population** of prospective patients that makes this a variable phenomenon of the type considered by statistics.

frequencies

One way to describe variability is to give the **frequencies** of the various responses in the population. An example would be the proportions of patients who (a) respond well, (b) do not respond, and (c) respond adversely to the remedy.

frequency distribution

A table of the possible responses and their frequencies is called a **frequency distribution.** Statistics is concerned with providing information about population frequency distributions. An example of a statistical application is the prediction of consumer response to a new product a business wants to market. If all members of the population of potential consumers were asked if they would use the product, it is unlikely that all would say "yes"; the responses would vary from person to person. The decision to market the product would be based on the proportion of "yes" responses. If the company determines that the product will be profitable if 30% of the potential customers react favorably, then the information required is whether or not the proportion of "yes" responses in the population is as large as 30%. This is a question involving a frequency distribution; as such, it is within the domain of statistics.

The Need for Sampling

If information is to be timely and cost-effective, it is seldom practical to poll an entire population. After extensive laboratory testing, the drug company wanting to market its new arthritis remedy would hire a few physicians specializing in the treatment of arthritis to perform clinical trials of the drug on selected patients. In this way, the drug would be tried on a relatively

sample

small number of prospective users. This subcollection (or **sample**) would be selected and the trials carried out according to statistically acceptable methods. Although the complete testing process for a drug can take several years, the use of clinical trials makes it possible to check the effectiveness of a drug on humans reasonably quickly. In this way, useful remedies can be made available in a timely fashion.

Businesses must get product acceptability information quickly. For this reason, it is common practice to hire a company specializing in market surveys to test the acceptance of the product. Such companies use statistical sampling methods to select individuals to test and give opinions about the product. Because a relatively small number of prospective customers are actually contacted, information can be obtained quickly and at reasonable cost.

Other examples of the use of samples can be found in nearly all areas of human activity. Politicians hire political polling organizations to evaluate their chances of election or the popularity of their stands on various issues. Virtually all the scientific understanding we have of variable phenomena in nature comes from experiments or observations on samples of the phenomena. The government obtains unemployment figures, food price indices, and nearly all other statistics on the United States population by sampling.

Only the decennial census, which provides information for congressional apportionment, is carried out on the entire population. But even as the census is being carried out, a great deal of additional information is obtained from samples of individuals. This information is available more quickly and at far less cost than if it were obtained from everyone in the population. A surprising bonus is that this information is also generally more accurate than the data from the complete census, because of errors inherent in such massive data-gathering efforts.

What Is Statistical Inference?

A familiar statement of the result of a political poll is that, say, 30% of the voters favor candidate A, 40% favor candidate B and 30% are undecided. Occasionally, the possibility of an error of, say, 2 percentage points will be acknowledged. In fact, it is almost always the case that a relatively small number of prospective voters have been contacted for the information, and the figures quoted are from this sample. However, the implication of the statement is that the figures apply to all prospective voters, not just to those actually polled; there has been an extrapolation of voter preferences from sample to population.

How is it possible to make such an extrapolation? Why should the frequencies in the sample and the population be related at all? Intuitively, it seems reasonable that a "properly chosen" sample should show frequencies that approximate those in the population, but the basis for the value of the possible error is much less clear. How is it possible to quantify the possibility of error in approximating the population frequencies by using those of the sample?

statistical inference

random sampling

The answer is that reputable polling organizations use the methods of **statistical inference** that form the core of this book. Statistical inference is the science of extrapolating frequency information from samples to their respective populations. It is based on a particular scheme of sampling called **random sampling.** What random sampling is and why it works will be discussed next.

Representative Samples and Random Sampling

The ideal method of constructing a sample to make the above extrapolation process work would be to select individuals in such a way that the sample frequency distribution is exactly the same as that of the population. However, to do this, we would have to know the population frequency distribution. If we had this information, sampling would be unnecessary!

representative

The term **representative** is used somewhat loosely to describe sample frequency distributions that are approximately equal to their population counterparts. Is it possible to form anything approaching a representative sample without knowing the population frequency distribution? The answer is that a method of random sampling produces samples that are representative "with high probability." The degree of representativeness is related to the degree of probability in a way that we will study in subsequent chapters.

The prototype of the random sampling method is a lottery. For example, to draw a sample for a political poll by lottery we could have each member of the voting population submit a ticket with an identification number (for example, a Social Security number) written on one side and the candidate preference (for example, candidate A, candidate B, or undecided) on the other. All of the tickets would be put into a large basket and the contents of the basket thoroughly mixed. If the sample were to consist of 100 individuals, a blindfolded person would draw 100 tickets from the basket, remixing the tickets between drawings for good measure. The individuals corresponding to the Social Security numbers on these tickets would then become the members of the sample. The distribution of voter preferences in the population is simply the proportions of the tickets in the basket labeled "candidate A," "candidate B," and "undecided." The sample frequency distribution consists of the corresponding proportions for the 100 sampled tickets.

To see why representative samples are the most probable ones to be drawn, consider the drawing of the first ticket from the basket. Initially, 30% of the tickets are marked "candidate A," 40% "candidate B," and 30% "undecided." It follows that there will be a 30% chance that the selected ticket will be marked "candidate A," a 40% chance that it will be "candidate B," and a 30% chance that it will be labeled "undecided."

Now, just as repeated tosses of a coin with probabilities $\frac{1}{2}$ each of heads and tails will most likely produce each result about half the time, so repeated drawings of the first ticket—which could be simulated by returning the ticket to the basket after each drawing—will favor the 30%, 40%, 30% distribution of candidate preferences. Thus, if the sample could be made up entirely of first drawings, chance would favor representative samples.

Intuitively, the 30%, 40%, 30% distribution is the most likely one. We would not expect to see exactly these proportions for most samples. However, it is likely that the actual sample proportions will be close to these values. With a probability analysis to be given in a later chapter, we will be able to quantify the terms "close" and "likely."

There is a difference between drawing 100 tickets to form a sample and making 100 separate drawings of a single ticket. In the first case, no replacement of tickets is made; in the second, sampling with replacement is involved. However, if the population is large compared to the size of the sample, this difference has relatively little impact on the sample frequency distributions or probability analysis.

Overview of the Book

Definitions and descriptions of frequency distributions will occupy the rest of this chapter and chapter 2. Once we have the appropriate tools, we will be ready to look in chapters 3 and 4 at the kinds of problems statisticians most often deal with in practice. Exploratory or descriptive "solutions" to these problems will be given.

An introduction to probability and to the role it plays in statistical inference is given in chapters 5–7. Because statistical inference deals with variable phenomena that can be described probabilistically, statistical theory has a strong probability component. We will study a theory of probability that has been developed over more than three centuries to handle such variable phenomena.

Confidence intervals and hypothesis tests, the fundamental statistical inference tools, are introduced in chapters 8 and 9. These tools are then applied to the problems introduced in chapters 3 and 4.

The final part of the text, consisting of chapters 10 through 17, contains both the classical and a modern approach to statistical methodology. It is there that the inference "capstone" will be placed on the exploratory solutions found in the first section of the book.

◪ SECTION 1.2
Variables and Their Types

Suppose we were interested in gathering information about the population of students enrolled in an elementary statistics course at a particular university in a given year. Display 1.1 provides data for a sample of students from such a population at the University of New Mexico. The characteristics of interest are age, sex, college class, height, and weight. These characteristics

variables   are called **variables,** and the sample data will be used to provide information about the population frequency distributions of these variables.

quantified   A variable is a characteristic that can be **quantified** in some way. Standard methods of quantifying characteristics are classification, measurement, and counting. Thus, the sex variable in display 1.1 classifies the members of the sample into the classes or categories M (male) and F (female). Values of the height and weight variables are obtained by measurement processes, while the age variable "counts" the number of years since birth.

Variables are classified into types by the same criteria. Thus, sex is a

categorical variable   **categorical variable.** Height and weight, on the other hand, are **measure-**
measurement variables   **ment variables,** and age is a **counting variable.**
counting variable
One important reason for assigning types to variables is that the frequency distributions for the different variable types require different mathematical "models," thus different treatments. We will find a way to eliminate the counting variable as a separate category, but the dichotomy of classification variables and measurement variables will persist throughout the entire book. They will require different descriptive, exploratory, and inference methods. For this reason the first step in solving an inference problem is to identify the types of the variables involved. This will make it possible to focus the analysis on the appropriate collection of methods.

Display 1.1

Statistics class data

I.D. NUMBER	AGE (YEARS)	SEX[a]	CLASS[b]	HEIGHT (INCHES)	WEIGHT (POUNDS)
1	28	M	Jun.	72	180
2	18	F	Fr.	66	114
3	46	F	Soph.	65	115
4	18	F	Fr.	62	122
5	55	M	Grad.	71	205
6	18	F	Fr.	69	125
7	20	M	Soph.	66	133
8	19	F	Fr.	63	105
9	19	F	Soph.	63	110
10	18	M	Fr.	71	175
11	21	M	Soph.	72	160
12	20	F	Soph.	67	133
13	19	M	Fr.	67	145
14	20	M	Soph.	73	163
15	24	F	Grad.	67	180
16	20	M	Jun.	65	125
17	20	M	Jun.	67	120
18	20	F	Jun.	66	125
19	19	F	Soph.	60	110
20	20	M	Soph.	68	140
21	18	F	Fr.	68	133
22	19	F	Fr.	65	107
23	18	F	Fr.	64	140
24	29	F	Sen.	64	125
25	26	F	Soph.	63	125
26	18	F	Fr.	70	170
27	19	F	Soph.	54	120
28	22	F	Jun.	64	110
29	27	F	Soph.	66	125
30	19	F	Jun.	65	120
31	18	M	Fr.	65	130
32	21	M	Jun.	70	175
33	19	F	Soph.	67	120
34	19	M	Soph.	66	175
35	19	M	Soph.	72	155
36	20	F	Soph.	57	82

[a] In this column M = male and F = female.
[b] In this column Fr. = freshman, Soph. = sophomore, Jun. = junior, Sen. = senior, and Grad. = graduate.

■ SECTION 1.3

Categorical Variables and Their Frequency Distributions

values

A categorical variable represents a characteristic with a fixed number of classes or categories, such as the sex and college class variables in display 1.1. The **values** of the variable are codes or names for these classes. The classes must satisfy the condition that each member of the population falls in one and only one class. This property is clearly satisfied for the sex variable. However, the categories of the class variable in display 1.1 do not account for special students, such as high school students taking college courses and students not working for a degree, who do not fall in one of the regular categories. This violates the condition that the classes must exhaust all possibilities; everyone must be in one of the categories. To solve this problem, the categories *freshmen* through *graduate* could be augmented with a sixth category labeled *other*.

An example of a set of classes that violates the condition that each individual fall in only one category is a religion variable with the categories *Protestant, Catholic, Baptist, Jewish* and other. Baptists would fall in both the *Baptist* and the *Protestant* classes.

EXERCISE

1.1   Suppose that the following additional characteristics were obtained for the 36 students of display 1.1. Define variables to quantify these characteristics and give their types (categorical or measurement). Give the units of measurement for measurement variables and list the possible values of categorical variables. If there are many, give only a few typical values. Group categories likely to have few or no occurrences in the sample into a single category.

(a)   religion

(b)   waist size

(c)   zodiac sign

(d)   birth month

(e)   hour of birth

(f)   amount of money on his/her person at time of survey

(g)   body temperature at survey time

(h)   employment status

Frequencies and Frequency Distributions

absolute frequency

The **absolute frequency** *for a given category and a given collection of individuals is simply the number of individuals in the collection falling into that category.* Thus, the absolute frequency of males for the collection of students represented in display 1.1 is the number of males, 14.

relative frequency

The **relative frequency** *is the absolute frequency divided by the number of individuals in the collection.* Since there are 36 students in the sample of display 1.1, the relative frequency of males would be

relative frequency = $\frac{14}{36}$ = .389

It is usually more convenient to represent relative frequencies by percentages—100 times the fractional version. Thus, we could say that the sample of display 1.1 is 38.9% male.

frequency distribution

The **frequency distribution** *of a variable is a list of all possible values of the variable along with their relative frequencies for the group in question.* Since the sample of display 1.1 is 38.9% male, it follows that the relative frequency of females is $100 - 38.9$ or 61.1%. Thus, the frequency distribu-

frequency table

tion of the sex variable can be given in a **frequency table** as follows.

Frequency distribution of sex variable for display 1.1

CATEGORY	FREQUENCY
Male	38.9%
Female	61.1%
	100.0%

Constructing Frequency Tables

Frequency distributions are commonly constructed using computers nowadays. However, distributions for small data sets can be easily and quickly constructed by hand, using a tally process. As an example, the frequency table for the *class* variable of display 1.1 is shown in display 1.2. The categories of the variable are entered first, then a tally mark for each individual is entered opposite the appropriate category, in one pass through the data set. By grouping the tallies in groups of five, they are easily counted, producing the absolute frequency (absolute *f*) column. The sum of this column is the

sample size

sample size, $n = 36$. The relative frequency column is then computed by dividing absolute frequencies by sample size and converting the resulting fractions to percentages.

Display 1.2

Frequency table for the class variable of display 1.1

CATEGORY	TALLY	ABSOLUTE *f*	RELATIVE *f* (%)
Freshman	⊬⊬⊬ ⊬⊬⊬ I	11	30.56
Sophomore	⊬⊬⊬ ⊬⊬⊬ ⊬⊬⊬	15	41.67
Junior	⊬⊬⊬ II	7	19.44
Senior	I	1	2.78
Graduate	II	2	5.56
Other		0	0.00
		$n = 36$	100.01[a]
			(sum of *f* column)

[a] Does not sum to 100% because of rounding.

Bar Graphs

bar graph

The **bar graph** is a commonly used graphical display of the information in a categorical variable frequency distribution. A bar graph for the class variable of display 1.1 is given in display 1.3. Note that the horizontal axis lists the categories of the variable and the vertical scale represents relative fre-

Display 1.3

Bar graph for the class variable of display 1.1

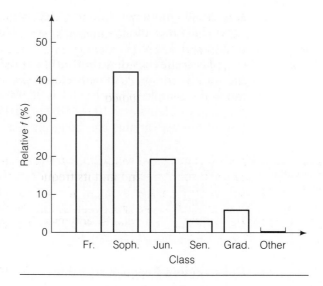

quency. Bars of heights equal to the appropriate relative frequencies are drawn over the category codes. Note that the bar over the *other* category is of height 0. Bar graphs can be produced by most statistical computer packages. We will use computerized bar graphs to compare categorical variable distributions in chapters 3 and 4.

Forming Categorical Variables By Grouping

grouping

Grouping is a process for forming a categorical variable by combining values of another variable into a prescribed collection of classes. For example, this process can convert a categorical variable into another categorical variable with fewer classes. In a later chapter we will have occasion to form a new variable from the class variable of display 1.1 by combining the categories *senior, graduate* and *other* into one. The values of the new variable could be coded Fr., Soph., Jun., and Sen.+, where Sen.+ represents the combined category. The frequency distribution of the new variable would be the same as that of the old class variable for the first three categories, and the relative frequency for the category Sen.+ would be the sum of the frequencies, $2.78 + 5.56 + 0.00 = 8.34\%$.

Grouping can be used to convert a counting variable into a categorical one. Suppose that a variable *number of brothers and sisters (siblings)* had been included in display 1.1. A counting variable can have values 0, 1, 2, and so on, theoretically without limit. Since a number of siblings larger than 7 is likely to be rare, little loss of information would result from combining all numbers beyond 6 into a single category. Thus, the grouped variable would have the eight values 0, 1, 2, . . . , 6 and 7+, where 7+ represents the combined category.

The age variable in display 1.1 has been previously described as a counting variable, but it would be possible to convert it into a categorical variable by grouping. However, the nature of age as a time measurement and the fact that a wide range of ages is observed in the display make it more natural and useful to treat age as a measurement variable. Each counting variable we will encounter will either be reduced to a categorical variable by grouping or will be dealt with as a measurement variable. Thus, only these two types will be studied.

EXAMPLE 1.1

In this example we will form a grouped age variable for the first 15 individuals of display 1.1 and find its frequency distribution. The grouped variable will have the value T (teen) if age is less that 20 (written as age <20), Y (young adult) if age is between 20 and 40 (20 ≤ age <40) and O (older adult) if age is 40 or more (age ≥ 40). The original ages and values of grouped age are given in the following table.

I.D. NUMBER	AGE	GROUPED AGE
1	28	Y
2	18	T
3	46	O
4	18	T
5	55	O
6	18	T
7	20	Y
8	19	T
9	19	T
10	18	T
11	21	Y
12	20	Y
13	19	T
14	20	Y
15	24	Y

The frequency table for the grouped age variable is as follows.

GROUPED AGE	ABSOLUTE f	RELATIVE f (%)
T (teen)	7	46.7
Y (young adult)	6	40.0
O (older adult)	2	13.3
	15	100.0

The relative frequency for teens, for example, was computed by the expression $100 \times \frac{7}{15} = 46.666 \ldots \%$, which has been rounded to 46.7%. If this subsample is representative of the student population in beginning statistics, we would conclude that the largest proportion of these students are still in their teens. This information would be useful to teachers responsible for designing elementary statistics courses.

◢

EXERCISES

1.2   Form a grouped class variable for the first 15 individuals from display 1.1 by combining the *Jun.*, *Sen.*, and *Grad.* categories into a single class called *Upper+*. Give a table of values for the original and the grouped variable, as in the last example, and find the frequency distribution for the grouped variable.

1.3   The first 15 individuals in display 1.1 will be classified as

L (light) if his or her weight is ≤115 pounds,

M (medium) if 115 < weight ≤175 pounds, and

H (heavy) if weight >175 pounds.

Use these categories to define a grouped weight variable. Give a table for weights and for grouped weights and the frequency table for grouped weights.

◢ SECTION 1.4

Measurement Variables and Their Frequency Distributions

measurement variable   The values of a **measurement variable** are obtained by applying a measurement process to individuals in a sample. This process requires a measuring instrument that assigns numerical values to the physical characteristic under study. Thus, a yardstick marked in inches could be used to measure students' heights. Weights can be obtained by the weighing device we call a scale.

A measurement variable will be defined for the entire population; indeed, it is the population distribution of the variable that is of interest. However, the measurements will actually be made only on the members of the sample. This is one of the primary economies achieved by sampling.

It is assumed that a measurement variable can take on any value in some interval of a number line. This assumption implies that measurements can be made with infinite accuracy, since measurement values with arbitrarily long decimal expansions are possible. Although this is never true of real measurement processes or for real populations, the assumption makes possible a convenient mathematical theory for measurement variables that leads to useful and accurate practical applications.

Sample Frequency Distributions

The idea of a sample frequency distribution for a measurement variable is somewhat more complex than in the case of a categorical variable because there are many possible ways to form such a distribution. We will restrict ourselves to the following procedure, which narrows the number of possibilities but still does not lead to a single "solution." An interval of the number line that contains all of the values of the measurement variable in question is subdivided into a number of intervals *of equal length*, called *classes*. Once this has been done, the absolute and relative frequencies of variable values falling into these classes are calculated, and a frequency table is formed, just as for categorical variables. The result is a sample frequency distribution for the variable.

The reason it is possible to obtain many different frequency distributions is that there are many ways to divide the original interval into subintervals. The numbers of observations falling in the different classes will depend on the number of classes chosen and the positioning of the class endpoints or boundaries. Since there is relatively little difference in the shapes of the frequency distributions when the numbers and positions of the class intervals are not too different, it is convenient to have a method for quickly obtaining a few representative frequency distributions. The stem and leaf plot, devised by J. W. Tukey, does just that. We will see that the stem and leaf plot provides not only a convenient table in which to store the sample data values, but also a useful picture of the data.

SECTION 1.5

Stem and Leaf Plots

stem
leaf

The stem and leaf plot is best introduced by an example; the weight data of display 1.1 (p. 9) will be used for this purpose. The first two digits of weight are designated the **stem** and the last digit the **leaf.** The possible stem values are aligned vertically from smallest to largest, and a vertical line is drawn beside this column of numbers. The appearance of the weight plot at this point is as follows:

```
 8 |
 9 |
10 |
11 |
12 |
13 |
14 |
15 |
16 |
17 |
18 |
19 |
20 |
```

Each observation provides both a stem and a leaf. The leaves are entered on the plot opposite their stem values as they are read from the data display. The first four weights of display 1.1 (180, 114, 115, and 122) are entered in the plot as follows:

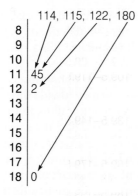

The completed plot is given in display 1.4.

Display 1.4

Stem and leaf plot for weight data of display 1.1

8	2
9	
10	57
11	45000
12	25505550500
13	3330
14	500
15	5
16	03
17	5055
18	00
19	
20	5

Frequency Tables and Histograms from Stem and Leaf Plots

Other examples of stem and leaf plots will be given presently, but first it is useful to note that a conventional frequency table and a graphical display called a *histogram* can be easily constructed from the stem and leaf plot.

class limits   Each stem defines a class of the frequency distribution; the **class limits** are the smallest and largest values possible for that class. Thus, the class limits defined by stem value 8 are 80 and 89; for stem 10 they are 100 and 109; and so forth. The absolute frequency for any class is just the number of leaves opposite the corresponding stem. The weight frequency table is given in display 1.5.

The second column of display 1.5 is obtained as follows. Recall that we consider the possible values of a measurement variable to be all values in an interval of the number line. The recorded values are then thought of as being rounded versions of the actual measurements. For example, the weight 82 pounds represents the rounded value of a weight between 81.5 and 82.5 pounds; the recorded weights between the class limits 80 and 89

Display 1.5

Frequency table for weight data of display 1.4

CLASS LIMITS	CLASS BOUNDARIES	FREQUENCY (f)	RELATIVE f (%)
80–89	79.5–89.5	1	2.78
90–99	89.5–99.5	0	.00
100–109	99.5–109.5	2	5.56
110–119	109.5–119.5	5	13.89
120–129	119.5–129.5	11	30.56
130–139	129.5–139.5	4	11.11
140–149	139.5–149.5	3	8.33
150–159	149.5–159.5	1	2.78
160–169	159.5–169.5	2	5.56
170–179	169.5–179.5	4	11.11
180–189	179.5–189.5	2	5.56
190–199	189.5–199.5	0	.00
200–209	199.5–209.5	1	2.78
		$n = 36$	

represent actual weights between 79.5 and 89.5 pounds; and so on. These values are made the endpoints of the class intervals in the frequency table. These endpoints, called **class boundaries,** are given in display 1.5 for the purpose of graphing the histogram.

class boundaries

histogram

The **histogram** is the classical graphical representation of a measurement variable frequency distribution. The histogram for the weight data tabulated in display 1.5 is given in display 1.6. Note that this graph resembles

Display 1.6

Histogram of weight variable, data from display 1.5

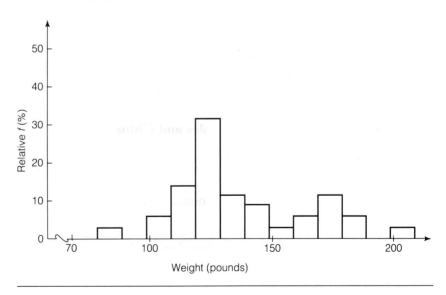

the bar graph for categorical variables, except that the sides of the rectangles are joined and the points on the horizontal axis where they come together are precisely the class boundaries given in display 1.5. The heights of the rectangles are, as in the bar graph, the relative frequencies of the classes.

The purpose of the histogram is to give information about the shape of the frequency distribution. We will see in the next subsection that this information is also contained in the stem and leaf plot.

EXERCISES

1.4   The following data are the ages of the first 40 registrants for the Sixth Annual Rio Grande Valley Bicycle Tour.

23	29	40	28	15
22	46	39	22	17
26	33	35	49	20
36	25	15	31	17
43	54	36	30	30
40	27	24	20	28
22	37	17	39	42
17	22	9	26	29

Make a stem and leaf plot for these data.

1.5   Refer to exercise 1.4. Identify the class limits and class boundaries for the resulting grouped frequency distribution. Calculate the relative frequencies.

1.6   Graph the histogram for the frequency distribution constructed in exercise 1.5.

Some Properties of the Stem and Leaf Plot

By rotating display 1.4 (p. 16) 90° and comparing it with display 1.6, we see that the leaves of the stem and leaf plot form a picture of the frequency distribution that is virtually identical to the histogram. See display 1.7 for

Display 1.7

Graphic representations for the weight variable of display 1.1.
(a) Histogram.
(b) Stem and leaf plot, tilted 90°.

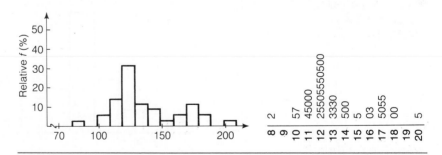

this comparison. The stem and leaf plot thus provides a useful and quickly formed picture of the distribution. This graphical representation is quite adequate for exploratory purposes. The histogram can be reserved for written reports and published data summaries.

Stem and leaf plots are often drawn by computers these days. However, when exploratory analyses are done by hand, the stem and leaf plot has important features beyond its pictorial qualities. First, since the class intervals of the frequency distribution are automatically determined by the choice of the stems, the otherwise annoying decisions of how many classes to choose and where to make the class boundaries are avoided. A second advantage from the viewpoint of hand calculation is that, unlike the histogram, all the original numbers are preserved in the plot. Since it is a compact summary of the data, the stem and leaf plot is a more useful starting point for further calculations than is the original data array.

Finally, some of the calculations we will make require that the data be put in increasing order. With a stem and leaf plot, it will not be necessary to write out the ordered data, since the plot itself provides a critical step in the process, namely the ordering of the stems. All that remains is to order the leaves within each stem row. For small- to moderate-size data sets, this is an easy task. Applications will be given in chapter 2.

Increasing the Scope of the Stem and Leaf Plot

Extensions of the stem and leaf plot to cover new data problems and more diverse frequency distributions will be given in a series of examples in this subsection.

EXAMPLE 1.2

In this example we will see how to increase the number of stems when the stem and leaf plot is too concentrated. Display 1.8 shows the stem and leaf plot of the height data from display 1.1, using the first digit as the stems and the second as the leaves. The plot has too few stem values to give a useful picture of the height distribution. Moreover, there are too many leaves in the 6 stem line to order the data easily.

Display 1.8

Stem and leaf plot of height data from display 1.1

```
5 | 47
6 | 65296337775760885443465576
7 | 21123002
```

To solve this problem, each stem line can be used twice, the leaves 0, 1, 2, 3, 4 recorded on the first line and 5, 6, 7, 8, and 9 on the second. The result of this modification is shown in display 1.9(a). Note that the stem lines

Display 1.9

Modified stem and leaf plots of height data. (a) Two lines per stem version (* indicates leaves 0–4; · leaves 5–9). (b) Five lines per stem version (* indicates leaves 0–1; t leaves 2–3; f leaves 4–5; s leaves 6–7; · leaves 8–9).

(a)
```
5* | 4
 · | 7
6* | 23304434
 · | 659677757688565576
7* | 21123002
```

(b)
```
 f | 4
 s | 7
 · |
6* | 0
 t | 2333
 f | 55544455
 s | 6777766676
 · | 988
7* | 1100
 t | 2232
```

two lines per stem

corresponding to leaves 0–4 are coded with an asterisk and those corresponding to 5–9 with a dot. This plot is said to have **two lines per stem.**

The two lines per stem modification is equivalent to doubling the number of classes in the frequency distribution, making each class half as long as before. Where the intervals for the plot of display 1.8 were 10 inches long, those for display 1.9(a) are 5 inches long. The next natural refinement would be to subdivide each interval into five equal parts, each one-fifth the original length. This is equivalent to subdividing each stem into five lines and recording leaves 0 and 1 for the first line, 2 and 3 for the second, and so on. This modification of the height stem and leaf plot is shown in display 1.9(b). The code * is used for leaves 0–1, **t** for 2–3, **f** for 4–5, **s** for 6–7, and · for 8–9. Note the correspondence between these codes and the first letter of the spelled-out versions of the leaf values (**t** = twos and threes, for

five lines per stem

example). This version of the stem and leaf plot is said to have **five lines per stem.**

Data with decimal points and more than three digits of accuracy are dealt with in the next example.

EXAMPLE 1.3

The capacitances of a sample of mylar capacitors are given in display 1.10. These data contain decimal points and a greater number of significant digits than we have encountered thus far. The strategy for dealing with decimal points is simply to ignore them in the formation of the plot. This is equivalent to converting the data to integers by moving the decimal point. For the capacitor data, the decimal is moved four places to the right. A notation to the effect that the decimal must be moved back four places in order to regain the original units is then given along with the plot.

Display 1.10

Measured capacitances of .6 μF rated mylar capacitors

.6309	.6241	.6359	.6320
.6116	.6585	.6399	.6301
.6428	.6458	.6483	.6397
.6329	.6266	.6430	.6362
.6180	.6544	.6521	.6390
.6296	.6328	.6380	.6499
.6215	.6300	.6445	.6320
.6262	.6404	.6471	.6459
.6402	.6449	.6394	.6214
.6368	.6611	.6390	.6441

Data courtesy of R. R. Prairie, Sandia Corporation, Albuquerque, New Mexico.

The "activity" from observation to observation in the capacitor data begins in the second decimal place, suggesting that the first two digits should be used as stems. However, unless the data are rounded to three digits, this requires that the leaves contain two digits each. This problem is easily solved by simply separating the leaves in the stem and leaf plot by commas to distinguish them visually from the one-digit leaf version. The resulting plot is given in display 1.11(a).

Multiple stem lines can be combined with multiple leaves to spread the plot out. This is demonstrated in display 1.11(b). Display 1.11(b) is computer generated and provides the additional bonus of ordered leaves within each stem line.

Display 1.11

Stem and leaf plots of capacitor data. (a) One line per stem version. (b) Computer generated, two lines per stem version.

(a) To regain original units, move decimal four places to the left.

```
61 | 16, 80
62 | 41, 66, 96, 15, 62, 14
63 | 09, 59, 20, 99, 01, 97, 29, 62, 90, 28, 80, 00, 20, 94, 68, 90
64 | 28, 58, 83, 30, 99, 45, 04, 71, 59, 02, 49, 41
65 | 85, 44, 21
66 | 11
```

(b) To regain original units, move decimal four places to the left.

```
61* | 16
  . | 80
62* | 14,15,41
  . | 62,66,96
63* | 00,01,09,20,20,28,29
  . | 59,62,68,80,90,90,94,97,99
64* | 02,04,28,30,41,45,49
  . | 58,59,71,83,99
65* | 21,44
  . | 85
66* | 11
```

Note that the * lines contain leaves 00–49 while the · lines contain leaves 50–99. Similarly, for stem and leaf plots with five lines per stem and double leaves, the * lines would contain leaves 00–19, the t lines leaves 20–39, and so forth.

The following example shows how to deal with negative numbers.

EXAMPLE 1.4

The following data were extracted from a data set we will encounter in the exercises to chapter 3.

$$.12, 1.01, -.20, .15, -.30, -.07, .32, .27, -.32, -.17, .24$$

This sample has the previously unseen feature of negative numbers. The stem and leaf plot can be readily extended to account for this feature by adding lines for negative stems—including a line for the stem -0 if both negative and positive data occur in the same plot. The stem and leaf plot using double leaves and one line per stem is given in display 1.12(a). Since this plot shows relatively little detail, the five line per stem version is given in display 1.12(b). Note that the stem line symbols radiate in different directions from 0.

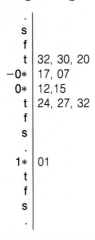

Display 1.12

**Stem and leaf plots for data containing negative numbers.
(a) One line per stem version. (b) Five lines per stem version.**

(a) To regain original units, move decimal two places to the left

```
-0 | 32, 30, 20, 17, 07
 0 | 12, 15, 24, 27, 32
 1 | 01
```

(b) To regain original units, move decimal two places to the left

```
      .
      s
      f
      t | 32, 30, 20
    -0* | 17, 07
     0* | 12,15
      t | 24, 27, 32
      f
      s
      .
     1* | 01
      t
      f
      s
      .
```

EXERCISES

1.7   Redo the stem and leaf plot of exercise 1.4 (p. 18) in the 2 lines per stem version. What are the class limits and class boundaries of the resulting grouped frequency distribution? What new details (if any) are seen in this expanded-scale distribution when compared to the one-stem plot?

1.8   Make a stem and leaf plot with one line per stem for the *age* variable of display 1.1 (p. 9).

1.9   Make a stem and leaf plot with five lines per stem for the *age* variable of display 1.1 (p. 9).

1.10   Refer to exercises 1.8 and 1.9. Compare the pictures of distributional shape provided by the two plots. What features of the frequency distribution are more clearly seen in the five lines per stem version? What are the class limits and boundaries of the grouped frequency distribution determined by this version of the stem and leaf plot?

1.11   The following data are the average heights of weld beads, relative to a given baseline, for welds used to secure cooling nozzles to the fuel rod assemblies in a certain nuclear reactor.*

3.45	−.37	−.85	.62
.94	.87	.24	4.11
.90	−.44	3.15	.98
−.43	1.35	1.23	.50
.52	−.34	2.11	.71

Make a stem and leaf plot for the data.

◪ SECTION 1.6

Population Frequency Curves and Their Shapes

frequency curves

The mathematical model of population frequency distributions utilizes curves that we will call **frequency curves.** Frequencies are then represented by the areas under these curves. Frequency curves are viewed as being smooth versions of histograms for entire populations. This representation has particular mathematical utility, but we will use frequency curves simply as a convenient way to picture and describe population frequency distributions.

The shapes of histograms and stem and leaf plots for representative samples will approximate the shapes of the frequency distributions for the corresponding populations. In fact, this is their primary function. The shape of the height stem and leaf plot of display 1.9(a) suggests the shape of the population frequency curve given in display 1.13. The population frequency of individuals with heights between 65 and 70 inches is illustrated by the gray area on the graph.

Display 1.13

Conception of the population frequency distribution for height, based on data of display 1.1

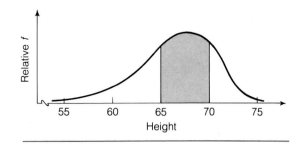

The Normal Distribution and Distribution Shapes

normal distribution

The frequency distribution that will play the central role in statistical inference is the so-called **normal distribution.** A picture of this celebrated distribution is given in display 1.14. Because of its importance, it is used as the

---

* Data courtesy of D. Sheldon and J. McKenzie, Sandia Corporation, Albuquerque, New Mexico.

standard by which other distributions are compared. In particular, certain information about the shapes of distributions will be used to describe important deviations from normality.

The three important regions of a distribution are indicated in display 1.14. They are the center of the distribution, which contains the bulk of the frequency, and the two tails, where the frequency trails off at the extremities of the curve.

Display 1.14

Frequency curve of the normal distribution

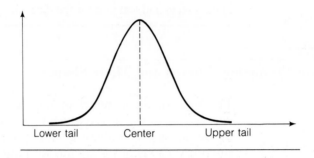

Lower tail   Center   Upper tail

The normal curve has a well-defined center of symmetry, marked by the dashed line in display 1.14. Frequency curves with this property are called **symmetric.** An example of a very nearly symmetric distribution is that of the capacitor data shown in display 1.11, page 21.

symmetric

skewed

Distributions that are not symmetric are called **skewed.** See display 1.15.

Display 1.15

Frequency curves for distributions.
(a) Skewed to the left.
(b) Skewed to the right.

(a)   (b)

Distributions for which the longer tail is to the left are said to be skewed to the left. This property is also seen in display 1.13. It is more common in practice to see distributions that are skewed to the right, as is the weight distribution shown in display 1.6, page 17.

The weight distribution appears to have another property not shared by the normal distribution, namely two humps or *modes*. The term for this property is **bimodal.** Multiple-mode distributions usually indicate the presence of two or more identifiable groups with different distributions in the population. The bimodality of the weight distribution arises from the rather different distributions of weights for men and women. A hint of this same feature is seen in the height distribution given in display 1.9(b) (p. 19).

bimodal

An important property of the normal distribution is a well-defined rate of decrease of frequencies from its center into the tails. Distributions for which this decrease is more rapid or abrupt are called **short-tailed.** Distribu-

short-tailed

long-tailed

tions for which the rate of decrease is less rapid, or for which appreciable frequency exists "abnormally" far from the center of the distribution, are called **long-tailed**. Illustrations of a short-tailed and a long-tailed distribution are given in display 1.16.

Display 1.16

Two nonnormal distributions.
(a) A short-tailed distribution.
(b) A long-tailed distribution.

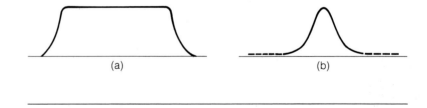

(a)   (b)

Long-tailed distributions create rather special problems for statistical inference. It will be important to deal with these problems, since long-tailed distributions occur frequently in practice. Unfortunately, the stem and leaf plot and histogram are not good diagnostic tools for identifying long-tailed distributions. We will need the box plot, to be introduced in the next chapter, for this purpose.

Tour of the Rio Grande Valley and Statistics Class Data Sets

Two data sets that will be used in exercises and examples throughout the text are given, for convenience of reference, at this point. Even medium-sized data sets can contain a wealth of information that requires much time and effort to extract. The two data sets in this subsection will be used to demonstrate this fact.

The Tour of the Rio Grande Valley is an annual bicycling tour sponsored by The New Mexico Wheelmen, a bicycling club based in Albuquerque. The data of display 1.17 are from the sixth annual tour, held in 1978.

Display 1.17

Data set for the sixth annual Tour of the Rio Grande Valley[a]

OBS	DIST[b]	TOWN[c]	SEX	AGE[d]	TIME[e]	SPEED[f]
1	100	1	M	23	4.9	20.2
2	100	0	F	22	8.3	12.0
3	100	1	M	26	8.0	12.4
4	100	0	M	36	5.7	17.6
5	100	1	M	43	6.1	16.5
6	100	1	M	40	.	.
7	100	1	M	22	.	.
8	100	1	M	17	7.9	12.6
9	100	1	M	29	9.0	11.1
10	100	1	F	46	8.1	12.3
11	100	1	M	33	4.8	21.1
12	100	1	M	25	8.0	12.5
13	100	0	M	54	8.8	11.4
14	.[g]	1	F	27	.	.
15	100	1	F	37	12.4	8.1
16	100	1	M	22	9.3	10.8

Display 1.17

(Continued)

OBS	DIST[b]	TOWN[c]	SEX	AGE[d]	TIME[e]	SPEED[f]
17	100	0	M	40	6.8	14.7
18	50	1	F	39	5.6	9.0
19	100	1	M	35	9.1	11.0
20	100	1	M	15	9.8	10.2
21	100	1	F	36	10.7	9.3
22	100	1	M	24	·	·
23	100	1	M	17	·	·
24	75	1	M	9	10.4	7.2
25	100	0	M	28	7.5	13.2
26	100	1	M	22	6.3	15.8
27	100	1	F	49	8.8	11.4
28	100	1	F	31	9.6	10.4
29	100	0	M	30	5.6	17.9
30	50	1	F	20	6.2	8.1
31	50	1	F	39	·	·
32	100	0	M	26	6.1	16.4
33	50	0	M	15	4.4	11.2
34	100	0	M	17	7.9	12.7
35	100	1	M	20	8.1	12.3
36	50	1	F	17	5.8	8.7
37	50	1	F	30	4.9	10.2
38	100	1	F	28	11.1	9.0
39	100	1	M	42	7.7	13.0
40	·	1	F	29	8.7	·
41	100	0	M	15	10.2	9.8
42	100	0	M	33	8.8	11.4
43	75	0	M	11	6.6	11.3
44	100	0	M	12	10.2	9.8
45	100	0	M	31	8.3	12.1
46	·	1	M	58	·	·
47	50	1	M	32	5.4	9.3
48	·	1	F	15	·	·
49	100	1	M	42	6.0	16.7
50	100	0	M	28	7.9	12.7
51	75	1	M	18	8.5	8.8
52	100	1	M	35	·	·
53	100	0	F	17	11.7	8.6
54	100	1	M	36	10.8	9.2
55	100	1	M	34	9.8	10.2
56	100	1	M	45	8.4	11.9
57	50	1	M	7	·	·
58	50	1	M	32	·	·
59	75	1	M	20	7.9	9.5
60	100	1	M	23	12.3	8.2
61	50	1	F	15	5.6	9.0
62	50	1	M	15	5.6	9.0
63	50	1	M	38	·	·
64	100	1	M	14	8.2	12.2
65	68	1	M	16	6.8	9.9
66	75	1	M	15	9.4	8.0

Display 1.17

(Continued)

OBS	DIST[b]	TOWN[c]	SEX	AGE[d]	TIME[e]	SPEED[f]
67	50	1	F	17	5.6	9.0
68	50	1	F	18	·	·
69	100	1	M	14	·	·
70	100	0	M	30	·	·
71	100	0	M	22	5.5	18.0
72	50	0	M	31	·	·
73	100	0	M	25	10.8	9.3
74	100	1	M	16	11.1	9.0
75	100	0	M	16	8.6	11.7
76	100	0	F	33	·	·
77	100	1	M	21	6.8	14.6
78	100	1	F	19	10.5	9.5
79	100	1	M	20	5.1	19.7
80	100	1	M	23	9.1	11.0
81	100	1	M	29	9.6	10.4
82	100	1	M	23	9.9	10.1
83	100	0	M	16	8.6	11.7
84	100	0	M	·	6.3	16.0
85	100	1	M	33	9.5	10.5
86	100	1	M	33	·	·
87	100	0	F	29	·	·
88	100	1	M	30	5.6	17.8
89	100	1	M	16	7.3	13.8
90	100	0	F	9	·	·
91	50	0	M	16	4.6	10.8
92	100	1	M	45	11.1	9.0
93	100	1	M	15	6.7	14.9
94	100	1	M	26	9.4	10.6
95	·	1	M	14	·	·
96	100	0	F	22	8.5	11.8
97	100	1	M	16	9.8	10.3
98	100	1	M	20	7.5	13.3
99	100	0	M	·	5.6	17.8
100	100	0	M	32	6.6	15.2

Data courtesy of W. Joseph, New Mexico Wheelmen.

[a] This data set is roughly a 25% sample of the data obtained for the over 400 bicyclists who signed up for Tour of the Rio Grande Valley held in April 1977. Riders had the option of three distances, 50, 75, and 100 miles, although, for various reasons, some riders reported other distances. Several riders who signed up either failed to show up on the morning of the ride or failed to check in at its completion. These no-shows have missing (·) time and speed values. Other missing speed values resulted from unspecified distances.

[b] DIST = distance ridden (in miles)

[c] TOWN = $\begin{cases} 1 \text{ if residence in Albuquerque} \\ 0 \text{ if residence outside Albuquerque} \end{cases}$

[d] Ages are reported in years.

[e] TIME = total time to complete ride (in hours).

[f] SPEED = DIST/TIME (in miles per hour)

[g] No-shows (missing values) are indicated by ·

Several values are missing because of no-shows. Missing values occur frequently in real data and they must be accounted for in some way so that any conclusions based on the data analysis will not be biased. We will investigate the no-shows in one of the exercises.

The entire data set for the class from which the data of display 1.1 were extracted is given in display 1.18. An additional variable gives the location in the classroom the students chose to occupy. We will study characteristics of those who choose the front of the class and compare them with the corresponding characteristics of those who choose to sit in the middle and the back of the classroom. There are some interesting differences, as we will see in the exercises.

Display 1.18

Complete class data set from which display 1.1 was extracted[a]

I.D. NO.	AGE (YEARS)	SEX	CLASS	HEIGHT (INCHES)	WEIGHT (POUNDS)	SEATING[b]
1	19	F	Soph.	64	112	1
2	30	F	Other	64	125	1
3	20	F	Soph.	64	120	1
4	19	F	Soph.	64	128	1
5	20	F	Fr.	65	110	1
6	27	F	Soph.	64	125	1
7	20	F	Jun.	62	110	1
8	19	F	Soph.	61	165	1
9	20	F	Jun.	68	140	1
10	30	F	Jun.	62	110	1
11	20	F	Jun.	62	110	1
12	30	M	Soph.	66	205	1
13	19	M	Soph.	75	175	1
14	21	F	Soph.	69	130	1
15	23	M	Soph.	69	135	1
16	22	M	Soph.	67	140	1
17	26	M	Jun.	75	180	1
18	39	F	Grad.	65	115	1
19	30	M	Soph.	72	175	1
20	19	F	Soph.	61	125	1
21	18	F	Fr.	64	120	1
22	19	F	Fr.	62	95	1
23	19	F	Fr.	63	112	1
24	28	M	Jun.	72	180	1
25	18	F	Fr.	66	114	1
26	46	F	Soph.	65	115	1
27	18	F	Fr.	62	122	1
28	55	M	Grad.	71	205	1
29	18	F	Fr.	69	125	1
30	20	M	Soph.	66	133	1
31	19	F	Fr.	63	105	1
32	19	F	Soph.	63	110	1
33	18	M	Fr.	71	175	1
34	21	M	Soph.	72	160	1
35	20	F	Soph.	67	133	1
36	19	M	Fr.	67	145	1
37	20	M	Soph.	73	163	1
38	24	F	Grad.	67	180	1
39	20	M	Jun.	65	125	1
40	20	M	Jun.	67	120	1

Display 1.18

(Continued)

I.D. NO.	AGE (YEARS)	SEX	CLASS	HEIGHT (INCHES)	WEIGHT (POUNDS)	SEATING[b]
41	20	F	Jun.	66	125	2
42	19	F	Soph.	60	110	2
43	20	M	Soph.	68	140	2
44	18	F	Fr.	68	133	2
45	19	F	Fr.	65	107	2
46	18	F	Fr.	64	140	2
47	29	F	Sen.	64	125	2
48	26	F	Soph.	63	125	2
49	18	F	Fr.	70	170	2
50	19	F	Soph.	64	120	2
51	22	F	Jun.	64	110	2
52	27	F	Soph.	66	125	2
53	19	F	Jun.	65	120	2
54	18	M	Fr.	65	130	2
55	21	M	Jun.	70	175	2
56	19	F	Soph.	67	120	2
57	19	M	Soph.	66	175	2
58	19	M	Soph.	72	155	2
59	20	F	Soph.	57	82	2
60	19	M	Soph.	72	155	2
61	19	M	Soph.	66	175	2
62	22	M	Soph.	72	165	2
63	27	M	Soph.	73	155	2
64	22	M	Soph.	68	155	2
65	19	M	Soph.	70	135	2
66	21	M	Soph.	74	185	2
67	19	M	Soph.	74	188	2
68	21	F	Sen.	69	130	2
69	21	M	Jun.	73	170	2
70	19	F	Fr.	64	125	2
71	18	F	Fr.	66	135	2
72	20	M	Soph.	71	160	2
73	19	F	Soph.	61	115	2
74	19	F	Soph.	64	114	2
75	18	F	Fr.	71	160	2
76	17	F	Fr.	66	130	2
77	22	F	Jun.	63	109	2
78	24	F	Soph.	63	115	2
79	19	M	Soph.	71	157	2
80	19	F	Fr.	66	127	2
81	20	F	Soph.	62	120	3
82	21	F	Jun.	67	150	3
83	21	M	Jun.	71	164	3
84	19	M	Fr.	73	170	3
85	22	M	Jun.	72	155	3
86	19	M	Soph.	69	135	3
87	20	M	Soph.	67	130	3
88	21	F	Fr.	68	135	3
89	21	F	Jun.	61	105	3

Display 1.18

(Continued)

I.D. NO.	AGE (YEARS)	SEX	CLASS	HEIGHT (INCHES)	WEIGHT (POUNDS)	SEATING[b]
90	20	F	Soph.	58	110	3
91	21	M	Jun.	66	160	3
92	19	M	Soph.	72	165	3
93	22	M	Jun.	72	180	3
94	19	M	Soph.	77	190	3
95	20	F	Jun.	65	114	3
96	24	F	Soph.	62	112	3
97	27	M	Jun.	66	140	3
98	18	F	Fr.	66	129	3
99	22	F	Soph.	64	108	3
100	19	M	Fr.	72	148	3
101	19	F	Soph.	67	110	3
102	19	F	Soph.	66	126	3
103	22	M	Grad.	72	190	3
104	19	F	Soph.	69	125	3
105	19	M	Soph.	74	180	3
106	19	M	Soph.	69	125	3
107	20	M	Soph.	73	170	3
108	19	M	Soph.	75	225	3
109	19	M	Soph.	73	170	3
110	19	M	Soph.	71	140	3
111	22	F	Sen.	65	135	3
112	22	M	Sen.	70	170	3
113	24	M	Sen.	68	140	3
114	27	M	Soph.	71	180	3
115	20	M	Soph.	66	125	3
116	26	M	Fr.	67	190	3

[a] Display 1.1 consists of the individuals with I.D. numbers 24 through 59. The height of individual 50 has been corrected from 54 to 64 inches.
[b] Seating = class setting location; 1 = front third of classroom; 2 = middle third of classroom; 3 = last third of classroom.

Chapter 1 Quiz

The purpose of the end-of-chapter quizzes is to pinpoint the main ideas and methods of the chapters. They can be used as study guides for chapter reviews.

1. What is the purpose of statistical inference?

2. What is a representative sample?

3. What are the differences between categorical and measurement variables?

4. Why is it important to identify the variable types in a statistical investigation?

5. What are the basic tabular and graphical representations of a categorical variable frequency distribution and how are they constructed?

6. What is grouping?

7. What is the mathematical idea of the values of a measurement variable?

8. What is a grouped frequency distribution for a measurement variable and why isn't there just one such frequency distribution?

9. How is a stem and leaf plot constructed?

10. How can a frequency table and a histogram be constructed from a stem and leaf plot?

11. What are the multiple lines per stem and multiple leaf extensions of the stem and leaf plot?

12. How does one deal with decimal points when constructing a stem and leaf plot?

13. What is the stem and leaf plot convention for dealing with data sets that contain both positive and negative numbers?

14. What is the mathematical "model" for population frequency distributions for measurement variables?

15. What frequency curve shapes are associated with the terms symmetric, skewed, bimodal, short-tailed, and long-tailed, respectively?

16. What is the assumed relationship between the stem and leaf plot of a representative sample and the frequency curve for the corresponding population?

SUPPLEMENTARY EXERCISES

1.12 To establish a standard for parachute design, a researcher recorded the following fill times (in seconds) of 27 standard parachutes, obtained under controlled test conditions:*

.59, .38, .47, .43, .44, .37, .43, .37, .27, .54, .39, .89, .48, .52,
.51, .49, .38, .38, .23, .44, .40, .36, .33, .82, .51, .44, .37

(a) Make a stem and leaf plot of these data. Are there any unusual features of this plot?

(b) What is the general shape of this distribution?

(c) Make a histogram of the frequency distribution.

1.13 An interesting data set from the work of F. Hampel will be used later to indicate that even small data sets with rather innocent-looking stem and leaf plots can, in fact, be somewhat strange. The data are as follows:

.0, .8, 1.0, 1.2, 1.3, 1.3, 1.4, 2.4, 4.6

(a) Construct a stem and leaf plot with 2 lines per stem for this data set.

(b) What is the shape of this distribution?

(c) Is anything unusual suggested by the stem and leaf plot?

(d) Take the square root of each data value (record to two decimal places and use both 2 lines per stem and double leaves) and repeat parts (a) and (b). How does this plot compare with that of (a)?

---

* Data courtesy of R. R. Prairie, Sandia Corporation, Albuquerque, New Mexico.

1.14   In a study to determine whether a particular method of swimming instruction re-
        duces tension, measurements of tension were made on several individuals before
        and after an instruction period. The differences between pre- and postmeasures are
        as follows:*

$$-1, 10, 2, -9, 8, 2, 2, 3, -20, -3, 11, 7, -11, 1, -9, -7, -3, -6$$

Make a stem and leaf plot of these data. Is there any indication that the instruction
method had an effect on tension?

Exercises 1.15–1.18 are based on the class data set given in display 1.18 (pp. 28–30).

1.15   (a)   Make a frequency table and bar graph for the *sex* variable for the front third of
               the class (seating = 1). Note: The use of *seating* to define subgroups of individ-
               uals designates it as a *classification variable*. The use of categorical variables for
               classification is common.

        (b)   Repeat part (a) for the second (seating = 2) and third (seating = 3) section of the
               class.

        (c)   Note any differences in the distributions of parts (a) and (b).

        (d)   To obtain a different look at the same comparison, make frequency tables and
               bar graphs of the *seating* variable for each sex. Now *sex* plays the role of the
               classification variable and *seating* is the variable whose distributions are of
               interest. Which sections of the classroom are preferred by the two sexes? Com-
               ment on possible reasons for what you find.

1.16   Repeat parts (a)–(c) of exercise 1.15 for the *class* variable.

1.17   Make separate stem and leaf plots for the *age* variable for the three sections of the
        class (seating = 1, 2, and 3). Compare and comment on the differences. (Note: The
        comparison is most easily made by putting the stem and leaf plots side by side with
        the scales aligned.) Comment on the shapes of the distributions.

1.18   (a)   Make separate stem and leaf plots for men and women for the following vari-
               ables: (i) *height* and (ii) *weight*. Compare the two distributions for each variable
               and comment on differences.

        (b)   Construct histograms from the stem and leaf plots of part (a).

*Exercises 1.19–1.22 are based on the tour of the Rio Grande Valley data set given in
display 1.17 (pp. 25–27).*

1.19   Make stem and leaf plots of the *speed* variable distributions for the three distances
        50, 75, and 100 miles. Discuss any differences seen.

1.20   (a)   Define a new variable, *showup*, with the value 0 for an individual if the value of
               the *time* variable is missing (·) and 1 if it is not. This variable indicates whether
               or not a rider showed up for the tour or logged in at the finish.

        (b)   Make a frequency table and bar graph for the *showup* variable.

---

* Data courtesy of K. Thomson, a former student in physical education, University of New
  Mexico, Albuquerque.

(c) Make separate frequency tables and bar graphs of the showup variable for men and women. How do they compare?

(d) Make separate frequency tables and bar graphs of the showup variable for people from in and outside of Albuquerque. Compare them.

(e) Make separate frequency tables and bar graphs of the showup variable for the three distances 50, 75, and 100 miles. Proportionately, for which distance did the fewest people show up?

(f) The *showup* variable provides a measure of dedication to bicycling, since the start time is 6:30 A.M. on a cold April morning. By summarizing parts (a) to (e), give the characteristics of the more dedicated cyclists.

1.21 Compare speed distributions for residents and out-of-towners by means of stem and leaf plots. Which is the speedier group, on the average? What might the reasons be for this result?

1.22 Compare speed distributions for men and women.

1.23 Scores on placement tests in arithmetic (20 points possible) and algebra (15 points possible) for students signing up for courses in the University of New Mexico Department of Mathematics and Statistics are given in the accompanying table.

(a) Make a stem and leaf plot of the arithmetic scores using the five lines per stem version of the plot. What is the shape of this score distribution?

(b) Repeat part (a) for the algebra score distribution. How are the shapes of the two distributions related? How do they differ?

(c) Do these exams appear to be too easy or too hard for this group of students? Why?

STUDENT NUMBER	ARITHMETIC SCORE (OUT OF 20)	ALGEBRA SCORE (OUT OF 15)
1	11	0
2	11	0
3	14	1
4	8	0
5	19	5
6	17	3
7	15	4
8	13	2
9	11	0
10	8	2
11	9	2
12	19	5
13	19	7
14	18	9
15	19	7
16	14	0
17	14	4
18	16	1
19	20	9

(table continued on next page)

STUDENT NUMBER	ARITHMETIC SCORE (OUT OF 20)	ALGEBRA SCORE (OUT OF 15)
20	16	4
21	14	2
22	5	0
23	16	5
24	19	9
25	14	3
26	16	1
27	6	0
28	16	5
29	2	0
30	14	7
31	13	0
32	19	1
33	20	14
34	16	7
35	18	7
36	17	7
37	18	5
38	17	1
39	19	6
40	19	12
41	10	5
42	16	5
43	9	0
44	15	7

1.24   The following data give one-day precipitation amounts (in inches) at Climax, Colorado, for the days in a given year for which the precipitation was measurable:*

.085	.030	.060	.080	.110	.100
.495	.260	.085	.220	.890	.045
.010	.020	.035	.055	.190	.015
.025	.065	.125	.010	.125	.020
.015	.010	.005	.070	.100	.030

Drop the decimal points and, using 2 lines per stem and two-digit leaves, make a stem and leaf plot of this data set. How would you describe the shape of this distribution? How would you describe the tail length?

1.25   The following data are extracted from the table of random numbers† to be given in chapter 5. One property of such tables is that, on the average, the relative frequency of each of the digits 0, 1, 2, . . . , 9 is one-tenth. Make a frequency table and a bar graph of the occurrence of these digits in the data set. Do the observed frequencies appear to support the theoretical frequencies of one-tenth each?

33839 40750 18898 61650 09970 47651 41205 65020 33537 01022
53070 61630 84434 05732 18094 71669 41033 82402 16415 83958

* *Source:* Mielke [23].
† *Source:* Rand Corporation [29].

Summary Measures for Measurement Variable Distributions

Introduction

parameters

Numerical quantities that summarize properties of a population frequency distribution are called **parameters.** Most of the questions asked in statistics can be posed in terms of the parameters of frequency distributions. The important problems of comparison and association for measure variables, to be introduced in the next two chapters, rely on a measurement of where the center of the distribution lies, called a **location parameter.** Since these are the only measurement variable problems we will consider in this book, we are especially interested in this type of parameter. Three such parameters, the *mean,* the *median,* and the *trimmed mean,* are introduced in this chapter.

location parameter

estimate

An **estimate** of a parameter is an "intelligent guess" at the true value of the parameter, computed using a sample from the population. The estimate is the result of applying a function or rule, called an **estimator,** to the sample. That is, by applying the "formula" for the estimator to the sample values, the estimate is obtained. For example, as we will see presently, the estimator to be called the *sample mean* is the expression or formula $(X_1 + X_2)/2$ for a sample of size 2. If a random sample produces the two numbers $X_1 = 3$ and $X_2 = 7$, then the estimate based on this sample (also called the sample mean, unfortunately) would be $(3 + 7)/2 = 5$. Estimators of the mean, median, and trimmed mean will be introduced in this chapter, and the methods for calculating their estimates will be illustrated.

estimator

Estimators are of great importance in statistical inference. They provide the primary information from samples for carrying out inferences about the corresponding population parameters. Estimators are also used to describe properties of sample frequency distributions; we are interested in the descriptive use of estimators here.

dispersion parameter

For location estimators, both the descriptive and inferential uses will require an associated measure of the *spread* or *dispersion* of the frequency distribution. A **dispersion parameter** measures how variable the population measurements are. The population *standard deviation* and *interquartile range* will be introduced in this chapter and used to clarify the definitions of the corresponding sample estimators.

The sample standard deviation and a similar quantity for the trimmed mean will be introduced in preparation for their use in statistical inference. The sample interquartile range will be needed in the definition of the box plot, an important exploratory tool that is also introduced in this chapter.

◢ SECTION 2.2

Measures of Location

With only occasional exceptions, a population parameter and its sample estimator are related in a simple way: The estimator is the same function of the sample frequency distribution that the parameter is of the population frequency distribution. This fact makes it convenient to present parameters and their estimators together. In some cases the definition of the parameter is the simpler one; seeing it first will help clarify the definition of the estimator.

The Population and Sample Means

The population mean has an intuitively useful physical interpretation. If relative frequency is equated to mass (for example, by cutting a picture of the frequency curve out of a piece of metal), then the population mean is the center of mass—the point at which the distribution would balance on your finger. See display 2.1.

Display 2.1

Illustration of the physical interpretation of the mean

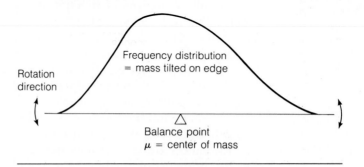

Frequency distribution = mass tilted on edge

Rotation direction

Balance point
μ = center of mass

Unfortunately, the precise definition of the mean requires mathematics beyond our scope. However, it is easily given for another representation of the population frequency distribution. Suppose there are N individuals in the population and that the values of the variable in question for these individuals are X_1, X_2, . . ., X_N. (The three dots stand for "and so on up to".) In practice, these values will not all be observed because we will not look at the entire population.

The symbol μ (lower case Greek letter mu) will be reserved for the population mean. Each member of the population carries relative frequency (mass) $1/N$. The **population mean** is then the sum, over the entire population, of the variable values times frequency:

population mean

$$\mu = X_1 \cdot \frac{1}{N} + X_2 \cdot \frac{1}{N} + \cdots + X_N \cdot \frac{1}{N}$$

$$= \frac{X_1 + X_2 + \cdots + X_N}{N}$$

For example, if the data of display 1.18 (p. 28) were viewed as having come from a population of size $N = 116$, then the mean of the *weight* variable would be

$$\mu = \frac{82 + 95 + \cdots + 225}{116}$$

$$= 140.86 \text{ pounds}$$

This value is shown on a stem and leaf plot of the population in display 2.2. (Think of the vertical line next to the stems as representing a scale with the stem values opposite the midpoints of the intervals represented by the stem lines. Thus, 13 would represent the value 135, 14 the value 145, and so forth. It follows that 140.86 would be nearly midway between the stems 13 and 14, as shown.)

Display 2.2

Stem and leaf plot of the weight data of display 1.18 viewed as a population, with the position of the population mean μ = 140.86 indicated. The population size is N = 116.

```
           8 | 2
           9 | 5
          10 | 55789
          11 | 0000000002224445555
          12 | 0000000255555555555556789
          13 | 00000333555555                μ = (82 + 95 + ··· + 225) / 116
 μ →      14 | 000000058
          15 | 0555557
          16 | 000034555
          17 | 000000555555
          18 | 00000058
          19 | 000
          20 | 55
          21 |
          22 | 5
```

sample mean

The **sample mean,** denoted by $\overline{X}$, is analogous to the population mean. The sample size is n and the relative frequency carried by each individual is $1/n$. Moreover, the variable values, X_1, X_2, . . ., X_n are now the values observed in the sample, so the sample mean can actually be computed. It is defined by the expression

$$\overline{X} = \frac{X_1 + X_2 + \cdots + X_n}{n}$$

For example, a random sample of size $n = 3$ from display 1.18 resulted in the I.D. numbers 11, 35, and 107. These correspond to weight values 110, 133, and 170 pounds, which thus represent a random sample of size 3 from the weight distribution for this population. The sample mean is

$$\overline{X} = \frac{110 + 133 + 170}{3} = 137.67 \text{ pounds}$$

Note that the sample mean has been rounded to two digits more than the accuracy with which the original data were given. While estimates will be carried to full accuracy in computations, we will follow the practice of rounding them to one or two digits of accuracy beyond that of the original data when reporting their values.

Sums with several terms are occasionally needed, so it is convenient to have a notation for them. The Greek capital letter sigma, Σ, is used for this purpose:

$$\Sigma X_i = X_1 + X_2 + \cdots + X_n$$

Thus, the sample mean can be written in the more compact form

$$\overline{X} = \frac{\Sigma X_i}{n}$$

where the sum is taken over all members of the sample.

EXAMPLE 2.1

Suppose the sample measurements consist of the numbers 1, 2, 3, 4, 5. The sample size is $n = 5$ and $X_1 = 1$, $X_2 = 2, \ldots, X_5 = 5$. A histogram of this sample is as follows.

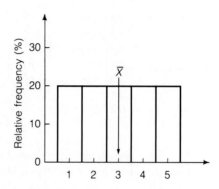

The sample mean is

$$\overline{X} = \frac{1 + 2 + 3 + 4 + 5}{5}$$

$$= \frac{15}{5}$$

$$= 3$$

Note from the histogram that the sample has a symmetric frequency distribution with center of symmetry at 3. An intuitively desirable property of a location measure is that it be at the center of symmetry for symmetric distributions. This is true of the mean as well as of the other location measures we will study.

As demonstrated in the next example, the stem and leaf plot is a convenient display in which to store data when the sample mean is calculated by hand.

EXAMPLE 2.2

The following stem and leaf plot contains the sample data used in example 1.4 (p. 21).

To regain original units, move decimal two places to the left.

```
-0 | 32, 30, 20, 17, 07
 0 | 12, 15, 24, 27, 32
 1 | 01
```

It is convenient, and correct, to compute the mean for the integer data provided in the plot and then move the decimal in the answer to arrive at the mean in the original data units. Thus, enter the data $-32, -30, -20, -17, -7, 12, 15, \ldots, 101$ in the calculator. It will produce the sample mean 9.5454 for these data. Now move the decimal two places to the left to find

$$\overline{X} = .0955$$

EXERCISES

2.1 Find the sample mean of the values 12, 14, 10, 17, 23, 10, 4.

2.2 Make a stem and leaf plot and find the sample mean of the numbers −1.23, 3.54, 7.15, −2.43, −4.22, 6.29, −3.55, −5.12.

2.3 The following data were attributed to F. Hampel in exercise 1.13 (p. 31). Make a stem and leaf plot and find the sample mean of these values.

.0, .8, 1.0, 1.2, 1.3, 1.3, 1.2, 2.4, 4.6

2.4 The stem and leaf plot of the age data from display 1.1 is as follows.

```
1*
 ·  888998998988998999
2*  010040000210
 ·  8967
3*

4*
 ·  6
5*
 ·  5
```

Compute the sample mean age. Indicate where the mean falls on the stem and leaf plot. Is this value a reasonable indicator of the center of frequency of the distribution? Why?

A Comparison Based on Sample Means

The most important use we will make of location measures is to determine "shifts" in frequency distributions relative to one another. For example, to ask whether men are taller than women *on the average* would, in statistical terms, be interpreted as asking whether the height frequency distribution for men is shifted toward large values relative to the distribution for women.

As an example, a similar question could be asked about the population of statistics students from which the data of display 1.1 were drawn. A useful method for visually comparing two distributions is to graph their stem and leaf plots back-to-back on the same scale. Back-to-back stem and leaf plots of the height distributions for men and women are given in display 2.3.

Display 2.3

Back-to-back stem and leaf plots of the height variable based on data of display 1.1 with observation 20 omitted

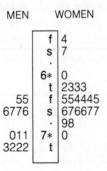

```
           MEN     WOMEN

                  f │ 4
                  s │ 7
                  · │
                 6* │ 0
                  t │ 2333
         55       f │ 554445
       6776       s │ 676677
                  · │ 98
        011      7* │ 0
       3222       t │
```

Intuitively, an *average* represents a value central to a collection of measurements. It is reasonable to use an average to measure distribution shifts, because if two distributions differ from one another only by a relative shift of a certain number of units, then the averages are shifted by the same number of units. Thus, a measure of the amount of shift can be obtained from the difference in the averages. The mean is the most commonly used average; the sample mean is also called the *arithmetic average*.

The sample means of the above height distributions are $\overline{X}_M = 69.00$ inches and $\overline{X}_W = 64.32$ inches. It follows that the men in the sample are taller than the women by an average of $\overline{X}_M - \overline{X}_W = 4.68$ inches. To extrapolate this information to the population, we will need a statistical inference procedure, to be developed in chapter 10.

The sex of the individual with I.D. number 20 was unreadable from the survey form, so the height for this person was omitted from the analysis. To avoid missing values in other computations, this individual is classified as male in display 1.1, based on these height stem and leaf plots; this individual's height (68 inches) is more central to the distribution for men than for women.

EXERCISE

2.5   The following ages are for a sample of students who took a statistics course ten years ago.

> 23, 18, 44, 19, 18, 33, 19, 19, 21, 23, 21, 18, 20, 18, 19, 19, 22, 22, 36, 18, 22, 19, 20, 21, 17

Use these ages in conjunction with the age data from display 1.1, repeated in exercise 2.4 (p. 39), to compare the average ages of recent statistics students with those of ten years ago. Form back-to-back stem and leaf plots of the samples. What is the average age difference? Do the data appear to indicate that students are older, on the average, now than they were ten years ago?

The Population and Sample Medians

population median

Another useful measure of location is the median. The **population median,** denoted by m, is a number such that half of the population measurements are less than or equal to and half are greater than or equal to this value. See display 2.4.

Display 2.4

Population frequency distribution with the median indicated

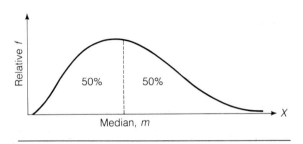

Again viewing the individuals of display 1.18 as a population of size $N = 116$, it is seen from the stem and leaf plot of the weight data given in display 2.5 that the median is $m = 133$. Half of the frequency (58 observations) occurs for weights less than or equal to this value, while half occurs for weights greater than or equal to this value. Comparing displays 2.2 and 2.5, note that the mean and median do not have the same values for this population; they measure the "center" of the distribution somewhat differently.

sample median

The **sample median,** denoted by M, is also intended to divide the sample frequency distribution into equal parts. If the sample size is an odd number, the sample median is the middle value among the observations (listed in increasing order). When the sample size is even, the median is the average of the two middle observations.

Thus, if the sample consists of the five (already ordered) numbers 1, 2, 3, 4, and 5, the median is the middle observation: $M = 3$. If the sample contains the four numbers 1, 2, 3, and 4, then the median is the average of the two middle observations, 2 and 3: $M = (2 + 3)/2 = 2.5$.

Display 2.5

Stem and leaf plot of the weight data of display 1.18 viewed as a population, with the position of the population median, *m* = 133, indicated

```
        8 | 2
        9 | 5
       10 | 55789
       11 | 0000000002224445555
       12 | 00000002555555555555556789
m →    13 | 00000333555555
       14 | 000000058
       15 | 0555557
       16 | 000034555
       17 | 000000555555
       18 | 00000058
       19 | 000
       20 | 55
       21 |
       22 | 5
```

50% = 58 observations

Sample medians are easily hand-calculated by taking advantage of the ordering property of stem and leaf plots. First compute the *median location* by the expression

$$\text{median location} = \frac{n + 1}{2}$$

where n is the sample size. Then, starting from either extreme of the stem and leaf plot, count in order a number of values equal to the median location. When the sample size is odd, the median location will be a whole number and the value reached will be the median. When the sample size is even, the median location will end with the decimal .5. This alerts you to take the average of the values corresponding to the integer part of the median location and this number plus one. For example, if the median location were 16.5, the median would be the average of the 16th and 17th ordered values in the stem and leaf plot.

EXAMPLE 2.3

The stem and leaf plot of the first ten completion times (in hours) for bicycle riders choosing the 100-mile option in the tour of the Rio Grande Valley (see display 1.17, p. 25) is as follows.

To regain original units, move decimal one place to the left.

```
4 | 98
5 | 7
6 | 1
7 | 9
8 | 3010
9 | 0
```

Since the sample size is 10, the median location is

$$\text{median location} = \frac{10 + 1}{2} = 5.5$$

Thus, to find the median, count six observations in order from either end of the stem and leaf plot, then take the average of the fifth and sixth observations. Again, we can ignore the decimal point until the last step, then insert it in the computed median.

Starting from the smallest value (the top of the stem and leaf plot) and counting up, the first six ordered values are 48, 49, 57, 61, 79, and 80. The median of the integer values is then $(79 + 80)/2 = 79.5$. Now, move the decimal one place to the left to obtain the median completion time

$M = 7.95$ hours

Check that the same answer is obtained by starting the count with the largest value (at the bottom of the stem and leaf plot) and ordering the values in decreasing order: 90, 83, 81, and so on.

Negative numbers can cause some confusion in ordering the data. The following example indicates the proper procedure.

EXAMPLE 2.4

The stem and leaf plot of the data used for the mean calculation in example 2.2 is repeated here.

To regain original units, move decimal two places to the left.

```
-0 | 32, 30, 20, 17, 07
 0 | 12, 15, 24, 27, 32
 1 | 01
```

The sample size is 11, so the median location is $(11 + 1)/2 = 6$. The median will be the sixth value in order from either end of the stem and leaf plot. Starting from the smallest value, the first 6 values in order are −32, −30,

-20, -17, -7, and 12. The median of the integer values is 12. Now, move the decimal two places to the left to obtain the median in the original units:

$$M = .12$$

Check this value by counting in decreasing order, starting from the largest value in the stem and leaf plot.

EXERCISES

2.6 Find the sample median of the values 12, 14, 10, 17, 23, 10, and 4.

2.7 Find the sample median of the numbers

$$-1.23, 3.54, 7.15, -2.43, -4.22, 6.29, -3.55, -5.12$$

2.8 Calculate the sample median of the Hampel data given in exercise 2.3 (p. 39). How does this value compare with the sample mean as a measure of the center of frequency?

2.9 Compute the sample median of the age data given in exercise 2.4 (p. 39). Locate this value on the stem and leaf plot and comment on any differences between the median and the mean.

SECTION 2.3

Measures of Dispersion

The two major measures of spread or dispersion are the interquartile range and the standard deviation. The standard deviation will be paired with the mean in statistical inference, and the interquartile range will be used in conjuncton with the median in exploratory and descriptive analyses. Since the interquartile range has a convenient intuitive interpretation not shared by the standard deviation, it is presented first.

The Population and Sample Interquartile Ranges

first quartile
third quartile

Just as the population median divides a population frequency distribution in half, the population quartiles subdivide it into quarters. The **first quartile,** q_1, subdivides the lower half of the distribution again in half, and the **third quartile,** q_3, does the same for the upper half. See display 2.6.

interquartile range

The median is the second quartile. The **interquartile range,** denoted by iqr, is then the distance between the first and third quartiles:

$$\text{iqr} = q_3 - q_1$$

Display 2.7 shows the quartiles and interquartile range for the weight distribution of display 1.18. The quartiles q_1 and q_3 enclose the middle 50% of the frequency and the length of the interval between these values, iqr, measures the dispersion of the distribution.

Display 2.6

Population quartiles and interquartile range

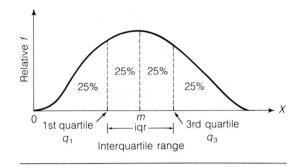

Display 2.7

Stem and leaf plot of the weight data of display 1.18 viewed as a population, with the quartiles ($q_1 = 120$ and $q_3 = 164.5$) and the interquartile range (iqr = 44.5) indicated

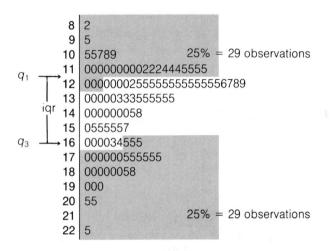

The interquartile range measures dispersion in the following way. Since the middle half of the distribution's frequency lies between q_1 and q_3, if the frequency is tightly concentrated around its center, the quartiles will be close together, and the iqr will have a small value. On the other hand, if the frequency is widely dispersed, the quartiles will be widely separated, leading to a large interquartile range. This phenomenon is illustrated in display 2.8.

The sample quartiles are defined as follows. The method used to define and compute the median is extended by defining the **quartile location** in terms of the median location:

quartile location

$$\text{quartile location} = \frac{\text{truncated median location} + 1}{2}$$

truncated median location

first sample quartile

The **truncated median location** is simply the median location, $(n + 1)/2$, *with the fraction .5 omitted if present.* (That is, we drop *down* to the lower whole number.) To find the **first sample quartile** Q_1, start with the smallest observation and count a number of observations equal to the quartile location in increasing order. If the quartile location is an integer, the value of

Display 2.8

Illustration of the
interquartile range for
a distribution with
small dispersion and
one with large
disperson

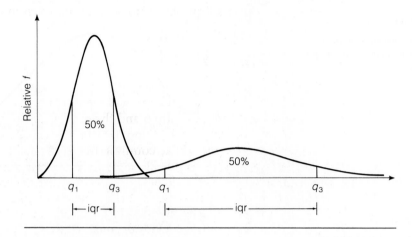

the observation arrived at is Q_1. If the quartile location ends in .5, Q_1 is taken to be the average of the two adjacent values, just as in the calculation of the median.

third sample quartile

sample interquartile range

The **third sample quartile** Q_3 is obtained in the same way by counting observations in decreasing order starting with the largest. The **sample interquartile range** is then

$$IQR = Q_3 - Q_1$$

EXAMPLE 2.5

We will calculate the interquartile range for the bicycling completion times given in example 2.3 (p. 43). As before, the stem and leaf plot will be used to organize the computation.

To regain original units, move decimal one place to the left.

```
4 | 98
5 | 7
6 | 1
7 | 9
8 | 3010
9 | 0
```

The median location was seen to be $(10 + 1)/2 = 5.5$. The truncated median location is then 5 and the quartile location is $(5 + 1)/2 = 3$.

The smallest three observations in order are 48, 49, and 57. The first quartile, in integer form, is then 57. Moving the decimal, the first quartile in the original units is $Q_1 = 5.7$ hours.

Starting from the largest value, the three observations in order are 90, 83, and 81. The third quartile is $Q_3 = 8.1$ hours.

Finally, the sample interquartile range is $IQR = 8.1 - 5.7 = 2.4$ hours.

EXAMPLE 2.6

The interquartile range will be calculated for the sample data last seen in example 2.4 (p. 43). The stem and leaf plot is as follows.

To regain original units, move decimal two places to the left.

```
-0 | 32, 30, 20, 17, 07
 0 | 12, 15, 24, 27, 32
 1 | 01
```

The median location was seen to be 6 for these data. The truncated median location is also 6 and the quartile location is $(6 + 1)/2 = 3.5$.

The first quartile is found by taking the average of the third and fourth observations in order, counting from the smallest. The first four observations are -32, -30, -20, and -17. The first quartile for the integer data is then $(-20 + -17)/2 = -18.5$. Moving the decimal two places to the left, $Q_1 = -.185$. Similarly, counting down to the third and fourth observations, starting from the largest, the third quartile is the average of 27 and 24: $(27 + 24)/2 = 25.5$. With decimal adjusted, $Q_3 = .255$. The sample interquartile range is then IQR $= .255 - (-.185) = .440$.

EXERCISES

2.10   Find the sample interquartile range of the values 12, 14, 10, 17, 23, 10, and 4.

2.11   Find the sample interquartile range of the numbers -1.23, 3.54, 7.15, -2.43, -4.22, 6.29, -3.55, and -5.12.

2.12   Find the sample interquartile range for the Hampel data: .0, .8, 1.0, 1.2, 1.3, 1.3, 1.4, 2.4, and 4.6.

2.13   Find the sample interquartile range of the age data:

```
1* |
 · | 888998998988998999
2* | 010040000210
 · | 8967
3* |
 · |
4* |
 · | 6
5* |
 · | 5
```

The Population and Sample Standard Deviation

population standard deviation

The **population standard deviation** will be denoted by the lower case Greek letter sigma, σ. For a population of size N whose members have values X_1,

$X_2, \ldots, X_N$ for a given variable, the standard deviation is defined by the expression

$$\sigma = \sqrt{\frac{\Sigma(X_i - \mu)^2}{N}}$$

The calculation of the standard deviation for the weight variable of display 1.18 is indicated in display 2.9.

Display 2.9

Calculation of the standard deviation of the weight variable for display 1.18 viewed as a population. The interval from $\mu - \sigma = 113.12$ to $\mu + \sigma = 168.62$ is shown. It contains $71/116 = 61.2\%$ of the distribution.

```
               8 │ 2
               9 │ 5
              10 │ 55789
 μ − σ ┬11 │ 0000000002224445555
              12 │ 00000002555555555555556789
              13 │ 00000333555555
 μ ──▶ 14 │ 000000058
              15 │ 0555557
         ┌16 │ 000034555
 μ + σ ┴17 │ 000000555555
              18 │ 00000058
              19 │ 000
              20 │ 55
              21 │
              22 │ 5
```

$\mu = 140.86$

$$\sigma = \sqrt{\frac{(82 - 140.86)^2 + (95 - 140.86)^2 + \cdots + (225 - 140.86)^2}{116}}$$

$$= 27.75$$

population variance

The square of the standard deviation, σ^2, is called the **population variance.** The variance and standard deviation contain the same dispersion information, and both measures are used in inference. However, the standard deviation has the added attraction of being in the same units as the original measurement.

In the mass analogy used to explain the population mean, the standard deviation is a measure of dispersion called the *moment of inertia.* Aside from the square root, the definition multiplies each frequency $1/N$ by the square of the distance of the measurement value, X_i, from the mean, μ. These products are then summed over the entire population.

From this definition, it is evident that if most measurement values were close to the mean, indicating small dispersion, the squared terms would be small and the resulting standard deviation would also be small. On the other hand, widely dispersed measurements should lead to a large standard deviation. But how can a particular value of the standard deviation be interpreted? What does it mean for a population to have standard deviation $\sigma = 3$, for example?

Recall that if the interquartile range has the value iqr = 3, this means that the middle 50% of the frequency spans a range of 3 units. We could hope for a similar interpretation for the standard deviation: perhaps a certain proportion of the frequency distribution is within one standard deviation of the mean (between $\mu - \sigma$ and $\mu + \sigma$). Unfortunately, this is not the case for all distributions.

A somewhat deceptive practice is to "calibrate" the standard deviation by means of the normal distribution. About 68% of a normal distribution is within one σ of the mean. However, this depends on the shape of the normal curve—especially on the tail length—and does not hold true for all distributions. For example, the one-σ interval around the mean for the weight distribution shown in display 2.9 contains only 61.2% of the total frequency. As will be demonstrated in a supplementary exercise, nearly all of the frequency for suitably long-tailed distributions can be within one σ of the mean, while short-tailed distributions can have almost no frequency in this range. It is this sensitivity to distribution shape that makes the standard deviation a poor descriptive measure. We will not discover the real importance of the standard deviation until we begin our study of probability and statistical inference in the second section of the text. However, it is convenient to define and give examples of the sample standard deviation at this point.

sample standard deviation

The **sample standard deviation,** denoted by s, is defined for the sample frequency distribution in near (but not perfect) analogy with the population version:

$$s = \sqrt{\frac{\Sigma (X_i - \overline{X})^2}{n - 1}} \qquad (1)$$

As before, n is the sample size and the X_i's are the observed values of the variable. Note that the population mean in the definition of σ has been replaced by the sample mean, and the population size, N, has been replaced by the sample size n *minus* 1. The $n - 1$ divisor is introduced to provide the **sample variance,** s^2, with a desirable statistical property called *unbiasedness.* This property will be discussed in chapter 7. For now, we will simply accept this divisor as part of the definition of the sample standard deviation.

sample variance

For those who compute the sample standard deviation by hand, it is useful to have a more convenient (but algebraically equivalent) definition of the standard deviation:

$$s = \sqrt{\frac{\Sigma X_i^2 - [(\Sigma X_i)^2/n]}{n - 1}} \qquad (2)$$

Only the three quantities n, ΣX_i, and ΣX_i^2 are required to compute this version of the standard deviation. This economy is particularly suited to hand calculators, which have rather limited memory space; the sums are formed as the data are entered, thus requiring only three memory locations. For reasons of numerical accuracy, expression (1), or something quite similar, is the one usually used to calculate the standard deviation with computers. While expressions (1) and (2) are *algebraically* equivalent (i.e., one can be derived from the other by purely algebraic steps), they can lead to different numerical answers. This fact will be illustrated in a supplementary exercise at the end of the chapter.

EXAMPLE 2.7

In this example, the standard deviation of the $n = 5$ numbers 1, 2, 3, 4, 5 will be found, using first formula (1), then formula (2).

The variable values are $X_1 = 1$, $X_2 = 2$, . . ., $X_5 = 5$. As was seen earlier, the value of $\overline{X}$ for these numbers is 3. Thus, $X_1 - \overline{X} = 1 - 3 = -2$, $X_2 - \overline{X} = 2 - 3 = -1$, and so forth. It follows that

$$\Sigma (X_i - \overline{X})^2 = (-2)^2 + (-1)^2 + 0^2 + 1^2 + 2^2$$
$$= 4 + 1 + 0 + 1 + 4$$
$$= 10$$

Then

$$\frac{\Sigma (X_i - \overline{X})^2}{n - 1} = \frac{10}{4}$$
$$= 2.5$$

Finally, s is the square root of this number.

$$s = \sqrt{2.5}$$
$$= 1.58$$

To use the computing formula (2), we first calculate

$$\Sigma X_i = 1 + 2 + 3 + 4 + 5$$
$$= 15$$

and

$$\Sigma X_i^2 = 1^2 + 2^2 + 3^2 + 4^2 + 5^2$$
$$= 1 + 4 + 9 + 16 + 25$$
$$= 55$$

Then

$$\Sigma X_i^2 - \frac{(\Sigma X_i)^2}{n} = 55 - \frac{15^2}{5}$$
$$= 10$$

Dividing this by $n - 1 = 4$ produces $10/4 = 2.5$. Taking the square root, we again obtain

$$s = 1.58$$

A note of warning is in order for those who compute standard deviations using hand calculators. Most calculators with this feature have buttons for both the *sample* standard deviation (usually designated s or σ_{n-1}) and the *population* standard deviation (designated σ or σ_n). Be sure to use the appropriate option. The two give similar but not identical answers because of the division by $n - 1$ in the sample standard deviation and by N in the population version.

◪
EXERCISES

2.14 Find the sample standard deviation of the numbers 12, 14, 10, 17, 23, 10, and 4.

(a) using the defining expression (1), page 49

(b) using the computing expression (2)

2.15 Find the sample standard deviation of the numbers -1.23, 3.54, 7.15, -2.43, -4.22, 6.29, -3.55, and -5.12.

2.16 Find the sample standard deviation for the Hampel data: .0, .8, 1.0, 1.2, 1.3, 1.3, 1.4, 2.4, and 4.6

2.17 Find the standard deviation of the age data:

```
1*
 ·  888998998988998999
2*  010040000210
 ·  8967
3*
 ·
4*
 ·  6
5*
 ·  5
```

2.18 Refer to the bicycling completion time data last given in example 2.5 (p. 46). Calculate the standard deviation of the integer data given in the stem and leaf plot, then move the decimal in the answer to obtain the standard deviation of the data in the original data units.

2.19 Follow the directions of exercise 2.18 to find the standard deviation of the data given in the following stem and leaf plot.

To regain original units, move decimal two places to the left.

```
-0 | 32, 30, 20, 17, 07
 0 | 12, 15, 24, 27, 32
 1 | 01
```

◪ SECTION 2.4
The Box Plot

The box plot is a graphical display, devised by J. W. Tukey, that provides both diagnostic and descriptive information about a sample. As we have seen, the histogram and the stem and leaf plot excel at providing shape information about the center of a frequency distribution but tell us little about the tails. The box plot also gives useful information about location, dispersion, and skewness but is designed to provide pictorial information about the tails as well. The two types of displays together give complementary pictures of a frequency distribution that will serve all of our descriptive and diagnostic needs.

Interpretation of the Box Plot

A box plot of the ages of statistics students from display 1.1 (p. 9) is given in display 2.10.

Display 2.10

Box plot of age data from display 1.1

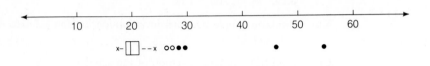

The box, for which the display was named, contains the middle 50% of the sample.

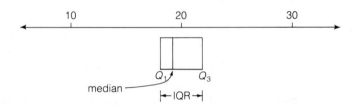

The left edge of the box is the first quartile, Q_1, and the right edge is Q_3. The line inside the box is at the sample median. It follows that this line pinpoints the "center" of the distribution, while the length of the box is a measure of dispersion—the interquartile range.

The remaining details in the box plot provide information about the tails of the distribution. *Tail lengths* are described by comparing features of the distribution at hand relative to the tails of a normal distribution. A remarkably consistent property of sample distributions, observed by the noted statistician C. P. Winsor, is that the variation of frequency in the center of the distribution can be closely approximated by that of a normal distribution. Because the sample quartiles depend only on the middle 50% of the frequency, this observation suggests that a "yardstick" for evaluating tail length can be based on a normal distribution with the same quartiles as the sample quartiles. The following scaling of the box plot uses this idea implicitly.

A scale factor of 1.5 interquartile ranges is used to establish three regions on either side of the box in a box plot. Dashed lines are drawn to the most extreme observations in the first or adjacent regions, which extend 1.5 IQRs from the box.

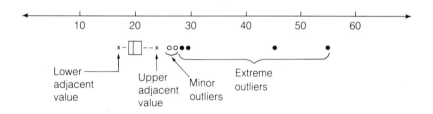

minor outliers

extreme or major outliers

Observations falling in the second region on either side, which extends from 1.5 to 3 IQRs from the box, are called **minor outliers** and are drawn as open circles on the plot. Observations falling in the third region, outside of the three-IQR limit, are called **extreme** (or **major**) **outliers** and are drawn as solid circles.

The presence of outliers signals a long-tailedness in the underlying population frequency distribution that can adversely affect the inference methods we will learn in later chapters. A rough calibration of the "seriousness" of different kinds of outliers can be based on how frequently they occur for normal distributions: If observations were sampled at random from a normal distribution, an outlier (of either kind) would be seen only about seven times in every 1000 draws. Extreme outliers would be seen only about twice in every million draws. It follows that the sample shown in display 2.10 would be an extremely unusual one indeed if the population age distribution were normal; in this sample of only 36 observations, we find two minor and three extreme outliers. We can anticipate problems with the application of traditional inference methods to these data.

Information concerning the symmetry of the distribution can also be obtained from the box plot. The box plot of a symmetric distribution would show the median line in the center of the box, equally long dashed lines, and roughly an equal scatter of outliers, if any, on both sides of the box. The age distribution shows a strong skewness to the right.

The box plot does not show the locations or sizes of peaks and troughs in frequency distributions, however. We need a stem and leaf plot or histogram to provide that information. One important use of the box plot is described in the next subsection.

Using Box Plots to Screen Data for Errors

Sometimes the outliers in a box plot represent errors in the data. By correcting these errors, we can avoid future analysis problems. A case in point is the box plot of the height data from display 1.1, given in display 2.11.

Two women's heights show up as outliers. One is a legitimate height for a rather short woman, but the other is a recording error. The woman with I.D. number 27 misrecorded her height of 5′4″ as 54″ instead of 64″. This correction was made in the data of display 1.18. Of course, errors that

Display 2.11

Box plot of the height data from display 1.1

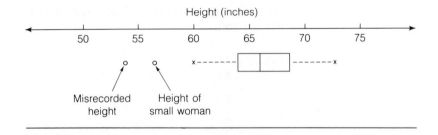

fall in the central part of the distribution will not be detected by the box plot, but they will have little effect on the analysis in any event, so their detection is generally not critical.

The use of box plots for descriptive purposes will be illustrated in succeeding chapters. Outlier detection is a diagnostic function of considerable importance. The fact that it is possible to carry out both functions at the same time makes the box plot a powerful exploratory tool.

Computer programs for graphing box plots are not universally available, so it is useful to have an organized method for computing and drawing them by hand. The next subsection will give such a method.

Making Box Plots By Hand

The data required to graph a box plot can be conveniently displayed in a pair of computing forms. The steps of the calculation are organized by the process of filling out these forms. The procedure will be illustrated by deriving the information needed to graph the box plot of the age data given in display 2.10.

We begin with the age stem and leaf plot.

```
1*
 ·  888888889999999999
2*  000000001124
 ·  6789
3*
 ·
4*
 ·  6
5*
 ·  5
```

The information required to plot the box itself is collected in the first form.

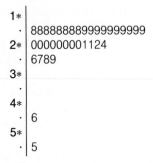

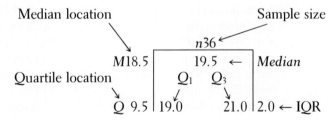

First, the median location and quartile location are calculated and written beside the form. These are used in conjunction with the stem and leaf plot to calculate the median and quartiles, which are entered in the form as shown. Q_1 is then subtracted from Q_3 to obtain the IQR.

A second form is now drawn below the first, with the scale factor 1.5 × IQR entered at the top.

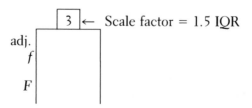

| | 3 | ← | Scale factor = 1.5 IQR |

adj.
f
F

fences

This form is filled out in two steps. First, the boundaries or **fences** between the three regions on both sides of the box are computed. The fences on the left side are obtained by *subtracting* the scale factor from Q_1 to obtain $f = 19.0 - 3 = 16$, then subtracting it again from f to obtain $F = 16 - 3 = 13$. The fences to the right of the box are obtained by *adding* the scale factor to Q_3: $f = 21 + 3 = 24$, and $F = 24 + 3 = 27$. The fences are entered in the form as shown in the next diagram.

```
          | 3 |
        ┌─────────┐
        │  Space  │
      f │ 16   24 │
        │  Space  │
      F │ 13   27 │
        │  Space  │
```

In the final step, the adjacent values and outliers are entered into the form from the stem and leaf plot. The two adjacent values are the most extreme observations in the first regions on each side of the box. The first region to the left of the box extends from the first quartile, $Q_1 = 19$, down to the first fence, $f = 16$. Thus, the lower adjacent value will be the value from the stem and leaf plot between 16 and 19 that is closest to 16. This value is 18. The adjacent value is entered in the form as shown in the next diagram.

```
              | 3 |
           ┌─────────┐
     adj.  │ 18      24 │
        f  │ 16      24 │
           │ None    Two │ 26, 27
        F  │ 13      27 │
           │ None    Four │ 28, 29, 46, 55
```

Since 18 is also the smallest value in the stem and leaf plot, there will be no observations in the other two regions below the box. To indicate this fact, "None" is written in the form in the spaces reserved for the numbers of minor and extreme outliers lying to the left of the box.

The upper adjacent value is the number from the stem and leaf plot between the third quartile, $Q_3 = 21$, and the fence, $f = 24$, closest to 24. We will always put an observation that falls *on* a fence in the less extreme region (i.e., the region closer to the box.) Since one of the observations is equal to 24, this is taken to be the adjacent value. Note that this adjacent value is not the largest observation in the data set, so we must continue to catalog the outliers.

Minor outliers are values from the stem and leaf plot lying between the fences, $f = 24$ and $F = 27$. Thus, since observations falling on fences are put in the less extreme region, both 26 and 27 will be classified as minor outliers. Since there are two minor outliers, "Two" is written inside the form, and their values are listed as shown.

Extreme outliers are observations lying beyond the fence $F = 27$. There are four of them, with values 28, 29, 46, and 55. At this point, we are ready to graph the box plot.

Refer to display 2.10. The first step in drawing the box plot is to construct a (convenient) scale that encompasses the entire range of data—here, from 18 to 55. A scale in 10-year intervals from 10 to 60 does nicely. The box itself is drawn first, with its edges at the quartiles 19 and 21 and the median line at 19.5. Dashed lines are drawn from the box to the adjacent values, "x'ed" at 18 and 24. Finally the minor outliers are plotted as open circles and the extreme outliers as solid ones.

Display 2.12

Stem and leaf plot and computing form for the box plot of height data given in display 2.11

```
5*   4
 ·   7
6*   23304434
 ·   659677757688565576
7*   21123022
 ·
```

```
                         n36
            M18.5 ┌──────────────────┐
                  │        66         │
            Q9.5  │ 64          68.5  │ 4.5
                  └──────────────────┘
```

```
                  ┌────────────┐
                  │    6.75     │
        adj. │ 60 └────────────┘ 73
           f │ 57.25              75.25
      54, 57 │ Two                None
           F │ 50.50              82.00
                       None
```

A useful observation for data containing decimals is that the decimal point can be ignored throughout the entire box plot construction and then simply reinserted in the final scale of the plot. Had we measured ages in ten-year units, so that a 24-year-old is 2.4 units of age, for example, the box plot in the appropriate scale would be obtained simply by inserting a decimal point in the scale values of display 2.10: 10 would become 1.0, 20 would be 2.0, and so forth.

EXAMPLE 2.8

The box plot for the height data given in display 2.11 was graphed from the computing form in display 2.12. Use the stem and leaf plot to verify the details of the computation.

EXERCISES

2.20   Make a box plot of the observations 12, 14, 10, 17, 23, 10, and 4. Use it to describe the shape of the distribution.

2.21   Make a box plot of the data $-1.23, 3.54, 7.15, -2.43, -4.22, 6.29, -3.55$, and -5.12. Use the plot to describe the shape of the distribution.

2.22   Construct a box plot of the Hampel data: .0, .8, 1.0, 1.2, 1.3, 1.3, 1.4, 2.4, and 4.6. Describe the shape of the distribution with special reference to outliers.

2.23   Make a box plot of the following bicycling completion time data:

To regain original units, move decimal one place to the left.

```
4 | 98
5 | 7
6 | 1
7 | 9
8 | 3010
9 | 0
```

Describe the shape of the distribution.

2.24   Make a box plot of the data given in the following stem and leaf plot.

To regain original units, move decimal two places to the left.

```
-0 | 32, 30, 20, 17, 07
 0 | 12, 15, 24, 27, 32
 1 | 01
```

Describe the shape of the distribution.

◢ SECTION 2.5

Resistant Measures and the Trimmed Mean*

Our interest in outliers stems from their harmful influence on traditional methods of statistical inference. This influence shows up at the descriptive level in the way outliers affect the values of important estimators, such as the sample mean and standard deviation.

To illustrate this influence, we can calculate estimates first in the presence of outliers, then in their absence. Differences in the estimates can then be attributed to the outliers. The following table gives values of the sample mean and standard deviation for the age data of display 1.1. The box plot of ages in display 2.10 (p. 52) showed four major and two minor outliers. We will assess the effect of both types of outliers by computing the estimates with no outliers, with minor outliers only, and with all outliers present.

VALUES OF	WITHOUT OUTLIERS	WITH MINOR OUTLIERS ONLY	WITH ALL OUTLIERS
$\bar{X}$	19.40	19.84	22.03
s	1.354	2.187	7.648

Note that the standard deviation is the more highly influenced statistic. It nearly doubles in value from the influence of the minor outliers and increases by more than a factor of 6 under the influence of all outliers. The extreme outliers have by far the greatest effect. The mean is shifted outside of the middle 50% of frequency by the outliers, and its value as a measure of location or distribution center becomes somewhat questionable for this distribution.

The following table repeats the same analysis for the sample median and interquartile range.

VALUES OF	WITHOUT OUTLIERS	WITH MINOR OUTLIERS ONLY	WITH ALL OUTLIERS
M	19.0	19.0	19.5
IQR	2.0	1.5	2.0

resistant estimators

Except for a curious drop from 2 to 1.5, the IQR remains unaffected by the presence of the outliers. The median increases by only .5 when all outliers are included. Estimators whose values are affected very little by the presence of outliers in the data are called **resistant estimators**. It follows that the sample median and interquartile range are resistant estimators, but the sample mean and standard deviation are not.

Robust methods of statistical inference are designed to counteract the influence of outliers in samples. To achieve robustness, resistant estimators

---

* The trimmed mean and standard deviation are used only in chapter 12 on robust statistical methods.

will be substituted for more traditional estimators such as the mean and standard deviation in the inference procedures we will consider.

In chapter 7 we will see that the essential connection between a population parameter and its sample estimator, which is required in statistical inference, is made by the so-called sampling distribution of the estimator. It follows that an important requirement of an estimator is that a good approximation to its sampling distributions must be readily available. Unfortunately, this is not the case for the median. Although it is useful at the descriptive and exploratory levels of statistics because of its resistance, the sample median is rarely used in inference. Other resistant estimators with more easily approximated sampling distributions are needed; an example of such an estimator is given in the next subsection.

The Trimmed Mean

trimmed mean

trimming fraction

An alternative location estimator that is both resistant and has a tractable sampling distribution is the **trimmed mean.** The trimming process that produces this estimator is an appealing one from the viewpoint of eliminating outliers: A percentage of the observations are removed from each end of the ordered sample. The percentage, called the **trimming fraction,** is made large enough to include at least all extreme outliers among the observations so removed. The trimmed mean is then the ordinary mean of the remaining observations.

Since the trimming fraction can be selected to be any percentage between 0 and 50, there are a number of different possibilities. The trimming fraction is usually taken to be a multiple of 5; the ones most often seen in practice are between 5% and 25%.

We will use a stem and leaf plot of the age data to illustrate the calculation of the 10% trimmed mean.

```
1*
 ·  | 888888889999999999
2*  | 000000001124
 ·  | 6789
3*
 ·  |
4*
 ·  | 6
5*
 ·  | 5
```

Recall that the sample size is $n = 36$. When an application of the trimming fraction leads to a fractional number, the strategy is to round the number to be trimmed *up* to the next larger whole number. Since the 10% trimming fraction leads to $.1 \times 36 = 3.6$, the number of values to be trimmed from each end of the sample here is four. These observations are underlined in the above stem and leaf plot.

The 10% *trimmed sample* then consists of the values not underlined.

trimmed sample size

The **trimmed sample size,** n_T, is the number of observations in the trimmed sample. Since $2 \times 4 = 8$ observations have been trimmed,

$$n_T = 36 - 8 = 28$$

The 10% sample trimmed mean is the ordinary sample mean of the trimmed sample, or

$$\overline{X}_T = 20.11$$

To satisfy the additional property of having a sampling distribution that can be conveniently approximated, the trimmed mean will have to be paired with the "right" measure of dispersion. This measure will be called the *trimmed standard deviation*—but it is not the standard deviation of the trimmed sample, as is seen next.

The Trimmed Standard Deviation

Tukey and McLaughlin [34] showed that the statistical theory built around the summary values n, $\overline{X}$, and s can be applied without change to obtain robust procedures if these summary values are replaced by n_T, $\overline{X}_T$, and the following definition of the trimmed standard deviation s_T.

Winsorized sample   A sample called the **Winsorized sample** is constructed by replacing the values removed to form the trimmed sample by the values next in line for trimming. Thus, for the age data, the four underlined values 28, 29, 46, and 55 in the above stem and leaf plot would be replaced by the next value in decreasing order, 27. The four 18s that were trimmed would reappear as 18s, since the next value in order is also 18. The stem and leaf plot of the Winsorized sample, with replacements boxed, is as follows.

```
1*
 ·  | 8888 88889999999999
2*  | 010040000210
 ·  | 67 7777
```

trimmed standard deviation   Now the *Winsorized standard deviation*, s_W, is the ordinary standard deviation of the Winsorized sample. The **trimmed standard deviation** is then defined by the expression

$$s_T = \sqrt{\frac{(n-1)s_W^2}{n_T - 1}}$$

where n is the original sample size and n_T is the trimmed sample size.
For the age data we calculate $s_W = 3.072$. Thus, since $n - 1 = 36 - 1 = 35$, and $n_T - 1 = 28 - 1 = 27$, we find

$$s_T = \sqrt{\frac{35 \times 3.072^2}{27}}$$
$$= 3.50$$

EXERCISES

2.25 (a) Calculate the 15% trimmed mean and standard deviation of the sample 1, 2, 3, 4, 5.

(b) Calculate the 25% trimmed mean and standard deviation of the sample of part (a).

2.26 Find the 5% trimmed mean and standard deviation of the data 12, 14, 10, 17, 23, 10, and 4. How do these values compare with the ordinary mean and standard deviation?

2.27 Find the 10% trimmed mean and standard deviation of the sample -1.23, 3.54, 7.15, -2.43, -4.22, 6.29, -3.55, and -5.12.

2.28 Calculate the 10% trimmed mean and standard deviation for (a) the sample consisting of the age data, given in the stem and leaf plot on page 54, with extreme outliers omitted, and (b) the age data with all outliers omitted.
Use these calculations to verify the following table.

VALUES OF	WITHOUT OUTLIERS	WITH MINOR OUTLIERS ONLY	WITH ALL OUTLIERS
$\bar{X}_T$	19.21	19.33	20.11
s_T	1.14	1.24	3.50

Discuss the resistance of the trimmed mean and standard deviation to the outliers of the age sample. How does this resistance compare to that of the mean and standard deviation? To the median and the interquartile range?

The Effect of Varying the Trimming Fraction—The Mean and Median as Trimmed Means

The sample mean involves no trimming, so it could be considered the trimmed mean with a 0% trimming fraction. The sample median, on the other hand, is the average of the one or two middle observations in the sample, which can be viewed as a trimmed mean with all other observations trimmed. For this reason, the median is sometimes called the 50% trimmed mean.

It is logical that the degree of resistance (insensitivity to outliers) of the trimmed mean should increase from that of the ordinary mean to that of the median as the trimming fraction varies from 0 to 50%. This is the case, as the comparison of tables requested in exercise 2.28 suggests. However, it is important to note that resistance generally increases very rapidly with the trimming fraction, so that a 10% or 20% trimmed mean will provide good results for most problems encountered in practice.

The reason for this rapid increase is that the *sizes* of the outliers are more of a problem in real data than are the number of outliers. Once the outliers have been trimmed, their sizes no longer play a role in the value of the trimmed mean. Thus, a modest trimming fraction can lead to a substantial improvement in performance.

A case in point is the following example, which contains one of the author's favorite data sets. The context has been changed, but the data set appeared in a published report of real data along with summaries including means and standard deviations. Unfortunately, the analysis did not include a screening device such as a box plot, and a rather obvious error made the summary of this data set meaningless.

EXAMPLE 2.9

Annual incomes for a sample of households from a fictitious town

In an attempt to obtain an average income on which to base a uniform tax, we imagine that the following sample was obtained from a particular town.

HOUSEHOLD	ANNUAL INCOME (×$1000)
1	7
2	1110 (retired billionaire living on dividends)
3	7
4	5
5	8
6	12
7	0 (unemployed)
8	5
9	2
10	2
11	46
12	7

The following box plot suggests that the sample mean (and standard deviation) are in for a lot of trouble.

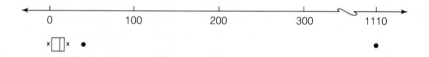

This trouble is confirmed by the values

$$\overline{X} = \$100,920 \quad \text{and} \quad s = \$318,010$$

With the middle 50% of the incomes in the range from $3500 to $29,000, it is unlikely that the "average" wage earner will be happy about being taxed on the basis of a $100,920 average income. The median income of M = $7000 seems more appropriate. The influence of the outliers on the mean is very strong.

We now show that a 10% trimming fraction is sufficient to counteract the outlier effect. 10% of the sample size ($n = 12$) rounds up to 2; two observations will be trimmed from each end of the sample. The following stem and leaf plot shows the trimmed observations. Note the use of the HI (high) category to avoid distorting the scale of the plot.

Stem and leaf plot of
household income data
with values to be
trimmed for a 10%
trimmed mean
underlined

```
0 | 775805227
1 | 2
2 |
3 |
4 | 6
HI| 1110
```

The trimmed sample size is $n_T = 12 - 4 = 8$. The trimmed mean is

$$\overline{X}_T = \$6625$$

The Winsorized sample is obtained by replacing the underlined values 0 and 2 by 2 on the low side and 46 and 1110 by 12 on the high side. The standard deviation of the Winsorized sample (in thousands of dollars) is then

$$s_W = 3.793$$

Thus, the trimmed standard deviation is

$$s_T = \sqrt{\frac{(12 - 1) \times 3.793^2}{8 - 1}} = 4.755$$

or \$4755.

The 10% trimmed mean, \$6625, is also a more reasonable average of the household incomes than is the sample mean. An "average" close to the median has been achieved with a reasonably small trimming fraction. The effect of replacing a traditional inference procedure by a robust procedure based on the trimmed mean will be demonstrated for this sample in chapter 12.

The following table illustrates how the trimmed mean and trimmed standard deviation vary with trimming fraction for these data.

TRIMMING FRACTION	0%	5%	10%	20%	50%
Trimmed Mean	100.92	10.1	6.625	6.50	7.00
Trimmed Standard Deviation	318.01	17.6	4.75	1.97	—

EXERCISE

2.29 The ordinary (0% trimmed) mean and standard deviation for the Hampel data—0.0, .8, 1.0, 1.2, 1.3, 1.3, 1.4, 2.4, 4.6—were found to be $\overline{X} = 1.556$ and $s = 1.302$, and the median was seen to be 1.3.

(a) Calculate the 10% trimmed mean and standard deviation for these data.

(b) Calculate the 20% trimmed mean and standard deviation for these data.

(c) Make a table like the one in the last example and compare the values of the 0%, 10%, 20%, and 50% trimmed means and the values of the 0%, 10%, and 20% trimmed standard deviations. Explain what you find in terms of the box plot computed in exercise 2.22 (p. 57). (Compute the box plot here if it was not previously obtained.)

Chapter 2 Quiz

1. What is a parameter?

2. What does a location parameter measure for a population frequency distribution? Give examples of location parameters.

3. What is the distinction between an estimator and an estimate of a parameter? Give examples of an estimator and of an estimate.

4. What does a dispersion parameter measure? Give examples of dispersion parameters.

5. What is the physical interpretation of the population mean?

6. How is the sample mean computed? Show such a computation and check your result using the automatic feature of your hand calculator.

7. What is the intuitive idea of an average?

8. How does the population median measure location?

9. How is the sample median computed? Give an example.

10. What are the first and third (population) quartiles? What is the population interquartile range?

11. How is the sample interquartile range computed? Give an example.

12. On what property of a frequency distribution does the statement, "About 68% of the frequency lies within one standard deviation of the mean," depend?

13. What is the relationship between standard deviation and variance?

14. How is the sample standard deviation computed? Show such a computation and check your result using the automatic feature of your hand calculator.

15. What do the various parts of a box plot for a sample represent? Describe the interpretation of each feature. In particular, describe how the plot "calibrates" outliers.

16. How is a box plot constructed? Give an example demonstrating the use of the computing form.

The following questions refer to the optional section 2.5.

17. What is meant by the influence of outliers on an estimator? What is a resistant estimator? Give examples of a resistant and a nonresistant estimator.

18. What is a trimmed mean? Why are such estimators used in statistics?

19. How are the trimmed mean and trimmed standard deviation calculated? Give an example for a 10% trimming fraction.

20. How do the mean and median fit in to the family of trimmed means?

21. What effect does varying the trimming fraction have on the resistance of the trimmed mean and trimmed standard deviation for long-tailed distributions?

SUPPLEMENTARY EXERCISES

2.30 The Department of Meteorology of the University of Stockholm has, for many years, monitored various chemical constituents of the atmosphere from several stations located throughout Sweden. A typical summary table for the year 1973, obtained from Station No. 80 located at Sjöängen, is given on page 66.

The chemicals are precipitated out of the atmosphere by rain and deposited on filters, from which the amount of each chemical, in milligrams per square meter of filter surface, can be measured. The amounts are given for each month M of the year (1 = January, 2 = February, etc.), along with the precipitation (PREC.) amounts, in millimeters.

Monthly data such as these often display interesting distributional shapes because of the natural yearly weather patterns, which tend to create a month-to-month association or dependence among the data values. We examine the distributional shapes for the various variables of this data set in this and the next few problems.

(a) Make a box plot for the precipitation (PREC.) data.

(b) Calculate $\overline{X}$ and s for the precipitation data.

(c) Using the box plot and the accompanying stem and leaf plot, describe the shape of the precipitation distribution.

(d) Describe the tail lengths of the precipitation distribution.

(e) Compare the values of $\overline{X}$ and M for the precipitation data and try to explain any differences in these quantities in terms of the descriptions of parts (c) and (d).

2.31 Repeat steps (a)–(e) of exercise 2.30 for the sulfur (S) data.

2.32 Repeat steps (a)–(e) of exercise 2.30 for the chlorine (CL) data.

2.33 Repeat steps (a)–(e) of exercise 2.30 for the magnesium (MG) data.

2.34 Repeat steps (a)–(e) of exercise 2.30 for the potassium (K) data.

2.35 The parachute fill times from exercise 1.12 are as follows.

0.59	0.54	0.23
0.38	0.39	0.44
0.47	0.89	0.40
0.43	0.48	0.36
0.44	0.52	0.33
0.37	0.51	0.82
0.43	0.49	0.51
0.37	0.38	0.44
0.27	0.38	0.37

(a) Make a box plot of the data and describe any unusual features of the distribution.

(b) Find $\overline{X}$, s, M, and IQR. Comment on the reasons for the difference in the values of M and $\overline{X}$.

(c) Select the trimming fraction (multiple of 5) that just eliminates the outliers, and calculate the trimmed mean and standard deviation.

(d) Comment on the differences in the location measures of parts (b) and (c) and on the differences in the ordinary and trimmed standard deviations.

PRECIPITATION STATION NR 80 SJÖÄNGEN   (LONG 14.30)   (LAT 58.77)   (ALT 128)

ST	YEAR M	STATION NR M	PREC. MM	S	CL	NO3-N MG/	NH4-N SQ M	NA	K	MG	CA	PH	ALC MICRO-EQU/L	CONDUCT. MICRO-S/CM
80	1973	1	35	55	9	23	18	5	3	12	7	4.1	-87	34
80	1973	2	25	30	18	16	5	11	1	2	9	4.1	-75	34
80	1973	3	12	25	7	5	12	5	1	1	5	4.0	-82	55
80	1973	4	36	43	24	22	16	14	3	2	11	4.2	-46	19
80	1973	5	81	135	26	36	70	15	19	9	31	4.3	-43	18
80	1973	6	19	38	11	8	9	6	11	2	12	4.2	-66	35
80	1973	7	55	63	14	18	30	8	4	2	12	4.4	-43	25
80	1973	8	63	93	52	38	47	21	16	3	26	4.2	-65	39
80	1973	9	69	64	30	19	23	12	25	3	15	4.4	-42	25
80	1973	10	23	17	21	7	6	12	2	2	6	4.4	-31	20
80	1973	11	52	34	65	17	14	36	3	6	8	4.4	-29	18
80	1973	12	35	34	45	12	9	28	2	5	5	4.3	-51	29
SUM			505	631	322	221	259	173	90	49	147			

Source: Data from the European Atmospheric Chemistry Network, submitted on request from the Department of Meteorology, University of Stockholm, Arrhenius Laboratory, S-106 91 Stockholm, attention Dr. Lennart Granat.

2.36 Make a box plot and find $\overline{X}$ and s for the swimming tension data of exercise 1.14, which is repeated below. Do the location measures appear to indicate that swimming instruction had any effect on tension? (Recall that the observations represent the difference in tension before and after instruction.)

$$-1, 10, 2, -9, 8, 2, 2, 3, -20, -3, 11, 7, -11, 1, -9, -7, -3, -6$$

2.37 The arithmetic and algebra scores for the first 20 students from exercise 1.23 are given below. Make stem and leaf plots and box plots of both data sets. What information about the distributions is most easily obtained from the box plots? What is most easily seen from the stem and leaf plots?

I.D. NUMBER	ARITHMETIC SCORES	ALGEBRA SCORES
1	11	0
2	11	0
3	14	1
4	8	0
5	19	5
6	17	3
7	15	4
8	13	2
9	11	0
10	8	2
11	9	2
12	19	5
13	19	7
14	18	9
15	19	7
16	14	0
17	14	4
18	16	1
19	20	9
20	16	4

2.38 Make a box plot of the precipitation data from exercise 1.24, repeated below. How would you describe the tail length of this distribution? Its shape?

OBS	PRECIP	OBS	PRECIP
1	0.085	16	0.080
2	0.495	17	0.220
3	0.010	18	0.055
4	0.025	19	0.010
5	0.015	20	0.070
6	0.030	21	0.110
7	0.260	22	0.890
8	0.020	23	0.190
9	0.065	24	0.125
10	0.010	25	0.100
11	0.060	26	0.100
12	0.085	27	0.045
13	0.035	28	0.015
14	0.125	29	0.020
15	0.005	30	0.030

2.39 A stem and leaf plot of the nuclear reactor weld data from exercise 1.11 is given below.

To regain original units, move decimal two places to the left.

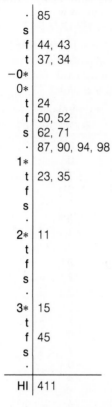

```
        ·  85
        s
        f  44, 43
        t  37, 34
     −0∗
      0∗
        t  24
        f  50, 52
        s  62, 71
        ·  87, 90, 94, 98
      1∗
        t  23, 35
        f
        s
        ·
      2∗  11
        t
        f
        s
        ·
      3∗  15
        t
        f  45
        s
        ·
    ──────────
      HI  411
```

(a) Make a box plot of these data.

(b) Use the box plot to select a trimming fraction that just eliminates the outliers, and calculate the trimmed mean and standard deviation for these data.

(c) Calculate the ordinary sample mean and standard deviation for the data, and comment on the differences between these values and the quantities calculated in part (b).

2.40 This exercise will demonstrate that it is possible to construct population frequency distributions for which the proportion of frequency within one standard deviation of the mean can be arbitrarily small or arbitrarily large. Both populations will consist of $N = 100$ individuals with "measurement" values -2, -1, 1, and 2. The distributions will be symmetric, so the population means will be $\mu = 0$.

(a) For the first population, 49 individuals have value -2, 49 have value 2, and one each have the values -1 and 1. Calculate the population standard deviation σ and show that only 2% of the population has values in the range from $-\sigma$ to σ.

(b) For the second population, 49 individuals have value -1, 49 have value 1 and the other 2 have values 2 and -2, respectively. Show that 98% of the frequency for this population is within one standard deviation of the mean.

2.41† The purpose of this exercise is to demonstrate that the two expressions for standard deviation on page 49 can lead to different results if the precision of the observations is close to the precision of the computer. Suppose you are using a calculator or computer that has only four digits of accuracy. What this means is that any number with more than four places of accuracy will have the rightmost digits rounded to four places. For example, the number 123456 would lose the 5 and 6 and the 4 would be rounded to 5. The relative position of the decimal point would be retained, however, so arithmetic operations on this number would have the same results as if the number 123500 were used.

(a) Carry out the calculation of s on the numbers 1001, 1002, and 1003, using formula (2), page 49. Round numbers to four places as though the four-digit calculator described above were being used. (Any time an operation leads to a number of more than four-place accuracy, it will be rounded to four places as described.)

(b) Repeat the calculation of s on the four-digit calculator for these same three numbers, but now use formula (1).

(c) Describe where in the calculations the use of four places of accuracy leads to differences in the values found for s in parts (a) and (b).

---

† The dagger denotes more difficult exercises.

CHAPTER 3

The Comparison Problem—An Exploratory View

Introduction

comparison
association

A unifying theme of this book is that, despite the variety of measures already introduced and yet to be introduced, there are really only two kinds of problems to consider. We call them problems of **comparison** and **association.** Exploratory and traditional descriptive methods will be applied to the comparison problem in this chapter and to the association problem in chapter 4.

The former is simply the problem of comparing the frequency distributions of a given variable for two or more populations based on (random) samples. The goal of a full-scale comparison analysis is to answer the following three questions:

1. Are there any differences (in distributions)?
2. What is the nature of the differences?
3. How large are the differences?

These are questions about the population frequency distributions, to be answered on the basis of samples. Put in more precise terms, question 1 asks whether the sample contains sufficient information to conclude that differences exist in the population frequency distributions. This question will require statistical inference to answer. Question 3 is concerned with the sizes of the differences indicated by the data. This is also an inference question. What, then, is the purpose of an exploratory analysis?

An exploratory analysis serves both a descriptive and a diagnostic function and will come first in the overall strategy for attacking both comparison and association problems. It shows clearly the nature of any differences in distributions (question 2) and often gives an indication of whether the differences are real—that is, whether they actually are properties of the population frequency distribution. The diagnostic features of the box plot help in deciding whether the standard inference analyses need to be supplemented with so-called robust procedures. The inference procedure or procedures will then be applied to "confirm" the indications of the exploration and to quantify the distributional differences. In this way, exploration and inference work hand in hand as mutually supportive activities in the solution process.

A comparison method based on a useful display, called a *schematic diagram* by its inventor, J. W. Tukey, will be applied to measurement variable problems in the next section. Comparisons based on bar graphs will be made for categorical variables in section 3.3.

◤ SECTION 3.2

Measurement Variable Comparisons

For measurement variable comparisons, we are (principally) interested in *location differences* among the distributions. This is the reason for our interest in location parameters such as the mean and median. The "scaling" for differences in location to be made in statistical inference depends both on the dispersions of the distributions and the sizes of the samples. A rough assessment at the exploratory level can be based on dispersions alone. Thus, if two sample distributions differ substantially in location *relative to their dispersions*, it is likely that these differences will be "confirmed" for the populations at the inference stage.

The box plot is ideal for assessing such location differences because of its elegantly simple representations of location (the median line) and dispersion (the box length). Box plots of all distributions to be compared, drawn on the same scale, provide the most useful exploratory tool available for

schematic diagrams   measurement variable comparison problems. Such plots are called **schematic diagrams.**

The schematic diagram is also a useful diagnostic tool. Differences in dispersions, indicated in the diagram by differences in box lengths, will have to be accounted for in an inference analysis. Extreme skewness and the presence of outliers will require special attention as well. These features are also easily seen in the diagram.

Some of the uses of the schemtic diagram will be illustrated in the following examples. Other examples will be seen in the exploratory analyses that precede the inference procedures in later chapters.

EXAMPLE 3.1   In this example we will compare a physical feature of modern-day Englishmen with the corresponding feature of some of their ancient countrymen. The Celts were a vigorous race of people who once populated parts of England. It is not entirely clear whether they simply died out or merged with other people who were the ancestors of those who live in England today. A goal of this study might be to shed some light on possible genetic links between the two groups.

The study is based on the comparison of maximum head breadths (in millimeters) made on unearthed Celtic skulls and on a number of skulls of modern-day Englishmen. The data and schematic diagram are given in display 3.1.

Display 3.1

Data and schematic diagram for skull measurements. (a) Data. (b) Schematic diagram

ENGLISH		CELTS	
141	158	133	136
148	150	138	131
132	140	130	126
138	147	138	120
154	148	134	124
142	144	127	132
150	150	128	132
146	149	138	125
(a) 155	145		

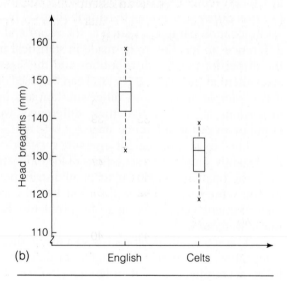

(b)

Note that the schematic diagram simply consists of box plots for the two samples, graphed side by side, with a common scale.

These are what might be called "textbook" data sets. The distributions appear to differ in location relative to dispersion, as indicated by the fact that the boxes do not overlap on the scale. It is a sure bet that the classical inference procedure will "confirm" a substantial difference between the two distributions. We will verify this in chapter 10.

Outliers are absent, and only two minor diagnostic features are interesting. The classical inference procedure prefers normal distributions, which are symmetric. The head breadth distributions appear to be slightly skewed; the median lines are not quite centered in the boxes, and the dashed lines for the Celts are of unequal length. Moreover, samples from normal distributions tend to have adjacent (dashed) lines of a length roughly equal to that of the box. The Celts' distribution appears to be mildly short-tailed. The influence of such "abnormalities" on the classical inference procedure diminishes rapidly with increasing sample size, as will be demonstrated in chapter 7. For the sample sizes of these data sets, the influence is negligible.

The following example takes fuller advantage of the diagnostic features of box plots.

EXAMPLE 3.2

This example is based on data that have been rigged to illustrate the point that conclusions based on traditional summaries of data, consisting primarily of sample means, can be misleading. An exploratory analysis is required to call attention to features of the data not revealed by the summary and to arrive at the correct interpretation. As we will see in subsequent examples and exercises, the effects shown are not uncommon in real data sets. Thus, an exploratory analysis is an essential feature of all statistical studies.

The following data are final exam scores for a professor's statistics course over a three-year period.

YEAR 1		YEAR 2		YEAR 3	
88	43	54	45	59	47
49	90	48	59	51	64
31	34	36	39	0	32
86	54	53	50	58	55
41	28	45	33	45	0
26	48	31	47	0	50
52	40	49	43	53	42
39	89	42	57	41	62
46	22	46	27	50	0
40	32	44	37	44	0
37	35	41	40	38	36
58	45	51	46	56	49
97		63		68	

Source: Wainer [35].

A calculation of sample means showed the following downward trend in average performance.

	YEAR 1	YEAR 2	YEAR 3
$\overline{X}$	50	45.04	40

The professor was understandably distressed by this trend, since his perception was that student performance (and his teaching of the course) had actually improved. The schematic diagram of the data, given in display

3.2, tells the real story. A check of records showed that the outliers in year 1 were scores for statistics majors who had previously taken more advanced courses and were trying to improve their grade-point averages. The proctor for the final exam in year 3 informed the professor that five students were sick with the flu and left early without being able to do any work on the problems.

After eliminating the "unusual" scores represented by the outliers, the following trend in means is observed.

	YEAR 1	YEAR 2	YEAR 3
$\overline{X}$	40	45.04	50

This trend follows closely the trend in medians in the box plots.

Display 3.2

Box plots for final exam scores; sample means indicated by $\otimes$

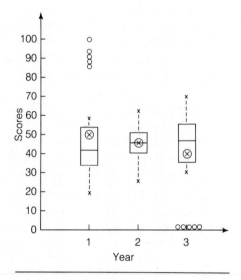

It should be pointed out that discarding data before an analysis, as we did in the last example, must be justified carefully and convincingly. The old adage "you can prove anything with statistics" owes a great deal to the clandestine selection or doctoring of data by unscrupulous (or unknowing) investigators. The justification in example 3.2 is that the individuals represented by the outliers are not from the "target population" for the problem, namely the population of non-statistics majors (for whom the course was designed) who are healthy enough to finish the final exam.

EXERCISES

3.1   In an attempt to justify a new position for statistics in the Department of Mathematics and Statistics at the University of New Mexico, the author accumulated data on the average numbers of student credit hours taught by each member of the faculty, by specialty, over a three-year period. The results are shown in the accompanying table. Does it appear that the author was able to make the case that the statistics faculty was overworked in comparison to the other groups in the department? (Make a schematic diagram.) Do there appear to be extensive variations in teaching loads between specialties? Is anyone seriously underworked or overworked relative to his or her colleagues in the same specialty?

STATISTICS	APPLIED MATH	PURE MATH AND MATH EDUCATION	
89	118	197	203
135	144	139	92
168	108	104	90
70	48	174	90
100	110	182	121
190	110	181	101
145	85	50	98
195	107	44	
	135		
	147		
	99		

3.2   Data from a study of the motivational effects of different kinds of instructions in experimental games are given below. We will deal with the inference questions in exercise 10.30.

COMPETITIVE INSTRUCTIONS	INDIVIDUALISTIC INSTRUCTIONS
32	24
19	23
8	13
19	25
9	26
15	25
17	6
16	26
21	22
24	13
10	7
27	16

Data courtesy of Richard J. Harris, Department of Psychology, University of New Mexico.

Make a schematic diagram of these data and describe any differences noted in the distributions.

3.3   The following data are from a study of the political conservatism of college students. Large scores indicate more conservative individuals. Use a schematic diagram to describe the relative conservatism of this sample of students with the three indicated majors.

ANTHROPOLOGY		MATHEMATICS		PSYCHOLOGY	
58	55	56	56	59	57
58	56	57	54	60	60
59	56	58	55	61	63
57	57				

3.4 The following data were used in an experiment to determine the best way to cut sheets of insulating material so as to retain impact strength. The values are strengths in foot-pounds. Make a schematic diagram of the data and describe any differences in the distributions.

LENGTHWISE SPECIMENS	CROSSWISE SPECIMENS
1.15	0.89
0.84	0.69
0.88	0.46
0.91	0.85
0.86	0.73
0.88	0.67
0.92	0.78
0.87	0.77
0.93	0.80
0.95	0.79

Data courtesy of the Statistical Laboratory, University of California, Berkeley.

3.5 The following data are measurements of total dissolved solids in water obtained from two wells, Lee 1 and Lee 2, in the Ambrosia Lake Mining District near Grants, New Mexico.

DATE	LEE 1	LEE 2
3/78	528	554
10/78	562	580
1/79	517	480
5/79	548	461
12/79	612	466
3/80	536	475
6/80	572	491
9/80	611	494

Data courtesy of the Environmental Protection Agency, State of New Mexico, Santa Fe.

(a) Make a schematic diagram of the two data sets. Do the plots indicate a difference in total dissolved solids for the two wells?

(b) Note that the data occur in pairs. An inferential analysis to be carried out in chapter 10 will utilize the differences, Lee 1 value minus Lee 2 value, for each date. Compute these differences and make a box plot of them. A zero average for the differences would support the contention that the average levels of total dissolved solids are the same for the two wells. Does the plot support this contention? Compare the dispersion of the differences with the dispersions of the two data sets in part (a). What effect has the formation of differences had on this property?

3.6   The following data are thicknesses in ten-thousandths of an inch from a study of the break strengths of seven types of starch films. Make a schematic diagram and describe any differences in thicknesses suggested by the data. Describe any other features of the data seen in the box plots.

STARCH	WHEAT	RICE	CANNA	CORN	POTATO	DASHEEN	SWEET POTATO
	5.0	7.1	7.7	8.0	13.0	7.0	9.4
	3.5	6.7	6.3	7.3	13.3	6.0	10.6
	4.7	5.6	8.8	7.2	10.7	7.1	9.0
	4.3	8.1	11.8	6.1	12.2	5.3	7.6
	3.8	8.7	12.4	6.4	11.6	6.2	
	3.0	8.3	12.0	6.4	9.7	5.8	
	4.2	8.4	11.4	6.9	10.8	6.6	
		7.3	10.4	5.8	10.1	6.6	
		7.5	9.2		12.7		
		7.8	9.0		13.8		
		8.0	12.5				

Data courtesy of the Statistical Laboratory, University of California, Berkeley.

SECTION 3.3
Comparison Problems for Categorical Variables

two-way table

The comparison of the frequency distributions of a categorical variable is greatly aided by a two-dimensional tabular array called a **two-way table**. The observed (absolute) sample frequencies are recorded in the table, and all necessary information for both descriptive and inferential purposes are computed directly from it.

Display 3.3 is such an array that contains data from an evaluation of a baby-food advertisement by a large producer. Mothers' responses to an ad

Display 3.3

Two-way table of responses to baby-food survey made in three types of living environments

RESPONSE	ENVIRONMENT (population)[a]			ROW MARGINALS (total responses)
	UR	SU	RU	
Yes	30	19	21	70
No	38	17	28	83
COLUMN MARGINALS (sample sizes)	68	36	49	153

Data courtesy of Francis Wall, private consultant, Albuquerque, New Mexico.
[a] UR = urban; SU = suburban; RU = rural.

for a particular product were evaluated using samples from three living environments, urban (UR), suburban (SU), and rural (RU). The mothers responded either "yes" or "no" to the question of whether the ad persuaded them to adopt the company's product.

marginal totals

The table of observed frequencies is augmented by a row and column of **marginal totals,** which represent the sums of column entries and row entries, respectively. The column totals are the sample sizes for the three regions, and the row totals represent the "yes" and "no" responses for the *pooled sample* formed by combining the data for all three regions. The number in the lower right-hand corner of the table is the sum of either the row marginal or column marginal frequencies and represents the sample size of the pooled sample.

The object of the study is to determine whether there are differences in the distributions of responses for the three regions. By frequency distributions we mean *relative* frequency distributions. It is the intent of the study to generalize the results of the analysis to populations in the three living environments, and the information to be generalized will be about relative frequencies. We are thus concerned with comparing the proportions of "yes" responses for the three environments. The first step in the analysis is to form a table of relative frequencies from the above two-way table, by dividing each absolute frequency by the marginal total (the sample size) for its column. Each number is then multiplied by 100 to represent the frequencies in percent. The resulting table is given in display 3.4.

Display 3.4

Relative frequency distributions (in percent) of the response variable for the data of display 3.3

RESPONSE	LOCATION			POOLED SAMPLE AVERAGE
	UR	SU	RU	
Yes	44.1	52.8	42.9	45.8
No	55.9	47.2	57.1	54.2
Total frequencies	100	100	100	100

We see from this table that 44.1% of the urban mothers were persuaded to adopt the manufacturer's product, as opposed to 52.8% of the suburban and 42.9% of the rural mothers. The column labeled AVERAGE is the relative frequency distribution for the pooled sample; it represents an all-region average to which the other distributions can be compared. For example, the average "yes" response is 45.8%. The suburban "yes" response is above the average for all regions, and the urban and rural responses are below.

A graphical presentation of this comparison would clearly be useful. We used bar graphs in chapter 1 to represent frequency distributions of categorical variables, and it is natural to extend them to the comparison problem. A principle of graphical display is that objects to be compared should be graphed close together, on the same scale. Since we want to compare the frequencies of each response across regions, the bars for the

comparison bar graph   different regions are graphed together. A **comparison bar graph** of the baby-food survey response data is given in display 3.5.

Display 3.5

Comparison bar graphs for variable response data

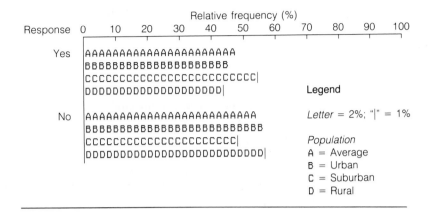

Note that the differences in "yes" responses are clearly seen in the graph. The picture makes it easier to visualize the magnitudes of the differences as well as the overall sizes of the responses represented by the bar graph of the average frequency distribution. The overall success of the advertisement is impressive; it persuaded 45.8% of the mothers to adopt the product. This fact may outweigh the differences from region to region, which are relatively small.

EXAMPLE 3.3

In this example we will look at a comparison of political party preference by voting location in a student senate election held at the University of New Mexico. Display 3.6 gives the election results, tabulated by voting location and candidate party affiliation. The relative frequencies are given below the observed frequencies.

A comparison bar graph of these distributions is given in display 3.7.

If we assume that students vote at places convenient to at least one of their centers of campus activity, the table and graph may tell us something of the voting patterns of various categories of students. For example, the president's garage is along the route to the largest free parking lot near campus. Students voting there are likely to be commuters unwilling to pay the fee to park on the central campus. The distinction is not sharp, however, since the Geology building is also on a major route to off-campus parking and draws commuting voters as well as geology students.

As the Average bar graph shows, the overall vote most strongly favors USDA candidates, with Student Voice and Unaffiliated candidates winning only about half this proportion of the total vote. The vote for the other two parties was less than 10% each, which suggests that differences in voting patterns by location will be of interest only for the top three parties.

Display 3.6

Observed and relative frequency table for party votes by voting location; relative frequencies are in percent

		VOTING LOCATION					
PARTY VOTED FOR	FREQUENCY	GEOLOGY BUILDING	PRESIDENT'S GARAGE	STUDENT UNION BUILDING	LA POSADA[a]	ENGINEERING CENTER	TOTAL (Average)
USDA[b]	Obs.	640	776	1628	537	436	4017
	Rel.	39.8	49.5	45.9	46.8	62.8	46.9
How and Why	Obs.	78	47	170	51	23	369
	Rel.	4.9	3.0	4.8	4.4	3.3	4.3
Student Voice	Obs.	390	348	776	232	107	1853
	Rel.	24.3	22.2	21.9	20.2	15.4	21.6
Progressive	Obs.	112	60	278	131	26	607
	Rel.	7.0	3.8	7.8	11.4	3.7	7.1
Unaffiliated	Obs.	387	336	695	197	102	1717
	Rel.	24.1	21.4	19.6	17.2	14.7	20.1
Total		1607	1567	3547	1148	694	8563 = total votes

Data courtesy of the *New Mexico Daily Lobo*, University of New Mexico, Albuquerque.
[a] A student dormitory dining hall.
[b] United Students for Democratic Action.

Display 3.7

Comparison bar graph for the student election data of display 3.6

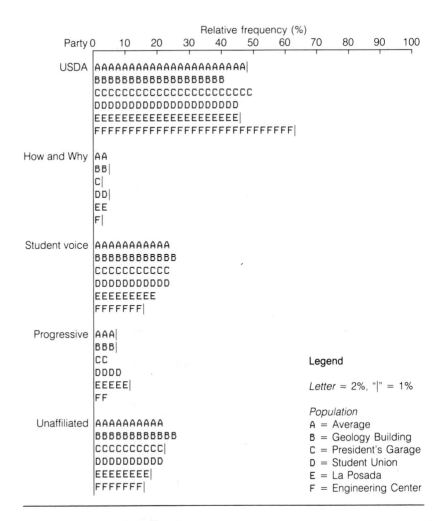

USDA was most strongly supported by those students voting at the Engineering Center, receiving over 60% of the vote at that location. The other parties received somewhat less than the average voting percentage at this location. Since USDA had a relatively conservative platform and list of candidates, this might support the traditional view of engineers as being conservatively oriented. Except for a somewhat lighter than average vote for USDA among those voting at the Geology building, there appear to be few other differences of any interest. However, as we will see in chapter 13, the huge total vote in this example makes some of the more subtle differences *statistically significant*—that is, indicative of real differences in population frequencies. For example, some of the seemingly trivial differences in voting frequencies for the Progressive party are of this nature. This will allow us to make the point that statistically significant differences need not be interesting differences in any practical sense. Statistical inference does not free us from the necessity of making practical judgments about the results of our analyses.

EXERCISES

3.7   The distributions of the *showup* variable from the bicycle tour data of display 1.17 for those who chose the 100- and 50-mile options and for those who rode other distances or for whom a distance wasn't given (*other*) are represented in the accompanying table.

VALUES OF	DISTANCE		
SHOWUP VARIABLE	100	50	OTHER
No-show	11	6	5
Show	62	10	6

Construct a relative frequency table and a comparison bar graph. Discuss possible reasons for what you see in the graph.

3.8   Data concerning the perceived need for post-secondary education for individuals at various education levels are shown below. Make a frequency table and comparison bar graph of the data and describe the difference in perceived need seen for the five groups.

EDUCATION LEVEL	NEED TO UPGRADE EDUCATION	
	LOW	HIGH
Less than high school	198	192
High school diploma	386	481
Some college	803	924
Bachelor's degree	453	301
Graduate degree	383	126

Data courtesy of G. Mallory, Albuquerque Urban Observatory, *Survey of Adult Needs for Postsecondary Education Programs in the Albuquerque Metropolitan Area, 1980.*

3.9   A sample of death certificates of individuals over 20 years of age in the Southwest who are known to have died by suicide during the period 1978 to 1981 revealed the following age and ethnicity information.

AGE		ETHNICITY	
	ANGLO	HISPANIC	NATIVE AMERICAN
20–29	34	25	11
30–39	26	9	2
40–49	23	6	1
50–59	22	6	0
over 59	21	4	0

(a)   Make a frequency table and comparison bar graph that simplifies the comparison of suicide rates for the three ethnic groups. Describe your findings.

(b)   Make a frequency table and comparison bar graph that simplifies the comparison of suicide rates for the different age groups. Describe your findings.

3.10   In an experiment to determine what kind of vegetation is most likely to be associated with archaeological sites in the Southwest, 98 sites were randomly selected from a population of known sites and 68 "non-sites" (i.e., locations known not to contain archaeological finds) were selected. The vegetation was classified for each location into one of the categories *juniper, salt-desert,* and *other.* The results of the classification are given in the following table.

	SITE TYPE	
VEGETATION	SITE	NON-SITE
Juniper	14	1
Salt-desert	64	52
Other	20	15

Describe the difference in vegetation for sites and non-sites by means of a frequency table and comparison bar graph.

3.11   In a study of drug prescription prices in a large city, a physician was interested in determining whether a pharmacist's choice of filling a prescription with a brand name drug or a generic one depended on the number of different brands of that drug available in the pharmacy. The physician had the same prescription filled at a random sample of 39 pharmacies in the city. The following table summarizes his findings.

TYPE USED	NUMBER OF AVAILABLE BRANDS		
	1	2	3 or more
Brand Name	6	17	4
Generic	2	2	8

Data courtesy of L. Chilton, M.D., Department of Pediatrics, University of New Mexico School of Medicine.

Describe with a frequency table and comparison bar graph the differences in uses of brand name and generic drugs for the three categories of available brands.

3.12   An archaeologist hypothesized that the manufacturing quality of pots found at sites with a high density (HD) of artifacts should be different from that of pots found at low density (LD) sites, because of the use of pots in transporting trade goods and the likelihood that HD sites were trade centers. To test this hypothesis, potsherds from the two types of sites were rated on a three-point scale as being of low, medium, or high quality. The ratings of a number of pots from the two types of sites are given in the following table.

	SITE TYPE	
POT RATING	HD	LD
Low	5	8
Medium	12	8
High	7	5

Form a frequency table and comparison bar graph to describe the differences in ratings for the two site types.

Chapter 3 Quiz

1. What is the comparison problem?

2. What questions are to be answered in the solution to a comparison problem? Give the "detailed" versions as well as the short versions of these questions.

3. What is a schematic diagram? Give an example.

4. What features of a schematic diagram are important for demonstrating location differences in distributions?

5. What features of a schematic diagram are primarily of diagnostic importance?

6. What is the form of the two-way table of observed frequencies for categorical variable comparison problems? Give an example.

7. How are the relative frequency distributions formed from the two-way table? What is the function of the average relative frequency distribution?

8. How is a comparison bar graph formed from the relative frequency distributions? How is it interpreted? Give an example.

◢
_____

SUPPLEMENTARY EXERCISES

3.13 Groundwater quality is affected by the geological formations in which it is found. A study of two water quality parameters, sodium (Na) ion concentration and chlorine (Cl) ion concentration, was carried out to determine whether differences in quality existed for the water tables on the east and west sides of the Rio Grande River in Bernalillo County, New Mexico. The data from several wells on each side of the river (in milliequivalents) are given in the accompanying table.

WEST SIDE		EAST SIDE			
NA	CL	NA	CL	NA	CL
6.10	.58	1.43	.34	1.10	.46
4.50	.29	1.22	.24	1.00	.12
4.31	.32	2.50	1.03	1.10	.39
6.10	.55	1.45	.68	1.65	.65
5.00	.56	.90	.29		
6.04	.62	1.60	1.14		
5.92	.61	2.70	.34		
5.92	.63	3.90	.46		
6.70	.52	2.30	.53		
6.12	.53	2.60	.40		
4.50	.29	1.46	.33		
4.17	.37	1.90	.37		
2.80	.40	2.10	.40		
4.60	.62	1.65	1.64		
2.90	.44	.90	2.06		
4.51	.25	.70	.37		
5.04	.24	1.18	.40		
4.92	.21	1.20	.45		
		1.40	.30		

Data courtesy of C. Rail, New Mexico Environmental Health Department, Albuquerque.

(a) Construct comparison (side-by-side) stem and leaf plots and a schematic diagram for sodium (Na) for the two sides of the river. Use them to describe and compare the sodium ion concentration frequency distributions.

(b) Repeat the steps of part (a) for the chlorine (Cl) ion concentration.

3.14 An investigator wanted to determine the effect the home occupants had on monthly demand for natural gas in a New Jersey community in which some homes were equipped with instruments to study natural gas consumption. He randomly paired homes in which there was a change in occupants with homes of the same type, size, and block position but having the same occupants throughout the study period. For each home, the difference in average monthly consumption before and after the occupant change was calculated for both the home experiencing the change and its twin. Since the pairs of homes were matched as closely as possible for a number of factors known to influence energy consumption, the differences seen should be largely attributable to the change of occupant and thus should provide information about how much the occupant contributed to the natural gas consumption equation. The differences in the accompanying table are the postmove minus premove averages. Compare the distributions of consumption differences for the two groups of homes by means of stem and leaf plots and a schematic diagram.

	NATURAL GAS CONSUMPTION DIFFERENCES	
HOME PAIR	FOR HOMES WITH CHANGE IN OCCUPANT	FOR HOMES WITH NO CHANGE IN OCCUPANT
1	4.4	−36.5
2	−19.4	3.8
3	−20.5	−7.6
4	−3.0	5.3
5	−30.7	−13.0
6	12.1	−13.5
7	−19.0	.5
8	−19.3	−27.7
9	−61.7	−8.1
10	6.2	25.3
11	−37.0	−36.5
12	−11.1	−36.5
13	12.7	−6.7
14	−6.7	−4.1
15	22.7	−14.0
16	−35.4	7.7
17	−42.2	−14.9
18	−7.7	−5.3
19	−8.8	−21.3
20	19.3	−11.7
21	−77.5	−10.0
22	12.1	−10.7
23	−16.3	−10.7
24	−41.9	−14.4
25	−9.0	11.8

Source: Mayer [22].

(a) What appears to be the principal characteristic of the new occupants insofar as natural gas consumption is concerned? Does there appear to be much difference in average change in consumption over the premove and postmove periods for the two groups?

(b) This study was made in the period during the mid-1970s in which the energy crisis was emphasized by rapidly escalating natural gas prices. Is there any indication that either group of home owners was responsive to this situation? Describe.

(c) Because these are paired data, they can be analyzed inferentially by taking the pairwise differences ("with change" minus "without change") of the two data columns. Make a box plot of these differences and comment on their dispersion in comparison to the dispersions of the original data as seen in the schematic diagram constructed for part (a).

3.15 This exercise provides an illustration of the kind of quality control practiced in a clothing manufacturing plant. Levi-Strauss manufactures clothing from cloth supplied by several mills. The data used in this exercise were obtained from the quality control department of the Levi plant in Albuquerque, New Mexico. In order to maintain the anonymity of the five mills supplying cloth for this particular time period, we have coded them A, B, C, D, and E. A measure of wastage due to defects in cloth and so on is called *run-up*. It is quoted as percentage of wastage per week and is measured relative to computerized layouts of patterns on the cloth. Since the people working in the plant can often beat the computer in reducing wastage by laying out the patterns by hand, it is possible for run-up to be negative.

From the viewpoint of quality control, it is desirable not only that the run-up be small but also that the quality from week to week be fairly consistent. These results would be reflected in both small average run-up and small variability. We will look for the mill with the best balance of performance in both categories over the given time period. The data are given in the accompanying table.

MILL A		MILL B		MILL C		MILL D		MILL E	
.12	.03	1.64	.63	1.21	.98	1.15	.59	2.40	2.23
1.01	.35	−.60	.90	.97	.65	1.02	1.30	−.37	.31
−.20	−.08	−1.16	.71	.74	.57	.38	.68	.82	1.68
.15	1.94	−.13	.43	−.21	.51	.83	1.45	.92	1.13
−.30	.28	.40	1.97	1.01	.34	.66	.52	−.93	1.23
−.07	1.30	1.70	.30	.47	−.08	1.02	.73	.80	1.69
.32	4.27	.38	.76	.46	−.39	.88	.71	1.58	
.27	.14	.43	7.02	.39	.09	.27	.34		
−.32	.30	1.04	.85	.36	.15	.51	.07		
−.17	.24	.42	.60	.96		1.12			
.24	.13	.85	.29						

Data courtesy of Levi-Strauss Corp., Albuquerque, New Mexico.

(a) Make two schematic diagrams as follows. In the first the scale for the box plots will encompass the full range of the data to demonstrate the magnitude of the outliers. (This diagram is said to be scaled to the extremes of the data.) The second diagram will be a blowup of the boxes and adjacent values of the first in

a scale appropriate to show the differences in the main parts of the distributions. (This diagram is said to be scaled to the adjacents.)

(b)   Which mill has the best average run-up performance (smallest median run-up)?

(c)   Which mill has the most consistent quality, as measured by small IQR?

(d)   If both items in (b) and (c) are considered important, and if infrequent but
unusual quality variations (outliers) are undesirable, which mill demonstrates
the best overall quality?

3.16   Refer to the data of display 1.1 (p. 9).

(a)   Compare the height distributions for men and women by means of a schematic
diagram.

(b)   Repeat part (a) for the weight distributions.

(c)   Repeat part (a) for the age distributions.

3.17   Recently, 168 prospective lawyers took the New Mexico bar exam. Of this group, 30
were members of minority groups—5 Native Americans, 23 Hispanics, and 2 blacks.
Of the total, 127 passed the exam. Nine of these were minority group members—1
Native American, 7 Hispanics, and 1 black.*

(a)   Form a two-way table of exam performance (passed, failed) by ethnic group
(Native American, Hispanic, black, Anglo) and fill in the cell frequencies from
the information given above.

(b)   Make a relative frequency table and a comparison bar graph for these data.
Comment on any performance differences seen among the groups.

(c)   Do you feel that the results *prove* that bar exams discriminate against minori-
ties? Why?

3.18   Are men or women better at elementary statistics? This question can be studied by
using the final grades for a recent statistics course. The data are given in the accompanying table. Note that in order to save space, we have listed the population
indicators (men and women) vertically, while the values of the variable of interest,
grades, have been listed horizontally.

	GRADE				
	A	B	C	D	F
Men	6	21	18	5	5
Women	10	29	13	6	4

(a)   Construct a relative frequency table and comparison bar graph for the data.

(b)   Comment on the difference in grade distributions.

3.19   The accompanying data were extracted from the tour of the Rio Grande Valley
bicycle data set given in display 1.17 (p. 25). They represent the average speeds, in
miles per hour, for riders selecting the three distances 50, 75, and 100 miles.

----

* Data courtesy of R. Griego, Department of Mathematics and Statistics, University of New
Mexico, Albuquerque.

50 MILES	75 MILES	100 MILES	
9.0	7.2	20.2	12.5
8.1	11.3	12.0	11.4
11.2	8.8	12.4	8.1
8.7	9.5	17.6	10.8
10.2	8.0	16.5	14.7
9.3		12.6	11.0
9.0		11.1	
9.0		12.3	
9.0		21.1	

(a) Using a schematic diagram, compare the speed distributions for the three distances. Comment on the tail lengths and shapes of the distributions.

(b) Compute the sample mean and standard deviation for each distribution. Plot the sample means on the corresponding box plots constructed in part (a) (use the symbol $\otimes$). How well do the means and medians correspond in this example? Explain any differences in terms of the distribution shapes.

3.20 One hypothesis states that sudden infant death syndrome (SIDS) is a lung disease that is influenced by the amount of available oxygen in the air. This conjecture can be tested by studying the death rates due to SIDS at different altitudes. New Mexico is well suited for this study in that it has centers of population ranging in altitude from 3400 feet to about 7400 feet. In the accompanying table, altitudes in the range 3400–4500 feet are categorized as low, 4600–6500 feet as medium, and 6600–7400 feet as high. Are there any indications of increased SIDS death rates at higher altitudes? How useful is the standard comparison bar graph here?

	ALTITUDE		
SIDS VARIABLE	LOW	MEDIUM	HIGH
Deaths due to SIDS	27	38	12
Others	6,601	10,512	2,511
Total births (two-year period)	6,628	10,550	2,523

Data courtesy of E. S. Sweeney and James T. Weston, Department of Pathology, University of New Mexico School of Medicine, Albuquerque.

3.21 Make a two-way table and comparison bar graph of the *class* variable distributions for the three seating locations (front third, middle third, back third) of the complete class data set given in display 1.18 (p. 28). Identify and comment on any interesting differences in these distributions.

3.22 A current hypothesis in medical circles is that conditions such as alcoholism, and the likelihood of alcoholics developing cirrhosis of the liver, are at least in part genetically determined. The accompanying data isolate the distributions of a single antigen (genetic marker), B12, from a more extensive study. The controls for both ethnic groups are nonalcoholics with no liver disease.

POSSESS B12	ANGLO		
	ALCOHOLIC, NONCIRRHOSIS	ALCOHOLIC, CIRRHOSIS	CONTROL
Yes	3	14	15
No	20	13	24
	HISPANIC		
Yes	5	16	3
No	12	29	26

Data courtesy of R. T. Rada, M. D., Department of Psychiatry, University of New Mexico School of Medicine, Albuquerque.

(a) Analyze the entire table (consisting of all six B12 distributions) for interesting differences by making a comparison bar graph. Interpret your results.

(b) Analyze separately the subtables for the two ethnic groups Anglo and Hispanic. What differences, if any, occur in the two analyses? Does B12 appear to distinguish the same things for the two ethnic groups? What is being distinguished for Anglos? For Hispanics?

3.23 The accompanying data were entered as evidence in a lawsuit brought against a company for failure to follow equal-rights employment practices involving a minority group. The nature of the business is such that blocks of workers are periodically laid off when the work load is low and rehired when new contracts bring in more work. At issue was possible discrimination in the selection of the workers to be retained during the slow periods. In the accompanying table, zero or one layoffs for an employee were classified as low, 2–3 as medium, and 4 or more as high.

	LAYOFFS		
	LOW	MEDIUM	HIGH
Minority	95	13	19
Nonminority	120	8	0

Data courtesy of R. Griego, Department of Mathematics and Statistics, University of New Mexico, Albuquerque.

(a) Compare the layoff distributions for these two groups. Is there any indication of discriminatory layoff practices by this company?

(b) Refer to part (a). If this indication were too large to be attributable purely to chance, would this prove that the company is discriminating against its minority employees? Discuss.

3.24 A study of the blood groups of SIDS and non-SIDS children was made in an early effort to determine the cause of the sudden infant death syndrome. If substantial differences exist, blood grouping could be used, in part, to predict which children are potential SIDS victims and a program to monitor these children could then be initiated to prevent their deaths. The data for a sample of 135 SIDS and 135 non-SIDS (controls) are given in the accompanying table.

BLOOD GROUP	SIDS	CONTROLS
O	69	55
A	40	71
B	17	7
AB	9	2
	135	135

(a) Analyze this table for interesting differences in the blood group distributions by using a relative frequency table and a comparison bar graph.

(b) Assuming that the blood group distribution of the combined sample (right margin of your completed table) is reasonably representative of the frequency of occurrence of the various blood groups in the population of all children, how useful would you judge blood group information to be for predicting SIDS?

3.25 An associate dean of the College of Arts and Sciences at the University of New Mexico ran a survey to compare the study habits of freshmen and sophomores based on samples of individuals coming to the college office for consultation on various matters. Among the questions asked was how many hours per week the student studied for all of his or her courses. The results for this question are summarized in the accompanying table.

NUMBER OF HOURS OF STUDY PER WEEK	FRESHMAN	SOPHOMORE
1–10	48	3
11–26	73	15
over 26	29	6

Data courtesy of Arts and Sciences Dean's Office, University of New Mexico, Albuquerque.

(a) What are the more interesting differences in these study time distributions?

(b) Assuming a standard 9- to 15-hour course load, what proportion of these students spend at least 3 hours studying outside class for every hour in class?

(c) Discuss the appropriateness of the arts and sciences' office as a place to obtain a representative sample for looking at student study habits.

3.26 One theory of the formation of the solar system states that all solar system meteorites have the same evolutionary history and thus have the same cooling rates. By a delicate analysis based on measurements of phosphide crystal widths and phosphide-nickel content, the cooling rates, in degrees Celsius per million years, were determined for samples taken from seven meteorites, named in the accompanying table after the places they were found.

METEORITE	COOLING RATES
Walker County	.69, .23, .10, .03,[a] .56, .10, .01,[b] .02,[b] .04,[b] .22
Lombard	.10, .15, .45, .50, .38, .45, .51, .61, 1.00, .84
Uwet	.21, .25, .16, .23, .47, 1.2, .29, 1.1, .16
Quillagua	.60, .23, .15, .33, .32, .58, .52, .42, 2.2, 1.4
Coahuila	5.6, 4.9, 3.3, 1.8, 4.8, 3.6, 3.2, 1.4, 1.6, 3.2, 8.2
Tocopilla	5.6, 2.7, 6.2, 2.9, 1.5, 4.0, 4.3, 3.0, 3.6, 2.4, 6.7, 3.8
Hex River	7.4, 5.6, 3.0, .64, 8.0, 7.4, 10.0

Data courtesy of R. Randich, Sandia Corporation, Albuquerque, New Mexico.
[a] Sample corroded.
[b] Estimated.

(a)   After making a schematic diagram of the distributions of cooling rates for the seven meteorites, discuss the reasonableness of the theory mentioned above.

(b)   What, if any, natural groups are there among the meteorites, based on cooling rates? [Answer this question both with and without the questionable (footnotes *a* and *b*) data values for Walker County.]

3.27   Not all measurement variable comparison problems are automatically concerned with location, and it is always necessary to assure that the analysis method fits the problem. A case in point, given by Enrick [11], involves the evaluation of a new pneumatic, continuous torque wrench used in assembling covers to bodies of gasoline pump meters. Management expected the new wrench to produce more uniform torque readings than those of the currently used pneumatic impact wrench. Product specifications call for a torque of 6 foot pounds with a tolerance of ±2. Otherwise, loose fits will permit leakage, while excessively tight fits will cause cover distortion and, again, leakage. The cost savings, from a reduction in the number of service calls to replace or repair meters in the field, would more than outweigh the expense of the new wrench and its somewhat higher maintenance costs. The two wrenches were each tested on 24 meters, with each meter using four screws, so that a total of 96 torque readings resulted. These are displayed in the accompanying back-to-back stem and leaf plot, with each entry x rounded to the nearest foot pound. Tolerance limits are indicated by the dashed lines.

```
        OLD WRENCH                              NEW WRENCH

                    x │  2 │
              xxxxxxx │  3 │ xx
              ------- │----│--
          xxxxxxxxxxx │  4 │ xxxx
        xxxxxxxxxxxxx │  5 │ xxxxxxxxxx
        xxxxxxxxxxxxx │  6 │ xxxxxxxxxxxxxxxxxxxxxxxxxxxxxxxxxxxx
        xxxxxxxxxxxxx │  7 │ xxxxxxxxxxxxxxxxxxxxxxxxxxxxx
         xxxxxxxxxxxx │  8 │ xxxxxxxxxxxx
          ----------- │----│--------
                 xxxx │  9 │
                   xx │ 10 │
```

(a) Make a schematic diagram for the two sets of torque readings. How likely do you think it would be that a comparison of means would indicate a difference in the wrenches?

(b) A tempting alternative comparison is to check for the equality of dispersions. On the basis of the box plots, which wrench produces the more variable torque readings?

(c) Using (a) and (b) together, we can begin to make a case for the new wrench. However, this approach does not really address the important question of whether the new wrench will significantly reduce the number of service calls the company has to make. Why not attack the question directly by looking at the relative proportions of out-of-tolerance readings for the two wrenches?

Set up a two-way table of a grouped tolerance variable, called *tol*, for the two wrenches, with categories for *in-tolerance* and *out-of-tolerance* readings. Use a relative frequency table to compare them. What are the relative percentages of out-of-tolerance readings for the two wrenches? Which wrench would be preferred on the basis of this analysis? (We will see in chapter 13 that the indicated difference is highly significant.)

3.28   The increase in food prices during the period 1974–1980 is a well-documented fact both from governmental statistics and personal experience. It is of interest to study how the food purchases for a typical household responded to these price increases. Did a person simply continue to buy the same items at their increased prices, or were purchasing strategies changed? It is known that in many families certain nonessential and more expensive types of food were often given up. Also, cheaper brands were substituted where available. However, it was necessary to pay the increased prices of items the household members viewed as essential. The data in the accompanying table are the prices of items (in dollars) from cash register receipts obtained from two shopping trips made by the same person to the same grocery store in May 1974 and April 1980.

MAY 1974			APRIL 1980			
2.05	.46	.25	4.35	3.19	1.99	.46
1.32	.79	.50	4.80	1.66	2.99	.56
.50	.79	.56	.68	1.66	.99	.94
1.92	.79	.56	.68	2.85	.72	2.12
2.17	1.00	.25	1.93	1.63	.39	2.12
2.10	1.00	.56	.83	6.89	.39	1.65
2.23	1.00	.79	.79	.56	.46	
.80	1.00	2.12	.79	.89	.46	
.95	.25	2.12	.79	.89	.46	
.69	.25	5.55	1.19	1.63	.46	

(a) Use a schematic diagram to compare the distributions of item prices for these two trips. What would be reasonable conjectures concerning changes in the grocery shopping strategy for this purchaser?

(b) Assume that these data are from typical shopping trips. Is it likely that the (population) distributions of item prices for items purchased by this shopper in 1974 or in 1980 are normal? Describe the shapes of the distributions.

3.29   Differences in the head breadth distributions of example 3.1 were convincing largely because the boxes in the schematic diagram were well separated. This implies that the differences between medians (location measures) were large compared to the dispersions of the measurements (box lengths). The difference in distributions will not be so convincing in this example. Statistical inference will be needed to draw any conclusions. The data are from a study of the rate of degradation of insulin by the liver cells of rats under two different experimental conditions.

Make a schematic diagram of the data and comment on both the diagnostic features of the distributions and their relative locations.

EXPERIMENTAL CONDITIONS	
1	2
30.2	19.5
34.2	40.4
33.7	24.1
7.8	12.9
23.9	18.5
9.9	25.0
31.3	29.5
27.3	6.6
17.0	
26.7	
21.4	

3.30   While testosterone is basically a male hormone, women also produce it in measureable amounts. Because of this, Purifoy and Koopmans [27] asked whether personality traits, such as those linked with the achievement of high-level occupational positions, are associated with womens' testosterone levels. To study this question, 47 women were grouped into 3 general occupational categories: housewives (with no other occupations outside the home), secretaries and office workers (which includes clerks, lab technicians not requiring an advanced degree, etc.), and professional career women (including physicians, lawyers, college professors, business executives, etc.). The free testosterone coefficient (FTC), a measure of the amount of biologically active testosterone in the blood, was obtained for each woman. The relationship between testosterone levels and occupational category was then determined by comparing the FTC distributions for the three occupational groups.

The FTC data from [27], in milligrams per 100 milliliters of blood (called mg%), are given below. Make a schematic diagram of the data. Does there appear to be any difference in testosterone distribution for these occupation levels? If so, what is the nature of the difference?

HOUSEWIVES		SECRETARIES AND OFFICE WORKERS		PROFESSIONAL CAREERS			
.8	2.3	1.1	2.3	1.2	2.9	3.4	5.2
1.1	2.3	1.8	2.3	1.5	3.0	3.5	5.3
1.2	2.3	2.0	2.4	2.1	3.1	3.6	5.7
1.4	3.9	2.1	2.5	2.2	3.2	3.6	6.1
1.8	4.2	2.1	5.9	2.2	3.3	4.8	6.2
2.0		2.2		2.7	3.4	4.9	7.8

An Exploratory Look at Association

Introduction

The focus in an *association* problem is on a single population and on the relationship among selected characteristics or variables of the population. Each member of the population (potentially) contributes a value for each variable. The pattern of frequency of the different combinations of values then represents the association among the variables.

For example, in a technologically advanced society such as the United States, it seems reasonable that the variables X = number of years of schooling and Y = annual income would be associated. We would expect to find large values of both X and Y occurring together more frequently than, say, large values of X and small values of Y. The actual population frequencies of the various combinations of variable values, called the *joint frequency distribution*, can then be used to give quantitative meaning to the association.

The kinds of questions we will ask in association problems are

1. Is there an association among the variables?
2. What is the nature of the association?
3. How large is the association?

Again, because we will only see a sample from the population, statistical inference will be needed to address the first and third questions. The exploratory methods to be studied in this chapter will aid both in answering the second question and in diagnosing troublesome features of the data that could cause problems with inference. As with comparison problems, exploration will again be the first step in an overall statistical problem-solving strategy.

We begin with the study of association for categorical variables. By defining the idea of *conditional distributions*, association problems can be solved with the comparison methods of the last chapter. Measurement variable association problems, on the other hand, require something new—the important methodology of *regression*. This topic will be introduced in section 4.3.

◪ SECTION 4.2

Association for Categorical Variables

The raw material for studying the association between two categorical variables is, as in the comparison problem, a two-way table of observed (absolute) frequencies. The table for the *sex* and *class* variables for the sample of display 1.1 (p. 9) is given in display 4.1. The positions of data in the table are called *cells*. The number in each cell is the number of individuals in the sample who show the corresponding values of the two variables. For example, there are three male freshmen in the sample of display 1.1.

Display 4.1

Two-way table of observed and marginal frequencies for *sex* and *class* variables of display 1.1

SEX	CLASS FRESHMAN	SOPHOMORE	JUNIOR	SENIOR	GRADUATE	SEX MARGINAL
Male	3	6	4	0	1	14
Female	8	9	3	1	1	22
Class Marginal	11	15	7	1	2	36 = total frequency

---

contingency table
marginal frequencies

For categorical variables, association is sometimes called *contingency*, and the table of display 4.1 is called a **contingency table.** The row and column sums are the **marginal frequencies,** as shown in the display. Note that the total frequency, which is the sum of either the row or column marginals, is the sample size, $n = 36$.

Joint Distributions

sample joint frequency distribution

All of the quantities required to analyze the association between two categorical variables will be calculated directly from the contingency table. For example, the **sample joint frequency distribution** is obtained by dividing the absolute frequencies in the body of the table by the sample size. Display 4.2 contains the joint frequency distribution of *sex* and *class*, computed from display 4.1. We see, for example, that 8.3% of the class were freshman men; the largest percentage, 25%, were sophomore women; there were no senior men, and so on.

marginal distributions   Display 4.2 also gives the **marginal distributions** of the two variables. These distributions are formed by dividing the marginal frequencies from the contingency table by the sample size. These are the ordinary frequency distributions, seen in chapter 1, calculated as though the other variable were not present. Thus, the statistics class was 38.9% male, sophomores made up 41.7% of the class, and so forth.

Display 4.2

Joint and marginal frequency table (%) of sex and *class* variables of display 1.1

		CLASS				
SEX	FRESHMAN	SOPHOMORE	JUNIOR	SENIOR	GRADUATE	SEX MARGINAL
Male	8.3	16.7	11.1	0.0	2.8	38.9
Female	22.2	25.0	8.3	2.8	2.8	61.1
Class Marginal	30.6	41.7	19.4	2.8	5.6	100.0

Note also that the marginal frequencies are the sums of the joint frequencies in the corresponding row or column.

As mentioned in the introduction, the goal is to obtain information about the population joint frequency distribution of the variables. Logically, the way to do this is to work with the corresponding sample distribution. In fact, this is not the way it is done in practice. The analysis of joint frequency distributions is made indirectly, and the connection is not very transparent. The reason for this is that another representation exists for association information that is intuitively more appealing and that also allows us to use the comparison techniques developed for categorical variables in the last chapter. This representation requires the idea of conditional distributions, to be taken up next.

◢

EXERCISES

4.1   Group the weight variable of display 1.1 (p. 9) into three categories, ≤120, 121–140, and ≥141 pounds, and the height variable into the categories ≤64, 65–67, and ≥68 inches. Construct the contingency table for these grouped variables.

4.2   From the table constructed in exercise 4.1, form the table of joint and marginal frequency distributions.

4.3   Use the table of exercise 4.2 to answer the following questions. What percentage of the students in the sample weigh 120 pounds or less and are 64 inches tall or shorter?

What percentage of them weigh 120 pounds or more and are 68 inches tall or more? What proportion is less than or equal to 120 pounds? What proportion is between 65 and 67 inches?

4.4   In an arthritis clinic, patients are asked to evaluate the degree of activity of their disease on a scale from 1 to 3, with 1 indicating the least activity (severity) and 3 the greatest. At the same time, a clinical rating on the same scale is made by a physician. Data for a sample of patients are given in the following contingency table.

	PATIENT'S RATING		
PHYSICIAN'S RATING	1	2	3
1	31	14	4
2	7	5	5
3	2	9	12

(a)   Form the table of joint and marginal distributions for these data.

(b)   Based on the table of part (a), determine the percentage of the patients whose ratings agree with those of the physician.

(c)   For what percentage of the sample did the patient's assessment of activity exceed the rating given by the physician?

4.5   A sample of bicycle riders from the data for the Tour of the Rio Grande Valley (display 1.17 p. 25), was taken to determine whether there is an association between the age of the rider and the average speed with which he or she made the tour. The intent was to determine whether bicyclists slow down after they reach the age of 30.

	SPEED (MPH)		
AGE	<11	11–15	>15
≤17	14	9	0
18–29	11	11	5
≥30	10	9	7

(a)   Make a table of joint and marginal distributions.

(b)   What is the proportion of cyclists with speeds of less than 11 mph? What proportion is 17 years of age or less?

(c)   What proportion have speeds of less than 11 mph and are 17 years old or younger? What proportion have speeds of less than 11 mph and are 30 or older?

Conditional Distributions

The idea of an association between variables implies that the values of one variable are related to the values of the other in a frequency sense; the frequencies of the values of the first variable would be different for different values of the second. This suggests that we should be able to describe

*conditional
distributions*

association by comparing distributions, if the appropriate frequency distributions are used.

The appropriate frequency distributions are the **conditional distributions** of one variable *given* values of the other. For example, the conditional
distribution of the *class* variable given *sex* = male for the data of display 1.1
is the relative frequency distribution of the class variable *among men*. To
calculate this distribution from the contingency table of display 4.1, the
frequencies of the class categories in the *sex* = male row are divided by the
total number of men, 14, found in the column of marginal sums. The
conditional distribution of *class* given *sex* = female is obtained by dividing
the frequencies in the *sex* = female row by the number of females, 22.
These distributions are given in display 4.3. Except for rounding errors, the
frequencies in each row total 100%.

Display 4.3

**Conditional
distributions of *class*
given *sex* and marginal
distribution of *class* for
data of display 1.1**

SEX	FRESHMAN	SOPHOMORE	JUNIOR	SENIOR	GRADUATE	TOTAL
			CLASS			
Male	21.4	42.9	28.6	0.0	7.14	100.0
Female	36.4	40.9	13.6	4.55	4.55	100.0
Class
Marginal	30.6	41.7	19.4	2.8	5.6	100.0

An indication that *sex* and *class* are associated is seen, for example, in
the difference in conditional frequencies of freshmen. The frequency of
freshmen among women is 36.4%, while it is only 21.4% among men. Other
minor differences are seen for the other class categories as well. The comparison bar graph in display 4.4 gives a visual indication of the extent of the
differences. The marginal *class* distribution, shown in displays 4.3 and 4.4, is
an average* of the conditional distributions of *class* and can be used as a
basis of comparison for these conditional distributions. Thus, the proportion of freshmen among males is below the average for the two sexes, while
the proportion of freshmen among females is above the average.

To gain a different perspective on their association, it will often be
useful to form the conditional distributions of the variables in reverse order.
For example, displays 4.5 and 4.6 show the conditional distributions of the

* This average is what is known as a *weighted average*, in which the frequencies of the conditional distributions are multiplied by weights equal to the *sex* marginal frequencies rather
than the $1/n$ weights of the usual arithmetic average.

Display 4.4

Comparison bar graphs of conditional distributions of *class* given *sex* and marginal distribution of *class* for data of display 1.1

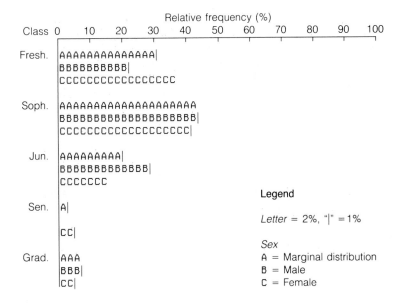

Display 4.5

Conditional distributions of *sex* given *class* and marginal distributions of *sex* for data of display 1.1

			CLASS			
SEX	FRESHMAN	SOPHOMORE	JUNIOR	SENIOR	GRADUATE	SEX MARGINAL
Male	27.27	40.00	57.14	0.00	50.00	38.89
Female	72.73	60.00	42.86	100.00	50.00	61.11
Total	100.00	100.00	100.00	100.00	100.00	100.00

sex variable given *class* for the data of display 1.1. Among freshmen, the proportion of men (3/11 = 27.3%) is seen to be well below the average provided by the marginal frequency (38.9%), while it is well above this average among juniors. When a variable has two values, only the bar graph for one value is needed, since the frequencies for the second value are the complements of (100% minus) the frequencies for the first. It would thus be possible to delete the portion of display 4.6 for males or for females without loss of information.

Note that the widest variation in frequencies is in the senior class, which contains 0% men. However, since there was only one senior in the class, this large difference should not carry much weight in measuring the amount of association between the variables. The inference procedure we will consider in chapter 13 automatically makes this kind of adjustment.

Display 4.6

Comparison bar graphs of conditional distributions of *sex* given *class* and marginal distribution of *sex* for data of display 1.1

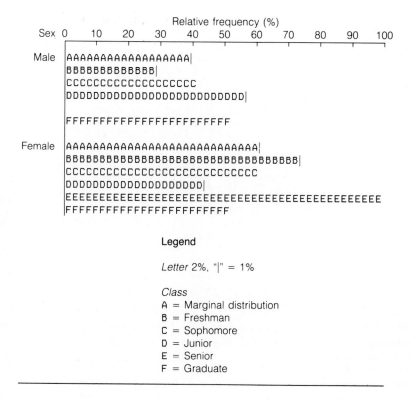

Legend

Letter 2%, "|" = 1%

Class
A = Marginal distribution
B = Freshman
C = Sophomore
D = Junior
E = Senior
F = Graduate

EXERCISES

4.6 Refer to the grouped *height* and *weight* variables considered in exercises 4.1–4.3. Compute the conditional distributions of *weight* given *height* and the marginal distribution of *weight*. Put this information into a conditional distribution table like the one in display 4.3 and make a comparison bar graph. Describe the nature of the association between *height* and *weight* in terms of these distributions. How do the frequencies of the categories of *weight* vary for increasing values of *height*?

4.7 Repeat the analysis of exercise 4.6, interchanging *height* and *weight*. What new interpretations, if any, are gained by interchanging variables?

4.8 Refer to the arthritis rating data of exercise 4.4. Form the table of conditional distributions of patients' ratings given physician's ratings and the marginal distribution of patients' ratings. Make a comparison bar graph of these distributions and describe the association seen in the graph.

4.9 Make a conditional distribution table and comparison bar graph of speeds given age groups for the data of exercise 4.5. Describe the trend in speed distributions with age. Does the "slowing down with age" hypothesis seem to be substantiated by the data?

The State of No Association—Independence

independence

If association is described in terms of differences in conditional distributions, it would seem reasonable to define the state of no association or **independence** to mean that these distributions are all the same. However, an important point to remember is that independence is a *population* concept, defined in terms of conditional frequency distributions for the population. These distributions could be obtained from a contingency table of population frequencies (if such were available) in the same way the sample distributions were computed from display 4.1. With this in mind, independence can be defined as follows.

Variables X and Y are said to be **independent** *if the population conditional distributions of Y given X are the same for every value of X.*

Some aspects of this important concept will be illustrated in the next example.

EXAMPLE 4.1

The contingency table for independent variables X and Y for a hypothetical population of 1000 individuals is given in the following table.

	Y				
X	1	2	3	4	TOTAL
1	60	40	20	80	200
2	240	160	80	320	800
Total	300	200	100	400	1000 = N

First form the conditional distributions of Y given X for both values of X by dividing the frequencies in each row of the table by the appropriate marginal total in the last column. The marginal frequency distribution of Y is similarly obtained from the marginal frequency row. It is seen that the conditional distributions are equal to the marginal distribution and have relative frequencies 30%, 20%, 10%, and 40%, respectively, for the values Y = 1, 2, 3, and 4. The equality of the conditional distributions establishes the independence of the two variables.

It is readily verified that the conditional distributions of X given Y are also all equal to one another and to the marginal distribution of X. Thus, the independence of X and Y is equivalent to the equality of the conditional distributions of either variable given the other.

The bar graphs for the population conditional distributions of Y given X and the marginal distribution of Y are given on the next page.

We would expect the bar graphs for a random sample from this population to be close to this form but not identical to it, because of sampling variation. With samples, there will always be the problem of trying to decide whether observed deviations from this form indicate a real association be-

Marginal distribution of Y (A) and conditional distribution of Y given X for X = 1 (B) and X = 2 (C), when X and Y are independent variables

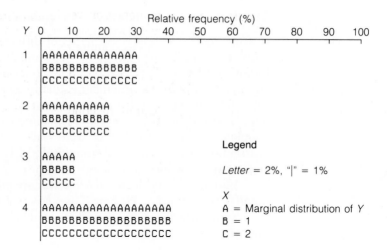

tween the variables or are just indicative of the variation likely to be seen from one random sample to another. One of the functions of inference is to help with this decision.

SECTION 4.3

Association for Measurement Variables: Regression

For measurement variables, the association problem takes on new richness and complexity. As before, all relevant information is contained in the joint frequency distribution of the variables. The important aspects of this information are best seen from the conditional distributions of one variable given the other. We will use the model for measurement variable distributions from chapter 1 to describe these conditional distributions. The conditional distributions will be represented by frequency curves and conditional probabilities by areas under these curves.

For variables X and Y, the decision to look at the conditional distributions of Y given X can be viewed as selecting a directed relationship to study, in which Y is the *dependent* variable and X the *independent* variable. This terminology originates in so-called designed experiments, in which the experimenter can freely or independently select one or more values of the X variable. The goal of the experiment is then to determine the response to the input values as measured by the dependent variable Y. The information about the response is contained in the conditional distributions of Y for the given values of X. The terminology has come to be used in many situations dealing with conditional distributions despite its unfortunate conflict with the meaning of "independence" in the last section. Other commonly used terms for Y and X are *response* variable and *explanatory* (or *predictor*) variable, respectively.

Conditional frequency curves for a hypothetical student population are shown in display 4.7 for measurement variables height (X) and weight (Y). (Imagine the curves coming out of the figure toward you in a third dimension.) The locations of the curves illustrate the fact that there are proportionately more students in the 120-to-150-pound range among those who are 60 inches tall than there are among those who are 72 inches tall; there is an upward shift in the conditional frequency distributions of weight with increasing height.

Display 4.7

Hypothetical conditional frequency distributions of weight (Y) given height (X) for heights x = 60 inches and x = 72 inches; relative frequencies for weights between 120 and 150 pounds given by shaded areas

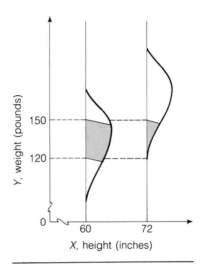

Only two distributions are shown in display 4.7, but in the mathematical idealization of a measurement variable, X has values that range over a continuum. Since there is a distribution attached to each value of X, it follows in theory that there are infinitely many conditional distributions. The conditional frequency curves show the flow and change of the Y variable frequency with increasing values of X.

For most practical purposes, it suffices to describe the flow of the *centers* of the conditional distributions. One of the location measures introduced in chapter 2 can be used for this description. The mean is one possibility. The traditional **regression curve** is the curve swept out by the means of the conditional distributions of Y for increasing values of X. An illustration of such a curve is given in display 4.8.

regression curve

Another possibility would be a *regression curve of medians*, which uses the median in place of the mean as the measure of distributional center. We can then study the population regression curve on the basis of a (random) sample by constructing an approximation to it from the sample medians. We will call such a curve an **exploratory regression curve**. The resistance property of the median, which was discussed in chapter 2, makes this an appealing exploratory tool.

exploratory regression curve

The remainder of the chapter will be devoted to the construction of

Display 4.8

Hypothetical conditional frequency distributions of Y given X for four values of X, showing regression curve of means

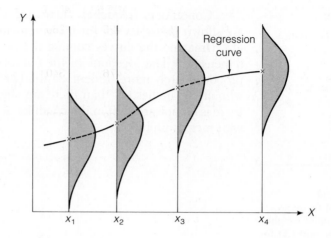

such curves and to another important exploratory display called the *scatter diagram*. As well as providing descriptive pictures of regression curves, scatter diagrams will give diagnostic information we will need in the study of linear regression starting in chapter 14.

Exploratory regression curves require a substantial amount of computation for large data sets and are then best carried out on a computer. Unfortunately, they are not implemented in many standard computer packages. However, schematic diagrams are becoming increasingly popular and are now included in several well-known packages. The next subsection shows how to use these diagrams to form exploratory regression curves in an important special case.

A Regression Method for Data Grouped at Discrete X Values

When data are grouped at a few values of the independent variable, schematic diagrams can be used to plot the exploratory regression curve. Since the box plots in schematic diagrams are printed at equal spacings, if the X values are also equally spaced then the regression curve can be drawn directly from the schematic diagram; the distance between boxes is simply set equal to the number of units separating the data groups. The following example illustrates this method.

EXAMPLE 4.2

Display 4.9 lists the (coded) scores on a verbal ability test for children of ages 3 to 10. The objective of this example is to chart the trend in the childrens' verbal ability with age. All children in the 3-to-5 age groups were either preschoolers or kindergartners, while the 6-year-olds were in first grade, the 7-year-olds in second grade, and so on.

Because the ages are equally spaced, the median regression curve can be obtained by joining the medians in a schematic diagram, as shown in display 4.10.

Display 4.9

Coded verbal ability scores by age for preschool or kindergarten (3–5) and elementary school (6–10) children

AGE 3	AGE 4	AGE 5	AGE 6	AGE 7	AGE 8	AGE 9	AGE 10
68	255	425	260	376	324	418	428
35	202	370	182	350	366	372	366
145	317	380	292	304	322	387	386
173	327	476	340	328	398	416	320
170	297	410	330	310	448	408	404
190	100	358	324	326	372	400	414
225	448	338	378	342	344	398	396
340	412	373	172	284	390	402	399
123	228	377	296	332	434	436	412
228	192	467	312	332	350	416	436
	297	388		298	364	407	452
						388	

Preschool or kindergarten ├───────┤   ├─────────── Elementary grades ───────────┤

1   2   3   4   5

Data courtesy of M. Kartas, M.S., Department of Communicative Disorders, University of New Mexico, Albuquerque.

Display 4.10

Schematic diagram used in the calculation of the exploratory regression of example 4.2

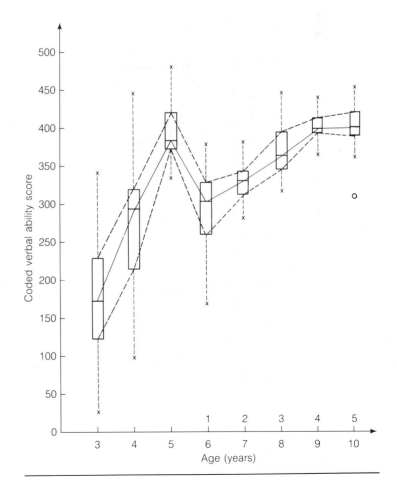

The boxes are one year apart, so the changes in medians from box to box are changes per year of age. It is seen from the shape of the curve that the rate of increase of verbal ability is greater for younger children than for older ones.

Information about trends in the variability of the data, which is of diagnostic interest, can be obtained by joining the quartiles with dashed lines as shown in the display. The distances between dashed lines are the interquartile ranges of the conditional distributions.

By removing the box plots from the background, the less cluttered regression curve of display 4.11 emerges.

Display 4.11

Regression curve of coded verbal ability score versus age for preschool or kindergarten children and children in grades 1–5

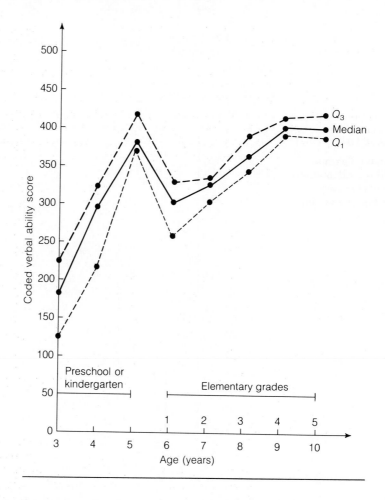

The regression curve shows a rather startling decrease in verbal ability as children enter elementary school. Before using this curve to condemn the public schools, it should be pointed out that the information for the various age groups was obtained from different groups of children; the same children were not followed through the 8 ages represented in the study. At the time of this study, parents in New Mexico had to pay in order to send their children to preschool and kindergarten. Thus, the children in the 3-to-5 age group were a "select group" relative to the general public school students in the 6-to-10 range.

The rather unique ability of this regression method to follow trends is well demonstrated by this example. Without previous warning that the features seen in display 4.11 existed, the traditional regression methodology could easily miss them. Thus, an exploratory regression is a useful first step in any regression analysis.

EXERCISES

4.10   A biologist studying the seasonal activity patterns of a species of field mice recorded the average distances traveled between captures by those mice trapped at least twice in a given month. The monthly capture data have been grouped by seasons in the accompanying table. The distances have been rounded to the nearest meter. Plot the exploratory regression of the median activity against season and indicate its variability by quartile curves. Describe the seasonal distributions, including the outliers. (Plot the outliers on the graph.) What is the season of peak activity? When is the variation in activity greatest?

FALL		WINTER		SPRING	SUMMER	
0	15	0	30	15	60	0
0	8	34	15	0	21	0
21	29	0	15	15	15	21
0	15	15	8	18	15	17
15	46	15	21	109	15	0
0	39	87	15	15	33	15
15	30	15	15	0	24	106
15	15	0	15	15	33	17
0	11	5	22	47	42	21
8	0	0	0	30	54	21
0		0	15	15	11	
0		15	15	34	32	
15		8	33	47	8	
21		0	21	42	71	
0		15	15	0	150	
34		47	0	22	18	
0		0	0	34	12	

Data courtesy of J. Scheibe, Department of Biology, University of New Mexico, Albuquerque.

4.11   In a study of occupation levels versus testosterone levels [27], Purifoy and Koopmans grouped 47 women into three general occupational categories: housewives (with no other occupations outside the home), secretaries and office workers (including clerks, lab technicians not requiring an advanced degree, etc.), and professional career women (including physicians, lawyers, college professors, business executives, etc.) These categories can be viewed as resulting from grouping an unobserved measurement variable. However, it is unlikely that this grouping preserves whatever scale that variable might have. Consequently, we cannot assume that the three groups are equally spaced in these units. We will be able to derive only directional information from this study, not rates of change information.

For the dependent variable, the free testosterone coefficient (FTC), a measure of the amount of biologically active testosterone in the blood, was obtained for each woman. The relationship between testosterone levels and occupational category can now be studied by forming a schematic diagram of the FTC distributions for the three occupational groups.

The FTC data from [27], in milligrams per 100 milliliters of blood, are given below.

HOUSEWIVES		SECRETARIES AND OFFICE WORKERS		PROFESSIONAL CAREERS			
.8	2.3	1.1	2.3	1.2	2.9	3.4	5.2
1.1	2.3	1.8	2.3	1.5	3.0	3.5	5.3
1.2	2.3	2.0	2.4	2.1	3.1	3.6	5.7
1.4	3.9	2.1	2.5	2.2	3.2	3.6	6.1
1.8	4.2	2.1	5.9	2.2	3.3	4.8	6.2
2.0		2.2		2.7	3.4	4.9	7.8

(a)   Make a schematic diagram of the data and join the medians of the three box plots with lines. Use this "curve" to describe the trend in testosterone levels with increasingly professional occupation levels.

(b)   Join the lower and upper quartiles with dashed lines, as in display 4.10, and describe any trend in dispersion.

(c)   If only the median and quartile lines drawn in parts (a) and (b) were to be shown on a graph, what important details seen in the schematic diagram would be lost?

A Regression Method for Ungrouped Variables

In the example of the last subsection, data were conveniently collected into groups of roughly equal sizes at equally spaced values of the independent variable. More frequently, the data will be given as pairs of numbers consisting of the X and Y values for each individual in a sample. Forming these data into groups at equally spaced X values is often not a good strategy, because of the differences in group sizes that can result. These differences can lead to a sample regression curve that gives a jumbled and inaccurate picture of the population regression.

The height and weight values of display 1.1 (p. 9) are an example of ungrouped regression data. A useful picture of the data is given by the

scatter diagram   **scatter diagram** of weight (Y) versus height (X) in display 4.12. Points corresponding to males are circled.

A scatter diagram is simply a graph of the individual paired data values as points. For example, the point (72, 180) indicated with an arrow represents the male with I.D. number 1 who is 72 inches tall and weighs 180 pounds.

Display 4.12

Scatter diagram of weight (Y) versus height (X) for data of display 1.1 (males circled)

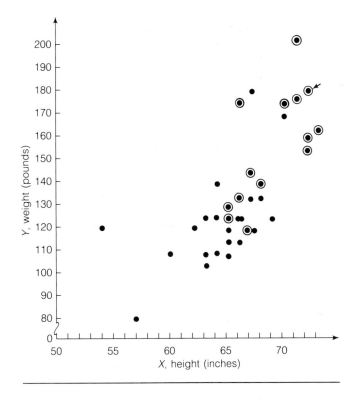

Note the heavy concentration of points near the middle of the plot, with diminishing numbers toward the extremes. An attempt to group these data at equally spaced heights would lead to a substantial variation in group size. To avoid this, the widths of the intervals are allowed to vary in order to put roughly equal numbers of points in each group.

cutpoints

It is convenient to use the scatter diagram to establish the **cutpoints** or boundaries between the groups. Slide a ruler, held vertically, across the plot from left to right and draw lines to establish groups as close as possible to the selected size. Once these groups have been chosen, the **exploratory regres-**

exploratory regression

sion is simply a plot of the points consisting of the X and Y group medians joined by straight lines.

An exploratory regression plot of the weight versus height data with cutpoints 63.5, 65.5, 67.5, and 70.5 inches is given in display 4.13. Note that all seven pairs of observations with X-values of 63 or less have been put into the first group, those with X-values larger than 63 but less than or equal to 65 are in the second group, and so forth.

In addition to the regression of medians, dashed lines joining the weight (Y) quartiles show the trend in dispersion for the data. An increase in dispersion is seen in the region of the plot where the data for the two sexes most overlap.

The cutpoints were initially chosen to produce five groups of about seven points each. (Multiple points with the same X-value have caused the

Display 4.13

Exploratory regression curve and quartile curves superimposed on scatter diagram of the height-weight data of display 1.1

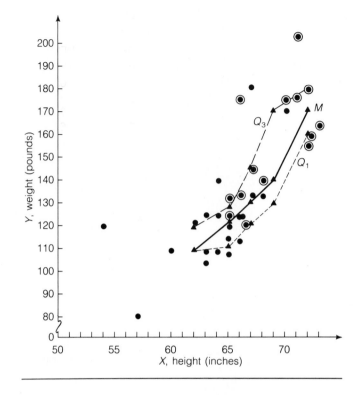

actual numbers to vary from group to group.) The tradeoff between the number of groups and the number of points per group must be taken into account in an exploratory regression. Too few groups can cause interesting features of the regression to be "averaged out" and missed. On the other hand, too few points per group can lead to unstable median estimates and the introduction of features that are purely artifacts of the data.

An ideal strategy to arrive at a satisfactory plot would be to calculate several regressions with different cutpoints and different numbers of groups. Methods of smoothing regression curves would also be useful. Clearly, however, these options are appealing only if they are computerized. A reasonable compromise is to compute a single curve using the rule of thumb (when the number of observations is sufficiently large) of choosing at least three groups and including at least five points in each group. This strategy will generally produce smooth regression plots that contain enough detail to show general trends and curvature. For example, an upward curvature in the regression of weight on height is seen in the five-group plot in display 4.13. This display suggests that we will have to go beyond a straight-line description of this regression. We will do so in chapter 14.

EXERCISES

4.12   The following data are the height and weight values from display 1.1, ordered by height. The cutpoints used to calculate the exploratory regression of display 4.13 are shown. Verify the calculations of the median and quartile values given in display 4.13.

			WEIGHT		
GROUP	NUMBER IN GROUP	HEIGHT MEDIAN	MEDIAN	Q_1	Q_3
54–63	7	62	110	107.5	121
64–65	8	65	122.5	112.5	127.5
66–67	10	66.5	129	120	145
68–70	5	69	140	133	170
71–73	6	72	169	160	180

HEIGHT	WEIGHT	
54	120	
57	82	
60	110	
62	122	
63	105	
63	110	
63	125	Cutpoint = 63.5
64	140	
64	125	
64	110	
65	115	
65	125	
65	107	
65	120	
65	130	Cutpoint = 65.5
66	114	
66	133	
66	125	
66	125	
66	175	
67	133	
67	145	
67	180	
67	120	
67	120	Cutpoint = 67.5
68	140	
68	133	
69	125	
70	170	
70	175	Cutpoint = 70.5
71	205	
71	175	
72	180	
72	160	
72	155	
73	163	

4.13 The following data are the long-term growth rates, Y, for children who achieved growth rate X during a trial period of treatment for dwarfism with a human growth factor, hGH.

X	Y	X	Y
.133	.202	.083	.158
.075	.123	.058	.094
.125	.150	.158	.206
.108	.165	.025	.067
.025	.077	.017	.060
.050	.092	.125	.150
.025	.098	.200	.254
.183	.227	.150	.215
.117	.167	.142	.181

Source: Rudman, Kutner, Goldsmith, and Blackston [30]. Copyright 1979, the Endocrine Society.

(a) Make a scatter diagram of the data and use it to choose cutpoints so as to put roughly six points into each of three groups.

(b) Construct an exploratory regression plot using the cutpoints chosen in part (a). Discuss the soundness of fitting these data with a straight line.

(c) Draw the quartile curves on the exploratory plot of part (b) and discuss the trend in dispersion of the Y values with increasing X.

4.14 The body weights (X) and heart weights (Y) of 15 female and 16 male cats are as follows.

FEMALES		MALES	
BODY WEIGHT (X)	HEART WEIGHT (Y)	BODY WEIGHT (X)	HEART WEIGHT (Y)
2.3	9.6	2.9	9.4
3.0	10.6	2.4	9.3
2.9	9.9	2.2	7.2
2.4	8.7	2.9	11.3
2.3	10.1	2.5	8.8
2.0	7.0	3.1	9.9
2.2	11.0	3.0	13.3
2.1	8.2	2.5	12.7
2.3	9.0	3.4	14.4
2.1	7.3	3.0	10.0
2.1	8.5	2.6	10.5
2.2	9.7	2.5	8.6
2.0	7.4	2.8	10.0
2.3	7.3	3.1	12.1
2.2	7.1	3.0	13.8
		2.7	12.0

(a) Make a scatter diagram of the data, circling the points for females.

(b) Use the scatter diagram of part (a) to find cutpoints, putting roughly six observations into each of five groups. Make an exploratory regression plot for the data and discuss the trend observed.

(c)   Draw the quartile lines in the plot and discuss the trend in dispersion. How is this trend related to the sexes of the cats in the various groups?

Chapter 4 Quiz

1. What is the association problem?

2. What questions are to be answered in the solution of an association problem?

3. What is the form of the two-way table of observed frequencies for a categorical variable association problem? Give an example.

4. What are the joint frequency distribution and the marginal frequency distributions? How are they formed from the two-way table? Give an example.

5. What are the conditional frequency distributions? How are they formed from the two-way table? Give an example.

6. How are the conditional frequency distributions used to study association?

7. How are comparison bar graphs used in the association problem?

8. When are two variables independent?

9. What is the meaning of the terms "dependent variable" and "independent variable" in an association problem?

10. What is the (traditional) regression curve?

11. What is an exploratory regression curve?

12. What is a scatter diagram?

13. How is an exploratory regression curve constructed when the data are grouped at equally spaced values of the independent variable?

14. What is the function of the quartile curves drawn on an exploratory regression graph?

15. How is an exploratory regression curve constructed for ungrouped data?

SUPPLEMENTARY EXERCISES

4.15   In a study of the relationship between diabetes and birth defects among the Pima Indians of Arizona (Comess et al. [8]), Pima mothers were classified by status: nondiabetic, prediabetic, (i.e., not currently having the disease but showing indications of developing it later), or diabetic. Among 1207 children born to these mothers, the observed frequencies of birth defects are given in the accompanying table.

MOTHER'S DIABETIC STATUS	CHILD'S BIRTH DEFECT STATUS	
	ONE OR MORE DEFECTS	NO DEFECTS
Nondiabetic	31	754
Prediabetic	13	362
Diabetic	9	38

(a) Obtain the joint distribution table for these data. What percentage of the children are born with birth defects? What percentage of the mothers are diabetic? What percentage of the children are born to diabetic mothers and have birth defects?

(b) Form the table and bar graphs of conditional distributions of birth defects given the mother's diabetic status. Comparing these distributions to the marginal distribution of birth defects, what can you say about the birth defect rate among children of diabetic mothers? Among children of nondiabetic and prediabetic mothers?

(c) Form the conditional distribution table and bar graphs of maternal diabetic status given the child's birth defect status. Among those children having at least one birth defect, which classification of diabetic status is most out of line with the corresponding marginal distribution?

(d) Would you be inclined to say that birth defects are associated with diabetic maternal status among the Pimas? (Recall that this is a sample, albeit a rather large one.)

4.16 This is a two-way analysis of a measurement variable. The accompanying data give a time history of the numbers of armed robberies in Albuquerque, New Mexico, by month from 1970 to 1974. By analyzing the table in both directions—that is, both horizontally and vertically—we can get an indication of the trends in armed robberies both by month of the year and by year over this five-year period.

(a) To obtain the trend by month of the year—and an indication of dispersion as well—obtain the medians and quartiles for each *row* of the table. Graph the median points at the 12 equally spaced month (time) values, and join the points by straight lines, as in display 4.11 (p. 106). Plot the 12 Q_1 points and join them by dashed lines. Repeat this step for the Q_3 points. (Note: A computer schematic diagram program may be used.) Describe the trend and variability of armed robberies by month. Give possible reasons for the highest incidence of robberies falling in the month shown by the curve.

| | | | YEAR | | |
MONTH	1970	1971	1972	1973	1974
Jan.	31	17	43	73	49
Feb.	29	36	24	39	43
Mar.	24	32	41	28	41
Apr.	21	33	25	64	36
May	16	32	48	42	38
June	16	25	75	59	29
July	15	24	76	55	46
Aug.	17	35	52	59	57
Sept.	21	53	60	33	36
Oct.	22	51	58	62	70
Nov.	28	46	26	77	52
Dec.	20	61	65	64	59

Data courtesy of J. Pedroncelli, Albuquerque Police Department.

(b) A large proportion of armed robberies are committed by people traveling through the city on the two major highways passing through Albuquerque. Correlate this information with the trend seen in the curve.

(c) To display the trend in armed robberies over the five-year period, repeat the calculations and graphs of part (a) for the columns of the table. Describe the trend in medians and in dispersion from the resulting graph.

(d) Late during this five-year period, a system was initiated that assigned more police officers to areas of the city shown by statistical analysis to be particularly robbery-prone during certain months. Does there appear to be any relationship between the robbery rate and the initiation of this system? What is it?

4.17 The relationship between grandparents living with the family in the home and the language predominantly spoken in the home was of interest in a recent study of English language use by young Native Americans. The accompanying table gives the observed frequency distribution of the pertinent variables. What are the differences in the conditional distributions of language use given grandparent status? What possible causes can you think of for these differences? Would any of these causes be established by the data if the effect were large enough to be accepted as real? Why?

	GRANDPARENT VARIABLE	
LANGUAGE USAGE	LIVE WITH FAMILY	LIVE ELSEWHERE
Native language predominant	10	23
English predominant	5	23

4.18 The accompanying data give hydrocarbon (HC) emissions at idling speed, in parts per million (ppm), for automobiles of various years of manufacture. The data were extracted via random sampling from an extensive study of the pollution levels from automobiles in current service in Albuquerque, New Mexico. The study was conducted over a period of time in the parking lot of a local shopping center. (What selection factors are likely to be operating to bias the sample because of the location of the test station?)

(a) Make a schematic diagram of the data and note any unusual features of the distributions.

PRE-1963	1963–1967		1968–1969		1970–1971		1972–1974	
2351	620	900	1088	241	141	190	140	220
1293	940	405	388	2999	359	140	160	400
541	350	780	111	199	247	880	20	217
1058	700		558	188	940	200	20	58
411	1150		294	353	882	223	223	235
570	2000		211	117	494	188	60	1880
800	823		460		306	435	20	200
630	1058		470		200	940	95	175
905	423		353		100	241	360	85
347	270		71		300	223	70	

Data courtesy of F. Wessling, College of Engineering, University of New Mexico, Albuquerque.

(b) What is the trend in average (median) HC levels as year of manufacture increases? Is there any apparent change in HC emissions coincident with the establishment of federal emission control standards in 1967–1968?

(c) What is the trend in dispersion with year of manufacture? Comment on the uniformity of car condition relative to HC emissions within the given time period.

(d) Do the automobiles seem to fall into natural performance groups by year of manufacture? If so, what are they?

4.19 The management of the National Public Radio (NPR) affiliate in Albuquerque commissioned a telephone survey of the Albuquerque metropolitan area to determine FM radio listeners' awareness of the existence of their station (or of NPR in general) and to assess the programming preferences of those individuals who regularly listened to the station. Demographic information including the sex, educational level, household income, and household size, along with a statement of awareness of NPR, was obtained from each of 136 respondents, making it possible to determine whether awareness is associated with any of these characteristics. More precisely, it is possible to obtain comparative profiles of those individuals who were and those who were not aware of NPR in the local community by looking at the conditional distributions of each demographic variable given the two values of the awareness variable. This analysis will be done in this and the next four exercises.

The two-way table of *sex* and *NPR awareness* is as follows:

SEX	NPR AWARENESS	
	AWARE	NOT AWARE
Male	81	8
Female	43	4

Source: The survey yielding these data was carried out by students at the University of New Mexico under the direction of S. Teaf, Department of Sociology, through the services of the Statistical Laboratory of the Department of Mathematics and Statistics. The summary and analysis of the data were carried out by S. Teaf and M. J. Pence.

Analyze the table for differences in NPR awareness for the two sexes; that is, form the conditional distribution of *awareness* given *sex*. Interpret your results.

4.20 Refer to exercise 4.19. The summary table of educational level and awareness from the NPR survey is as follows:

EDUCATIONAL LEVEL	NPR AWARENESS	
	AWARE	NOT AWARE
High school	22	5
Some college	27	2
College graduate	33	2
Postgraduate	39	2
Other	3	1

Analyze the association between educational level and NPR awareness by forming the conditional distributions of *NPR awareness* given *education level*. Interpret your results.

4.21   Refer to exercise 4.19. The summary table of household income and awareness from the NPR survey is as follows:

	NPR AWARENESS	
HOUSEHOLD INCOME	AWARE	NOT AWARE
Less than $10,000	25	4
$10,000–$14,999	23	1
$15,000–$19,999	16	1
$20,000–$29,999	30	2
$30,000 or more	20	2
No response	10	2

Analyze the association between income and NPR awareness. (Form NAR *awareness* given *income* conditional distributions.) Interpret your results.

4.22   Refer to exercise 4.19. The summary table of household size (number of people in the household) and awareness from the NPR survey is as follows:

HOUSEHOLD SIZE (NO. OF PEOPLE)	NPR AWARENESS	
	AWARE	NOT AWARE
1	26	3
2	43	5
3	22	3
4	24	1
5 or more	9	0

Analyze the association between household size and NPR awareness. (Form *NPR awareness* given *household size* conditional distributions.) Interpret your results.

4.23   Refer to exercises 4.19–4.22. Summarize the major differences in awareness of National Public Radio among individuals with the given characteristics.

4.24   (Computer project) An area of current interest is the field of gerontology, the study of the aging process. Modern laboratory techniques have made it possible to measure small differences in important hormone levels and thus to study the variations of these hormones with age. The hormone androstenedione in females is produced by both the adrenal glands and the ovaries, the latter contributing from 25% to 50% of the hormone in young women. This hormone is believed to have a connection with the ability to produce strong, healthy children (called reproductive capacity). Consequently, its progressive levels with age as well as its levels relative to other essential hormones are of considerable interest in the study of the maturation and aging process of women.

The accompanying data were obtained in part from a study involving a group of normal women ranging from 20 to 83 years of age (Purifoy et al. [28]). To extend the study to younger women, the researchers obtained androstenedione levels from the records of an endocrinologist, for 26 women in the age range 9–25 years. These 26 data points are recorded first in the accompanying table.

(a)   Make a scatter diagram of the data, circling the first 26 points. What, if anything, appears to be different about the data obtained from the endocrinologist's records, when compared to that for the normal women?

(b) Break the data up into eight groups by age, with the following cutpoints: 19.5, 23.5, 25.5, 33.5, 39.5, 47.5, 59.5. Do an exploratory regression analysis of the data based on these subdivisions, plotting both the median regression curve and the (dashed) quartile curves.

(c) Comment on the trends in both regression and dispersion (indicated by the quartile curves). Does the trend in median agree with what you would expect from your intuitive understanding of the term *reproductive capacity*? What influence does the endocrinologist's data have on the graph?

(d) What important features of the data would a straight-line regression miss? Would a straight-line regression provide a realistic representation of the maturation and aging process? For what ages would a straight line fitted to the entire data set provide unrealistic average androstenedione levels?

FROM ENDOCRINOLOGIST'S RECORDS			FROM NORMAL WOMEN			
(25, 105)	(22, 143)	(19, 151)	(33, 273)	(66, 32)	(46, 116)	(32, 130)
(17, 314)	(19, 224)	(25, 84)	(76, 45)	(22, 227)	(24, 186)	(38, 140)
(24, 220)	(22, 224)	(25, 157)	(40, 196)	(42, 196)	(39, 123)	(73, 74)
(25, 322)	(19, 140)	(12, 343)	(39, 175)	(56, 74)	(34, 137)	(61, 35)
(24, 525)	(25, 336)	(17, 140)	(39, 221)	(87, 56)	(29, 140)	(57, 60)
(24, 298)	(23, 144)	(22, 133)	(56, 102)	(28, 158)	(40, 165)	(62, 84)
(16, 133)	(25, 336)	(21, 489)	(31, 77)	(83, 84)	(44, 91)	(31, 133)
(25, 234)	(15, 87)	(21, 87)	(29, 165)	(53, 56)	(35, 165)	(66, 63)
(9, 67)	(15, 101)		(34, 147)	(35, 154)	(63, 84)	(51, 67)
			(40, 116)	(42, 108)	(56, 105)	(20, 189)
			(43, 176)	(53, 119)	(55, 116)	(42, 140)
			(73, 137)	(61, 84)	(51, 109)	(38, 172)
			(34, 109)	(20, 147)	(47, 98)	(59, 81)
			(31, 172)			

4.25 (Computer project) The discovery of the link between serum cholesterol level and heart disease has sparked a great deal of interest in determining cholesterol levels for various populations and how they vary with such other variables as sex and age. In a project carried out by the University of New Mexico School of Medicine, cholesterol levels were obtained from a total of 1509 volunteers during the 1975 and 1976 New Mexico state fairs. A booth was set up each year and blood was extracted from anyone volunteering for the study. The only reward for volunteering was that cholesterol levels were reported to each individual, with an indication of the currently available normal limits for their appropriate age and sex. Among the goals of the study (reported in Eaton et al. [10]) was the investigation of the variation of cholesterol levels with age for males and females. The following data were obtained by randomly subsampling the 1975 data set. Cholesterol levels are in milligrams per 100 milliliters of blood (mg%).

MALES				FEMALES		
(18, 158)	(38, 260)	(56, 174)	(18, 166)	(38, 230)	(42, 212)	(59, 147)
(26, 207)	(30, 265)	(54, 217)	(18, 123)	(36, 201)	(46, 181)	(65, 293)
(28, 251)	(38, 153)	(50, 263)	(28, 226)	(36, 162)	(47, 203)	(69, 291)
(22, 203)	(39, 180)	(54, 199)	(22, 177)	(38, 190)	(55, 229)	(60, 168)
(28, 217)	(34, 106)	(54, 225)	(23, 162)	(38, 178)	(59, 286)	(65, 167)
(26, 174)	(45, 242)	(57, 172)	(20, 114)	(30, 155)	(59, 239)	(66, 224)
(21, 190)	(47, 241)	(65, 245)	(23, 180)	(36, 211)	(59, 290)	(62, 201)
(23, 153)	(42, 305)	(60, 195)	(20, 253)	(34, 201)	(59, 292)	(66, 203)
(23, 146)	(49, 149)	(64, 212)	(28, 194)	(44, 229)	(55, 273)	(69, 271)
(25, 138)	(49, 165)	(65, 222)	(21, 139)	(44, 185)	(50, 199)	(67, 268)
(29, 191)	(44, 237)	(62, 215)	(22, 132)	(42, 257)	(55, 165)	(71, 257)
(31, 280)	(40, 182)	(63, 215)	(23, 183)	(44, 239)	(59, 246)	(72, 299)
(33, 160)	(47, 325)	(68, 250)	(29, 162)	(46, 148)	(50, 228)	(74, 214)
(35, 198)	(40, 241)	(70, 272)	(26, 141)	(48, 182)	(57, 253)	(70, 251)
(39, 215)	(57, 271)	(71, 219)	(38, 157)	(44, 226)	(50, 124)	(82, 220)
(38, 180)	(52, 287)	(80, 219)	(35, 150)	(42, 162)		
(36, 223)	(58, 260)					

Break the data for each sex into three groups by age at cutpoints 30.5 and 55.5 years and plot exploratory regression curves of cholesterol level versus age for the two sexes. Describe the trends in cholesterol with age and comment on any differences seen in these trends. What physical factors might account for these differences?

PART II

Statistical Inference—Concepts and Tools

Chapters 5 through 9 are intended to be transitional in nature, forming the bridge between the problems of statistics and their exploratory treatment given in part I and their inferential treatment given in part III. Statistical inference depends critically on the use of statistical models which are defined in terms of probabilities. Consequently, our first order of business will be to study the basic ideas from probability to be used in the later discussion of inference. The important normal distribution, which forms the basis of classical statistical inference, occupies much of chapters 6 and 7.

In chapters 8 and 9 we will get down to business with the two major forms of inference, confidence intervals and hypothesis tests. A glimpse of the important activity of experimental design will be given there. After completing these chapters we will be ready to return to the problems of part I, armed with the tools needed to evaluate the strength of our samples' evidence supporting the exploratory analyses.

CHAPTER 5

Probability

Introduction

There are two goals for this chapter. First, because probability is a rich and important intellectual discipline in its own right, a "cultural" introduction to the subject containing some of the more important probability ideas will be provided. Second, the necessary background will be given to support many of the statistical applications of probability in the following chapters.

Our treatment of probability will follow the ideas of the *frequentist* school, wherein probabilities are viewed as long-term relative frequencies. However, probability has a much broader scope, and we will touch briefly on other kinds of probabilities in this chapter.

Probability is the term used both for the numerical quantity that measures the likelihood of occurrence of a chance event and for the science that has been created to manipulate such quantities. A method for devising probabilities presumably first occurred to individuals involved in performing particular kinds of repetitive "chance experiments"—namely, gambling. (Names are unknown, because this activity and many of the ideas predate recorded history.) Since "high-probability" events happen frequently when such an experiment is repeated, while "low-probability" events happen infrequently, it occurred to them that a natural measure of the probability of an event is *its relative frequency (number of occurrences of the event divided by the number of trials) in repetitions of the experiment.*

One difficulty with this idea is that the relative frequency of an event can vary, both for different sequences of trials and for different numbers of trials. However, it was observed that as the number of trials increases, the relative frequencies settle down and approach the same value, no matter what the sequence of trials. This "limiting" value is then the natural choice for the probability of the event.

As an example, it is commonly accepted that the probability of *heads* for coins is .5. The author had members of a statistics class each toss a coin 50 times and record the tosses that turned up heads. The results for 116 students were combined, providing a coin toss experiment of 5800 trials. The relative frequencies of *heads* were then plotted against the number of

Display 5.1

Relative frequencies of _heads_ with increasing numbers _n_ of coin tosses; horizontal scale distorted (logarithmic)

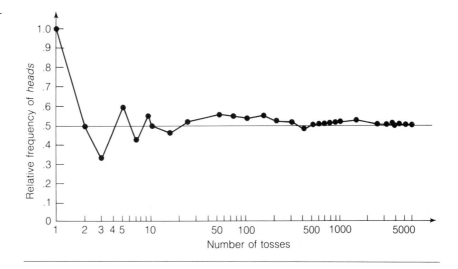

tosses, yielding the graph of display 5.1. Note how the relative frequencies settle down near .5 as the number of trials increases.

The frequency idea of probability was eventually extended to other areas of human experience in which repetitive random experiments occur naturally. Applications to human mortality were made in the 1600s; even now, life insurance companies base their rate structures on the observed frequencies of deaths within various groups of individuals.

The steady profits of gambling houses as well as the continuing financial health of insurance companies depend on probabilities calculated in this way. However, it soon became clear that it was impossible to calculate the probability of every relevant event by observing frequencies, and mathematical methods for computing and manipulating probabilities were devised.

Mathematical methods for manipulating probabilities make it possible to extend probability ideas to random experiments that are not naturally repetitive, such as the occurrence of a nuclear reactor accident. It is clearly not desirable to calculate the probability, say, that a lethal amount of radiation will escape from the reactor during an accident by physically enacting the accident. However, by modeling the relationships among reactor components mathematically and then using methods of the kind we will study for combining and refining probabilities, it is often possible to derive, at least approximately, the probabilities of interest.

When probabilities cannot be derived mathematically, another computational alternative has been made available by the development of the digital computer. This method takes advantage of the frequency concept of probability and the rapid computational speed of computers. The random experiment can be modeled or _simulated_ by a computer program. The computer can then be instructed to perform the experiment many times, keeping track of the numbers of occurrences of the events of interest. The

Monte Carlo simulation

final relative frequencies are then used to approximate the probabilities. Because the experiment produces random outcomes, this process is known as **Monte Carlo simulation,** after the famous gambling casino in Monte Carlo, Monaco.

The Monte Carlo method of computing probabilities is becoming increasingly important in statistics. The needs and interests of statisticians often surpass the present capabilities of mathematical derivation, and the only technology capable of keeping up with current needs is computer simulation. Some details of Monte Carlo simulation will be included in this chapter.

◢ SECTION 5.2
Random Experiments, Events, and Probabilities

We will consider the probability model for a random experiment in this section. Events will be defined, and an "algebra" for combining events will be discussed. The axioms for probability will then be given and some of the more important probability relationships derived.

Probability Models for Random Experiments

random experiment

outcomes

sample space

A probability model is designed to describe an experiment with a random outcome, which we will call a **random experiment.** When such as experiment is performed, we assume that it results in one of a well-defined collection of possible chance **outcomes.** For example, tossing a coin constitutes a random experiment that produces one of the outcomes *heads* or *tails.* The first component of a probability model is the collection, S, of outcomes which is called the **sample space** of the experiment. Thus, the sample space for the coin toss experiment can be represented symbolically as

$$S = \{heads,\ tails\}$$

This notation represents a collection by specifying its components in braces { }.

Another random experiment is the toss of a simple six-sided die with faces marked with from one to six dots. We can model this experiment by a sample space S consisting of the six outcomes *one dot, two dots, . . . , six dots.* Symbolically

$$S = \{one\ dot,\ two\ dots,\ .\ .\ .\ ,\ six\ dots\}$$

events

Subcollections of outcomes are called **events.** For example, the event "even number of dots" for the die experiment is the collection consisting of the outcomes *two dots, four dots, six dots:*

$$\text{"even number of dots"} = \{two\ dots,\ four\ dots,\ six\ dots\}$$

An event is said to *occur* if one of the outcomes contained in the event occurs when the random experiment is performed.

Our goal will be to formulate rules for calculating and manipulating the probabilities of events. Before discussing probabilities, however, it is necessary to see how events can be combined and changed to form the new events that will be needed later.

An Algebra of Events

not A

complement

New events will be formed by applying one or more of the three operations *not*, *or*, and *and*. If A is an event, then the event (**not** *A*) consists of the outcomes in S that are not in A. Thus, (not A) occurs in a random experiment only if A does not occur. (Not A) is called the **complement** of A.

For example, if A is the event "an even number of dots" in the die experiment, then

(not A) = {*one dot, three dots, five dots*}

That is, the complement of "even number of dots" is "odd number of dots."

A useful display for picturing operations on events is the *Venn diagram*. This diagram represents S as a square and various events as enclosed regions inside the square. The Venn diagram of the event (not A) is given in display 5.2.

Display 5.2

Venn diagram showing the event (not *A*) (shaded)

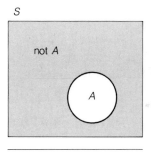

A *or* B

If A and B are events, then the event (**A or B**) consists of all outcomes that are in A or in B or in both A and B. Thus, the event (A or B) occurs in an experiment if *at least one* of the events occurs. The Venn diagram of this event is given in display 5.3.

Display 5.3

Venn diagram showing the event (*A* or *B*)

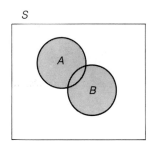

In the die experiment, if

A = {*four dots, five dots, six dots*}

and

B = {*two dots, four dots, six dots*}

then

(A or B) = {*two dots, four dots, five dots, six dots*}

A and B   The event (**A and B**) consists of all outcomes contained in both A and B. That is, this event occurs in a random experiment only if A and B occur simultaneously. The Venn diagram of (A and B) is given in display 5.4.

Display 5.4

Venn diagram showing the event (A and B) (shaded)

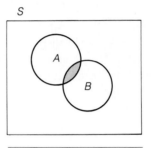

For events A and B that were defined above,

(A and B) = {*four dots, six dots*}

It would be convenient if an application of one of the above operations always produced an event, but this is not necessarily the case. Since S is a collection of outcomes, it is an event. Because S always occurs in a random experiment, it is sometimes called the *sure* event. But then (not S) contains no outcomes, so (not S) is not an event according to the current definition. This difficulty will be circumvented by extending the definition to include an event that contains no outcomes. This event, denoted by the symbol ∅, is called the empty or **null event.**

null event

Another situation leading to the null event is when the "and" operation is applied to events that have no outcomes in common. Such events are said to be **mutually exclusive.** Consequently, A and B are mutually exclusive if

mutually exclusive

(A and B) = ∅

An example of mutually exclusive events is A = {*two dots, four dots*} and B = {*one dot, three dots*}, in the die experiment. A Venn diagram of mutually exclusive events is shown in display 5.5.

Complex events can be built up by applying the three basic operations in combination and to more than two events at a time. For example, if A, B,

Display 5.5

Venn diagram showing (A and B) = ∅

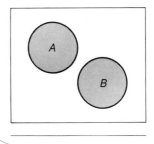

C, . . . are events, then (A or B or C or . . .) is the event "at least one of the events A, B, C, . . . occurs." The event (A and B and C and . . .) is the collection of outcomes common to all of the events, and so forth.

Events A, B, C, . . . are said to be *mutually exclusive* if no two of them have any outcomes in common. A Venn diagram of mutually exclusive events would show a series of non-overlapping regions. These ideas will be used to add the final component of probability to the model of a random experiment.

EXERCISES

5.1   A miniature card deck consists of the four cards king of clubs (KC), queen of clubs (QC), king of hearts (KH) and queen of hearts (QH). The deck is shuffled and one hand of two cards is dealt. The sample space for this game consists of the six distinct hands that can be formed from the four cards:

$$S = \{[KC, QC], [KC, KH], [KC, QH], [QC, KH], [QC, QH], [QH, KH]\}$$

(a)   Recall that clubs are black cards and hearts are red cards. Write down the outcomes in the following events, using the bracket notation.

A = "exactly one red card"
B = "at least one king"

(b)   For the events of part (a), write down in bracket notation the following events:

(1)   (not A)

(2)   (A or B)

(3)   (A and B)

(4)   (not B)

(5)   (A and (not B))

(c)   Which of the events of part (b) are mutually exclusive?

(d)   Show that the events (3) and (5) of part (b) satisfy the relation

((A and (not B)) or (A and B)) = A

5.2   Make a Venn diagram of two overlapping events A, B like the one in display 5.3.

    (a)   Identify the following events by shading the appropriate regions.

        (1)   C = (A and (not B))

        (2)   D = (B and (not A))

    (b)   Confirm, by identifying regions on the Venn diagram, that if C and D are the events found in part (a), then

        (1)   the three events C, D, and (A and B) are mutually exclusive

        (2)   (C or D or (A and B)) = (A or B)

Probability

The final component of a probability model consists of the probabilities themselves. To define probabilities for the model is to provide a method for assigning numerical values $P(A)$ for every event A to be formed from S. However these assignments are made, the probabilities will need to satisfy the following properties or axioms, which were first proposed by the Russian probabilist A. N. Kolmogorov.

    Kolomogorov Axioms

    1.   For every event A, $0 \leq P(A) \leq 1$

    2. $P(S) = 1$

    3.   If A, B, C, . . . are mutually exclusive events, then $P(A$ or B or C or . . .$) = P(A) + P(B) + P(C) + \cdots$

This is a rather modest set of requirements. Axiom 1 simply states that probabilities are between 0 and 1. This conforms with the relative frequency idea presented in the introduction, since relative frequencies are always in this range. Axiom 2 states that the sure event, S, will have probability 1.

addition law   Axiom 3, called the **addition law,** states that for mutually exclusive events the probability that at least one event will occur is equal to the sum of the probabilities of these events.

    So far we have considered only random experiments for which S consists of a finite number of outcomes. However, the above axioms cover all possible probability models, many of which have infinitely many outcomes. These models admit the possibility of infinitely many mutually exclusive events, and the addition law plays a unique and important role in shaping the properties of these models.

    A number of the most useful properties of probabilities follow (with little effort) from the axioms, as will now be demonstrated.

The Law of Complementation

Note that the events A and (not A) are mutually exclusive. Moreover,

    (A or (not A)) = S

By Axioms 2 and 3,

$$P(A) + P(\text{not } A) = P(A \text{ or } (\text{not } A)) = P(S) = 1$$

It follows by subtraction that

$$P(\text{not } A) = 1 - P(A)$$

law of
complementation

This expression is called the **law of complementation.** It has the useful consequence that, given the probability of an event A, the probability of (not A) can be calculated simply by subtracting $P(A)$ from 1.

For example, if we know that $P(A) = .35$, it follows immediately that

$$P(\text{not } A) = 1 - P(A) = 1 - .35 = .65$$

Note that the law of complementation also implies that

$$P(A) = 1 - P(\text{not } A)$$

Consequently, if $P(\text{not } A)$ is the easier probability to calculate, we can use this equation to calculate $P(A)$ indirectly.

An example of this kind of application is a random experiment consisting of the production of a computer chip. The chip can have one or more defects, any one of which would cause the chip to be rejected for use. If S contains the potential number of defects the chip can have, then $S = \{0, 1, 2, \ldots, N\}$, where N is the maximum number of possible defects. The probability that the chip will be rejected is the probability of the event $A = \{1, 2, \ldots, N\}$. Because A can also be written as $A = (\{1\} \text{ or } \{2\} \text{ or } \ldots \text{ or } \{N\})$, by axiom 3 the probability of A is the sum of the probabilities $P(1)$, $P(2)$, and so forth. Even if each term is reasonably easy to compute, the calculation of this sum could be rather tedious if N is very large. However, we note that $(\text{not } A) = \{0\}$. Consequently, the law of complementation makes it possible to calculate $P(A)$ by the substantially simpler expression $1 - P(0)$.

Another consequence of the law of complementation is that, since the null event, $\emptyset$, is the complement of S,

$$P(\emptyset) = 1 - P(S) = 0$$

The General Addition Law and Boole's Inequality

A second probability law, called the *general addition law* (for two events), is also a direct consequence of the axioms. If A and B are mutually exclusive, then axiom 3 yields

$$P(A \text{ or } B) = P(A) + P(B)$$

What can be said about $P(A \text{ or } B)$ when A and B are not mutually exclusive? As seen in display 5.6, the event (A or B) consists of three mutually exclusive parts: the event C, which consists of outcomes in A but not in B; the event D, consisting of outcomes in B but not in A; and the overlap event (A and B) (see exercise 5.2).

Display 5.6

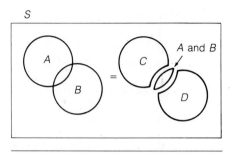

Since $A = (C$ or $(A$ and $B))$, axiom 3 yields

$$P(A) = P(C) + P(A \text{ and } B)$$

thus

$$P(C) = P(A) - P(A \text{ and } B)$$

Similarly,

$$P(D) = P(B) - P(A \text{ and } B)$$

Applying axiom 3 to the three events that make up $(A$ or $B)$ yields

$$P(A \text{ or } B) = P(C) + P(D) + P(A \text{ and } B)$$

By substituting the above expressions for $P(C)$ and $P(D)$ in this equation, we obtain

$$P(A \text{ or } B) = P(A) + P(B) - P(A \text{ and } B)$$

general addition law   This is the **general addition law.**

Note that if $(A$ and $B) = \emptyset$ (i.e., A and B are mutually exclusive), then $P(A \text{ and } B) = 0$ and the general addition law reduces to the special case of axiom 3.

One more general consequence will be given before we consider specific examples. It was seen in the derivation of the general addition law that $P(A) + P(B)$ exceeded $P(A$ or $B)$ by an amount equal to $P(A$ and $B)$. If we simply ignore this term, we get the inequality

$$P(A \text{ or } B) \leq P(A) + P(B)$$

Boole's inequality   This inequality, known as **Boole's inequality,** extends to any number of events:

$$P(A \text{ or } B \text{ or } C \text{ or } \ldots) \leq P(A) + P(B) + P(C) + \cdots$$

The importance of this inequality follows from the fact that probabilities of events formed by the "and" operation are often hard to calculate. When this is the case, although we cannot calculate exactly the probability that at least one of the events occurs, we will be able to find an upper bound for it. This is often sufficient.

For example, suppose that each of ten factors that would lead to a nuclear reactor's venting radiation into the atmosphere have been shown to have probability .001 of occurring in isolation from the others. Denote the events that these factors occur in a nuclear accident by A_1 to A_{10}. The probability of venting is the probability that at least one of these events occurs: (A_1 or . . . or A_{10}). By Boole's inequality, we can say that the probability of venting cannot exceed the sum of the probabilities of the separate factors:

$$P(\text{venting}) \le .001 + \cdots + .001 = 10 \times .001 = .01$$

The calculation of a more precise venting probability would require an extension of the general addition law, which requires the computation of the probabilities of all combinations of the A_i's (joined with the "and" operation). Because these probabilities depend on complex interactions among the factors, they may be impossible to compute in practice. Boole's inequality provides a practical upper bound for the venting probability that can be determined from the more easily computed individual probabilities.

The Law of Total Probability

Another useful probability rule is obtained as follows. Suppose that events $A_1, A_2, \ldots, A_m$ are mutually exclusive and exhaust S. Such a collection of events is called a **partition** of S. Then, if B is any other event, the events (A_1 and B), (A_2 and B), . . . , (A_m and B) are mutually exclusive, and

partition

$$B = ((A_1 \text{ and } B) \text{ or } (A_2 \text{ and } B) \text{ or } \ldots \text{ or } (A_m \text{ and } B))$$

See display 5.7 for an illustration of this property.

Display 5.7

Venn diagram showing the relationship $B = ((A_1 \text{ and } B) \text{ or } (A_2 \text{ and } B) \text{ or } (A_3 \text{ and } B))$ for three mutually exclusive events A_1, A_2, A_3 that exhaust all outcomes in S

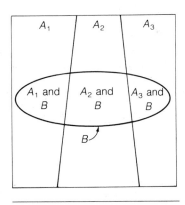

It follows from the addition law for probabilities that

$$P(B) = P(A_1 \text{ and } B) + P(A_2 \text{ and } B) + \cdots + P(A_m \text{ and } B)$$

law of total probability   This is called the **law of total probability**.

An important special case of this law occurs when the partition consists of A and (not A) for some event A. If B is any other event, we have

$$P(B) = P(A \text{ and } B) + P((\text{not } A) \text{ and } B)$$

This expression makes it possible to calculate, for example, the probability of $P((\text{not } A) \text{ and } B)$ from $P(B)$ and $P(A \text{ and } B)$:

$$P((\text{not } A) \text{ and } B) = P(B) - P(A \text{ and } B)$$

Note that when $B = S$, this expression reduces to the law of complementation.

EXAMPLE 5.1

Suppose that it is known that the probability of rain during a certain month is .15 and that this probability is .12 when the humidity is at least 60%. What is the probability of rain when the humidity falls below 60%?

If B = "rain during the month" and A = "humidity at least 60%" then (not A) = "humidity below 60%." We are asked to find $P((\text{not } A) \text{ and } B) = P(B) - P(A \text{ and } B)$. We are given that $P(B) = .15$ and $P(A \text{ and } B) = .12$. Consequently,

$$\begin{aligned} P((\text{not } A) \text{ and } B) &= P(B) - P(A \text{ and } B) \\ &= .15 - .12 \\ &= .03 \end{aligned}$$

EXERCISES

5.3   If $P(A) = .5$, $P(B) = .6$, and $P(A \text{ and } B) = .2$, find

(a) $P(A \text{ or } B)$

(b) $P(\text{not } A)$

(c) $P((\text{not } A) \text{ and } B)$

(d) $P((\text{not } A) \text{ or } B)$

(e) $P((\text{not } A) \text{ or } (\text{not } B))$ $= P(\text{NOT } (A \text{ and } B)) = 1 - P(A \text{ and } B)$

(f) $P(((\text{not } A) \text{ and } B) \text{ or } (A \text{ and } B))$

5.4   Let $P(A) = .7$ and $P(B) = .8$.

(a)   Is it possible for A and B to be mutually exclusive events? Explain.

(b)   Suppose $P((\text{not } A) \text{ and } B) = .3$. What is $P(A \text{ or } B)$?

5.5   The country is divided into four regions according to the probability of at least one severe earthquake (Richter scale 7.0 or greater) during the next year. The probability is .001 in region 1, .0005 in region 2, .0001 in region 3, and .00002 in region 4. What is the maximum probability of at least one severe earthquake in the country next year?

◢ SECTION 5.3

Finite and Equally Likely Probability Models

The topic of finite probability models is introduced in this section. These models will furnish the examples for the rest of this chapter and will be used to introduce general probability ideas in a simple setting. An especially important class of finite probability models are the *equally likely* models. These models will be shown to have useful applications to random sampling and Monte Carlo simulation.

Finite Probability Models

finite probability models

Probability models for random experiments with finitely many outcomes are called **finite probability models**. Probabilities for these models are assigned as follows. If $S = \{e_1, e_2, \ldots, e_N\}$, then a number (probability) p_i is associated with each e_i. These numbers must have the properties:

1. $p_i \geq 0$ for all i
2. $p_1 + p_2 + \cdots + p_N = 1$

The probability of an event A is then the sum of the probabilities associated with the outcomes in A. In symbols,

$$P(A) = \sum_{e_i \text{ in } A} p_i$$

Probabilities assigned in this way satisfy the Kolmogorov axioms, and thus they follow the probability laws discussed in the last section.

EXAMPLE 5.2

Readers familiar with the game Dungeons and Dragons will be aware of the possibility that dice need not be six-sided and regular like the ones used in the game of craps. Suppose a four-sided die has faces labeled with one through four dots, and these faces have relative areas $2:1:3:2$. If we assume that the probability of a given face is proportional to its area, the probabilities of the $N = 4$ outcomes are as follows:

OUTCOME (e_i)	PROBABILITY (p_i)
One dot	$\frac{2}{8}$
Two dots	$\frac{1}{8}$
Three dots	$\frac{3}{8}$
Four dots	$\frac{2}{8}$
	Sum = $\frac{8}{8}$ = 1

Now if A is the event "an even number of dots," then this event is composed of the outcomes *two dots, four dots,* and its probability is

$$P(A) = \tfrac{1}{8} + \tfrac{2}{8} = \tfrac{3}{8}$$

What is the probability of the event "an odd number of dots"? We can evaluate this probability in two ways. First, since the outcomes in this event have probabilities $\frac{2}{8}$ and $\frac{3}{8}$, by direct evaluation the probability of the event is $\frac{2}{8} + \frac{3}{8} = \frac{5}{8}$. But we also observe that this event is (not A), where $A =$ "an even number of dots." The law of complementation then implies that

$$
\begin{aligned}
P(\text{not } A) &= 1 - P(A) \\
&= 1 - \tfrac{3}{8} \\
&= \tfrac{5}{8}
\end{aligned}
$$

Equally Likely Probability Models

equally likely models

An important class of finite probability models are those for which the outcomes all have the same probabilities. They are called **equally likely models.** If there are N outcomes, since all probabilities must be equal and at the same time add up to 1, the outcome probabilities are necessarily

$$p_i = \frac{1}{N}$$

This expression leads to a simple formula for the probabilities of events. If N(A) denotes the number of outcomes in the event A, then the sum of the probabilities of the outcomes in A is simply N(A) times the probability of a single outcome, 1/N. It follows that the probability of A is

$$P(A) = \frac{N(A)}{N}$$

The importance of this expression is that it shows how to compute the probabilities of events for equally likely models simply by counting outcomes in events. First, one counts the total number of outcomes in S, N = N(S). Then one counts the number of outcomes in A, N(A). The ratio of these numbers is P(A).

EXAMPLE 5.3

The fairness (and legality) of the game of craps in gambling casinos depends on the regularity and symmetry of the six-sided dice used. The price of admission for the game and the odds on the various dice combinations are computed on the basis of the assumption that the dice are *fair*—that the probabilities of the six faces of each die are equal. We will look at the fair die probability model for the toss of a single die. It is an equally likely model with six outcomes, so the probability of each outcome is $\frac{1}{6}$.

Let A be the event "an even number of dots." What is the probability of A? We have seen that this event consists of the three outcomes *two dots, four dots,* and *six dots.* It follows that the probability of A is

$$P(A) = \tfrac{3}{6}$$

What is the probability that the number of dots will be even or greater than four? If A denotes "an even number of dots" and B = "more than four dots," the probability of interest is that of the event (A or B). A direct computation of the probability depends on an enumeration of the outcomes in the event. Since

(A or B) = {*two dots, four dots, five dots, six dots*}

its probability is $\frac{4}{6}$.

We will also compute the probability by means of the general addition law,

$$P(A \text{ or } B) = P(A) + P(B) - P(A \text{ and } B)$$

to show that it works in this context. We have seen that $P(A) = \frac{3}{6}$. The event B contains the two outcomes *five dots* and *six dots*. Consequently, $P(B) = \frac{2}{6}$. There is only one outcome common to A and B, namely *six dots*. Thus, $P(A$ and $B) = \frac{1}{6}$. The general addition law now yields

$$P(A \text{ or } B) = \tfrac{3}{6} + \tfrac{2}{6} - \tfrac{1}{6} = \tfrac{4}{6}$$

The Model for Simple Random Sampling: Lotteries

The lottery method of random sampling given in the introduction to chapter 1 was described as follows. Each member of the population submits a ticket with his or her I.D. number written on one side and various characteristics such as sex, height, and weight written on the other. The tickets are all placed in a basket and the basket is rotated several times to mix them thoroughly. For a sample of size one, one ticket is taken from the basket. The person whose I.D. number is on the selected ticket constitutes the sample and the various characteristics written on the reverse side are the randomly chosen values of the variables.

Suppose that the I.D. numbers are $1, 2, \ldots, N$. These numbers can be used to denote the outcomes in a probability model of this random experiment:

$$S = \{1, 2, \ldots, N\}$$

The intent of the lottery is that each individual in the population will have the same chance of being chosen. This implies that the appropriate probability model is an equally likely one.

The data of display 5.8, taken from one of the author's intermediate statistics classes, will be treated as a population of size $N = 20$ for demonstration purposes. Events can be defined in terms of the various characteristics possessed by the individuals. For example, the event "female" consists of all I.D. numbers for female students:

"female" $= \{2, 3, 4, 5, 10, 17, 20\}$

It follows, since there are seven outcomes in this event, that the probability that the sampled individual will be female is

$$P(\text{female}) = \tfrac{7}{20}$$

By the law of complementation, the probability of a male is then

$$P(\text{male}) = 1 - P(\text{female}) = \tfrac{13}{20}$$

What is the probability that the sampled student will be a senior, or a biology major? If A denotes the event "senior" and B denotes "biology major," we need $P(A$ or $B)$. The events are

$A = \{1, 3, 13, 18\}$
$B = \{2, 6, 19, 20\}$

Display 5.8

I.D.	AGE[a]	SEX	HEIGHT[b]	WEIGHT[c]	DISTANCE[d] FROM CAMPUS	NUMBER[e] OF SIBS	MAJOR	CLASS
1	30	M	65	135	4	2	Bus	Sen
2	29	F	67	120	1	2	Biol	Grad
3	24	F	66	137	10	4	Math	Sen
4	25	F	67	130	3	3	Econ	Grad
5	34	F	62	150	0	0	Econ	Grad
6	23	M	70	140	2	2	Biol	Grad
7	25	M	80	185	1	2	Geol	Grad
8	27	M	72	165	1	1	Math	Grad
9	31	M	66	135	1	5	Anth	Grad
10	19	F	67	135	1	1	Math	Soph
11	23	M	71	155	2	1	Engr	Soph
12	23	M	72	175	1	1	Engr	Jun
13	20	M	72	200	10	9	Engr	Sen
14	22	M	72	175	7	5	Engr	Jun
15	22	M	71	175	7	4	Engr	Jun
16	23	M	68	180	5	2	Engr	Jun
17	24	F	61	105	2	1	Anth	Grad
18	27	M	72	155	1	0	Econ	Sen
19	30	M	67	130	5	1	Biol	Grad
20	28	F	64	120	1	5	Biol	Other

[a] Age in years
[b] Height in inches
[c] Weight in pounds
[d] The distance in miles of the student's residence from UNM
[e] The number of brothers and sisters the student has

Note that these events are mutually exclusive. Thus

$$P(A \text{ or } B) = P(A) + P(B)$$
$$= \tfrac{4}{20} + \tfrac{4}{20} = \tfrac{8}{20}$$

Other examples will be given in the exercises at the end of the section.

Monte Carlo Simulation

How is it possible for a computer to toss a coin or to perform a lottery experiment of the kind described above? One can imagine a futuristic robot that physically carries out these tasks, but they can also be performed by the less human-like computers of today, by a process of simulation.

Monte Carlo simulation consists of modeling the outcomes and probabilities of a real experiment by means of a computer program. The computer performs the experiment by generating a "random" outcome according to the appropriate probabilities. At the heart of this process is the generation of what are known as **random numbers.**

random numbers

Display 5.9 gives random numbers extracted from one of the first computer-generated tables [29]. This table was published in 1955, before computer programs that generate random numbers were as commonly available as they are now. Today, tables are unnecessary, because random numbers can be produced by the computer as they are needed in a simulation. However, we will use display 5.9 to illustrate the methodology.

Display 5.9

Table of random numbers. Numbers are in groups of 5. Read the digits consecutively in book fashion—from left to right along the lines and down the table.

47505	02008	20300	87188	42505	40294	04404	59286	95914	07191
13350	08414	64049	94377	91059	74531	56228	12307	87871	97064
33006	92690	69248	97443	38841	05051	33756	24736	43508	53566
55216	63886	06804	11861	30968	74515	40112	40432	18682	02845
21991	26228	14801	19192	45110	39937	81966	23258	99348	61219
71025	28212	10474	27522	16356	78456	46814	28975	01014	91458
65522	15242	84554	74560	26206	49520	65702	54193	25583	54745
27975	54923	90650	06170	99006	75651	77622	20491	53329	12452
07300	09704	36099	61577	34632	55176	87366	19968	33986	46445
54357	13689	19569	03814	47873	34086	28474	05131	46619	41499
00977	04481	42044	08649	83107	02423	46919	59586	58337	32280
13920	78761	12311	92808	71581	85251	11417	85252	61312	10266
08395	37043	37880	34172	80411	05181	58091	41269	22626	64799
46166	67206	01619	43769	91727	06149	17924	42628	57647	76936
87767	77607	03742	01613	83528	66251	75822	83058	97584	45401
29880	95288	21644	46587	11576	30568	56687	83239	76388	17857
36248	36666	14894	59273	04518	11307	67655	08566	51759	41795
12386	29656	30474	25964	10006	86382	46680	93060	52337	56034
52068	73801	52188	19491	76221	45685	95189	78577	36250	36082
41727	52171	56719	06054	34898	93990	89263	79180	39917	16122
49319	74580	57470	14600	22224	49028	93024	21414	90150	15686
88786	76963	12127	25014	91593	98208	27991	12539	14357	69512
84866	95202	43983	72655	89684	79005	85932	41627	87381	38832
11849	26482	20461	99450	21636	13337	55407	01897	75422	05205
54966	17594	57393	73267	87106	26849	68667	45791	87226	74412
10959	33349	80719	96751	25752	17133	32786	34368	77600	41809
22784	07783	35903	00091	73954	48706	83423	96286	90373	23372
86037	61791	33815	63968	70437	33124	50025	44367	98637	40870
80037	65089	85919	73491	36170	82988	52311	59180	37846	98028
72751	84359	15769	13615	70866	37007	74565	92781	37770	76451
18532	03874	66220	79050	66814	76341	42452	65365	07167	90134
22936	22058	49171	11027	07066	14606	11759	19942	21909	15031
66397	76510	81150	00704	94990	68204	07242	82922	65745	51503
89730	23272	65420	35091	16227	87024	56662	59110	11158	67508
81821	75323	96068	91724	94679	88062	13729	94152	59343	07352
94377	82554	53586	11432	08788	74053	98312	61732	91248	23673
68485	49991	53165	19865	30288	00467	98105	91483	89389	61991
07330	07184	86788	64577	47692	45031	36325	47029	27914	24905
10993	14930	35072	36429	26176	66205	07758	07982	33721	81319
20801	15178	64453	83357	21589	23153	60375	63305	37995	66275

Display 5.9

(continued)

79241	35347	66851	79247	57462	23893	16542	55775	06813	63512
43593	39555	97345	58494	52892	55080	19056	96192	61508	23165
29522	62713	33701	17186	15721	95018	76571	58615	35836	66260
88836	47290	67274	78362	84457	39181	17295	39626	82373	10883
65905	66253	91482	30689	81313	01343	37188	37756	04182	19376
44798	69371	07865	91756	42318	63601	53872	93610	44142	89830
35510	99139	32031	27925	03560	33806	85092	70436	94777	57963
50125	93223	64209	49714	73379	89975	38567	44316	60262	10777
25173	90038	63871	40418	23818	63250	05118	52700	92327	55449
68459	90094	44995	93718	83654	79311	18107	12557	09179	28408
24810	61498	24935	86366	58262	44565	91426	86742	61747	79346
75555	42967	02810	16754	08813	40079	62385	21488	38665	94197
49718	05437	73341	17414	91868	24102	76123	67138	43728	43627
04194	83554	98004	14744	63132	75018	75167	24090	02458	78215
66155	24222	91229	63841	03271	56726	36817	51182	94336	20894
10801	76783	05312	30807	40006	04465	70163	17305	46414	76468
40422	75576	82884	27651	58495	87538	23570	66469	46900	95568
95572	12125	12054	17028	03599	73764	48694	85960	89763	58305
57276	49498	88937	08659	46840	83231	75611	94911	28467	67928
80566	94963	83787	87636	89511	64735	86699	66988	91224	72484
44517	55108	17435	33109	60343	46193	66019	43713	24097	52921
55424	87650	13896	90005	99458	20153	86688	13650	75201	79447
80506	78301	97762	16434	62430	28438	13602	63236	81431	75641
03646	54402	75413	39128	82975	73849	27269	73444	26120	06824
14537	53791	43951	51326	33274	54833	80802	66976	04878	35832
01644	33630	71247	59273	07811	33546	88628	06469	86257	39298
39387	94217	77995	54285	13354	84980	83590	63494	06036	18502
74962	49489	54662	93588	50466	55026	62458	06195	07995	71054
21165	45577	46383	38855	21561	89332	94248	09703	78397	38770
58519	95396	73607	72106	76597	85596	99075	39195	99605	66179
46982	79519	22294	15676	83484	98279	79200	02640	22501	43073
58463	67619	18006	05028	32441	83599	28915	05362	21612	64681
43055	00020	39254	68439	27399	24259	04641	50935	07112	55117
84073	38387	14337	90766	60436	65757	57590	17880	13776	35810
93542	37270	09361	62404	74056	52964	67372	81398	01482	97589
54467	20234	52813	85296	14542	73241	74848	39001	97598	76641
43608	42832	93917	67031	50220	94089	64858	27691	16719	99870
64808	01692	46424	64722	87162	06582	01452	14980	17397	07403
78703	93006	59651	48404	82284	66405	89818	00989	56112	78144
14886	70359	32158	30401	20829	22534	88848	07669	25100	48602
69280	61856	78974	91485	01583	11620	53740	32705	80391	56749
99680	99636	54107	79588	90845	21652	58875	13171	68531	18550
01662	21554	63836	41530	21864	81711	68921	61749	36051	78024
67852	69123	14280	17647	65125	82427	61594	32015	93473	05627
13911	67691	97854	89950	40963	06697	82660	69097	65284	49808

Display 5.9

(continued)

95822	09552	65950	34875	64250	41385	80133	70818	09286	30769
44068	24928	27345	34235	44124	06435	06281	43723	97380	76080
99222	66415	71069	62293	77467	35751	22548	23799	96272	58777
08442	61287	72421	35777	61079	42462	17761	94518	98114	74035
14967	60637	32097	28122	87708	19378	93372	23225	38453	80331
27864	15358	16499	91903	62987	98198	15036	23293	68241	44450
99678	16125	52978	79815	85990	18659	00113	93253	49186	25165
89143	79403	22324	54261	97830	42630	48494	09999	69961	39421
34135	31532	42025	83214	83730	28249	25629	11494	70726	45051
72117	97579	36071	29261	89937	78208	23747	56756	37453	51344
19725	76199	08620	22682	52907	25194	84597	93419	95762	14991
96997	66390	27609	41570	17749	23185	24475	56451	91471	33969
44158	67618	15572	95162	95842	08301	11906	68081	40436	58735
33839	40750	18898	61650	09970	47651	41205	65020	33537	01022
53070	61630	84434	05732	18094	71669	41033	82402	16415	83958

Source: Rand Corporation [29].

In the ideal, random numbers are generated as though the following die experiment were being carried out. A fair, ten-sided die with faces labeled 0 through 9 is rolled repeatedly, and the number on the upturned face is recorded after each roll. This experiment would generate a sequence of numbers such as those in display 5.9. Because the die is fair, the probability of each face is $\frac{1}{10}$. However, an additional feature is built into the generation process. At any given roll of the die, the probabilities of the ten outcomes are in no way influenced by the outcomes of previous rolls. The random digits are said to be *independent* or *independently generated*. (This concept will be treated more fully in section 5.6.)

The equally likely and independence properties of random numbers mean that not only do consecutive digits behave like the outcomes of rolls of a ten-sided die, but consecutive *pairs* of digits behave like independent rolls of a 100-sided die with faces labeled 00 through 99. Similarly, consecutive triples behave like the outcomes for a 1000-sided die, and so forth. Thus, random numbers can be used to represent the outcomes of a large class of die-toss games.

The next step in the design of a Monte Carlo simulation is to model the random experiment of interest by one of the die experiments. This is done by equating the *outcomes* of this experiment with *events* in the appropriate die-toss game. For example, to have a computer toss a fair coin, we proceed as follows. The outcomes of the coin toss are *heads* and *tails*, each occurring with probability $\frac{1}{2}$. The sample space of the ten-sided die-toss game is

$$S = \{0, 1, \ldots, 9\}$$

Equate the coin toss outcomes *heads* and *tails* with the following die-toss events.

"heads" = {0, 1, 2, 3, 4}

"tails" = {5, 6, 7, 8, 9}

Now, both events contain five outcomes and thus have probabilities $\frac{5}{10} = \frac{1}{2}$, as required for the fair-coin model.

To use the table of random numbers to toss a fair coin ten times, say, select a random starting point in the table (close your eyes and point) and read off ten consecutive numbers. If a number is 0, 1, 2, 3, or 4, record *heads*; otherwise, record *tails*. The resulting sequence of *heads* and *tails* is the Monte Carlo simulation of ten actual coin tosses.

EXAMPLE 5.4

To simulate the random selection of one individual from the population of display 5.8, we could proceed as follows. The sample space for the lottery is $S = \{1, 2, \ldots, 20\}$. We will use a 100-sided fair die for this simulation with faces labeled 00, . . . , 99. The die's sample space is then $S = \{00, 01, \ldots, 99\}$. Now, equate the outcomes of the lottery with the following die events:

"1" = {01, 21, 41, 61, 81}

"2" = {02, 22, 42, 62, 82}

. . . .

"19" = {19, 39, 59, 79, 99}

"20" = {00, 20, 40, 60, 80}

Each event contains five outcomes from the $N = 100$ outcomes of the die toss. Thus, each has probability $\frac{5}{100} = \frac{1}{20}$, the desired probabilities for the lottery draws.

To simulate the lottery selection of a sample of size one, pick a random starting place in display 5.9 and read a consecutive pair of digits. These digits form a number between 00 and 99. Locate this number in the events listed above. The I.D. number corresponding to this event is that of the sampled individual.

An alternative strategy for finding this I.D. number is the following. Divide the number formed by the random digits by 20; the appropriate I.D. number is then the remainder after this division. For example, suppose the two digits are 7 and 9. The remainder after dividing 79 by 20 is 19. The appropriate I.D. for the sample, according to the above scheme, is then 19.

EXERCISES

5.6   Two fair coins are to be tossed and the number of *heads* recorded. This suggests that the outcomes of the experiment are 0 *heads*, 1 *head*, and 2 *heads*. A natural tendency is to use the probabilities of $\frac{1}{3}$ for each outcome. However, probability models are intended to reflect accurately the "physical" probabilities (long-term relative frequencies) of events, and a somewhat more complicated sample space is needed in order to make the outcomes equally likely.

　　The two coins will be made distinguishable from one another, for example, by making one a dime and the other a penny. The single-toss outcomes, to be abbreviated as H and T, are then assembled in pairs to form the outcomes of the two-coin model, where the first letter in the pair refers to the dime and the second to the penny. The resulting sample space is

$$S = \{HH, TH, HT, TT\}$$

Now, the "outcomes" 0 *heads*, 1 *head*, and 2 *heads* are events in this sample space, with probabilities $\frac{1}{4}$, $\frac{1}{2}$, and $\frac{1}{4}$, respectively.

(a) To determine which of the two models, the $\frac{1}{3}, \frac{1}{3}, \frac{1}{3}$ or the $\frac{1}{4}, \frac{1}{2}, \frac{1}{4}$ model, best describes a real coin toss experiment, toss a pair of coins 100 times and tally the numbers of times 0 *heads*, 1 *head*, and 2 *heads* are seen. By the relative frequency definition, the probabilities of these events are approximated by the tallied values divided by 100. Which of the two models best represents the results of your experiment?

(b) Simulate the toss of two fair coins by using pairs of random numbers from display 5.9 as follows. If the first number is 0, 1, 2, 3, or 4, record H for the first coin (e.g. the dime); otherwise, record T. Record the outcome for the second coin in the same manner, using the second number. Tally the number of *heads*. Repeat this experiment 100 times by locating a random starting point in display 5.9 and reading off 100 consecutive pairs of random digits. (Cycle back to the beginning of the table if you run past the end.) Compare your Monte Carlo relative frequencies against the probabilities of the two models given above and with the results of part (a). Does the Monte Carlo experiment appear to give approximate probabilities that agree with the model best representing the real experiment?

5.7 The card game of exercise 5.1 involved dealing a hand of two cards from a deck consisting of the king and queen of clubs and the king and queen of hearts. The (equally likely) sample space is as follows:

$$S = \{[KC, QC], [KC, KH], [KC, QH], [QC, KH], [QC, QH], [QH, KH]\}$$

Let A denote the event "exactly one red card" and let B = "at least one king."

(a) Calculate $P(A)$.

(b) Calculate $P(B)$.

(c) Use the result of part (b) and the law of complementation to find the probability of the event "no kings."

(d) Calculate $P(A \text{ and } B)$.

(e) Use the general addition law and parts (a), (b), and (d) to find $P(A \text{ or } B)$. Verify this result by enumerating the outcomes in the event.

5.8 A random sample of size one is to be taken from the student population of display 5.8.

(a) List the outcomes of the following events (in bracket {} notation) and calculate their probabilities:

(1) "economics major"
(2) "female"
(3) "female economics major" (i.e., "female" and "economics major")
(4) "female or economics major"

(b) Calculate the probability of the event (4) of part (a), using the probabilities of events (1)–(3) and the general addition law.

(c) Calculate the probability that the selected student lives ten or more miles from campus. Then, without enumeration, calculate the probability that the student lives less than ten miles from campus, using the law of complementation. What is the advantage of using this law here?

(d) What outcomes do the events "business student" and "graduate student" have in common? What is the term used for such events? Calculate the probability that the selected individual is either a business student or a graduate student.

5.9   The Monte Carlo coin-toss method of this section can be extended to cover the tosses of biased coins. For example, a coin with *heads* probability equal to $0.4 = \frac{4}{10}$ can be constructed by equating *heads* with the outcomes 0, 1, 2, and 3 from the random digit sample space $S = \{0, 1, \ldots, 9\}$ and *tails* with the remaining outcomes.

Explain how to use display 5.9 to carry out 100 repetitions of a Monte Carlo simulation of the toss of a coin with the following *heads* probabilities. (Note: don't carry out the experiments.)

(a) .7

(b) .2

(c) .25   (Note: Use pairs of random digits.)

(d) .43

(e) .125   (How many random digits are needed here?)

Conditional Probability

In this section the concept of conditional probability will be introduced and illustrated. Several rules of probability will be derived from this idea. In particular, conditional probability will be used to define the independence of events, which plays an important role in the models used in statistics.

Conditional Probability

Suppose we were interested in the probability that a particular event B will occur in a random experiment. With no information about the potential outcome of the experiment, we would compute the probability of B as we did in section 5.3. Now, suppose that we are given the additional information that the outcome to be realized will be one of the outcomes in the event A. How should we modify the probability of B to account for this information? The (rational) answer is that we should now use *the conditional probability of* B *given* A, which is short for "the conditional probability that B will occur given that we know that A will occur." The notation for this probability is $P(B|A)$.

At first glance, this seems to be a rather artificial problem. How could we know that an event A is going to occur if the experiment has not yet been performed? In fact, there are many circumstances in which exactly this kind of information is available. Many random experiments consist of subexperiments or trials that are (or can be viewed as being) carried out one after the other.

For example, a sample of size 2 can be taken from a population by drawing the ticket for the first member of the sample, then drawing the second. Midway in the collection of the sample we will know all of the variable values for the first member of the sample. That information may influence the probabilities of the variable values for the second. For instance, if a sample of size 2 is drawn from the population described in display 5.8, the probability of obtaining a female on the first draw is $\frac{7}{20}$. Given the information that a female has been obtained on the first draw, which means that the event A = "female on first draw" has occurred, the probability of a female on the second draw will no longer be $\frac{7}{20}$; the change in the population caused by removing one of its members will affect the probabilities of subsequent draws.

An example will be used to illustrate the definition and computation of conditional probabilities. We will show that it is possible to simulate the toss of a coin with, say, a heads probability of $\frac{3}{7}$ by one roll of a ten-sided Monte Carlo die.

The die sample space is $S = \{0, 1, \ldots, 9\}$, and we must identify the outcomes *heads* and *tails* with events from this sample space. The identification will be as follows.

"heads" = {0, 1, 2}

"tails" = {3, 4, 5, 6}

"ignored" = {7, 8, 9}

After picking a random start in display 5.9, the "coin" is tossed by recording heads, if the random digit is 0, 1, or 2, or tails if it is 3, 4, 5, or 6. If the random digit is 7, 8, or 9, it is skipped and the process is repeated for the next digit.

Now, we know that in order for heads or tails to be recorded, the event A = {0, 1, 2, 3, 4, 5, 6} must occur in the random experiment. We know this before the experiment takes place because we will skip digits until it does occur. Thus, the appropriate probability of the event B = "heads" is the conditional probability of B given A.

It is easy to see how this conditional probability should be calculated. The sample space has been effectively reduced to the event A, which contains $N(A) = 7$ outcomes. Moreover, the outcomes of A are equally likely because they had this property in the original model. Now, $N(A \text{ and } B) = 3$ of these outcomes are also in the event B. Consequently, the counting method can be applied to obtain

$$P(B|A) = \frac{N(A \text{ and } B)}{N(A)} \tag{1}$$

$$= \tfrac{3}{7}$$

This counting solution is appropriate for any equally likely model. But how should conditional probability be defined when the model is not equally likely? An observation about the counting solution provides the answer.

Divide the numerator and the denominator of expression (1) by the total number of outcomes, $N = N(S)$. The resulting ratios are the ordinary probabilities $P(A \text{ and } B)$ and $P(A)$, computed by the counting method. Thus, (1) can be written in the alternative form

$$P(B|A) = \frac{P(A \text{ and } B)}{P(A)}$$

conditional probability *This is the expression used to define* the **conditional probability of B given**
of B given A **A**.

EXAMPLE 5.5

A fair coin is tossed twice by a friend, and you are told that *heads* is obtained on at least one toss. What is a rational value to adopt for the probability of two *heads*?

If A is the event "at least one head" and B = "two heads," then the appropriate choice would be $P(B|A)$.

For the equally likely sample space $S = \{HH, HT, TH, TT\}$, we have

$$A = \{HH, HT, TH\} \quad \text{and} \quad A \text{ and } B = \{HH\}$$

Thus,

$$P(B|A) = \frac{\tfrac{1}{4}}{\tfrac{3}{4}}$$

$$= \tfrac{1}{3}$$

The probability of two heads has changed from the "no information" probability of $P(B) = \tfrac{1}{4}$ to the more favorable probability of $\tfrac{1}{3}$ because of the given information.

Conditional probabilities behave like probabilities in all respects. Thus, for example, the laws of addition and complementation and Boole's inequality apply. The only possible difficulty in dealing with conditional probabilities surfaces when one tries to compute the conditional probability of B given A when the event A has probability 0. The ratio $P(A \text{ and } B)/P(A)$ is not defined in this case. This issue requires attention in more advanced treatments of probability. However, it is not important for the finite probability models that provide the examples in this chapter, and we will continue to ignore it.

A Multiplication Rule for Conditional Probabilities

By multiplying the expression defining $P(B|A)$ on both sides of the equals sign by $P(A)$, we obtain the important relation

$$P(A \text{ and } B) = P(A)P(B|A)$$

multiplication rule

This expression is called the **multiplication rule**. The reason for its importance is that we will often be interested in computing $P(A \text{ and } B)$, although the probabilities that are available or are easy to calculate are $P(A)$ and $P(B|A)$. The multiplication rule then makes it possible to compute $P(A \text{ and } B)$ indirectly.

For example, suppose that two individuals are to be sampled at random, *without replacement*, from the population of display 5.8. That is, two tickets are drawn one at a time, and the first is not put back before drawing the second. What is the probability that both ticketholders will be women? In order to calculate this probability directly by counting, we would have to create a new equally likely sample space consisting of all possible pairs of I.D. numbers—a space consisting of 190 distinct pairs of numbers! We would then have to count the pairs for which both I.D.s belong to women. Although special counting methods have been devised for doing this sort of thing, it is clear that counting could quickly become unattractive.

However, a simple solution to this problem can be found, using the multiplication rule. Let A be the event "woman on first draw" and B the event "woman on second draw." Since 7 of the 20 individuals are women,

$$P(A) = \tfrac{7}{20}$$

But, given that a woman was sampled on the first draw, the basket now contains only 19 tickets, and 6 of them belong to women. Thus,

$$P(B|A) = \tfrac{6}{19}$$

Finally, since "woman on both draws" is equal to the event $(A \text{ and } B)$,

$$P(\text{women on both draws}) = \tfrac{7}{20} \cdot \tfrac{6}{19}$$
$$= \tfrac{42}{380}$$

The multiplication rule can be extended to any number of events. For three events A, B, C, it is

$$P(A \text{ and } B \text{ and } C) = P(A)P(B|A)P(C|A \text{ and } B)$$

This expression could be used to calculate, for example, the probability of getting three women in a random sample of size three from the population of display 5.8.

Independent Events

independent

Events A and B are **independent** if the conditional probability of A given B is the same as the unconditional probability of A:

$$P(A|B) = P(A)$$

The multiplication rule for conditional probabilities is particularly simple in this case:

$$P(A \text{ and } B) = P(A)P(B)$$

product rule

This is known as the **product rule** for independent events.

Note that if the product rule holds, then the definition of conditional probability yields

$$P(A|B) = P(A \text{ and } B)/P(B) = P(A)P(B)/P(B) = P(A)$$

That is, events that satisfy the product rule are independent. It follows that the product rule can be used to establish the independence of events; if the probabilities of events satisfy the product rule, they are independent.

This fact can be used, for example, to show that the independence of events A and B implies the independence of A and (not B). By the law of total probabilities,

$$P(A \text{ and } (\text{not } B)) = P(A) - P(A \text{ and } B)$$

But since A and B are independent, the product rule holds. Thus,

$$\begin{aligned} P(A \text{ and } (\text{not } B)) &= P(A) - P(A)P(B) \\ &= P(A)(1 - P(B)) \\ &= P(A)P(\text{not } B) \end{aligned}$$

The last equation is a consequence of the law of complementation. It follows that A and (not B) are independent. It also follows by this same argument that (not A) and B are independent, as are (not A) and (not B). Thus, the independence of events is a "stronger" property than first meets the eye; if A and B are independent, then so are all combinations of these events and their complements.

More than two events A, B, C, . . . are said to be independent if the product rule holds for all combinations of these events and their complements. For example, the independence of A, B, and C implies that

$$P(A \text{ and } B \text{ and } C) = P(A)P(B)P(C)$$

and

$$P(A \text{ and } B \text{ and } (\text{not } C)) = P(A)P(B)P(\text{not } C)$$

Adding these two equations together and using the laws of total probabilities and complementation yields

$$P(A \text{ and } B) = P(A)P(B)$$

It follows, rather sensibly, that the independence of A, B, and C implies the independence of A and B.

The same argument can be used to establish the independence of the other two pairs of events. It follows that the independence of three events implies the independence of any two among the three. In general, the independence of any collection of events implies the independence of any subcollection.

One might imagine that this property can be turned around. That is, we might hypothesize that if all subcollections of a collection of events are independent, then the entire collection is independent. This, however, is not true in general. A simple example illustrates this phenomenon.

Two fair coins are tossed. Let

A = "a head on the first coin" = {HT, HH},

B = "a head on the second coin" = {TH, HH}

C = "different faces on the two coins" = {TH, HT}

It is easily established that all three pairs of events are independent. However, the three events are not, because $P(A$ and B and $C) = 0$ while $P(A)P(B)P(C) = \frac{1}{8}$.

◪
_____

EXERCISES

5.10   A bag contains three red and four white marbles. Two marbles are randomly selected, one at a time, from the bag, without replacement.

(a)   Use the multiplication rule to find the probability that both marbles are red.

(b)   What is the probability that the second marble drawn is a different color from the first?

(c)   What is the probability that the marbles are the same color?

5.11   Use the multiplication rule to calculate the probability that in a random sample of two individuals, sampled without replacement from the population of display 5.8,

(a)   both are men

(b)   one is a man and one a woman

5.12   If A and B are independent events with probabilities .3 and .5, respectively, find

(a) $P(A$ and $B)$

(b) $P(A$ or $B)$

(c) $P(A$ and (not B))

(d) $P(A$ or (not B))

(e) $P($(not A) and (not B))

(f) $P($(not A) or (not B))

_____

A Chain Rule for Conditional Probabilities*

chain rule

The law of total probability and the multiplication rule for conditional probabilities can be combined to obtain a useful **chain rule** for calculating the probabilities of certain events. If $A_1, A_2, \ldots, A_m$ form a partition of S

_____
* Optional section.

(that is, they are mutually exclusive events that exhaust S) and B is any event, then

$$P(B) = P(A_1)P(B|A_1) + P(A_2)P(B|A_2) + \cdots + P(A_m)P(B|A_m)$$

An application of this expression is the following. Suppose that we are not told the outcome of the first draw in a random sample of size two, without replacement, from the population of display 5.8. What is the probability that the second ticket belongs to a woman?

We have seen that the conditional probability of a woman on the second draw depends on the outcome of the first draw. However, we are not told this outcome. The only information available is that this is the second ticket drawn. We will show that this is not enough information to change the probability that a woman is drawn; the probability will be the same as on the first draw.

Let B = "woman on second draw," A_1 = "woman on first draw," and A_2 = "man on first draw." Note that A_1 and A_2 form a partition of S. Now, using the argument given earlier, the probabilities of these events are $P(A_1) = \frac{7}{20}$ and $P(A_2) = \frac{13}{20}$, and the conditional probabilities are $P(B|A_1) = \frac{6}{19}$ and $P(B|A_2) = \frac{7}{19}$. The chain rule then gives

$$P(B) = P(A_1)P(B|A_1) + P(A_2)P(B|A_2)$$

$$= \frac{7}{20} \cdot \frac{6}{19} + \frac{13}{20} \cdot \frac{7}{19}$$

$$= \frac{7}{20}\left(\frac{6}{19} + \frac{13}{19}\right)$$

$$= \frac{7}{20}$$

That is, the probability of a woman on the second draw, given no information about the outcome of the first draw, is exactly the same as the probability of a woman on the first draw.

This is true of any number of draws up to the point that the sample completely exhausts the population. Thus, if a sample of size 20 were taken without replacement from the population (of size 20) of display 5.8, and you were not told the outcomes of the first 19 draws, the probability of a woman on the twentieth draw would still be $\frac{7}{20}$. So would the probability of a woman on the fifth draw or the seventeenth—or any other. These conclusions can be established by extensions of the chain rule.

Bayes' Rule*

Bayes' rule addresses the following question. Suppose that we have established, by whatever means, probabilities for mutually exclusive and exhaustive events A_1, A_2, . . . , A_m. These probabilities, $P(A_1)$, $P(A_2)$, . . . , $P(A_m)$, are called the *prior* or *a priori* probabilities. Now, a random experiment is carried out, and an event B is observed. How should the occurrence of B influence the probabilities of the above events? The reasonable ap-

---

* Optional section.

proach is to compute the conditional probabilities of these events given B. These probabilities are called the *posterior* or *a posteriori* probabilities.

The probabilities likely to be available more or less directly from this experiment are the conditional probabilities of B given the various events A_1 through A_m. **Bayes' rule** shows how to convert these probabilities, along with the prior probabilities, into the posterior probabilities:

Bayes' rule

$$P(A_i|B) = \frac{P(A_i)P(B|A_i)}{P(B)}$$

where $P(B)$ is calculated from the chain rule,

$$P(B) = P(A_1)P(B|A_1) + P(A_2)P(B|A_2) + \cdots + P(A_m)P(B|A_m)$$

As an example of the use of Bayes' rule, suppose that before sampling from the population of display 5.8, we entertain one of the two "hypotheses" A_1 (that the population is made up equally of men and women) and A_2 (that the population is 80% male). We have no reason to prefer one "hypothesis" over the other, so our prior probabilities for them are $\frac{1}{2}$ each. Now, a random sample of size 2 is taken from the population (without replacement), and we observe the event B = "both are men." What are reasonable new values for the probabilities of A_1 and A_2 in the light of this information?

Bayes' rule will allow us to calculate the posterior probabilities of the A_i's given B if we can supply $P(B|A_1)$ and $P(B|A_2)$. But if "hypothesis" A_1 is correct, then the population consists of ten men and ten women, and the probability of drawing two men is

$$P(B|A_1) = \tfrac{10}{20} \cdot \tfrac{9}{19}$$
$$= \tfrac{90}{380}$$

On the other hand, if A_2 holds, then the population consists of sixteen men and four women, and

$$P(B|A_2) = \tfrac{16}{20} \cdot \tfrac{15}{19}$$
$$= \tfrac{240}{380}$$

It follows from the chain rule that

$$P(B) = P(A_1)P(B|A_1) + P(A_2)P(B|A_2)$$
$$= \tfrac{1}{2} \cdot \tfrac{90}{380} + \tfrac{1}{2} \cdot \tfrac{240}{380}$$
$$= \tfrac{1}{2} \cdot \tfrac{330}{380}$$

Then, Bayes' rule yields

$$P(A_1|B) = \frac{P(A_1)P(B|A_1)}{P(B)}$$
$$= \frac{\tfrac{1}{2} \cdot \tfrac{90}{380}}{\tfrac{1}{2} \cdot \tfrac{330}{380}}$$
$$= \tfrac{90}{330}$$

Similarly,

$$P(A_2|B) = \tfrac{240}{330}$$

The information from the sampling experiment has led to a downgrading of the probability of the "hypothesis" of half men and half women from $\frac{1}{2} = .50$ to $\frac{90}{330} = .27$. The 80% male "hypothesis" has, correspondingly, gone up in weight from .50 to .73.

Bayes' rule, which dates from the mid-1700s, is the basis for a theory of statistical inference different from the one we will study in this text. The above example illustrates how it can be used to modify assumptions about a population (the prior probabilities) on the basis of information from a sample.

From the viewpoint of those who hold the frequentist view of probability, this use of Bayes' rule has two disagreeable features. We have implicitly had to extend the probability model to include a "pseudo" random selection of the proportions of men and women in the population, in order to view the "hypotheses" A_1 and A_2 as events. In contrast, the viewpoint we will take of statistical inference is that the physical constitution of the population is fixed—it is simply unknown. The sample information will then be used in a different way to decide among various "hypotheses," such as A_1 and A_2.

Second, the selection of the prior probabilities is rather arbitrary. In the above example, the use of $\frac{1}{2}$ for the prior probabilities simply reflected a uniform lack of preference for the two options. Even in cases where it is reasonable to view the formation of the population as a random experiment, the probabilities will seldom be known. However, values must be assigned in order to carry out the computation of posterior probabilities, and the methods of doing so often lack solid support.

From the viewpoint of its adherents, the inadequacies of the Bayesian theory are counterbalanced by an elegant unification of both the ideas and the mathematics of inference. For further details, the reader is referred to the book by Box and Tiao [4].

EXERCISES

5.13 An urn experiment consists of selecting a ball at random from an urn labeled 1 and transfering it to an urn labeled 2, then drawing a ball at random from urn 2. Initially, urn 1 contained three white and two red balls, while urn 2 contained five white and two red balls.

   (a) Calculate the probability that the ball drawn from urn 2 is white.

   (b) If the ball drawn from urn 2 is red, what is the probability that the ball drawn from urn 1 was white?

5.14 Calculate the posterior probabilities of the two "hypotheses" A_1 and A_2, given in the above example of Bayes' rule, if the sample of size 2 yielded two women.

5.15 You are the second person to choose one carton of milk from a grocer's shelf originally containing three cartons. One of the three cartons is sour.

   (a) What is the probability that you get the sour carton given the information that the first one selected was fresh?

(b) What is the probability that you get the sour carton if no information is given about the first selection? Justify your answer with a probability calculation.

(c) Suppose that only one carton of the three is left when you get to the store. What is the probability that it will be the sour one? Justify your answer by extending the second chain rule to account for all possible states of the first two cartons chosen. For example, let A_1 = "fresh on first choice and fresh on second," A_2 = "fresh on first choice and sour on second," and so forth.

5.16 Two dairies each deliver three unmarked cartons of milk to your grocery store according to a particular schedule. On the day you shop, the probability is $\frac{1}{3}$ that the delivery was made by dairy 1 and $\frac{2}{3}$ that it was made by dairy 2. One of the cartons delivered by dairy 1 is sour and two delivered by dairy 2 are sour.

(a) The carton you buy is sour. What is the probability that the delivery was made by dairy 1?

(b) What would be the probability that the delivery was made by dairy 1 if the carton you bought were fresh?

5.17 Three urns are filled with colored balls as follows:

Urn 1 has three red and seven green balls
Urn 2 has five red and five green balls
Urn 3 has six red and four green balls.

A ball is drawn at random from urn 1. If this ball is red, a ball is drawn from urn 2; if it is green, a ball is drawn from urn 3.

(a) What is the probability that a red ball is obtained on the second draw?

(b) Suppose that a red ball is obtained on the second draw. What then is the (conditional) probability that the first ball drawn was red?

SECTION 5.5

Random Variables and Their Probability Distributions

The application of probability to statistical inference requires probability models for random experiments with numerical outcomes. These models are based on the concept of random variables and their probability distributions. This section is devoted to the introduction of these ideas. The important concept of expectation will also be defined and some of its properties illustrated.

Random Variables

The intuitive idea of a random variable is that of the *potential* value of a characteristic in a random experiment. For example, in a game consisting of tossing two coins, the random variable X might represent the number of *heads* to be realized when the game is carried out. Before the game, we can list the possible values of the variable as 0, 1, and 2 and talk about the probability that, say, the random variable will take on the value 1 when the game is played. Symbolically, the "event" that X takes on the value 1 can be written $(X = 1)$, and its probability will be denoted by $P(X = 1)$.

The values 0, 1, and 2 of X can be viewed as the outcomes of a new random experiment with probabilities $P(X = 0)$, $P(X = 1)$ and $P(X = 2)$. These values, together with their probabilities, constitute the *probability distribution* of the random variable. If we are interested only in the number of *heads* that occur in a coin toss game, the probability distribution of X summarizes all of the relevant information about the random experiment.

Other examples of random variables arise in the random selection of samples from a population by lottery. Consider the student population represented by display 5.8. Recall that to carry out the lottery, 20 tickets are put in a basket, each labeled with the I.D. number on one side and the values of the variables of interest on the other.

Before the lottery draw, we can speculate, for example, about the event that a female will be selected—that the Sex variable will have the value F. The lottery experiment has converted the ordinary variables into random variables, each with its own probability distribution. Moreover, the relationship of the probability distribution of a variable to its population frequency distribution is simple and direct: They are exactly the same. This fact has important implications for statistical inference.

Our interest in subsequent discussions will be focused on numerical-valued variables. This does not eliminate from consideration categorical variables such as the *sex*, *class*, and *major* variables, however, since numerical codes can be assigned to their values. For example, we will find it useful to code the *sex* variable 0 for males and 1 for females. In this way, it becomes a numerical-valued random variable.

A simple yet important idea makes it possible to graft the treatment of random variables onto the already developed probability ideas of previous sections. A random variable is viewed as a numerical-valued *function* defined for each outcome of a probability model. Thus, the variable that counts the number of *heads* in two tosses of a fair coin is defined on the sample space

$$S = \{HH, HT, TH, TT\}$$

If X denotes this random variable, its values are

$$X(HH) = 2$$
$$X(HT) = 1$$
$$X(TH) = 1$$
$$X(TT) = 0$$

Thus, the "event" $(X = 1)$ is an actual event in this model:

$$(X = 1) = \{HT, TH\}$$

The random variable probabilities are then "inherited" from the original probability model. For example, if fair coins are used in the coin-toss game, then

$$P(X = 0) = P(\{TT\}) = \tfrac{1}{4}$$
$$P(X = 1) = P(\{HT, TH\}) = \tfrac{1}{2}$$
$$P(X = 2) = P(\{HH\}) = \tfrac{1}{4}$$

A tabular display of the probability distribution of this variable is the following.

VALUE OF X	PROBABILITY
0	$\tfrac{1}{4}$
1	$\tfrac{1}{2}$
2	$\tfrac{1}{4}$
	1 = sum of probabilities

Note that the values of X exhaust all possible outcomes, in the sense that

$$(X = 0) \text{ or } (X = 1) \text{ or } (X = 2) = S$$

Moreover, these events are mutually exclusive. Consequently, the addition law for probabilities (Axiom 3) yields

$$P(X = 0) + P(X = 1) + P(X = 2) = P(S) = 1$$

This demonstrates a general property of probability distributions, namely that *the sum of the probabilities over all values of the variable is 1.*

As a second example, let Y denote the coded *sex* variable in a lottery drawn from the population of display 5.8. The ticket to be drawn from the basket will have the value $Y = 0$ written on it if the individual is a male and $Y = 1$ if female. The sample space for the lottery is the collection of I.D. numbers

$$S = \{1, 2, \ldots, 20\}$$

The value of Y for each outcome of this experiment is obtained from display 5.8. For example, the individual with I.D. number 1 is a male, so $Y(1) = 0$. Similarly, $Y(2) = 1$, $Y(3) = 1$, and so forth. The events $(Y = 0)$ and $(Y = 1)$ are as follows.

$$(Y = 0) = \{1, 6, 7, 8, 9, 11, 12, 13, 14, 15, 16, 18, 19\}$$
$$(Y = 1) = \{2, 3, 4, 5, 10, 17, 20\}$$

Since the probability model for a lottery is an equally likely one, the probability distribution of Y is

VALUE OF Y	PROBABILITY
0	$\tfrac{13}{20}$
1	$\tfrac{7}{20}$

We will largely ignore the function interpretation of random variables in our discussion of the properties of probability distributions. However, it is useful to know that a mathematically legitimate conception of random variables exists that supports these properties within the framework of the theory of probability we have studied—the theory based on Kolmogorov's axioms.

Discrete Variables and the Bernoulli Distribution

discrete

Random variables that take on only a finite number or a countable infinity of values are called **discrete**. These variables are also said to have *discrete distributions*. The probabilities of a discrete variable are concentrated at discrete points on a number line. In contrast, the probabilities of continuous variables, to be considered in the next chapter, are "smeared" continuously over intervals of the line and require a different computing method. A picture of a discrete distribution is given in display 5.10.

Display 5.10
Picture of the probability distribution of a discrete random variable *X* with values 1, 2, 3, 4. Probabilities are indicated by the heights of the lines. The sum of probabilities = 1

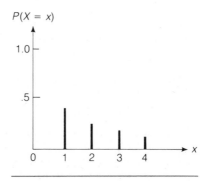

Bernoulli variable

A random variable that takes on only the values 0 and 1 is called an *indicator* or **Bernoulli variable**. The term "indicator" arises from the use of such a variable to indicate the occurrence of a particular event in a random experiment. Such a variable counts the number of times the event occurs; the count is 1 if the event occurs and 0 if it doesn't. Thus, the coded *sex* variable Y counts the number of females in the sample of size one from the population.

The name "Bernoulli" is given both to the variable and to the general form of its probability distribution. If the variable is denoted by X, and $P(X = 1) = p$, where p is a number between 0 and 1, then the Bernoulli probability distribution is

VALUE OF X	PROBABILITY
0	$1 - p$
1	p

In the context of random sampling from a population, p is both the population proportion of the characteristic of interest and the probability that an individual with this characteristic will be drawn. For example, the coded *sex* variable for the population of display 5.8 has a Bernoulli distribution with $p = \frac{7}{20}$.

The Bernoulli distribution is not a single distribution but rather a family of distributions "indexed" by the *parameter p*. That is, a different distribution is associated with each value of p between 0 and 1. This parameter will be of particular interest to us in statistical inference.

EXERCISES

5.18   Let X be the random variable that counts the total number of dots on two fair (six-sided) dice when they are tossed. An equally likely sample space for this experiment is constructed by making the dice distinguishable and recording in pairs the numbers of dots on each die. For example, $(1, 1)$ represents the outcome "one dot on the first die and one dot on the second." Then, $X(1, 1) = 2$.

(a)   Give the possible values of X.

(b)   For each possible value x, show the event $(X = x)$ in bracket notation.

(c)   Give the probability distribution of X in tabular form.

5.19   The sample space of the card game of exercise 5.1 (p. 127) is as follows.

$$S = \{[KC, QC], [KC, KH], [KC, QH], [QC, KH], [QC, QH], [QH, KH]\}$$

For each of the following random variables, (1) give the values of the variable for each outcome, (2) list in bracket notation the events corresponding to each value of the random variable, and (3) give the probability distribution of the random variable in tabular form.

(a) X = the number of kings in the hand

(b) Y = the number of red cards

(c) Z = the indicator of the event "at least one queen in the hand"

Expectation

Expectation is an important operation that assigns numerical values to random variables and their probability distributions. It is not hard to guess that the term "expectation" originated in a gambling environment. The idea behind the name is as follows.

Suppose that in a toss of two fair coins you win $1 for *heads* but nothing for *tails*. How much can you expect to win *per toss* "on the average"? Since probabilities model the long-term relative frequencies of events, in a large number of tosses two heads will be seen about $\frac{1}{4}$ of the time, one head about $\frac{1}{2}$ of the time and 0 heads $\frac{1}{4}$ of the time. Thus, in 100 tosses, you would win $2 in about 25 games, $1 in about 50 and nothing in 25. For the 100 tosses, your total winnings would be about

$$2 \cdot 25 + 1 \cdot 50 + 0 \cdot 25 \text{ dollars}$$

Put this on a per-toss basis by dividing by the number of tosses. After reducing fractions, the average winnings per toss is approximately

$$2 \cdot \tfrac{1}{4} + 1 \cdot \tfrac{1}{2} + 0 \cdot \tfrac{1}{4} = \$1$$

Now, as the number of tosses is increased, the relative frequencies will stabilize and approach their respective probabilities. Thus, this expression will become exact, and it follows that $1 is the average per-toss winnings to be expected as the number of tosses becomes arbitrarily large.

If X denotes the winnings for a given toss, the above expression suggests the following definition of the expectation of X.

$$E(X) = 2 \cdot P(X = 2) + 1 \cdot P(X = 1) + 0 \cdot P(X = 0)$$

expectation of X

More generally, if X denotes any discrete random variable, the **expectation of X** is defined to be

$$E(X) = \Sigma \, x \cdot P(X = x)$$

where the sum ranges over all possible values x of X.

EXAMPLE 5.6

Since a Bernoulli variable, X, takes the value 1 with probability p and 0 with probability $1 - p$, its expectation is

$$E(X) = 1 \cdot p + 0 \cdot (1 - p)$$
$$= p$$

Transformations of Variables and Their Expectations

A key problem in the statistical applications of probability is to find the probability distribution of a random variable formed by transforming other random variables. For example, given the probability distribution of X, we might require the distribution of $W = X^2$ or of $Y = 3X + 10$. Forming new random variables from old in this way is called **variable transformation**.

variable transformation

While finding the distribution of a transformed variable can be difficult, especially when random variables are of the continuous variety to be studied in the next chapter, the computation of its expectation is often much easier. One reason for this is that the expectation can be computed from the distribution of the original variable; it is not necessary to first derive the distribution of the transformed variable. Thus, the expectation of $W = X^2$ can be calculated from the distribution of X by the expression

$$E(X^2) = \sum_{\text{all } x} x^2 \cdot P(X = x)$$

For example, the expectation of the square of a Bernoulli variable is

$$E(X^2) = 1^2 \cdot p + 0^2 \cdot (1 - p)$$
$$= p$$

which is the same as the expectation of the variable itself. (This result is also easily seen from the fact that $W = X^2$ is equal to X in the Bernoulli case.)

linear transformation

The transformation $Y = 3X + 10$ is called a **linear transformation** of X, because only a sum and a constant multiple of X are involved. Linear transformations are of fundamental importance in statistics. Most of the statistical estimators we will be interested in are linear transformations of other variables. Thus, it is fortunate that expectations of linear transformations can be computed by following extremely simple rules.

These rules will make it possible to simplify the calculation of expectations of transformations that involve linear combinations of other transformations. Thus, for example, the rules will reduce the computation of the expectation of $W = 3X^2 + 10X + 4$ to the problem of computing the expectations of X and X^2.

rules of expectation

The **rules of expectation** are given here for transformations of a single variable. They will be extended to linear combinations of more than one random variable in section 5.6.

1. If $g(X)$ is a transformation of a random variable X and c is a constant, then

$$E(cg(X)) = cE(g(X))$$

2. If $g(X)$ and $h(X)$ are any two transformations of X, then

$$E(g(X) + h(X)) = E(g(X)) + E(h(X))$$

3. If c is a constant, then c can be viewed as a transformation of X that assigns probability 1 to c. Then

$$E(c) = c$$

These rules can be used to compute the expectations of complex expressions. For example, the expectation of $Y = 3X + 10$ can be found as follows. First, use rule 2 to write

$$E(Y) = E(3X + 10) = E(3X) + E(10)$$

Next, rule 1 yields $E(3X) = 3E(X)$, and rule 3 gives $E(10) = 10$. It follows that

$$E(3X + 10) = 3E(X) + 10$$

Similarly, the expectation of $W = 3X^2 + 10X + 4$ is

$$E(W) = 3E(X^2) + 10E(X) + 4$$

The following example justifies a computing formula that will be used in the next subsection.

EXAMPLE 5.7

We will show that

$$E(X - E(X))^2 = E(X^2) - E(X)^2$$

The rules can be applied to the expression $(X - E(X))^2$ by expanding the square algebraically. Since $(a + b)^2 = a^2 + 2ab + b^2$, we can write

$$(X - E(X))^2 = (X^2 - 2E(X)X + E(X)^2)$$

Thus, by the rules of expectation,

$$E(X - E(X))^2 = E(X^2 - 2E(X)X + E(X)^2)$$
$$= E(X^2) - 2E(X)E(X) + E(X)^2$$
$$= E(X^2) - 2E(X)^2 + E(X)^2$$
$$= E(X^2) - E(X)^2$$

◢

EXERCISES

5.20   Refer to exercise 5.18 (p. 155) and calculate the expected number of dots seen on two rolls of a fair die.

5.21   Refer to exercise 5.19 (p. 155) and calculate

(a) $E(X)$

(b) $E(Y)$

(c) $E(Z)$

5.22   The random variable Y has the following probability distribution.

VALUE OF Y	PROBABILITY
-1	$\frac{1}{4}$
0	$\frac{1}{4}$
1	$\frac{1}{2}$

(a) Calculate the expectation of Y.

(b) Calculate the expectation of $W = Y^2$.

(c) Find the expectation of $-3Y^2 + 5Y + 2$ using the rules of expectation.

The Variance of a Random Variable

variance

If X is a random variable with expectation $E(X)$, its **variance** is defined by the expression

$$\text{var}(X) = E(X - E(X))^2$$

The variance can be calculated using the probability distribution of X as follows.

$$\text{var}(X) = \sum_{\text{all } x} (x - E(X))^2 \cdot P(X = x)$$

For example, if X is the number of heads in two tosses of a fair coin, we have seen that $E(X) = 1$. The variance of X is then

$$\text{var}(X) = (2 - 1)^2 \cdot \tfrac{1}{4} + (1 - 1)^2 \cdot \tfrac{1}{2} + (0 - 1)^2 \cdot \tfrac{1}{4}$$
$$= \tfrac{1}{2}$$

computing formula for variance

A useful **computing formula for variance** was establishe

5.7:

$$\text{var}(X) = E(X^2) - E(X)^2$$

This expression reduces the computation of var(X) to the computati
the expectation of X and of X^2. These are usually easier to compute tha
the variance directly. Thus, if X is the number of *heads* in two tosses of a fa
coin, we have $E(X) = 1$ and

$$E(X^2) = 0^2 \cdot \tfrac{1}{4} + 1^2 \cdot \tfrac{1}{2} + 2^2 \cdot \tfrac{1}{4}$$
$$= \tfrac{1}{2} + 1 = \tfrac{3}{2}$$

The variance is then

$$\text{var}(X) = E(X^2) - E(X)^2$$
$$= \tfrac{3}{2} - 1^2$$
$$= \tfrac{1}{2}$$

as before.

EXAMPLE 5.8

Let X be a Bernoulli variable with $P(X = 1) = p$ and $P(X = 0) = 1 - p$. The variance of X will be computed using both the defining expression and the computing formula.

Since $E(X) = p$,

$$\text{var}(X) = (1 - p)^2 \cdot p + (0 - p)^2 \cdot (1 - p)$$
$$= (1 - p)^2 p + p^2(1 - p)$$

Factoring out a common factor of $p(1 - p)$ leaves

$$\text{var}(X) = p(1 - p)[1 - p + p]$$
$$= p(1 - p)$$

This expression is also quickly derived by means of the computing formula for variance by noting that X and X^2 have the same distribution, thus the same expectation p:

$$\text{var}(X) = E(X^2) - E(X)^2$$
$$= p - p^2$$
$$= p(1 - p)$$

A somewhat more complicated variance calculation that uses the computing formula to good advantage is the following.

EXAMPLE 5.9

A fair, ten-sided die has faces labeled 0 through 9. Let X denote the label value seen in a single toss of the die. What are E(X) and var(X)?

The expectation is

$$E(X) = 0 \cdot \tfrac{1}{10} + 1 \cdot \tfrac{1}{10} + \cdots + 9 \cdot \tfrac{1}{10}$$
$$= \tfrac{1}{10} \cdot [1 + 2 + \cdots + 9]$$

ıilarly,

$$E(X^2) = \tfrac{1}{10} \cdot [1^2 + 2^2 + \cdots + 9^2]$$

ıulas for evaluating sums and sums of squares are

$$+ 2 + \cdots + n = \frac{n(n + 1)}{2}$$

and

$$1^2 + 2^2 + \cdots + n^2 = \frac{n(n + 1)(2n + 1)}{6}$$

Setting $n = 9$, we find

$$E(X) = \frac{1}{10} \cdot \frac{9 \cdot 10}{2} = \frac{9}{2} = 4.5$$

$$E(X^2) = \frac{1}{10} \cdot \frac{9 \cdot 10 \cdot 19}{6} = 9 \cdot \frac{19}{6} = 28.5$$

The variance of X is then

$$\mathrm{var}(X) = 28.5 - (4.5)^2$$
$$= 8.25$$

Rules of Variance

The notation $\mathrm{var}(X)$ suggests that variance is viewed as an operation on a random variable similar to expectation. It is of interest to know whether "rules of variance" exist for linear transformations. That is, if we know $\mathrm{var}(X)$, what can be said about $\mathrm{var}(Y)$ if $Y = 3X + 10$, say? The following properties follow readily from applications of the rules of expectation.

If $g(X)$ is a transformation of a random variable X and c is any constant, then

1. $\mathrm{var}(cg(X)) = c^2 \mathrm{var}(g(X))$
2. $\mathrm{var}(g(X) + c) = \mathrm{var}(g(X))$

That is, multiplication of a random variable by a constant multiplies the variance by the square of the constant, but adding a constant does nothing to the variance.

Combining these relations, the variance of $Y = 3X + 10$ is seen to be $9\mathrm{var}(X)$.

EXAMPLE 5.10

If X is a random variable with $E(X) = \mu$ and $\mathrm{var}(X) = \sigma^2$, then the linear transformation

$$Z = \frac{X - \mu}{\sigma}$$

$$= \left(\frac{1}{\sigma}\right) X - \left(\frac{\mu}{\sigma}\right)$$

Z-score transformation   is called the **Z-score transformation**. This is a "standardizing" transformation, in that it converts X to a random variable Z with expectation equal to 0 and variance equal to 1. This follows easily from the rules of expectation and variance:

$$E(Z) = \left(\frac{1}{\sigma}\right) E(X) - \left(\frac{\mu}{\sigma}\right)$$

$$= \left(\frac{\mu}{\sigma}\right) - \left(\frac{\mu}{\sigma}\right)$$

$$= 0$$

$$\mathrm{var}(Z) = \left(\frac{1}{\sigma}\right)^2 \mathrm{var}(X)$$

$$= \frac{\sigma^2}{\sigma^2}$$

$$= 1$$

We will need the Z-score transformation in our discussion of the normal distribution in chapter 6.

Means and Variances for Probability Distributions and Population Frequency Distributions

The function definition of a random variable can be used to obtain the expectation of the random variable without first obtaining its probability distribution. If the outcomes of the sample space S on which X is defined are $e_1, e_2, \ldots, e_N$, and if the probabilities associated with these outcomes are $p_1, p_2, \ldots, p_N$, then

$$E(X) = \Sigma X(e_i) \cdot p_i$$

where this sum is over all outcomes in S.

For example, let X denote the number of dollars won in two tosses of a fair coin if $1 is paid for each *heads*. Since the outcomes HH, HT, TH, TT have probabilities of $\frac{1}{4}$ each,

$$E(X) = X(HH) \cdot \tfrac{1}{4} + X(HT) \cdot \tfrac{1}{4} + X(TH) \cdot \tfrac{1}{4} + X(TT) \cdot \tfrac{1}{4}$$
$$= 2 \cdot \tfrac{1}{4} + 1 \cdot \tfrac{1}{4} + 1 \cdot \tfrac{1}{4} + 0 \cdot \tfrac{1}{4}$$
$$= 1 \text{ dollar}$$

This is the same result obtained earlier using the probability distribution of X.

Expectations of functions of random variables can also be obtained in this way. For example,

$$\mathrm{var}(X) = \Sigma(X(e_i) - E(X))^2 \cdot p_i$$

These expressions for the expectation and variance have interesting implications for random sampling from a population. Suppose a random sample of size one is to be drawn from a population of N individuals by lottery. The lottery sample space is $S = \{1, 2, \ldots, N\}$ and $p_i = 1/N$ for each outcome (I.D. number) i. If the variable X has value $X(i) = x_i$ for the ith population member, then under random sampling X becomes a random variable with expectation

$$E(X) = \Sigma\, X(i) \cdot p_i$$

$$= x_1 \cdot \frac{1}{N} + x_2 \cdot \frac{1}{N} + \cdots + x_N \cdot \frac{1}{N}$$

This is exactly the same as the expression used to define the population mean of X in chapter 2. That is, under random sampling the expectation of X is equal to the mean of the population frequency distribution of X. We will use probability methods to arrive at statistical inferences for the expectation. However, it is the population mean that is of interest. Because they are the same quantity under random sampling, these inferences will also apply to the population mean.

mean of the random variable

The expectation of a random variable is also called the **mean of the random variable** and the symbol μ will be used frequently for $E(X)$ hereafter.

The equivalence of population and probability parameters also holds for the variance and standard deviation. For the lottery model,

$$\mathrm{var}(X) = \Sigma\, (X(i) - E(X))^2 \cdot \frac{1}{N}$$

$$= \frac{\Sigma(x_i - \mu)^2}{N}$$

This expression was used to define the population variance in chapter 2. Consequently, the symbol σ^2 for the population variance will also be used for the variance of the random variable. The square root of this quantity is the **standard deviation** of the random variable:

standard deviation

$$\sigma = \sqrt{\mathrm{var}(X)}$$

EXERCISES

5.23   A random variable X has the following probability distribution.

VALUE OF X	PROBABILITY
1	$\frac{1}{3}$
2	$\frac{1}{3}$
3	$\frac{1}{3}$

(a) Find the expectation of X.

(b) Compute the variance of X.

(c) Find the standard deviation of X.

(d) Find the expectation, variance, and standard deviation of Y = 2X + 8.

5.24   Refer to exercise 5.19 (p. 155), and calculate the variances and standard deviations of the variables

(a) X   (b) Y   (c) Z

5.25   Refer to exercise 5.18 (p. 155), and calculate the variance and standard deviation of the number of dots seen on two rolls of a fair die.

SECTION 5.6

Joint, Marginal, and Conditional Probability Distributions*

When two or more variables are observed simultaneously in a random experiment, extensions of the idea of a probability distribution are needed to handle the possible associations among the variables. Probability analogs of the joint, marginal, and conditional frequency distributions given in chapter 4 will be introduced for this purpose. The independence of random variables will be defined and used to derive the so-called *binomial* distribution. Means and variances of variables to be seen in statistical inference will be computed by extending the rules of expectation and variance given in the last section.

Joint and Marginal Probability Distributions

Let X and Y be random variables defined for a finite probability model. For example, in the toss of two fair coins, X might be the number of *heads* and Y the indicator of the event "opposite faces seen." The values of these variables on the sample space S = {HH, HT, TH, TT} are

$$X(HH) = 2 \qquad Y(HH) = 0$$
$$X(TH) = 1 \qquad Y(TH) = 1$$
$$X(HT) = 1 \qquad Y(HT) = 1$$
$$X(TT) = 0 \qquad Y(TT) = 0$$

joint probability distribution of X and Y   The **joint probability distribution of X and Y** is then the table of probabilities of the events ((X = x) and (Y = y)) (events joined by the *and* operation) for all combinations of the values x of X and y of Y. These probabilities will be written in the notation

$$P(X = x \text{ and } Y = y)$$

---

* The material in this section is not essential for the methods of subsequent chapters (if certain parameter relationships are taken on faith) and the section can be omitted if an abbreviated treatment of probability is desired.

Since, in the example, X has values 0, 1, and 2 and Y has values 0 and 1, the events whose probabilities are required are

$((X = 0)$ and $(Y = 0)) = \{TT\}$

$((X = 0)$ and $(Y = 1)) = \emptyset$

$((X = 1)$ and $(Y = 0)) = \emptyset$

$((X = 1)$ and $(Y = 1)) = \{HT, TH\}$

$((X = 2)$ and $(Y = 0)) = \{HH\}$

$((X = 2)$ and $(Y = 1)) = \emptyset$

If the coins are fair, the joint probability distribution of X and Y is as summarized in the following table.

		VALUES OF X	
VALUE OF Y	0	1	2
0	$\frac{1}{4}$	0	$\frac{1}{4}$
1	0	$\frac{1}{2}$	0

For any pair of discrete random variables, the events $((X = x)$ and $(Y = y))$ will form a partition, as x and y range over all possible values. It follows that

$$\sum_{\text{all } x,\, y} P(X = x \text{ and } Y = y) = 1$$

This property is seen to hold for the above table.

marginal distributions

An application of the law of total probability (p. 131) is the following. The individual probability distributions of X and Y are called **marginal distributions**. The reason for this terminology is that marginal probabilities can be obtained by summing the probabilities in the joint probability table along rows or columns into the margins of the table:

$$P(X = x) = \sum_{\text{all } y} P(X = x \text{ and } Y = y)$$

and

$$P(Y = y) = \sum_{\text{all } x} P(X = x \text{ and } Y = y)$$

That is, the marginal probability distribution of one variable is found by summing the joint probabilities over all values of the other variable. The probability table for the coin-toss game, augmented by the marginal probabilities, is as follows.

		VALUES OF X		
VALUE OF Y	0	1	2	Y MARGINAL
0	$\frac{1}{4}$	0	$\frac{1}{4}$	$\frac{1}{2}$
1	0	$\frac{1}{2}$	0	$\frac{1}{2}$
X Marginal	$\frac{1}{4}$	$\frac{1}{2}$	$\frac{1}{4}$	1

Conditional Probability Distributions

conditional probability distribution

If X and Y are random variables, then the **conditional probability distribution** of Y given $X = x$ is as follows. For each value y of Y,

$$P(Y = y | X = x) = \frac{P(X = x \text{ and } Y = y)}{P(X = x)}$$

Note that because summing $P(X = x \text{ and } Y = y)$ over all values of Y yields $P(X = x)$, we have

$$\sum_{\text{all } y} P(Y = y | X = x) = 1$$

In fact, conditional distributions have all the properties of probability distributions discussed in the last section.

EXAMPLE 5.11

The conditional distribution of Y given $X = 0$ for the table of joint and marginal distributions given above is computed as follows.

$$P(Y = 0 | X = 0) = \frac{P(X = 0 \text{ and } Y = 0)}{P(X = 0)}$$

$$= \frac{\frac{1}{4}}{\frac{1}{4}} = 1$$

$$P(Y = 1 | X = 0) = \frac{P(X = 0 \text{ and } Y = 1)}{P(X = 0)}$$

$$= \frac{0}{\frac{1}{4}} = 0$$

In tabular form, this distribution is

VALUE OF Y	CONDITIONAL PROBABILITY OF Y GIVEN $X = 0$
0	1
1	0

This and the conditional distributions of Y given $X = 1$ and $X = 2$ are given in the following table.

Conditional Distributions of Y given X and the Marginal Distribution of Y

VALUE OF Y	VALUES OF X 0	1	2	Y MARGINAL
0	1	0	1	$\frac{1}{2}$
1	0	1	0	$\frac{1}{2}$
Sum of Probs.	1	1	1	1

Thus, if it is known that the number of *heads* is 0 ($X = 0$), for example, then with probability 1 the faces are the same ($Y = 0$). The interpretations of the other two distributions are similar.

The conditional distributions of X given Y are somewhat more interesting. For example,

$$P(X = 0 | Y = 0) = \frac{P(X = 0 \text{ and } Y = 0)}{P(Y = 0)}$$

$$= \frac{\frac{1}{4}}{\frac{1}{2}}$$

$$= \frac{1}{2}$$

This and the remaining probabilities are given in the following table.

Conditional Distributions of X given Y and the Marginal Distribution of X

		VALUES OF X		
VALUE OF Y	0	1	2	SUM OF PROBS.
0	$\frac{1}{2}$	0	$\frac{1}{2}$	1
1	0	1	0	1
X Marginal	$\frac{1}{4}$	$\frac{1}{2}$	$\frac{1}{4}$	1

Thus, given the information that the faces are the same ($Y = 0$), the conditional probability of the number of *heads* is evenly split between 0 and 2. If $Y = 1$, on the other hand, $X = 1$ with probability 1.

multiplication rule   The multiplication rule for probabilities provides a **multiplication rule** that relates joint, marginal, and conditional probabilities:

$$P(X = x \text{ and } Y = y) = P(Y = y | X = x)P(X = x)$$

A similar equation holds for the conditional distribution of X given Y. These equations show that the association between X and Y described by the joint distribution is also described by the conditional distributions (each supported by a marginal distribution). As was seen in chapter 4, the association is easier to interpret when conditional distributions are used. For example, the association can be described by comparing the conditional distributions of Y given $X = x$ to the marginal distribution of Y for the various values of x. The differences in distributions represent the association.

EXERCISES

5.26 The sample space for the card game introduced in exercise 5.1 is

$$S = \{[KC, QC], [KC, KH], [KC, QH], [QC, KH], [QC, QH], [QH, KH]\}$$

Define random variables X and Y as follows.

X = the number of kings in the hand

Y = the number of red cards

(a) For each value x of X and y of Y, list the event $((X = x) \text{ and } (Y = y))$ in bracket ({}) notation.

(b)   Give the joint distribution of X and Y in table form.

(c)   Compute the marginal distributions of X and Y.

(d)   Find the conditional distributions of Y given X for each value of X. What is the conditional probability of no red cards given that the hand contains one king?

(e)   Find the conditional distributions of X given Y for each value of Y. What is the probability of exactly one king given that the hand contains no red cards?

Independent Random Variables

Let's take somewhat different variables for the random experiment consisting of the two tosses of a fair coin:

X = number of *heads* on the first toss

Y = number of *heads* on the second toss

The joint and marginal probability distribution table for these variables is as follows.

	VALUES OF X		
VALUE OF Y	0	1	Y MARGINAL
0	$\frac{1}{4}$	$\frac{1}{4}$	$\frac{1}{2}$
1	$\frac{1}{4}$	$\frac{1}{4}$	$\frac{1}{2}$
X Marginal	$\frac{1}{2}$	$\frac{1}{2}$	1

The conditional distributions of Y given X and the Y marginal are given in the following table.

	VALUES OF X		
VALUE OF Y	0	1	Y MARGINAL
0	$\frac{1}{2}$	$\frac{1}{2}$	$\frac{1}{2}$
1	$\frac{1}{2}$	$\frac{1}{2}$	$\frac{1}{2}$
Sum of Probs.	1	1	1

Note that the conditional distributions are the same and are equal to the marginal distribution of Y. No information about the distribution of Y is conveyed by X, in the sense that the distribution of Y to use if the value of X were given (the appropriate conditional distribution) is the same as the one we would use with no knowledge of X (the marginal distribution). This idea was discussed in the context of population frequency distributions in chapter 4. When the conditional frequency distributions of one variable were the same for all values of the other, we called the variables independent. The extension of this idea plays an extremely important role in probability: *Random variables for which the conditional probability distributions of one variable are the same for all values of the other are said to be (statistically)* **independent**.

independent

product rule for
independent random
variables

A property of independent variables that follows from the multiplication rule relating joint, marginal, and conditional distributions is

$$P(X = x \text{ and } Y = y) = P(X = x)P(Y = y) \qquad \text{for all } x \text{ and } y$$

This important relationship is called the **product rule for independent random variables**. Verify this relationship in the above table of joint and marginal distributions. It follows that if X and Y are independent, their joint distribution can be obtained from the marginal distributions alone.

It is also true that if the product rule holds for random variables X and Y, these variables are necessarily independent. It follows that the independence of random variables can be defined by the product rule.

Moreover, the definition of independence can be extended in this way to any number of random variables. Thus, random variables X, Y, Z, . . . are independent if

$$P(X = x \text{ and } Y = y \text{ and } Z = z \text{ and } . . .)$$
$$= P(X = x)P(Y = y)P(Z = z) . . .$$

for all values x, y, z, . . . of the respective variables. The product rule provides the most convenient starting point for performing probability calculations on independent variables, as will be seen shortly.

The Model-Building Application of Independence

The probability models most used in statistics are based on independent random variables. The accuracy of these models depends on being able to justify the assumption of independence. This justification is as follows.

The values of independent random variables are viewed as arising from *independent random experiments*. Random experiments are independent if the probabilities of the outcomes for any of them are not influenced by the outcomes of the others. Thus, in a pair of independent experiments, if one produces a value of a random variable X and the other a value of Y, then the (conditional) probability distribution of Y (given X) does not depend on the value observed for X. That is, independent random experiments produce independent random variables.

An example already considered is that of two tosses of a fair coin. The first toss leads to a number of *heads* X and the second to a number of *heads* Y. We verified above that the product rule held for the joint distribution of X and Y. That is, the equally likely model developed for this experiment led to the independence of X and Y.

However, intuitively, the independence of these variables should not depend on the fact that the coin was fair. Two tosses of a coin *should* be independent experiments, whatever the nature of the coin. This observation can be used as a starting point for the formulation of a model for the coin-toss experiment: We begin by *assuming* the independence of X and Y. Further calculations would then be based on this assumption.

This is how independence is used to construct probability models. Many random experiments in statistics consist of subexperiments or *trials* that are reasonably viewed as being independent. Repeated coin tosses are one example. Repeated sampling from a population is another, provided the sampling is done *with replacement*. Die tosses, including the tosses of a ten-sided Monte Carlo die, are independent trials.

It follows that an accurate probability model for a random experiment consisting of independent trials can be built up by using the product rule for independent random variables. An example is given in the next subsection.

The Binomial Distribution

Many experiments in probability consist of independent repetitions, or trials, of a two-outcome subexperiment. Repeated coin tosses are an example; the subexperiment has the two outcomes *heads* and *tails*. If the outcomes are coded 0 and 1, then the ith trial of the subexperiment can be viewed as resulting in the value of a Bernoulli random variable X_i. These variables will be independent because the trials are independent. Moreover, because the trials are repetitions of the same subexperiment, the variables will have the same distribution. That is,

$$P(X_i = 0) = 1 - p \quad \text{and} \quad P(X_i = 1) = p$$

where p is the same for all trials.

binomial distribution

Now, suppose we were interested in the total number of 1's to be seen in n repetitions of the experiment. Since the 1's occur randomly, this number will also be random and can be viewed as the value of a random variable Y. The probability distribution of Y is called the **binomial distribution**. This distribution has many applications in both probability and statistics. The parameters of the binomial distribution are the number of trials (or sample size) n and the probability p. As an illustration of an important use of independence, we will derive the binomial distribution for $n = 2$ trials.

The pair of numbers (x_1, x_2) will be used as shorthand notation for the event $((X_1 = x_1)$ and $(X_2 = x_2))$. A sample space that represents all possible outcomes of two trials of the above experiment is then

$$S = \{(0, 0), (0, 1), (1, 0), (1, 1)\}$$

Because the random variables are independent, the probabilities of the outcomes in this sample space can be computed from the equation

$$P(\{(x_1, x_2)\}) = P(X_1 = x_1 \text{ and } X_2 = x_2) = P(X_1 = x_1)P(X_2 = x_2)$$

for the possible values, 0 and 1, of x_1 and x_2. Using the distribution of X_i given above, we find

$$P(\{(0, 0)\}) = (1 - p) \cdot (1 - p) = (1 - p)^2$$
$$P(\{(0, 1)\}) = (1 - p) \cdot p = p(1 - p)$$
$$P(\{(1, 0)\}) = p \cdot (1 - p) = p(1 - p)$$
$$P(\{(1, 1)\}) = p \cdot p = p^2$$

Now, to calculate the probability distribution of Y, we must find the probability $P(Y = y)$ for each possible value y. But the values of y correspond to the following events in this sample space:

$$(Y = 0) = \{(0, 0)\}$$
$$(Y = 1) = \{(0, 1), (1, 0)\}$$
$$(Y = 2) = \{(1, 1)\}$$

Consequently, the distribution of Y is simply the probabilities of these events, which are the sums of the corresponding outcome probabilities. For example, $P(Y = 1) = P(\{(0, 1)\}) + P(\{(1, 0)\}) = (1 - p)p + p(1 - p) = 2p(1 - p)$.

Note that the factor 2 is the number of pairs, (x_1, x_2), with one 1 and one 0. Since each pair has the same probability $p(1 - p)$, the probability of the event $(Y = 1)$ is equal to $p(1 - p)$ multiplied by this factor.

The complete binomial distribution for $n = 2$ is given in the following table.

VALUE OF Y	BINOMIAL PROBABILITIES
0	$(1 - p)^2$
1	$2p(1 - p)$
2	p^2

The generalization of this expression to n trials is

$$P(Y = k) = \binom{n}{k} p^k (1 - p)^{n-k} \qquad k = 0, 1, \ldots, n$$

This is the binomial distribution for sample size n and parameter p. The number denoted by $\binom{n}{k}$, called a binomial coefficient, simply counts the number of distinct sequences of k 0's and $n - k$ 1's. In terms of factorials $(r! = r(r - 1) \ldots 2 \cdot 1)$, it can be evaluated by the expression

$$\binom{n}{k} = \frac{n!}{k!(n - k)!}$$

Thus, for example, in the above calculation we found the number of sequences (pairs) consisting of 1 zero and 1 one to be $\binom{2}{1} = 2!/1!1! = 2$.

It is possible to calculate the mean and variance of the binomial variable Y from its distribution. However, we will instead take advantage of a simple relationship that exists between Y and the individual Bernoulli variables of the n trials. The relationship is

$$Y = X_1 + X_2 + \cdots + X_n$$

This follows from the fact that the total number of 1's is the sum of the 1's for each trial. The expression $Y = X_1 + X_2 + \cdots + X_n$ is a linear transformation of independent random variables; as we will see shortly, special methods exist for calculating the mean and variance for such transformations.

EXERCISES

5.27   Use the binomial distribution with $n = 2$ and $p = \frac{1}{2}$ to calculate the probability of one *heads* in two tosses of a fair coin.

5.28   Two fair (six-sided) dice are tossed.

(a)   Use the binomial distribution to calculate the probability of "snake-eyes" (one dot twice). (Hint: Take $X_i = 1$ if one dot is seen, $X_i = 0$ otherwise.)

(b)   Calculate the probability that at least one of the dice shows one dot.

5.29   A random sample of size two is drawn, *with replacement*, from the population of display 5.8, page 136. Use the binomial distribution to determine the probability of

(a)   selecting one woman and one man

(b)   selecting no women

5.30   Let Y be the number of 1's in three independent Bernoulli trials, each of which has probability p of showing a 1. Derive the binomial distribution in this case and verify that it is of the form given above. (Note: the sample space will now be made up of triples (x_1, x_2, x_3) of 0's and 1's. First calculate the outcome probabilities using the independence of the three Bernoulli trials, then the probabilities of the events $(Y = y)$ for $y = 0, 1, 2,$ and 3.)

5.31   Use the binomial distribution to calculate the probability of obtaining two *heads* in three tosses of a fair coin.

5.32   Use the binomial distribution to calculate the probability of seeing three consecutive 0's in a table of random numbers such as that of display 5.9 (p. 137).

Expectation for Functions of More Than One Variable

The expectation of random variables of the form $W = f(X, Y)$ can be computed by (a) computing the probability distribution of W from the joint distribution of X and Y, then finding $E(W)$ as before, or (b) computing the expectation directly from the joint distribution of X and Y:

$$E(f(X, Y)) = \sum_x \sum_y f(x, y)P(X = x \text{ and } Y = y)$$

Because finding the distribution of a transformation of variables is often quite difficult, scheme (b) is usually much easier to carry out. An example of such a calculation, which also demonstrates the favored position of linear transformations, is as follows.

EXAMPLE 5.12

The joint and marginal distribution table for the random variables $X =$ number of *heads* and $Y =$ indicator of different faces ($Y = 1$ if the faces are different, $Y = 0$ if not) for two tosses of a fair coin is

VALUE OF Y	VALUES OF X 0	1	2	Y MARGINAL
0	$\frac{1}{4}$	0	$\frac{1}{4}$	$\frac{1}{2}$
1	0	$\frac{1}{2}$	0	$\frac{1}{2}$
X Marginal	$\frac{1}{4}$	$\frac{1}{2}$	$\frac{1}{4}$	1

We will find the expectation of the random variable $W = X + Y$. This is a linear transformation and, although X and Y are dependent variables, it will be seen that the addition law, $E(X + Y) = E(X) + E(Y)$, holds. The computation can be easily worked into a demonstration of this law, which will be extended and applied in the next subsection.

To find the expectation, for every possible pair of values of X and Y the value of $W = X + Y$ is multiplied by the joint probability of the pair. The resulting quantities are summed over all pairs. The following table contains this computation.

x	y	(1) $x + y$	(2) $P(X = x$ AND $Y = y)$	PRODUCT OF (1) AND (2)
0	0	0	$\frac{1}{4}$	0
0	1	1	0	0
1	0	1	0	0
1	1	2	$\frac{1}{2}$	1
2	0	2	$\frac{1}{4}$	$\frac{1}{2}$
2	1	3	0	0
				$E(W) = 1\frac{1}{2}$

We will arrange the computation a little differently. Instead of forming the sum $X + Y$ before taking the product, the values of X and Y will be multiplied by the joint probability first, and then the column sums will be formed:

x	y	(3) $xP(X = x$ AND $Y = y)$	(4) $yP(X = x$ AND $Y = y)$
0	0	$0 \cdot \frac{1}{4}$	$0 \cdot \frac{1}{4}$
0	1	$0 \cdot 0$	$1 \cdot 0$
1	0	$1 \cdot 0$	$0 \cdot 0$
1	1	$1 \cdot \frac{1}{2}$	$1 \cdot \frac{1}{2}$
2	0	$2 \cdot \frac{1}{4}$	$0 \cdot \frac{1}{4}$
2	1	$2 \cdot 0$	$1 \cdot 0$

Now, $E(W) =$ Sum of column (3) + Sum of column (4). Combine terms with the same values of x in the sum of column (3) as follows:

$$\text{Sum of (3)} = (0 \cdot \tfrac{1}{4}) + (0 \cdot 0) + (1 \cdot 0) + (1 \cdot \tfrac{1}{2}) + (2 \cdot \tfrac{1}{4}) + (2 \cdot 0)$$
$$= 0 \cdot (\tfrac{1}{4} + 0) + 1 \cdot (0 + \tfrac{1}{2}) + 2 \cdot (\tfrac{1}{4} + 0)$$

Verify from the above table of joint and marginal probabilities that the sums in parentheses are the marginal probabilities of X. Consequently, this sum is simply $E(X) = 1$. Similarly, the sum of column (4) is $E(Y) = \frac{1}{2}$, and it follows that $E(X) + E(Y) = 1\frac{1}{2}$. Thus we have shown by a method that can be applied in any example that $E(X + Y) = E(X) + E(Y)$.

The Mean and Variance for Linear Transformations of Independent Random Variables

A linear transformation of random variables X and Y is an expression of the form

$$W = aX + bY + c$$

where a, b, and c are constants. We will need extensions of the laws of expectation and variance, given earlier, to cover these transformations. For most applications, it will suffice to give the laws for independent variables X and Y. The effect of dependence on the law of variance will be discussed briefly in a later subsection.

law of expectation   The extension of the **law of expectation** is as follows.

$$E(aX + bY + c) = aE(X) + bE(Y) + c$$

As in the example of the last subsection, this property does not require that the random variables be independent; the law of expectation is valid whether X and Y are independent or dependent.

The **law of variance** for two or more random variables does depend on the nature of the relationship between X and Y, however. When the variables X and Y are *independent*, it has the following form:

$$\text{var}(aX + bY + c) = a^2\text{var}(X) + b^2\text{var}(Y)$$

The following *addition laws for expectation and variance of independent variables* are obtained by taking $a = b = 1$ and $c = 0$.

$$E(X + Y) = E(X) + E(Y)$$
$$\text{var}(X + Y) = \text{var}(X) + \text{var}(Y)$$

These expressions extend to sums of any number of variables: For independent variables $X_1, X_2, \ldots$

$$E(X_1 + X_2 + \cdots) = E(X_1) + E(X_2) + \cdots$$

and

$$\text{var}(X_1 + X_2 + \cdots) = \text{var}(X_1) + \text{var}(X_2) + \cdots$$

Somewhat unexpected things can happen for other linear combinations, however. For example, although

$$E(X - Y) = E(X) - E(Y)$$

the law of variance yields

$$\text{var}(X - Y) = \text{var}(X) + \text{var}(Y)$$

That is, the mean of the difference is the difference of the means, but the variance of the difference is the *sum* of the variances. The reason is that $X - Y = aX + bY + c$ with $a = 1$, $b = -1$, and $c = 0$. Thus,

$$\text{var}(X - Y) = 1^2 \text{var}(X) + (-1)^2 \text{var}(Y)$$
$$= \text{var}(X) + \text{var}(Y)$$

EXAMPLE 5.13

The mean and variance of a binomial variable Y are easily calculated from the addition laws. Recall that $Y = X_1 + X_2 + \cdots + X_n$ for a binomial variable with sample size n, where the X_i's are independent Bernoulli variables with parameter p. Earlier we calculated $E(X_i) = p$ and $\text{var}(X_i) = p(1 - p)$. It follows that

$$E(Y) = E(X_1) + E(X_2) + \cdots + E(X_n) = np$$

and

$$\text{var}(Y) = \text{var}(X_1) + \text{var}(X_2) + \cdots + \text{var}(X_n) = np(1 - p)$$

Other important applications of the addition laws are given in the next subsection.

The Mean and Variance of the Sample Proportion and Sample Mean

In statistical inference, we will not be interested in the binomial variable, Y, itself, but rather in this variable divided by sample size:

$$\hat{p} = \frac{Y}{n}$$

This ratio, called the *sample proportion*, is the proportion of 1's in the sample. It is an estimator (chapter 2, p. 35) of the population proportion, p. To apply this estimator in statistical inference, we will need its mean and variance. Note that $\hat{p}$ is a simple linear transformation of Y, bY, with $b = 1/n$. Consequently, we can use the mean and variance of Y found in example 5.13 to calculate these quantities:

$$E(\hat{p}) = \left(\frac{1}{n}\right) E(Y) = \left(\frac{1}{n}\right) np = p$$

and

$$\text{var}(\hat{p}) = \left(\frac{1}{n}\right)^2 \text{var}(Y) = \left(\frac{1}{n}\right)^2 np(1 - p)$$
$$= \frac{p(1 - p)}{n}$$

The standard deviation of $\hat{p}$ is then

$$\sigma_{\hat{p}} = \sqrt{\frac{p(1 - p)}{n}}$$

Another important statistical estimator is the sample mean $\overline{X}$ introduced in chapter 2. The sample mean is an estimator of the population mean of a measurement variable X. In the notation of chapter 2, if random variables $X_1, X_2, \ldots, X_n$ represent the (potential) values of the variable X for a random sample of size n, then the sample mean is

$$\overline{X} = \frac{1}{n}(X_1 + X_2 + \cdots + X_n)$$

For reasons to be discussed in chapter 7, it will be assumed that the X_i's are independent and have the same mean μ and variance σ^2.

By the addition laws for independent random variables, it follows that

$$E(\overline{X}) = \left(\frac{1}{n}\right) E(X_1 + X_2 + \cdots + X_n)$$

$$= \left(\frac{1}{n}\right) \cdot n\mu$$

$$= \mu$$

and

$$\text{var}(\overline{X}) = \left(\frac{1}{n}\right)^2 \text{var}(X_1 + X_2 + \cdots + X_n)$$

$$= \left(\frac{1}{n}\right)^2 \cdot n\sigma^2$$

$$= \frac{\sigma^2}{n}$$

The standard deviation of $\overline{X}$ is then

$$\sigma_{\overline{X}} = \frac{\sigma}{\sqrt{n}}$$

The Law of Variance When Variables Are Not Independent

If X and Y are dependent random variables, then the law of variance, $\text{var}(aX + bY + c) = a^2\text{var}(X) + b^2\text{var}(Y)$, has the added term, $2ab\text{cov}(X, Y)$, where $\text{cov}(X, Y)$ is the *covariance* of the variables. The covariance is a measure of the linear association between X and Y; that is, it is a measure of the tendency of the paired X, Y values to lie near a straight line (see chapter 14). An equivalent measure is the *correlation coefficient* $\rho = \text{cov}(X, Y)/\sigma_X\sigma_Y$, where the quantities in the denominator are the standard deviations of X and Y.

The correlation coefficient can have any value in the range from -1 to 1. Negative values of correlation imply a negative linear association between the variables. In this case the values of Y tend to decrease with increasing values of X, on the average. Positive correlation implies that the values of Y tend to increase, on the average, with increasing X. Curiously, dependent variables can be uncorrelated ($\rho = \text{cov}(X, Y) = 0$), as will be demonstrated in an exercise at the end of the section.

In chapter 10 we will be interested in the influence of the correlation of X and Y on $\text{var}(X - Y)$. Suppose X and Y have the same variance σ^2. If they were independent variables, we would have $\text{var}(X - Y) = \text{var}(X) + \text{var}(Y) = 2\sigma^2$, by the law of variance. However, if the variables are correlated, the term $-2\text{cov}(X, Y) = -2\rho\sigma^2$ would be added to this expression, yielding $\text{var}(X - Y) = 2(1 - \rho)\sigma^2$. Because the correlation ranges from -1 to 1, the actual value of $\text{var}(X - Y)$ could be as small as 0 or as large as $4\sigma^2$. Note that the smaller values of $\text{var}(X - Y)$ occur for positively correlated variables. In fact, if $\rho > \frac{1}{2}$ the variance of $X - Y$ will be smaller than the variance of either variable individually. The effect of correlation on the variance of $X - Y$ will have implications for the paired differences method, to be studied in chapter 10.

EXERCISES

5.33   Refer to exercise 5.26 (p. 166), in which the joint probability distribution of the random variables X = the number of kings in the hand and Y = the number of red cards was found for a simple card game.

(a)   Calculate the expectation of the random variable

$$W = X^2Y$$

(b)   Calculate the expectation of $Z = 2X + 3Y + 5$.

5.34   A random sample of size 5 is to be taken *with replacement* from the population of display 5.8. Let X be the height variable for the population. Random variables X_1 and X_2 will be the heights of the two individuals obtained in the sample.

(a)   Use the fact that the mean and variance of X under random sampling is the same as the population mean and variance of the height frequency distribution to calculate μ and σ^2 for this variable.

(b)   Find the mean, variance, and standard deviation of the sample mean $\overline{X} = (X_1 + X_2)/2$.

5.35   A random sample of size 2 is to be taken *with replacement* from the population of display 5.8. We are interested in estimating the proportion of individuals in the population who live at least ten miles from campus. Let p be this proportion.

(a)   Compute the population proportion p.

(b)   Find the mean and standard deviation of the sample proportion $\hat{p}$.

5.36   A fair coin is tossed twice. You win \$2 if the first toss comes up *heads* and \$3 if the second toss comes up *heads*. Let X and Y be the numbers of *heads* on the first and second tosses, respectively.

(a)   Write your winnings W in terms of X and Y.

(b)   Find the expectation of W.

(c)   Find the variance of W.

5.37   In a toss of two fair (six-sided) dice, you win $2 for every dot showing on the first die but lose $1 for every dot on the second. If W is your winnings for the two dice, find $E(W)$ and $\text{var}(W)$.

5.38   A computer contains three identical memory boards that have two (mutually exclusive) modes of failure. In failure mode 1 a board can be repaired, at a cost of $200. In failure mode 2 a board must be replaced, at a cost of $800. A failure of mode 1 occurs during a given time period with probability $\frac{1}{10}$, while a failure of mode 2 occurs with probability $\frac{1}{50}$. Assume that failures are independent and that a given board can have no more than one failure of either type during the time period. Let X denote the number of boards failing in mode 1 and Y the number failing in mode 2.

   (a)   The total number of boards failing during the given time period is $W = X + Y$. Find the expectation of W.

   (b)   Write the total cost of repairing or replacing memory boards for the given time period as a linear function of X and Y.

   (c)   Using the result of part (b), find the expected cost of repairing or replacing memory boards for the given time period.

5.39   A random sample of size n_X is taken from a population and the sample mean $\overline{X}$ of a variable X is found. A random sample of size n_Y is taken, independently, from a second population, and the sample mean $\overline{Y}$ of a variable Y is computed. The independence of the samples implies the independence of the random variables $\overline{X}$ and $\overline{Y}$. If the population mean and standard deviation of X are μ_X and σ_X, and the corresponding quantities for Y are μ_Y and σ_Y, then

   (a)   compute $E(\overline{X} - \overline{Y})$

   (b)   compute $\text{var}(\overline{X} - \overline{Y})$

5.40†   Random variables X and Y have the following joint distribution.

		X	
Y	−1	0	1
−1	$\frac{1}{8}$	0	$\frac{1}{8}$
0	0	$\frac{1}{2}$	0
1	$\frac{1}{8}$	0	$\frac{1}{8}$

   (a)   Find the marginal distributions of X and Y.

   (b)   Show that X and Y are dependent variables.

   (c)   Find $E(X)$ and $E(Y)$.

   (d)   Show that $E(W) = 0$, where $W = (X - E(X))(Y - E(Y))$. This expectation is the covariance $\text{cov}(X, Y)$, discussed in the last subsection. It follows that dependent variables can be uncorrelated. That is, variables can be associated without being linearly associated.

Chapter 5 Quiz

1.   What is the relative frequency concept of probability?

2.   How is a Monte Carlo simulation carried out on a computer?

3. What are the three components of a probability model?

4. What are the three basic operations on events? How are they depicted with Venn diagrams?

5. What is the bracket notation for events? Give an example.

6. When are two events mutually exclusive? When is a collection of more than two events mutually exclusive?

7. What are the Kolmogorov axioms for probability?

8. What is the law of complementation? Give an example of its use.

9. What is the general addition law of probability for two events? What form does this law take for mutually exclusive events? Give examples.

10. What is Boole's inequality? Why is it useful?

11. What is the law of total probability? Give an example of its use.

12. What is a finite probability model? How are the probabilities of events calculated for such a model?

13. What is an equally likely model? Give an example.

14. Why is it possible to calculate the probabilities for an equally likely model by counting outcomes in events?

15. What is the lottery model for simple random sampling from a population?

16. What are the die-toss games that model the generation of random numbers? How are these games used in Monte Carlo simulations?

17. What is the intuitive idea behind the conditional probability of one event given another?

18. What is the expression that defines the conditional probability of B given A?

19. What is the multiplication rule? How is it used? Give an example.

20. When are events independent?

21. What is the product rule for independent events?

22. If A and B are independent events, what can be said about the events (not A) and B? About (not A) and (not B)?

Answer the following two questions only if the starred sections (p. 147–151) were covered.

23. What is the chain rule for conditional probabilities? Give an example of its use.

24. What is Bayes' rule? Give an example of its use.

25. What is the intuitive idea of a random variable?

26. What is the purpose of the function definition of a random variable?

27. What is the probability distribution of a (discrete) random variable? Give an example.

28. What is a Bernoulli random variable? What is the Bernoulli probability distribution?

29. What is the intuitive idea of expectation?

30. What is the definition for the expectation of a discrete random variable? Give an example.

31. What is a transformation of a random variable? How can the expectation of a transformed random variable be calculated? Give an example of such a calculation?

32. What is a linear transformation? What are the laws of expectation? Use these laws in an example.

33. What is the definition of the variance of a random variable? What is the standard deviation?

34. What is the computing formula for variance? Use this formula in an example.

35. What are the rules of variance for transformations of a random variable? Use these rules in an example.

36. What is the Z-score transform? What are its expectation and variance?

37. What is the relationship between the population frequency distribution of a variable and the probability distribution of that variable under (lottery) random sampling? What does this imply about the relationship between the population mean and variance of the variable and its expectation and variance calculated from the random sampling probability model?

38. Why is the relationship requested in question 37 an important one from the viewpoint of statistical inference?

The following questions are for section 5.6.

39. What is joint probability distribution of two discrete random variables X and Y? Give an example.

40. What are the marginal distributions of X and Y? Give examples.

41. What is the conditional distribution of Y given $X = x$? Give an example.

42. When are discrete random variables independent? Give an example.

43. What is the product rule for independent random variables?

44. How is the independence of random variables used to build probability models for random experiments made up of independent trials? Give an example.

45. How is the expectation of a random variable $W = f(X, Y)$ calculated from the joint distribution of X and Y? Give an example.

46. What is a linear transformation of random variables X and Y? What are the laws of expectation for such transformations? What are the extensions of these laws to sums of more than two random variables?

47. What role does the concept of independent variables play in the laws of expectation?

48. What are the laws of variance for linear transformations of independent random variables X and Y? What are the extensions to sums of more than two variables?

49. What are the expressions for the mean and standard deviation of the sample proportion? How are these expressions derived from the laws of expectation and variance?

50. What are the expressions for the mean and standard deviation of the sample mean? How are these expressions derived from the laws of expectation and variance?

◢

SUPPLEMENTARY EXERCISES

5.41   Confirm by drawing Venn diagrams the following relationships between the various operations on events (called the DeMorgan laws)

(a)   (not $(A$ and $B)$) = ((not A) or (not B))

(b)   (not $(A$ or $B)$) = ((not A) and (not B))

5.42   Let $P(A) = .2$ and $P(B) = .1$.

(a)   How large can $P(A$ or $B)$ be?

(b)   How small can $P((\text{not } A)$ and $(\text{not } B))$ be?

5.43   The probability that a given individual will catch a cold during the winter is .30, and during the spring and fall the probability is .10 for each season. The probability of a cold in the summer is .05. If the probability that a person has more than one cold during any season is 0, what is the expected number of colds this individual will have during a given year?

5.44   A bag contains three red and four white marbles. Two marbles are selected at random from the bag.

(a)   Construct an equally likely sample space for this experiment. (Hint: Make the marbles distinguishable by writing the numbers 1, 2, 3 on the red marbles and 4, 5, 6, 7 on the white ones.)

(b)   What is the probability that both marbles drawn are red?

(c)   Find the probability of getting a white and a red marble. (Hint: "One white and one red" = "white on first draw and red on second" or "red on first and white on second.")

5.45   Three skiers, wearing numbers 1, 2, and 3 on their backs, are started on a cross-country race in random order.

(a)   Write down an equally likely space for this random experiment. (Hint: Use ordered triples, with the positions in the triple corresponding to the starting positions first, second, and third, and fill the positions with the skiers' numbers 1, 2, and 3.)

(b)   What probability model is suggested by the statement "started in random order"? Use this model to calculate the probability that skier 1 is started first, skier 2 second, and skier 3 third.

(c) Calculate the probability that skier 1 is started first.

(d) Calculate the probability that the starting position of exactly one skier coincides with the number on his back.

(e) Calculate the probability that the starting positions of one or more skiers coincide with the numbers on their backs.

5.46 Four identical components are connected "in parallel" in an electronic device. "In parallel" means that the device will operate if at least one of the components works. If the components fail independently with the same failure probability .1, what is the probability that the device will operate?

5.47 A code consisting of the numbers 1, 2, 3, and 4 is sent in random order, where all orderings are equally likely.

(a) What is the probability that the first digit sent is 1?

(b) What is the probability that all digits will be sent in their proper order (i.e., 1 sent first, 2 sent second, etc.)?

(c) What is the probability that no digits will be sent in their proper order?

(d) What is the probability that exactly three digits will be sent in proper order?

5.48 Use the multiplication rule (p. 145), extended to three events, to calculate the probability that in a random sample of size 3 from display 5.8 (p. 136), without replacement,

(a) all three are women

(b) two are men and one is a woman

5.49 Let $P(A) = .5$, $P(B) = .6$, and $P(A \text{ and } (\text{not } B)) = .2$.

(a) Are A and B independent events?

(b) Find $P(A \text{ or } B)$.

5.50 The space shuttle has two on-board computers to control landing, each with probability .01 of failing during a mission. If failures are independent, what is the probability that a computer will be available to control the landing on a given mission?

5.51 The variable Y is to be used as an indicator of the event "student lives at least ten miles from campus" in a random sample of size 1 from display 5.8 (p. 136). Find the probability distribution of this variable.

5.52 A random variable X has the following probability distribution.

VALUE OF X	PROBABILITY
1	$\frac{3}{10}$
2	$\frac{5}{10}$
3	$\frac{2}{10}$

(a) Calculate $E(X)$.

(b) Use the rules of expectation for linear transformations to calculate $E(-2X + 2)$.

(c) Calculate the expectation of $Z = 3X^2 + 4X + 8$.

5.53   A random sample of size 1 is to be taken from the population of display 5.8 (p. 136). Let X represent the distance of the selected individuals' residence from campus.

(a)   Find the expectation of X.

(b)   Find the variance and standard deviation of X.

(c)   What is the relationship between the mean and standard deviation of the random variable found in parts (a) and (b) and the mean and standard deviation of the population frequency distribution of the Distance from campus variable?

(d)   Convert the Distance from campus variable from miles to kilometers and give the mean and standard deviation in the new units. (Note: 1 mile = 1.609 kilometers. Make the computation without actually converting the original distances.)

5.54   A tie maker sells three kinds of ties, A, B, and C, for $5.00, $6.00, and $10.00, respectively. Customers have been observed to select tie A with probability .40, to select B with probability .60, and to select C with probability .50. Assume that they make selections among kinds of ties independently and buy no more than one tie of each kind. (They may buy more than one kind of tie, or no ties at all, however.) It costs the tie maker $1.00 to make each tie, regardless of type.

(a)   What is the expected net profit per customer to the tie maker?

(b)   What is the variance of the net profit per customer?

(Hint: Write the net profit per customer as a linear function of X = number of ties of type A, Y = number of type B, Z = number of type C sold to a "randomly selected" customer.)

Probability Distributions for Measurement Variables— The Normal Distribution

Introduction

This chapter deals with probability models for random variables obtained by a measurement process. These models differ from the ones treated in the last chapter in that measurement variables are viewed as taking on, potentially, any value in an interval of a number line. Because of this property, these random variables are called **continuous**.

continuous

A curious mathematical fact is that it is not possible to assign positive probabilities to every point in an interval if they are to add up to 1. Consequently, probabilities must be defined differently for continuous variables than they were for the discrete variables treated in chapter 5. The mathematical model for a population frequency distribution, introduced in section 1.6, will be adopted for this purpose. Events will be formed from intervals on the number line, and probabilities will be the areas under a *probability curve* over these intervals.

The idea of every point on a number line being a possible value of a variable seems a bit absurd. The number of points on a number line is infinite, while real populations, no matter how large, are finite in size. Thus, only a finite number of values of the variable are possible in reality.

However, it is important to realize that we are dealing with a *mathematical model* of a probability distribution—not with an actual distribution. While we want this model to reflect the true probabilities accurately, we also want to be able to deal with it using the available mathematical tools.

An overriding consideration in the use of the area-under-a-curve model is that the available mathematical tools are very powerful indeed.

The catch is that they involve concepts from the calculus—and beyond. This will put the development of the theory beyond our reach. However, by taking a few things on faith, we will be able to deal with others using the general probability properties presented in the last chapter. This will provide us with quite an adequate basis for the discussion of statistical inference.

Because probabilities are areas under curves in the continuous model, we will be able to manipulate them easily by drawing pictures of the events and areas and by applying the general laws of probability given in chapter 5. Most of our effort in this chapter will be devoted to the *normal distribution*. This distribution will play a key role in the large sample introduction to statistical inference, to be given in the next three chapters. We will learn to use the tables at the end of the book to compute specific normal probabilities. Later applications of this famous distribution will be anticipated in examples and exercises in this chapter.

Before introducing the normal distribution, we will give a general introduction to continuous variables and their distributions.

◢ SECTION 6.2

Continuous Variables, Events, and Continuous Distributions

Let X denote a continuous variable. Events will be defined in terms of X, as in chapter 5. Thus, although the event $(X \le x)$ will represent the outcomes in some unmentioned sample space for which X takes values less than or equal to x, we now view it as representing the interval of numbers among the possible values of X to the left of x. See display 6.1.

Display 6.1

Interval of a number line representing the possible values of a continuous random variable X, showing the event (X ≤ x)

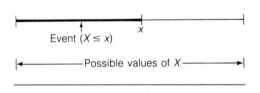

Other events will be built up from such intervals by the operations "and," "or," and "not."

probability curve

The probability distribution of X is specified by a curve, called a *probability density curve*, or more simply, a **probability curve**. Probabilities of events will be the areas under this curve over the events. Thus, the probability $P(X \le x)$ of the event of display 6.1 is depicted in display 6.2.

continuous distributions

Note that in order for probabilities to be nonnegative and add up to 1, the probability curve will never dip below the horizontal axis, and the total area under the curve will be 1. Probability distributions defined in this way are said to be **continuous distributions**.

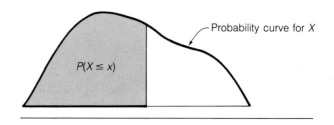

Display 6.2

Probability curve for the random variable X and the area representing $P(X \leq x)$

Probability curve for X

$P(X \leq x)$

Applying the Laws of Probability to Continuous Distributions

The laws of probability carry over completely to events and probabilities for continuous variables. For example, note that the event $(X > x)$ is the complement of $(X \leq x)$:

$$(X > x) = \text{not}(X \leq x)$$

Thus, by the law of complementation

$$P(X > x) = 1 - P(X \leq x)$$

This fact is also easily seen from display 6.2, since the total area under the curve is 1 and the areas above these two events clearly add up to this total.

Another useful fact is that we need not worry about whether the event whose probability we are after contains the boundaries of the intervals or not. For example, the events $(X \leq x)$ and $(X < x)$ have the same probability. To see this, note that $(X < x)$ and $(X = x)$ are mutually exclusive events, and

$$(X \leq x) = ((X < x) \text{ or } (X = x))$$

Thus, by the addition law of probability,

$$P(X \leq x) = P(X < x) + P(X = x)$$

But the area definition of probability requires that $P(X = x) = 0$ for all points x. It follows that $P(X \leq x) = P(X < x)$.

Another example of the use of the addition law allows us to calculate $P(a < X \leq b)$ for any numbers $a < b$ from the probabilities of the events $(X \leq a)$ and $(X \leq b)$. Write

$$(X \leq b) = ((X \leq a) \text{ or } (a < X \leq b))$$

The two events on the right are mutually exclusive, so by the addition law we have

$$P(X \leq b) = P(X \leq a) + P(a < X \leq b)$$

Thus, by subtraction,

$$P(a < X \leq b) = P(X \leq b) - P(X \leq a)$$

This fact can be seen using areas under a probability curve, as shown in display 6.3.

Display 6.3

Graphical demonstration of the equation
$P(a < X \leq b) =$
$P(X \leq b) - P(X \leq a)$

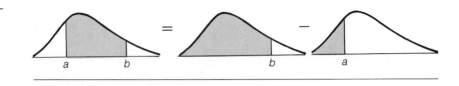

Thus, if tables giving probabilities for events of the form $(X \leq x)$ were available, the probabilities for events of a variety of other forms could be calculated by operations based on the laws of probability. Moreover, by drawing pictures of the events and their probabilities it is easy to see how to carry out the computations. This fact will be especially helpful in our dealings with the normal distribution, since most of the computational effort will involve manipulating probabilities into forms for which the normal tables can be used.

EXAMPLE 6.1

A random variable X is said to have the *uniform distribution on the interval from 0 to 1* if its possible values all lie between 0 and 1 and its probability curve is as shown in display 6.4.

Display 6.4

Probability curve for the uniform distribution on the interval from 0 to 1 and the area representing the probability that a variable X with this distribution is $\leq x$

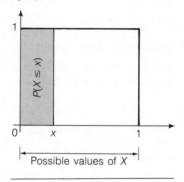

Because the area of a rectangle is base-length times height, probabilities for the uniform distribution are easy to calculate. For example, among the possible values of X, the event $(X \leq x)$ is the interval from 0 to x, which has length x. Since the height of the curve is 1,

$$P(X \leq x) = x \cdot 1 = x$$

Thus, we find that $P(X \leq .5) = .5$, for example.

The event $(a < X \leq b)$ is the interval from a to b, which has length $b - a$. If both a and b are in the interval from 0 to 1, it follows that the probability of this event is also $b - a$. This is a special case of the relationship derived just before this example.

Computers can be programmed to generate random numbers with (approximately) uniform distribution in the interval from 0 to 1. Such numbers are called *uniform random numbers*. What is the probability that a randomly generated number X will be between 0 and .3 or between .7 and .9?

By the addition law,

$$P((0 < X < .3) \text{ or } (.7 < X < .9)) = P(0 < X < .3) + P(.7 < X < .9)$$
$$= (.3 - 0) + (.9 - .7)$$
$$= .5$$

EXERCISE

6.1   If X has the uniform distribution on the interval from 0 to 1, identify on a picture (display 6.4) and calculate the probabilities of the following events.

(a) $(X \leq .7)$

(b) $(X > .6)$

(c) $((X < .3) \text{ or } (X > .7))$

(d) $((X < .3) \text{ and } (X > .7))$

(e) $(\text{not}(X > .4))$

(f) $(.4 \leq X \leq .8)$

(g) $(\text{not}(.4 \leq X \leq .8))$

The Mean, Median, Standard Deviation and iqr of a Continuous Probability Distribution

The interpretations of the common location and dispersion parameters for a continuous probability distribution are analogous to the interpretations of these parameters for population frequency distributions given in chapter 2. However, since we are now restricted to the continuous distribution model, it is no longer possible to *define* the mean μ and standard deviation σ without the calculus. This will not be a problem in what we intend to do, since an intuitive understanding of these parameters will suffice. It is enough to know that the mean and standard deviation of a continuous distribution behave like the mean and standard deviation of a frequency distribution, as described in chapter 2, or like the mean and standard deviation of a discrete probability distribution, as defined in the last chapter.

Recall that both the mean μ and median m are measures of where the "center" of the distribution is. In particular, for a symmetric distribution, both measures will be at the point of symmetry. Thus, for the uniform distribution, which is symmetric around .5, we have $\mu = m = .5$.

The interquartile range, iqr, is the difference between the third and first quartiles, as before. For the uniform distribution, $q_1 = .25$ and $q_3 = .75$, so iqr $= .5$.

The calculation of the standard deviation for the uniform distribution is strictly a calculus exercise and cannot be reasoned from a picture. It is $\sigma = 1/\sqrt{12} = 0.289$.

It will be seen that the mean and standard deviation are the key parameters of the normal distribution. We will use them in the process of calculating probabilities for this distribution.

SECTION 6.3

The Normal Distribution

The normal distribution is not a single probability distribution, but rather a family of distributions indexed by parameters μ and σ. This means that there is a different normal curve associated with each number μ and *positive* number σ. A picture of one such curve is given in display 6.5.

Display 6.5

Graph of normal probability curve with mean μ and standard deviation σ, where σ is shown as the distance from μ to the inflection point on the curve.

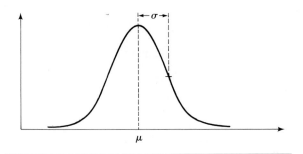

For obvious reasons, normal curves are sometimes said to be *bell-shaped*. The parameters μ and σ are the mean and standard deviation of the distribution, respectively. The manner in which μ controls the location of the distribution and σ its dispersion is demonstrated in display 6.6.

Display 6.6

Normal distribution curves. (a) With means $\mu_2 > \mu_1$ but same σ. (b) With standard deviations $\sigma_2 > \sigma_1$ but same μ.

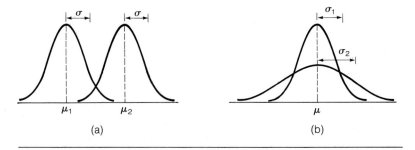

(a)   (b)

Note that an increase in μ moves the distribution to the right, whereas an increase in σ flattens and spreads out the distribution.

The first problem we will consider is that of finding probabilities of events for a random variable X that has a normal distribution with given mean μ and standard deviation σ. Normal probabilities, for our purposes, will come from tables. This suggests that we need a table for every possible value of μ and σ. Fortunately, it suffices to have a table for one distinguished member of the normal family of distributions, the so-called **standard normal distribution**, which has $\mu = 0$ and $\sigma = 1$.

standard normal distribution

The link between a normal random variable X with arbitrary mean μ and standard deviation σ and a variable Z with the standard normal distribution is provided by the **Z-score transformation** introduced in example 5.10 (p. 160):

Z-score transformation

$$Z = \frac{X - \mu}{\sigma}$$

We saw in the example that Z has mean 0 and standard deviation 1. The second fact needed is that *linear transformation of normal variables are normal*. Since the Z-score transformation is linear, Z has a normal distribution and is thus a standard normal variable.

The first step in a strategy to compute probabilities for X is to carry out the appropriate Z-score transformation. Thus, to calculate the probability of the event $(X \leq x)$, the Z-score transformation is carried out on both sides of the inequality:

$$\frac{X - \mu}{\sigma} \leq \frac{x - \mu}{\sigma}$$

The left-hand side is the standard normal variable Z. Thus, in terms of Z the above event becomes

$$\left(Z \leq \frac{x - \mu}{\sigma} \right)$$

Now, this event and the original event are "equal" in the sense that they actually represent the same event in the underlying probability space. It follows that they have the same probability:

$$P(X \leq x) = P\left(Z \leq \frac{x - \mu}{\sigma} \right)$$

Because of this equation (and others like it), normal probabilities for X can be obtained from probabilities for Z.

EXAMPLE 6.2

The normal random variable X has mean $\mu = 3$ and standard deviation $\sigma = 4$. What is the probability that X will take on a value less than or equal to 7?

Performing the Z-score transformation on both sides of the inequality $X \leq 7$ we find that

$$P(X \leq 7) = P\left(Z \leq \frac{7 - 3}{4} \right)$$

$$= P(Z \leq 1)$$

All that remains is to find the probability $P(Z \leq 1)$. This is the subject of the next subsection.

Example 6.3 further illustrates the property that linear transformations of normal variables are normal. The following example depends on material from section 5.7.

EXAMPLE 6.3

If X and Y are normal variables, then the linear transformation

$$W = aX + bY + c$$

for constants a, b, and c is also normal. This is true whatever the association between X and Y may be. In the special case of independent variables, the laws of expectation and variance given in chapter 5 can be used to determine the mean and variance of W. In this way, the distribution of W will be completely determined, and the normal tables can then be used to calculate probabilities for W.

Suppose, for example, that X and Y are independent normal variables with means and standard deviations given by

$$\mu_X = 2, \ \mu_Y = 5, \ \sigma_X = 3 \text{ and } \sigma_Y = 4$$

If $W = 3X + 2Y - 4$, how can $P(5 < W < 15)$ be calculated? By the rules of expectation and variance,

$$E(W) = 3E(X) + 2E(Y) - 4$$
$$= 3 \cdot 2 + 2 \cdot 5 - 4 = 12$$

and

$$\text{var}(W) = 3^2 \text{var}(X) + 2^2 \text{var}(Y)$$
$$= 9 \cdot 3^2 + 4 \cdot 4^2 = 145$$

Thus, $\sigma_W = \sqrt{145} = 12.04$.

It follows that the Z-score transformation can be applied to W, yielding

$$P(5 < W < 15) = P((5 - 12)/12.04 < Z < (15 - 12)/12.04)$$
$$= P(-.581 < Z < .249)$$

This probability can be calculated by using the method to be given in the next subsection.

Calculating Normal Probabilities Using Table 1A

Table 1A in the Appendix is designed for the direct calculation of probabilities, denoted Pr in the table, of the form

$$P(0 < Z < z)$$

See display 6.7.

Display 6.7

Standard normal probability curve showing the probability $P(0 < Z < z)$ **tabulated in table 1A**

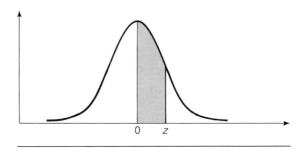

To use the table, round the number z, for which the probability is desired, to two decimal places and enter the row corresponding to the first two digits and the column corresponding to the last digit. The number in that row and column is the desired probability.

For example, to find $P(0 < Z < 1.236)$, round 1.236 to 1.24 and enter the table at the row labeled 1.2 and the column labeled .04. The desired probability is .3925. Greater accuracy can be obtained by a process of interpolation in the table. However, such accuracy is seldom warranted in practice, and we will not pursue the topic of interpolation.

Note that because the standard normal curve is symmetric around 0 and has total area equal to 1, $P(Z \geq 0) = .5$. Table 1A includes additional values of $P(0 < Z < z)$ for z ranging from 3.5 to 4.9, in order to show how rapidly the area of .5 is approached as z increases. This extension will also be used to justify the box plot "calibration" given in chapter 2.

EXAMPLE 6.4

The probability calculation begun in example 6.2 can now be completed. The probability of interest was seen to be $P(Z \leq 1)$. As we have seen, this probability is the same as $P(Z < 1)$. To put this in the form for which table 1A can be used, we observe that the event $(Z < 1)$ can be expressed in the form

$$(Z < 1) = ((Z \leq 0) \text{ or } (0 < Z < 1))$$

The events on the right-hand side of the equality are mutually exclusive, so

$$P(Z < 1) = P(Z \leq 0) + P(0 < Z < 1)$$

This relationship is given pictorially in display 6.8.

Display 6.8

Pictorial representation of the relationship $P(Z < 1) =$ $P(Z \leq 0) +$ $P(0 < Z < 1)$

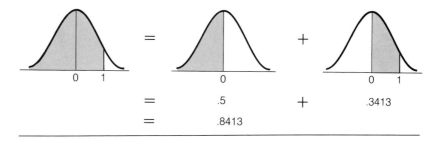

The second term is exactly of the form that can be obtained from table 1A. For $z = 1 = 1.00$, read the probability $.3413$ from the table. Thus,

$$P(Z < 1) = .5 + .3413$$
$$= .8413$$

It follows that for a normal variable X with mean $\mu = 3$ and standard deviation $\sigma = 4$,

$$P(X \leq 7) = .8413$$

Other normal probabilities can be calculated in a similar fashion. A somewhat more complicated example follows.

EXAMPLE 6.5

Suppose X is a normal variable with mean $\mu = 5$ and standard deviation $\sigma = 2$. Compute the probability $P(3 < X < 8)$.

The first step is to Z-score transform all three terms in the event $(3 < X < 8)$. The resulting probability in terms of the standard normal variable Z will then be the same as the probability of interest:

$$P(3 < X < 8) = P\left(\frac{3 - 5}{2} < \frac{X - 5}{2} < \frac{8 - 5}{2}\right)$$
$$= P(-1 < Z < 1.5)$$

Next, write the event $(-1 < Z < 1.5)$ as follows:

$$(-1 < Z < 1.5) = ((-1 < Z \leq 0) \text{ or } (0 < Z < 1.5))$$

The events on the right-hand side are mutually exclusive, thus excluding the boundary $z = 0$ in the first event,

$$P(-1 < Z < 1.5) = P(-1 < Z < 0) + P(0 < Z < 1.5)$$

Refer to display 6.9. Table 1A can be entered directly to obtain $P(0 < Z < 1.5) = .4332$. To evaluate the first term on the right, we appeal to the symmetry of the standard normal curve around 0 to conclude that

$$P(-1 < Z < 0) = P(0 < Z < 1)$$

Thus, $P(-1 < Z < 0) = .3413$. Summing the tabled probabilities, we find that

$$P(-1 < Z < 1.5) = .3413 + .4332 = .7745$$

Finally,

$$P(3 < X < 8) = .7745$$

The strategy for the calculation of normal probabilities can be summarized as follows. First, transform the event in a normal variable X to an "equal" event in the standard normal variable Z by using the Z-score transform. Second, write this event in terms of mutually exclusive events whose probabilities can be read either directly or indirectly from table 1A. This step

Display 6.9

Pictorial representation of $P(-1 < Z < 1.5) =$ $P(-1 < Z < 0) +$ $P(0 < Z < 1.5)$

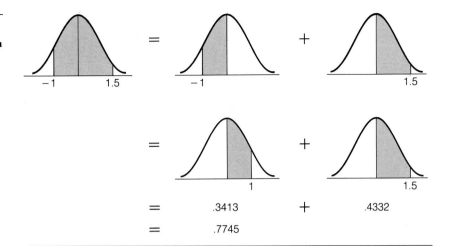

$$= \quad .3413 \quad + \quad .4332$$

$$= \quad .7745$$

may require using the symmetry of the standard normal curve. To avoid confusion, *draw a picture of the events and their probabilities,* as in displays 6.8 and 6.9. Then, by adding and/or subtracting the probabilities of the various pieces, the desired probability can be obtained.

EXAMPLE 6.6

If X is a normal random variable with mean $\mu = -1$ and standard deviation $\sigma = 3$, what is the probability that X will take on a value smaller than -8?
From the Z-score transform, we find that

$$P(X < -8) = P\left(Z < \frac{-8 - (-1)}{3}\right)$$

$$= P(Z < -2.33)$$

Now, write the event $(Z < 0)$ as

$$(Z < 0) = ((-2.33 < Z < 0) \text{ or } (Z \le -2.33))$$

Display 6.10

Pictorial representation of the calculation of $P(Z \le -2.33)$ for example 6.6

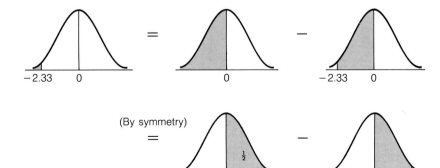

From the addition law, we see that

$$P(Z < 0) = P(-2.33 < Z < 0) + P(Z \leq -2.33)$$

or, by subtraction,

$$P(Z \leq -2.33) = P(Z < 0) - P(-2.33 < Z < 0)$$

See display 6.10.

By symmetry, $P(Z < 0) = P(Z > 0) = .5$ and $P(-2.33 < Z < 0) = P(0 < Z < 2.33)$. This probability is found from table 1A to be .4901.

It follows that

$$
\begin{aligned}
P(X < -8) &= P(Z < -2.33) \\
&= .5 - .4901 \\
&= .0099
\end{aligned}
$$

EXERCISES

6.2   Let X be a normal random variable with mean $\mu = 3$ and standard deviation $\sigma = 5$. Draw pictures of the standard normal areas and calculate the following probabilities.

(a) $P(X < 8)$

(b) $P(X < 3)$

(c) $P((X < 4) \text{ or } (X > 7))$

(d) $P(X > 20)$

(e) $P(X < -10)$

(f) $P(-10 < X < 20)$

6.3   Let X be a normal random variable with mean $\mu = -6$ and standard deviation $\sigma = 3$. Draw pictures of the standard normal probabilities and find

(a) $P(X < -5)$

(b) $P(X > 2)$

(c) $P(-4 < X < 2)$

*6.4   Random variables X and Y are independent and normally distributed, with means $\mu_X = -2$, $\mu_Y = 3$ and standard deviations $\sigma_X = 3$, $\sigma_Y = 1$. If $W = 3X + Y + 10$, find

(a) $P(8 < W < 14)$

(b) $P(W > 16)$

(c) $P(W < 6)$

*6.5   An electronic component will fail if the maximum voltage, X, that it receives exceeds a critical threshhold. The threshhold, Y, for a component randomly selected

---

* Depends on material from section 5.6.

from a lot is normally distributed with a mean of 10 volts and a standard deviation of 1.3 volts. The maximum voltage at the component is independent of Y and normally distributed, with a mean of 8 volts and a standard deviation of 1 volt. What is the probability that the randomly selected component will fail?

6.6† A random variable X has mean $\mu = 3$ and standard deviation $\sigma = 5$. The true mean is unknown to the experimenter, who has reason to conjecture that the mean is $\mu_0 = 1$.

(a) Calculate $P((X - 1)/5 \geq 1.645)$. (Note: The Z-score transform is $Z = (X - 3)/5$. First convert the event of interest to one involving X only on the left-hand side of the inequality, then apply the Z-score transform.)

(b) Calculate the probability of part (a) if the true mean of X is $\mu = 8$.

(c) Repeat the calculation of parts (a) and (b) for a mean μ of 13. Of 18. Of 23.

(d) Find the probability of part (a) if, in fact, the true mean is $\mu = 1$.

(e) Plot the probabilities calculated in parts (a) through (d) against the true means. For example, the result of the calculation of part (b) would be plotted as a point on the graph, with horizontal coordinate equal to $\mu = 8$ and vertical coordinate equal to the calculated probability. Join the plotted points with a smooth curve. This curve will be seen in chapter 9 to be the *power curve* for a 5% one-sided test of the hypothesis that the mean of a normal distribution has the hypothetical value μ_0, taken to be 1 in this exercise.

Events Defined in Terms of the Absolute Value

The absolute value $|x|$ of a number x is the positive magnitude of the number. Thus, $|5| = 5$ and $|-5| = 5$, for example. The distance between two numbers can be expressed in terms of the absolute value. Thus, the distance between 3 and 5 is $|3 - 5| = |-2| = 2$. Sensibly, this also is the distance between 5 and 3, since $|5 - 3|$ is also 2.

The distance between a number x and a number μ in units of a positive quantity σ is $|x - \mu|/\sigma = |(x - \mu)/\sigma|$. It follows that if X is a random variable with mean μ and standard deviation σ, then the absolute value of the Z-score transform

$$|Z| = \left|\frac{X - \mu}{\sigma}\right|$$

is a measure of the distance of X from its mean in units of its standard deviation. An event of the form $(|Z| \leq z)$ then represents the values of the variable X that differ from the mean μ by z or fewer standard deviations:

$$(|Z| \leq z) = (|X - \mu| \leq z\sigma)$$
$$= (\mu - z\sigma \leq X \leq \mu + z\sigma)$$

Similarly, the event $(|Z| \geq z)$ represents the values of X that differ from μ by z or more standard deviations.

The absolute value of the Z-score transform will be of considerable

---

† The symbol † will indicate more difficult exercises.

interest in our treatment of statistical inference because of this distance-measuring interpretation. Consequently, it will be important to be able to calculate the probabilities of events defined in terms of the absolute value. When the variable X has a normal distribution (and Z thus has a standard normal distribution), the calculations can be based on table 1A.

EXAMPLE 6.7

If Z has a standard normal distribution, calculate $P(|Z| \leq 2.3)$.

Since $|Z| \leq 2.3$ only if $Z \leq 2.3$ and $Z \geq -2.3$, we have the following relationships among events:

$$(|Z| \leq 2.3) = (-2.3 \leq Z \leq 2.3)$$
$$= ((-2.3 \leq Z \leq 0) \text{ or } (0 < Z \leq 2.3))$$

By the symmetry of the standard normal curve, the last two events have the same probability and can be read directly from table 1A:

$$P(|Z| \leq 2.3) = 2P(0 < Z < 2.3)$$
$$= 2(.4893)$$
$$= .9786$$

EXAMPLE 6.8

If Z has the standard normal distribution, compute $P(|Z| \geq 1.36)$.

Note that the event $(|Z| \geq 1.36)$ is equal to the event

$$((Z \geq 1.36) \quad \text{or} \quad (Z \leq -1.36))$$

The two components of this expression are mutually exclusive and, by the symmetry of the normal curve, have the same probability. Thus,

$$P(|Z| \geq 1.36) = 2P(Z \geq 1.36)$$

But

$$P(Z \geq 1.36) = .5 - P(0 < Z < 1.36)$$
$$= .5 - .4131$$
$$= .0869$$

Thus,

$$P(|Z| \geq 1.36) = 2(.0869)$$
$$= .1738$$

A pictoral representation of this calculation is shown in display 6.11.

Display 6.11

Pictorial representation of the computation for example 6.8

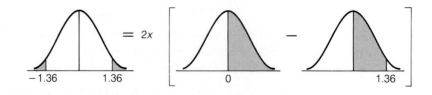

Note that this probability could also have been calculated by means of the law of complementation and a computation of the type given in example 6.7, since $(|Z| \geq z) = (\text{not}(|Z| < z))$.

EXERCISES

6.7   Let Z be a standard normal variable. Relate the probabilities of the following events to probabilities obtainable from table 1A by pictures (as in display 6.11) and calculate the probabilities.

(a) $(|Z| \leq 2.45)$

(b) $(|Z| \geq 1.96)$

(c) $(|Z| \leq 1.96)$

(d) $(|Z| \geq 4.2)$

(e) $(1.04 < |Z| < 2.56)$

6.8   Let X be a normal random variable with mean $\mu = -6$ and standard deviation $\sigma = 3$. Find

(a) $P(|X + 6| > 5)$

(b) $P(|X + 6| < 8)$

(c) $P(|X + 4| > 2)$ (Hint: Write $(|X + 4| > 2) = ((X + 4 > 2) \text{ or } (X + 4 < -2)))$

6.9†   Let X be a normal random variable with mean μ and standard deviation $\sigma = 5$. It is hypothesized that the value of μ is $\mu_0 = 1$. To "test" this hypothesis, the distance from X to 1 in units of σ is measured by $|(X - 1)/5|$.

(a)   Calculate the probability $P(|(X - 1)/5| \geq 1.96)$ if, in fact, the true mean is $\mu = 3$.

(b)   Calculate the probability of part (a) if the true mean of X is $\mu = 8$.

(c)   Repeat the calculation of parts (a) and (b) for a mean μ of 13. Of 18. Of 23.

(d)   Find the probability of part (a) if, in fact, the true mean is $\mu = 1$.

(e)   Plot the probabilities calculated in parts (a) through (d) against the true means. See the instructions for exercise 6.6(e), page 195.

The "Inverse" Problem—Use of Table 1B

Thus far, we have considered the problem: Given an event for a normal variable, find its probability. These probabilities have been found in table 1A. However, many of the more important problems in statistical inference are of the "inverse" variety: given a probability (and the "form") of an event, find the event that has that probability. For example, if Z is a standard normal variable and the event of interest has the form $(Z \geq z)$, we might want to find the number z for which $P(Z \geq z) = .05$.

Solving "inverse" problems using table 1A is rather awkward. For this reason, table 1B, given in the appendix, has been constructed to make the solution of these problems direct and convenient for the forms of events we will encounter in the next three chapters. For example, the third column of table 1B is seen to contain probabilities for events of the form $(Z \geq z)$. Scan down this column until the probability .05 is found. The number $z = 1.645$ opposite this probability in the last column is the value of z for which $P(Z \geq z) = .05$.

The columns of table 1B are labeled with the applications in which they will be used in chapters 8 and 9. For example, the second and third columns will be used to find what will be called *critical values* for the hypothesis tests of chapter 9. The first column will be used to determine critical values for confidence intervals. A preview of this use is given in the next example.

EXAMPLE 6.9

If X is a normal random variable with mean μ and standard deviation σ, for what value of z is the following equation true?

$$P(X - z\sigma \leq \mu \leq X + z\sigma) = .95$$

The event will be reduced to more familiar terms by algebraic manipulation. First note that the inequality $X - z\sigma \leq \mu$ is equivalent to $X \leq \mu + z\sigma$. Similarly, $X + z\sigma \geq \mu$ is equivalent to $X \geq \mu - z\sigma$. Thus,

$$(X - z\sigma \leq \mu \leq X + z\sigma) = (\mu - z\sigma \leq X \leq \mu + z\sigma)$$
$$= (-z\sigma \leq X - \mu \leq z\sigma)$$
$$= (-z \leq \frac{X - \mu}{\sigma} \leq z)$$
$$= (|Z| \leq z),$$

where we have used the Z-score transform $Z = (X - \mu)/\sigma$. Thus, to solve the problem, we need only find the value z for which

$$P(|Z| \leq z) = .95$$

Find .95 in the first column of table 1B and read the number $z = 1.960$ opposite this value in the last column.

EXAMPLE 6.10

If X is a normal variable with mean μ and standard deviation σ, then an important quantity in hypothesis testing is the probability of seeing a value of X farther from μ than z, where the distance from X to μ is measured in units of the standard deviation, σ:

$$P\left(\left|\frac{X - \mu}{\sigma}\right| > z\right) = P(|Z| > z)$$

This probability will be called the *P-value* in chapter 9. When z is given, the probability can be obtained from table 1A. However, it is also possible to obtain bounds for the P-value using table 1B.

For example, suppose we want to find bounds for $P(|Z| > 2.35)$. Scan down the last column in table 1B, headed "critical value," until the two consecutive values bounding 2.35 are found. The bounds for the probability are then the probabilities in the same rows as these values and in the column labeled $P(|Z| \geq z)$. Thus, since 2.326 and 2.576 bound the number 2.35, the probabilities .02 and .01 bound $P(|Z| > 2.35)$:

$$.01 < P(|Z| > 2.35) < .02$$

The more accurate calculation from table 1A is

$$P(|Z| > 2.35) = 2(.5 - .4906)$$
$$= .0188$$

The less precise information given by the bounds will generally be sufficient for hypothesis testing purposes.

The Interquartile Range of the Normal Distribution and the Calibration of the Box Plot

If X is a normal random variable with mean μ and standard deviation σ, then because its distribution around μ is symmetrical, the first and third quartiles of this distribution are of the form

$$q_1 = \mu - z\sigma$$

and

$$q_3 = \mu + z\sigma$$

Since the quartiles bound the middle half of the distribution, the number z must be determined so that the area under the probability curve between q_1 and q_3 is .5. After Z-score transformation, z must satisfy the equation

$$P(|Z| \leq z) = .5$$

From table 1B, column $P(|Z| \leq z)$, the solution is seen to be $z = .674$. It follows that the interquartile range of X is

$$\text{iqr} = q_3 - q_1 = 2(.674\sigma) = 1.348\sigma$$

An important application of the box plot is to call attention to outliers. Observations are classified as outliers if they fall more than 1.5 interquartile ranges from the ends of the box. In chapter 2 the claim was made that if the distribution were normal, an outlier would be seen about 7 times in 1000— that is, with probability about .007. We are now in a position to justify this statement, at least as an approximation.

The approximation arises from taking the normal distribution quartiles equal to the sample quartiles rather than the quartiles of the population from which the sample was drawn. This approximation is adequate for practical purposes.

With the quartiles of the normal distribution at the edges of the box, the X values more than 1.5 iqr below the lower edge are those for which

$$X < q_1 - 1.5 \text{ iqr} = \mu - .674\sigma - 1.5 \cdot 1.348\sigma = \mu - 2.696\sigma$$

Similarly, the values more than 1.5 iqr above the box satisfy the inequality $X > \mu + 2.696\sigma$. After applying the Z-score transform, the probability of an outlier is then

$$P(|Z| > 2.696) = 2(.5 - .4965)$$
$$= .007$$

as stated.

An extreme outlier, on the other hand, lies more than 3 interquartile ranges from the box. It was claimed that if the distribution were normal, an extreme outlier would be seen about twice in every million "draws" from the population. By the same process as above, the probability that X lies more than 3 iqr's below q_1 or above q_3 is

$$P(|Z| > 4.718) = 2(.5 - .4999987)$$
$$= .0000026$$

The number 4.718 has been rounded to 4.7 in order to use table 1A. If we were able to calculate the probability for the more accurate value, this probability would round to .000002, or 2 per million.

EXERCISES

6.10 Let Z have the standard normal distribution. In each (lettered) part, draw a picture of the event and its probability. Then find the z values requested in each subpart.

(a) Find the value of z for which $P(|Z| \le z)$ equals: (i) .95, (ii) .99, (iii) .90, (iv) .999.

(b) Find the value of z for which $P(Z \ge z)$ equals: (i) .05, (ii) .01, (iii) .005.

(c) Find the value of z for which $P(Z \le -z)$ equals: (i) .01, (ii) .10, (iii) .0005.

(d) Find the value of z for which $P(|Z| \ge z)$ equals: (i) .50, (ii) .05, (iii) .01.

6.11 If Z has the standard normal distribution, find bounds for the following probabilities.

(a) $P(|Z| \le 1.59)$

(b) $P(Z \ge 3)$

(c) $P(|Z| > 5)$ (Note: one bound will be 0)

(d) $P(Z < -2.435)$

6.12 If X has a normal distribution with mean μ and standard deviation σ, find the number z for which

$$P(X - z\sigma \le \mu \le X + z\sigma)$$

is equal to

(a) .90

(b) .95

(c) .99

Chapter 6 Quiz

1. What is a continuous random variable?

2. How are probabilities represented for continuous random variables?

3. What position on the measurement axis does the mean μ of a normal distribution occupy relative to the probability curve of the distribution?

4. What is an interpretation of the standard deviation σ of a normal distribution in terms of the probability curve?

5. What is the distribution of the Z-score transform of a normal variable?

6. What is the strategy for calculating the probabilities of events for a normal random variable X?

7. What is the purpose of drawing pictures of events and their probabilities (areas) in a probability calculation?

8. How can the probabilities of events, defined in terms of the absolute values of a standard normal variable, be calculated? Give examples for both $P(|Z| < z)$ and $P(|Z| > z)$.

9. For what kinds of problems is table 1B designed to provide solutions? Give examples.

SUPPLEMENTARY EXERCISES

6.13   Staff salaries at a certain university are normally distributed with mean $15,000 and standard deviation $1250. If a staff member is chosen at random, what is the probability that his or her salary will

(a)   exceed $18,000?

(b)   differ from the mean by more than $2000?

(c)   differ from the mean by more than two standard deviations?

6.14* Random variables X and Y are independent and normally distributed, with means $\mu_X = 7$ and $\mu_Y = 3$. The standard deviations of both variables are equal to $\sigma = 6$.

(a)   Calculate $P(X - Y > 4)$.

(b)   Calculate $P(X > Y)$.

(c)   Calculate $P(X < Y - 1)$.

---

* Depends on material from section 5.6.

6.15   During a given time period, the purchase price of U.S. dollars in British pounds is approximately normally distributed and averages 2.5 dollars to the pound with a standard deviation of .08 dollars. If a traveler changes 100 pounds into dollars during that time period and the transaction costs him $15,

(a)   what is the probability that he will be left with between $230 and $245?

(b)   what is the probability that he will be left with less than $225?

6.16   Let X be a normal random variable with mean μ and standard deviation σ.

(a)   Find the probability that X differs from μ by at least 3.2 standard deviations. Write the event in X symbolically and give the corresponding event in terms of the Z-score transformed variable X.

(b)   Find the probability that X lies in the interval from $\mu - 1.645\sigma$ to $\mu + 1.645\sigma$. Write the event in terms of the absolute value of the Z-score transform of X.

(c)   Find the probability of the event $(X - 1.96\sigma \leq \mu \leq X + 1.96\sigma)$.

6.17   The original version of the box plot had a scaling factor of $1.0 \cdot IQR$ rather than the currently used value of $1.5 \cdot IQR$.

(a)   What was the probability of an outlier (of any kind) for a box plot with the original scaling factor?

(b)   What was the probability of an extreme outlier, using the original scaling factor?

Sampling Distributions—The Link Between Samples and Populations

Introduction

We are now ready to forge the link that will allow us to generalize information from samples to the populations from which they are drawn. A basic assumption is that sampling is carried out at random. (See chapter 5 for specifics.) Because of this, the probability methods of chapters 5 and 6 will play a key role in the generalization process.

The generalized information will be about one or more parameters of a variable's population frequency distribution. For example, the physicians' question about the effectiveness of a new arthritis medication might be posed as a question about the mean of a measured characteristic. Such a characteristic might be the increase in the amount an affected joint can be moved after treatment. If the mean for treatment with the new medication "significantly" exceeds the mean for the standard treatment, then a case could be made for adopting the medication.

Alternatively, if no convenient measurement is available but an improvement in overall patient condition can be distinguished for either treatment, then one might agree to adopt the medication if the *proportion* of patients who show improvement is "significantly" larger than the proportion of patients who do when given the standard treatment. Here, the parameter of interest is a population proportion—the proportion of (prospective) patients improved by treatment.

In this and the next two chapters we will deal exclusively with inference problems involving population means, μ, and proportions, p. These are the most commonly used parameters in statistics, and the inference

203

procedures for them are of basic importance. Moreover, by restricting attention to these two parameters in relatively simple contexts, we will be able to isolate the more important ideas of statistical inference.

The starting point in the inference process is the formation of sample *estimators* of the parameters μ and p. The sample mean, $\overline{X}$ (introduced in chapter 2), will be used to estimate the population mean and the sample proportion, $\hat{p}$ (introduced in chapter 5), will be used to estimate the population proportion.

Since samples are drawn randomly, the sample observations are random variables. Estimators are functions of these observations, so they will be random variables as well. It follows that the estimators will have probability distributions. These probability distributions, called the *sampling distributions* of the estimators, are what make statistical inference possible; they are the key ingredient required to make inferences and to measure the quality of these inferences.

We will study the sampling distributions of the sample mean and proportion in this chapter. They will be seen to depend on factors such as the size of the sample and the distribution of the observations from which they are formed. Thus, it is not really correct to talk about *the* sampling distribution of an estimator; there is actually a different sampling distribution for each combination of circumstances.

In a given problem it will be necessary to know the sampling distribution, at least approximately, for whatever combination of circumstances the problem presents. Because the underlying distribution of observations will be unknown in practice, this would be impossible without a remarkable property of such estimators as the sample mean and proportion: Regardless of the form of the underlying distribution of observations, as sample size increases, the distributions of both estimators approach a normal distribution. This property makes it possible to use the normal distribution to approximate the sampling distributions of these estimators when sample sizes are "large." We will derive the appropriate forms of the normal approximations in preparation for their use in the *large sample theory* of statistical inference, to be studied in chapters 8 and 9.

An additional factor that influences a sampling distribution is the size of the population—or, more accurately, the size of the population relative to the size of the sample. The nature of this influence will be discussed in the next section. The theory of statistical inference in the remainder of the text will be based on the assumption that sample size is negligible relative to population size. Modifications appropriate when samples form appreciable fractions of their populations will be given in optional sections at the ends of this and the next two chapters.

The Infinite Population Model for Random Samples

If X is the variable of interest in a problem, then a random sample of size n results in a collection of random variables X_1, X_2, . . . , X_n, where X_i represents the value of X for the ith individual sampled from the population. The *infinite population model* is said to hold for the sample if the X_i's are (i) independent and (ii) identically distributed. Property (ii) means that the probability distributions of these variables are the same and equal to the probability distribution of X, which, under simple (lottery) random sampling, is identical to the population frequency distribution of X.

This is a model of independent trials (chapter 5, p. 168), in which a value of X is observed at each trial. While this model should represent random sampling *with* replacement rather well, it is not obvious that it will provide a good representation of sampling *without* replacement. We found it necessary to use conditional probabilities to analyze this form of sampling in chapter 5. Since sampling without replacement is the norm in practice, it seems curious that the infinite population model would be used as the basis of a theory of statistical inference.

An important justification for the use of the infinite population model is that the "mathematics" of independent variables is much simpler than that of dependent variables. However, this justification would not be very compelling unless the infinite population model provides a good approximation to a more accurate model involving dependent variables. That it does so can be argued as follows.

The dependence among observations in sampling without replacement arises from the fact that, as each sample member is drawn, the composition of the population is changed, thus the probabilities of X values for the next draw will be different from those for the undisturbed population. Moreover, the change in the distribution will depend on the individuals previously drawn, thus on the values of X previously seen. It follows that the X_i's are not independent random variables.

However, if the population is large, its composition is changed very little by removing one individual—or by removing n individuals if n is small compared to the population size. In this case, the measurements will be *nearly* independent and identically distributed. The larger the population relative to sample size, the more closely the model of independent and identically distributed variables approximates the actual distribution of X's. For an infinite population the approximation would be perfect; hence the term *infinite population model.*

By contrast, a *finite population model* would be required when the size of the sample size is not a negligible fraction of the population size. This situation occurs occasionally in practice. For example, in the field of Quality Assurance, in order to make very sure of the quality of a shipment (population) of items to be used in manufacturing expensive or vital products, it may be necessary to sample and test a large fraction of the shipment. Even when quality is exceptionally important, testing a sample rather than the entire shipment is preferable if testing is very expensive or if items are destroyed in the process. Details of how the finiteness of the population affects the sampling distributions of the sample mean and sample proportion will be given in the optional section 7.5.

■ SECTION 7.3

Sampling Distribution of the Sample Mean

Let X be a measurement variable with mean μ and standard deviation σ. In section 5.6 (p. 175), we saw that, for a random sample of size n, the mean and standard deviation of $\overline{X}$ are

$$\mu_{\overline{X}} = \mu$$

and

$$\sigma_{\overline{X}} = \frac{\sigma}{\sqrt{n}}$$

unbiased

An estimator whose mean is equal to the parameter being estimated is said to be **unbiased**. Intuitively, the sampling distribution of an unbiased estimator is centered at the "true" value of the parameter of interest. The expression for the mean of $\overline{X}$ tells us that this estimator is an unbiased estimator of the population mean.

Unbiasedness is a sufficiently important property that estimators are sometimes modified to achieve it. For example, in chapter 2 (p. 49), the divisor for the sample variance, s^2, was taken to be $n - 1$ instead of n in order to make it an unbiased estimator of the population variance, σ^2.

standard error

The standard deviation of an estimator is called its **standard error**. Since standard deviation measures the variation or dispersion of a variable from its mean, the second expression, $\sigma_{\overline{X}} = \sigma/\sqrt{n}$, tells us that as sample size increases, the dispersion of values of $\overline{X}$ around μ becomes smaller. As a consequence, the probability of $\overline{X}$ being "close" to μ becomes larger with increasing sample size; it becomes less and less likely that we will see values of $\overline{X}$ "far away" from μ. In this sense, $\overline{X}$ becomes a more informative indicator of the value of μ as n is increased.

Knowing the mean and standard deviation of $\overline{X}$ is not enough for the purposes of statistical inference. We must also know its probability distribution, or at least have a good approximation of it. If X has a normal distribution, then since linear transformations of normal variables are again normal (chapter 6), the distribution of $\overline{X}$ is also normal. Unfortunately, from a practical viewpoint this is not a very useful fact, since normal distributions rarely occur in practice. However, a remarkable property of an estimator involving sums of random variables, such as the sample mean, is that the distribution of the estimator "tends" toward a normal distribution with increasing sample size. Moreover, this happens whatever the underlying distribution of observations may be. Thus, a normal distribution can be used to approximate the sampling distribution of $\overline{X}$ for "large" samples.

This remarkable property can be established mathematically. The mathematical result is called the *Central Limit Theorem*. However, since its proof requires mathematics beyond our scope, we will instead illustrate this phenomenon with a Monte Carlo simulation.

Monte Carlo Demonstration of the Central Limit Theorem

A computer program was written to sample randomly from the population consisting of the first 100 students of display 1.18 (p. 28–30). In order to simulate random sampling from an infinite population, sampling was carried out with replacement. The variable X was taken to be the Weight variable. The population mean and standard deviation of this variable are

$$\mu = 138.4 \text{ pounds}$$

and

$$\sigma = 26.44 \text{ pounds}$$

The program carried out the following steps. Forty random samples of size $n = 5$ were drawn from the population and the sample mean was calculated for each sample. These 40 sample means then represent a random sample of size 40 from the sampling distribution of $\overline{X}$ for $n = 5$. The mean of this sample should approximate the theoretical population mean, $\mu_{\overline{X}} = \mu = 138.4$ pounds, and the sample standard deviation should approximate the standard deviation of the sampling distribution (the standard error), $\sigma_{\overline{X}} = \sigma/\sqrt{5} = 26.44/\sqrt{5} = 11.83$.

The sampling process was repeated to obtain samples of size 40 from the sampling distributions of the mean for $n = 10, 30, 50,$ and 100. The means and standard deviations of these samples should also be near the values predicted by theory. The observed means from the samples and the theoretical values are given in the following table.

n	MEAN OBSERVED	MEAN THEORETICAL	STANDARD ERROR OBSERVED	STANDARD ERROR THEORETICAL
5	135.7	138.4	11.77	11.83
10	138.2	138.4	9.05	8.36
30	138.7	138.4	4.72	4.83
50	138.8	138.4	3.53	3.74
100	138.3	138.4	2.78	2.64

More importantly, we should be able to observe the tendency toward normality of the sampling distributions in stem and leaf plots of these samples. This tendency should hold whatever the underlying distribution of observations may be. In chapter 1 the underlying frequency distribution of weights was observed to be bimodal. Display 1.7 (p. 18) shows a histogram and stem and leaf plot of a sample from this distribution. However, as seen from display 7.1, the bimodality is quickly lost in the sampling distributions as n increases.

Display 7.1

Stem and leaf plots of sample means for samples of sizes $n = 5$, 10, 30, 50, and 100 from the first 100 weights of display 1.18. Units = weights $\times$ 10.

$n = 5$

```
11*
  . | 68,88
12* | 08,08,14,14,42
  . | 50,64,76,78,78,96,98
13* | 04,10,10,16,24,24,40,48
  . | 60,62,74,80
14* | 00,06,14,20
  . | 50,52,96
15* | 00,20,30,36,44
  . | 50
16*
  . | 56
```

$n = 10$

```
11*
  . | 85
12* | 24,48
  . | 63,69,79
13* | 02,13,19,20,31,32,40,45,49
  . | 59,74,75,77,82,84,90,92,93,96,96
14* | 03,16,18,24,41,47,48
  . | 53,70,87
15* | 02,14
  . | 76
16* | 28
  .
```

$n = 30$

```
12* | 37
  . | 88
13* | 20,25,27,36,48
  . | 53,55,57,58,66,70,75,76,76,83,92,93,96,99
14* | 00,00,04,05,08,11,14,14,16,26,30,32,33,41,41,41,42
  . | 55,57
```

$n = 50$

```
13* | 13,28,28,28,44
  . | 53,54,56,58,61,65,67,67,69,75,75,77,79,82,85,87,88,94,97
14* | 06,06,06,12,13,13,16,17,22,29,30,32,33,40,44
  . | 51
```

$n = 100$
(note expanded scale)

```
13*
  t | 21,29
  f | 46,49,55,57,59,59
  s | 60,61,65,69,76,77,78,79,79
  . | 80,80,80,84,85,85,87,87,88,88,89,94,94,95
14* | 00,02,09,19,19
  t | 28,33,37
  f | 52
  s
  .
```

Already for $n = 10$ the sampling distribution displays the symmetry that is a characteristic of the normal distribution, although there appears to be too much frequency in the tails.

The reason for the term "central" in the name of the central limit theorem is that the middle or central part of the distribution tends most rapidly to that of a normal distribution. The tails follow more slowly or, put in another way, require larger sample sizes to conform to normality. This phenomenon is more easily seen in the schematic diagram of the Monte Carlo samples given in display 7.2.

Display 7.2

Schematic diagram of sample means for samples of sizes $n = 5$, 10, 30, 50, and 100 from the first 100 weights of display 1.18

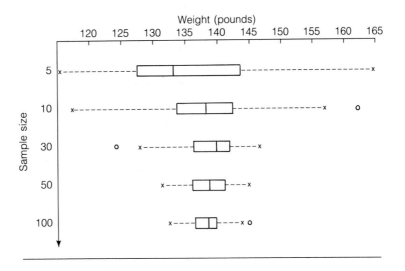

It was observed in chapter 2 that samples from a normal distribution tend to have box plots with dashed (adjacent) lines of roughly the same length as the box, and only about seven outliers in every 1000 observations. Even the sampling distribution for $n = 100$ has not completely reached this condition. However, calculations based on the normal distribution give acceptable approximations when sample sizes are 30 or more if the underlying distribution is not too far from normal. The closer to normal this distribution is, the better the approximation.

rule of thumb

We will assume, as a **rule of thumb**, *that it is reasonable to use the normal approximation for the sampling distribution of* $\overline{X}$ *when sample sizes, n, are 30 or larger.* That is, for the sample mean, the division point between the so-called *large sample theory* of statistics and the *small sample theory* is taken to be $n = 30$. In this and the next two chapters we will deal exclusively with the large sample theory of statistical inference for $\overline{X}$.

The normal approximation will not be applied directly to the distribution of $\overline{X}$. The next two subsections deal with the quantities actually used in practice.

The Z-score Transform Applied to $\overline{X}$

Recall that the Z-score transform of a random variable X with mean μ and standard deviation σ is

$$Z = \frac{X - \mu}{\sigma}$$

The result of applying this transformation to a normal variable was seen in chapter 6 to produce a *standard* normal variable. Applied to a variable with approximately a normal distribution, it will produce a variable with an approximate standard normal distribution. *Consequently, for $n \geq 30$, the Z-score transform of the sample mean,*

$$Z = \frac{\overline{X} - \mu}{\sigma/\sqrt{n}}$$

will have (approximately) a standard normal distribution.

EXAMPLE 7.1

A measurement variable X has mean $\mu = 9$ and standard deviation $\sigma = 5$ in a population. If a random sample of size $n = 75$ is taken from the population, what is the probability that the sample mean $\overline{X}$ will have a value between 8.7 and 9.5?

Since $n > 30$, this question can be answered approximately by a calculation based on the normal distribution. The mean of $\overline{X}$ is $\mu = 9$ and the standard deviation is $\sigma/\sqrt{n} = 5/\sqrt{75}$. Ignore the fact that an approximation is being used. Apply the Z-score transformation $Z = (\overline{X} - \mu)/(\sigma/\sqrt{n})$ and use table 1A, as in chapter 6:

$$P(8.7 \leq \overline{X} \leq 9.5) = P\left(\frac{8.7 - 9}{5/\sqrt{75}} \leq Z \leq \frac{9.5 - 9}{5/\sqrt{75}}\right)$$

$$= P(-.52 \leq Z \leq .87)$$

$$= .1985 + .3078$$

$$= .506$$

The approximate nature of the calculation can now be stated in the conclusion: The probability that $\overline{X}$ is between 8.7 and 9.5 is *approximately* .506.

As another example, we can use the normal approximation to demonstrate that the distribution of $\overline{X}$ becomes more concentrated around μ with increasing sample size. A more precise statement of this fact is that for any positive number c, the probability that $\overline{X}$ and μ differ by less than c units will increase to 1 as sample size increases. That is, $P(|\overline{X} - \mu| < c)$ will approach 1 as n becomes large. This effect is depicted in display 7.3. A numerical illustration is given in the next example.

EXAMPLE 7.2

Suppose that the population mean of a variable X is μ and its standard deviation is $\sigma = 5$. Random samples of size $n = 30, 50, 100, 500,$ and 1000 are drawn from the population. For each n, calculate the probability that the sample mean differs from μ by an amount less than .5.

Display 7.3

Illustration of the concentration of the distributions of $\overline{X}$ around μ for increasing sample size, n. The shaded areas are $P(|\overline{X} - \mu| < c)$ for the different sample sizes.

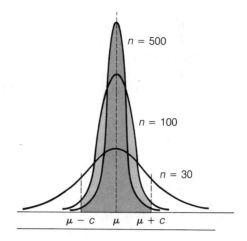

The event whose probabilities are to be calculated for different sample sizes is

$$|\overline{X} - \mu| < .5$$

The Z-score transform is applied by dividing both sides of this inequality by $\sigma_{\overline{X}} = 5/\sqrt{n}$. Thus,

$$P(|\overline{X} - \mu| < .5) = P\left(|Z| < \frac{.5}{5/\sqrt{n}}\right)$$
$$= P(|Z| < .1\sqrt{n})$$
$$= 2P(0 < Z < .1\sqrt{n})$$

where Z has (approximately) a standard normal distribution. The calculation now proceeds as in chapter 6. A table of these probabilities, computed from table 1A, is as follows. Note that the probabilities approach 1 as n increases.

n	$.1\sqrt{n}$	$P(|\overline{X} - \mu| < .5)$
30	.548	.418
50	.707	.522
100	1.00	.683
500	2.24	.975
1000	3.16	.998

The phenomenon demonstrated in example 7.2 will be important in statistical inference. The viewpoint, however, will be different. The population mean μ will be unknown and the value of $\overline{X}$ computed from actual sample data will be used to pinpoint the unknown value of μ. For sufficiently large sample size, the sample means computed from randomly selected samples will be "near" the value of μ with "high probability." By "near" we will mean $|\overline{X} - \mu| < c$ for some small (positive) number c. Thus, for the sample data at hand, unless we were unlucky enough to get one of the samples for which $\overline{X}$ is not "near" μ, we will have succeeded in pinpointing μ by the value of $\overline{X}$ to within $\pm c$ units.

The normal approximation to the distribution of $\overline{X}$ allows us to quantify and relate the terms "near" and "high probability." An application based on this idea is given in the next example.

EXAMPLE 7.3

Let X be a variable whose population frequency distribution has unknown mean μ but known standard deviation $\sigma = 10$. What is the smallest sample size needed to guarantee that the sample mean $\overline{X}$ differs from μ by less than .5 with probability at least .95? ("Near" means $|\overline{X} - \mu| < .5$, and "high probability" means .95 in this problem.)

Put symbolically, the problem is to find the smallest sample size n for which

$$P(|\overline{X} - \mu| < .5) \geq .95$$

We will initially ignore the fact that n is a whole number and that the normal distribution is an approximation. The standard deviation of $\overline{X}$ is $\sigma/\sqrt{n} = 10/\sqrt{n}$. Thus, applying the Z-score transform $Z = (\overline{X} - \mu)/(\sigma/\sqrt{n})$ yields

$$P(|\overline{X} - \mu| < .5) = P\left(|Z| < \frac{.5}{10/\sqrt{n}}\right)$$

$$= P(|Z| < .05\sqrt{n})$$

The desired sample size is the smallest n for which this probability is greater than or equal to .95. We will first try to find the value of n for which

$$P(|Z| < .05\sqrt{n}) = .95 \tag{1}$$

Except for the expression $.05\sqrt{n}$, this is a familiar problem from chapter 6. If we set $z = .05\sqrt{n}$, we are asked to find the number z such that

$$P(|Z| < z) = .95$$

for a standard normal variable Z. This problem was seen to have only one solution, namely $z = 1.960$.

Here is the key step. Since there is only one such solution, in order for equation (1) to hold, it is necessary that

$$.05\sqrt{n} = 1.960$$

Once this equation is established, the rest of the solution simply consists of solving the equation for n:

$$\sqrt{n} = \frac{1.960}{.05}$$

thus,

$$n = \left(\frac{1.960}{.05}\right)^2$$

$$= 1536.64$$

We now reintroduce the fact that n must be a whole number. Since $P(|Z| < .05\sqrt{n})$ increases with n, the smallest sample size for which this probability is greater than or equal to .95 is found by rounding to the next larger whole number, $n = 1537$. This is the solution to the problem—within the accuracy of the normal approximation.

The computation in example 7.3 will reappear as the solution of the sample size design problem for confidence intervals in chapter 8.

EXERCISES

7.1   A variable X has mean $\mu = 20$ and standard deviation $\sigma = 4$. If a random sample of size $n = 50$ is taken from the population, what is the (approximate) probability that the sample mean, $\overline{X}$, will have a value between 19.3 and 20.7?

7.2   For a given time period, the mean diameter of .3 millimeter (mm) bolts manufactured in a machine shop was actually .31 mm due to a misadjustment of the machine on which the bolts are made. If the standard deviation of the diameters for the population of bolts produced during this period is .2 mm and a random sample of $n = 50$ bolts is drawn from the population, how likely is it that

(a)   the sample mean diameter will exceed .3 mm?

(b)   the sample mean diameter will be within .01 mm of the true mean diameter?

(c)   the sample mean will differ from the true mean diameter by an amount exceeding .03 mm?

7.3   If X is a random variable with (unknown) mean μ and standard deviation $\sigma = 5$ in a large population,

(a)   given a sample of size $n = 30$, find the probability that the sample mean for this variable differs from the population mean by an amount less than .1 (that is, calculate $P(|\overline{X} - \mu| < .1)$)

(b)   Repeat the calculation of part (a) for $n = 50, 100, 500,$ and 1000. What happens to the probabilities as n increases?

Estimated Standard Errors and Student's t Statistic

From the viewpoint of statistical inference concerning an unknown mean μ, a disagreeable feature of the Z-score transform quantity $(\overline{X} - \mu)/(\sigma/\sqrt{n})$ is that it contains the parameter σ as well as μ. If the value of μ is unknown in a practical problem, it is likely that the value of σ will be unknown as well. Applications will require that the Z-score transform of $\overline{X}$ *not* contain any unknown parameters other than μ. To rid this expression of σ, a natural solution is to replace this parameter with a sample estimate. This is where the sample standard deviation, s, enters into statistical inference.

standard error   The standard error of $\overline{X}$ was defined to be its standard deviation $\sigma/\sqrt{n}$. An estimated version of this standard error, also (unfortunately) called the **standard error**, simply replaces the unknown parameter by its sample estimator:

$$SE_{\overline{X}} = \frac{s}{\sqrt{n}}$$

To avoid confusing the two, we will use the symbol SE to denote the estimated versions of standard errors while retaining the symbol σ for the unestimated kind.

Student's one-sample t statistic

The Z-score transform version of $\overline{X}$ with the estimated standard error replacing $\sigma/\sqrt{n}$ is an exceptionally important quantity in statistical inference known as **Student's one-sample t statistic:**

$$t = \frac{\overline{X} - \mu}{s/\sqrt{n}}$$

Our needs for the sampling distribution of $\overline{X}$ will be met by knowing the probability distribution of the t statistic. For large samples $(n \geq 30)$, its distribution is approximately standard normal, just as was the distribution of Z. This is not the case for small samples, however, as will be seen in chapter 10. The added difficulties with small samples will be postponed by considering only the large sample theory in this and the next two chapters.

EXAMPLE 7.4

A measurement variable X has unknown mean μ and standard deviation σ in a population. If a random sample of size $n = 45$ is taken from the population, what is the probability that $t = (\overline{X} - \mu)/(s/\sqrt{n})$ takes on a value between -1.32 and 1.84?

Since the sample size exceeds 30, the standard normal approximation to the distribution of t is valid. Consequently, the probability $P(-1.32 \leq t \leq 1.84)$ can be approximated by the standard normal probability $P(-1.32 \leq Z \leq 1.84)$. Using table 1A, we find the solution to be $.4066 + .4671 = .8737$.

An illustration of how the distribution of Student's t statistic is used in statistical inference is given in the next example.

EXAMPLE 7.5

Suppose that a random sample of size $n = 36$ is to be drawn from a large student population for which the mean age μ is of interest. For what value z is $P(|t| \leq z) = .95$, where t is Student's one-sample t statistic?

Since $n > 30$, the distribution of t can be approximated by the standard normal distribution. Consequently, we find $z = 1.960$ from table 1B.

Now, suppose that the data of display 1.1 (p. 9) represent the results of such a random sample. In chapter 2 we calculated the mean and standard deviation of the age variable to be $\overline{X} = 22.03$ years and $s = 7.648$ years. The value of the t statistic for this sample is then

$$t = \frac{\overline{X} - \mu}{s/\sqrt{n}}$$

$$= \frac{22.03 - \mu}{7.648/\sqrt{36}}$$

$$= \frac{22.03 - \mu}{1.275}$$

It follows that unless we were unlucky enough to have drawn one of the 5% of samples for which $|t| > 1.960$, we will have

$$|t| = \left| \frac{22.03 - \mu}{1.275} \right| \leq 1.960$$

or

$$|22.03 - \mu| \leq 1.960(1.275) = 2.50$$

With some algebra, this inequality becomes the two inequalities

$$22.03 - 2.50 \leq \mu \leq 22.03 + 2.50$$

or

$$19.53 \leq \mu \leq 24.53$$

We would infer that the mean age of the population is between 19.53 years and 24.53 years on the basis of this sample. This interval of values, which will be shown to be a confidence interval for μ in the next chapter, links the population mean with the sample; it represents the information about μ contained in the sample. The quality of this information is (in part) measured by how unlikely it is to draw a sample for which $|t| > 1.960$. Similarly, the number 1.960 enters into the computation of the interval bounds and thus plays a role in determining how closely the population mean is pinpointed by these bounds. The role played by the distribution of the t statistic is in forming the critical link between the 5% probability value and the number 1.960.

EXERCISE

7.4   A random sample of size $n = 80$ is to be taken from a large population, and the value of a measurement variable X with unknown population mean μ is to be obtained for each member of the population. If $\bar{X}$ denotes the sample mean of these measurements, s is their standard deviation, and $t = (\bar{X} - \mu)/(s/\sqrt{n})$, then

(a)   what is the (approximate) probability that $-1.5 < t < 1.7$?

(b)   for what value z is $P(|t| \leq z) = .99$?

◢ SECTION 7.4

Sampling Distribution of the Sample Proportion

In this section, the random variables $X_1, X_2, \ldots, X_n$ in a random sample of size n are Bernoulli variables. That is, they are categorical variables that have only two values, coded 0 and 1, with probability p of taking on the value 1. This model will apply, for example, to n tosses of a coin with *heads* probability p. Other important applications are to random samples of size n

from infinite populations for which the population proportion, p, of some characteristic (for example, "being female") is of interest.

The estimator for p is the sample proportion $\hat{p}$, which is simply the number of members of the sample that have the characteristic of interest divided by sample size. If Y denotes the number of sample members with this characteristic, then

$$\hat{p} = \frac{Y}{n}$$

The variable Y was seen to have the *binomial distribution* in chapter 5, and the sampling distribution of $\hat{p}$ can be based on this distribution for small sample sizes. However, we will restrict attention to large sample problems for categorical variables in this text and, because Y is a sum of random variables (chapter 5, p. 170), the Central Limit Theorem will provide a useful normal approximation to the distribution of Y, thus to the distribution of $\hat{p}$.

To use this approximation, we will need the mean and standard error of $\hat{p}$ derived in chapter 5:

$$\mu_{\hat{p}} = p$$

and

$$\sigma_{\hat{p}} = \sqrt{\frac{p(1-p)}{n}}$$

The first expression shows that $\hat{p}$ is an unbiased estimator of p. Note also the square root of n factor in the denominator of the standard error. This factor, as in the case of the sample mean, guarantees that the distribution of $\hat{p}$ becomes concentrated around p as sample size increases.

A Monte Carlo Justification of the Normal Approximation for the Sampling Distribution of $\hat{p}$

rule of thumb

The normal approximation to the distribution of $\hat{p}$ will be adequate for practical purposes when n is "large enough." An appropriate **rule of thumb** in this case is that *the normal approximation can be used for the distribution of $\hat{p}$ if the smaller of the numbers*

$$\frac{np}{1-p} \quad \text{and} \quad \frac{n(1-p)}{p}$$

is 9 or larger.

Thus, for example, if $p = .1$ then the smaller of

$$\frac{n(.1)}{(1-.1)} \quad \text{and} \quad \frac{n(1-.1)}{.1}$$

is

$$n(.1)/.9 = 0.111n$$

In order for the inequality $.111n \geq 9$ to be satisfied, we would have to take $n \geq 9/.111 = 81$.

As a second example, if $p = .8$, then the smaller of $n(.8)/(1 - .8)$ and $n(1 - .8)/.8$ is $n(.2)/.8 = .25n$. In order for the inequality $.25n \geq 9$ to hold, n would have to be at least $9/.25 = 36$.

This rule of thumb suggests that larger sample sizes are required for values of p near 0 and 1 than for p near .5. An indication of how well the rule works will be given in the following Monte Carlo demonstration.

A computer was programmed to "toss" a biased coin with *heads* probability $p = .1$. If the Bernoulli variable X has value 1 for *heads* and 0 for *tails*, a bar graph of the distribution of X would be as follows.

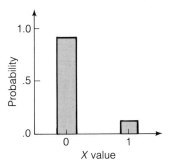

Note that this distribution, which represents the population frequency distribution for the sampling experiment, is badly skewed.

Forty samples each of sizes $n = 5, 10, 30, 50,$ and 100 were generated, and the sample proportion was calculated for each sample. The resulting proportions represent samples of size 40 from the sampling distributions of $\hat{p}$ for these five sample sizes and $p = .1$. If the normal approximation is to be valid, we would expect to see a tendency toward normality in these distributions with increasing sample size. Support for the above rule of thumb would be shown by good agreement between the sampling distribution and a normal distribution for $n \geq 81$.

Displays 7.4 and 7.5 give stem and leaf plots and a schematic diagram of these distributions. The unbiasedness of the estimator is supported by the fact that the distribution centers are quite close to .1. (For $n = 5$, the skewness of the distribution causes the mean and median to be different. The sample mean is actually close to .1.) Note also the decrease in dispersion with increasing n due to the square root of n factor in the standard error.

The symmetry and other features of a normal distribution become evident as early as $n = 30$. It appears that the above rule of thumb provides reasonable guidance for the normal approximation.

Display 7.4

Stem and leaf plots of sample proportions for samples of sizes $n = 5$, 10, 30, 50, and 100 for $p = .1$. (Note the scale expansions between $n = 10$ and $n = 30$ and between $n = 30$ and $n = 50$.) Units = 100 × value.

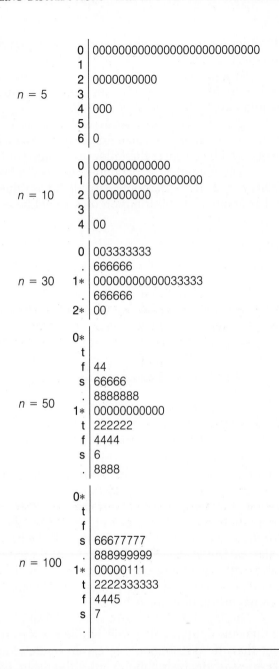

```
            0 | 0000000000000000000000000000
            1 |
            2 | 0000000000
   n = 5    3 |
            4 | 000
            5 |
            6 | 0

            0 | 000000000000
            1 | 00000000000000000
  n = 10    2 | 000000000
            3 |
            4 | 00

            0  | 003333333
            .  | 666666
  n = 30   1* | 00000000000033333
            .  | 666666
           2* | 00

           0* |
            t  |
            f  | 44
            s  | 66666
            .  | 8888888
  n = 50   1* | 00000000000
            t  | 222222
            f  | 4444
            s  | 6
            .  | 8888

           0* |
            t  |
            f  |
            s  | 66677777
            .  | 888999999
  n = 100  1* | 00000111
            t  | 2222333333
            f  | 4445
            s  | 7
            .  |
```

The Z-score Versions of the Sample Proportion

By applying the Z-score transform to $\hat{p}$, we obtain a statistic that has approximately a standard normal distribution for values of n satisfying the above rule of thumb. We will use the symbol V to denote the version of the Z-score transformation for which a value of p is assumed given:

$$V = \frac{\hat{p} - p}{\sqrt{p(1 - p)/n}}$$

Display 7.5

Schematic diagram of
sample proportions for
samples of sizes $n = 5$,
10, 30, 50, and 100
and $p = .1$

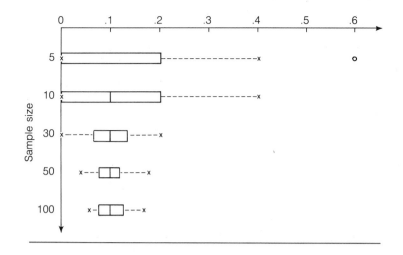

Note that if the value of p is unknown, then the standard error $\sigma_{\hat{p}} = \sqrt{p(1 - p)/n}$ will be unknown as well. In this case, its value will be estimated from the sample by replacing p in the expression for the standard error by $\hat{p}$:

$$SE_{\hat{p}} = \sqrt{\frac{\hat{p}(1 - \hat{p})}{n}}$$

We will also need the version of the Z-score transform of $\hat{p}$ for which the estimated standard error $SE_{\hat{p}}$ appears in the denominator. The symbol W will be reserved for this quantity:

$$W = \frac{\hat{p} - p}{SE_{\hat{p}}} = \frac{\hat{p} - p}{\sqrt{\hat{p}(1 - \hat{p})/n}}$$

The variable W also has approximately a standard normal distribution for the values of n satisfying the above rule of thumb.

EXAMPLE 7.6

A fair coin is tossed 50 times. How likely is it that the proportion of *heads* will deviate from one half by more than .15?

We are asked to calculate the probability that

$$|\hat{p} - .5| > .15 \tag{1}$$

when $n = 50$ and $p = .5$. The standard error of $\hat{p}$ is

$$\sigma_{\hat{p}} = \sqrt{\frac{.5(1 - .5)}{50}} = .0707$$

If both sides of the expression (1) are divided by the standard error, the left-hand side becomes $|V|$ and the right-hand side is $.15/.0707 = 2.121$. Thus, the problem reduces to calculating the probability

$$P(|V| > 2.121)$$

The rule of thumb for the normal approximation requires $n \geq 9$. Since $n = 50$, we can safely approximate this probability by $P(|Z| > 2.121)$, where

Z is a standard normal random variable. From table 1A we calculate

$$P(|Z| > 2.121) = 2(.5 - P(0 < Z \le 2.121))$$
$$= 2(.5 - .4830)$$
$$= .0340$$

This is the approximate probability of the event of interest; that is, the probability that the proportion of *heads* will deviate from $p = .5$ by more than .15 is .034.

EXERCISES

7.5   If the proportion of individuals in a population who have a given characteristic is $p = .35$ and a random sample of size $n = 80$ is selected, find the probability that the sample proportion $\hat{p}$ falls between .24 and .42.

7.6   A researcher suspects that 2% of a population carries a certain gene. How large a sample would be needed in order to use the normal approximation to the distribution of the sample gene proportion if the suspicion is correct?

7.7   A computer manufacturer claims that his latest computer will run without failures 95 hours out of every 100 that it is in operation, on the average. A potential buyer plans to run the computer for a number of hours to check this claim. He will estimate the true proportion of "up-time" hours by calculating their proportion for the time of his test.

(a)   How many hours should the test be run in order to use the normal approximation to the distribution of the sample proportion if the manufacturer's claim is correct?

(b)   If the computer is run the (smallest) number of hours found in part (a) and the manufacturer's claim is correct, what is the approximate probability that the sample proportion of "up-time" hours will be smaller than .93?

◢ SECTION 7.5

The Finite Population Correction Factor*

When is a population too small to use the infinite population model? What changes are required in the sampling distributions of $\overline{X}$ and $\hat{p}$ in this case? When sample sizes are large enough for the normal approximations to hold, the answers to these questions are rather simple.

Since normal distributions are determined by their means and standard deviations, changes in the sampling distributions can affect only the

---

* Optional section.

means and standard errors. The means and standard errors for $\bar{X}$ and $\hat{p}$ were derived in chapter 5 using the laws of expectation and variance for linear transformations of independent variables. When the population is not large relative to sample size, the sample variables are no longer independent, as argued in section 7.2. Because the law of expectation is not affected by dependence among the variables, the means of $\bar{X}$ and $\hat{p}$ will be the same for finite as for infinite populations. It is the standard errors that change, and they differ from the infinite population versions in a very simple and intuitively reasonable way.

Let N denote the size of the population and n the size of the sample. The finite population standard errors of $\bar{X}$ and $\hat{p}$ are obtained by multiplying the infinite population standard errors by the factor

$$\sqrt{(N - n)/(N - 1)}$$

finite population correction factor

which is called the **finite population correction factor**. That is,

$$\sigma_{\bar{X}} = \sqrt{\frac{N - n}{N - 1}} \frac{\sigma}{\sqrt{n}}$$

and

$$\sigma_{\hat{p}} = \sqrt{\frac{N - n}{N - 1}} \sqrt{\frac{p(1 - p)}{n}}$$

The estimated finite population standard errors are obtained from the infinite population versions, $SE_{\bar{X}}$ and $SE_{\hat{p}}$, in exactly the same way.

sampling fraction

An analysis of the correction factor shows that it is intimately related to the **sampling fraction**, n/N, which is the proportion of the population to be included in the sample. With a bit of algebra we find that

$$\sqrt{\frac{N - n}{N - 1}} = \sqrt{\frac{N}{N - 1}} \sqrt{1 - \frac{n}{N}}$$

For even modest population sizes, the ratio $N/(N - 1)$ is very close to 1, so the dominant term in the correction factor is the expression $\sqrt{1 - (n/N)}$, the square root of one minus the sampling fraction. This shows that the finite and infinite population standard errors will be different to an interesting degree only when the sampling fraction is appreciably different from 0. For example, even when the sample is half the size of the population ($n/N = .5$) the finite population standard error is still .71 times the infinite population version. For this reason, unless the sampling fraction is large, not much damage will be done to inference procedures in chapters 8 and 9 by ignoring the correction entirely.

On the other hand, we will see that the finite population correction factor plays an important role in the calculation of sample sizes for finite populations. Without it, calculations of sample sizes based on infinite population methods can lead to silly results such as samples larger than the populations from which they are to be drawn! The modifications required for finite populations will be given in optional sections at the ends of chapters 8 and 9.

EXAMPLE 7.7

Suppose the student population of example 7.5 (p. 214) actually consisted of only $N = 1000$ students. How would this fact affect the bounds for the population mean age computed in the example?

Since the sample size is $n = 36$, the value of the finite correction factor is

$$\sqrt{\frac{N-n}{N-1}} = \sqrt{\frac{1000-36}{1000-1}}$$

$$= .982$$

It follows that the finite population standard error of $\overline{X}$ is

$$SE_{\overline{X}} = .982 \left(\frac{7.648}{\sqrt{36}}\right)$$

$$= .982(1.275)$$

$$= 1.252$$

The effect of replacing the infinite population standard error by this quantity in the denominator of the t statistic and completing the calculations made in example 7.5 is to reduce the range of μ values from $19.53 \leq \mu \leq 24.53$ to $19.57 \leq \mu \leq 24.48$. The fact that the sampling fraction is small in this problem, only .036 or 3.6%, leads to a small difference in results.

EXAMPLE 7.8

A sample of 100 students is to be selected at random from a population of 500 students in order to estimate the proportion of freshmen. If the actual number of freshmen in the population is 200, how likely is it that the sample proportion will differ from the population proportion by more than 5%?

The true proportion of freshmen is $p = 200/500 = .4$. Thus, we are asked to calculate the probability of the event

$$|\hat{p} - .4| > .05 \tag{1}$$

The sampling fraction is $n/N = 100/500 = .2$. Consequently, the finite population standard error of $\hat{p}$ is

$$\sigma_{\hat{p}} = \sqrt{\frac{N-n}{N-1}} \sqrt{\frac{p(1-p)}{n}}$$

$$= \sqrt{\frac{400-100}{400-1}} \sqrt{\frac{.4(1-.4)}{100}}$$

$$= .0425$$

(Note that without the correction factor, the standard error would have been $\sqrt{.4(1-.4)/100} = .0490$.)

Divide both sides of expression 1 by the finite population standard error. The problem is then to calculate the probability

$$P\left(|V| > \frac{.05}{.0425}\right) = P(|V| > 1.176)$$

Based on the normal approximation, this probability is seen from table 1A to be approximately

$$2(.5 - .3810) = .238$$

Without the finite population correction, this probability is $P(|V| > .05/.0490) = P(|V| > 1.02) = .308$. The correction has made a noticeable difference in this example; the infinite population value of .308 overstates the more accurate value .238 calculated using the correction.

EXERCISES

7.8   Repeat the calculations of example 7.7 for a population of size $N = 100$.

7.9   If the population considered in exercise 7.5 (p. 220) contained only 150 members, what would be the probability that the sample proportion of individuals with the given characteristic is between .24 and .42?

7.10   A random sample of 100 television sets from a manufacturer's shipping lot of 500 sets is to be tested for defective channel selectors. If 50 of the sets in the lot actually have defective selectors, what is the probability that the sample proportion of TVs with defective selectors will be between 9% and 11%?

Chapter 7 Quiz

1.   What are the parameters and estimators dealt with in this and the next two chapters?

2.   What is the sampling distribution of an estimator?

3.   What probability distribution approximates the sampling distribution of the estimators of this chapter for large sample sizes?

4.   What is meant by the infinite population model for a random sample from a population?

5.   Why is the infinite population model a good approximation for the distribution of observations when sampling is done without replacement from a large population?

6.   What is meant by an unbiased estimator?

7.   What is the standard error of an estimator?

8.   What is the sampling distribution of the sample mean for a measurement variable X with mean μ and standard deviation σ when the sample size, n, is large?

9.   What is Student's one-sample t statistic? What is the approximate distribution of this statistic for large sample sizes?

10.   If p denotes the population proportion of some characteristic, then what is the definition of the sample proportion $\hat{p}$? What is the approximate sampling distribution of $\hat{p}$ for large sample sizes?

11. What is the rule of thumb for determining when a sample size is large enough to use the normal approximation for the distribution of $\hat{p}$?

12. What are the two versions of the Z-score transform of $\hat{p}$ that will be needed in statistical inference?

The following questions are for the optional section 7.5

13. What is the finite population correction factor?

14. What is meant by the sampling fraction?

15. Under what circumstances will the finite population correction factor be needed in an inference problem?

16. What is the relationship between the standard errors of the sample mean for finite and infinite populations?

17. What is the relationship between the standard errors of the sample proportion for finite and infinite populations?

SUPPLEMENTARY EXERCISES

7.11 If X is a random variable with mean μ and standard deviation $\sigma = 8$ in a large population, find the smallest sample size for which the probability $P(|\overline{X} - \mu| < .5)$

   (a) is at least .95

   (b) is at least .90

   (c) is at least .99

7.12 If μ is the unknown mean of a variable X in a large population, and a random sample of 100 individuals is taken from this population,

   (a) find the value z for which $P(|t| \le z) = .90$, where t is Student's one-sample t statistic.

   (b) If the sample mean and standard deviation are $\overline{X} = 18$ and $s = 3$, follow the steps of example 7.5 (p. 214) to find bounds for μ that will hold unless the selected sample is among the 10% of possible samples for which $|t| > z$ (found in part (a)).

7.13 Suppose that 52% of the people in a population are women, and a random sample of size $n = 75$ is drawn from the population. What is the probability that the proportion of women in the sample will be between 50.5% and 53.5%?

7.14 The units in a random sample of size 100 from a manufacturer's monthly production of television sets are to be tested for defective channel selectors. If 10% of the sets in the monthly production actually have defective selectors, what is the probability that the proportion of TVs in the sample with defective selectors will be between 9% and 11%?

Confidence Intervals—Large-Sample Theory

Introduction

In the introductions to chapters 3 and 4 we stated that the goals of statistical inference were to answer certain questions about population frequency distributions: Is there a difference in distributions? What is the nature of the difference? How large is the difference? These questions will be phrased in terms of population parameters, and answers to the third question will be based on *confidence intervals* for these parameters.

In this chapter we will study confidence intervals for the population mean and population proportion when sample sizes are "large." Suppose, for example, that a medical researcher has concluded that a new arthritis medication is more effective than one previously used by showing that a random sample supports a greater mean joint mobility for individuals taking the new medication than for individuals on the old medication. This information is likely to be only part of what is wanted in the study. It may also be desirable to "pinpoint" the mean mobility in order to decide, for example, whether the improvement is sufficiently great to establish the new medication as medically or economically important.

A confidence interval is a device for pinpointing the *true value* of a parameter within a range of values determined from a random sample. Statistical inference utilizes *statistical models*—families of population frequency distributions, where each distribution is associated with a value of a parameter. Examples of statistical models we have seen are the Bernoulli distribution of chapter 5, for which the parameter is the population proportion, p, and the normal distribution of chapter 6, for which the parameter of interest is the population mean, μ. By the **true value** of a parameter, we mean that value for which the frequency distribution of the statistical model in use best approximates the actual or true distribution of the observations.

true value

A traditional notational convention is to use the same symbol to represent both a parameter (the "index" that can assume any of the possible parameter values) and the true value of the parameter (a fixed but unknown number). Thus, μ would be used to denote any of the possible values of the mean joint mobility in the arthritis study; at the same time, the confidence interval symbolism $L \leq \mu \leq U$ means that the true value of μ lies between the limits L and U. Confidence intervals for the population mean and population proportion will be given in section 8.2.

An experiment in which statistical inference is applied should be planned in a preliminary design stage. An important aspect of *experimental design* is the selection of a sample size. Sample size selection depends on the concept of evaluating statistical procedures—an idea of fundamental importance in statistical inference. A method of sample size selection based on confidence intervals will be discussed in section 8.3.

Finite population adjustments in the calculation of both confidence intervals and sample sizes will be given in the optional section 8.4.

◢ SECTION 8.2

Confidence Intervals for the Population Mean and Population Proportion

A confidence interval is a range of values of a population parameter that will be assumed to contain the true parameter value. The lower bound, L, and upper bound, U, of the interval are computed from a random sample from the population. Consequently, before sampling, L and U are random quantities. For useful choices of these bounds, there is always a positive probability that the confidence interval will "miss" the true parameter value. An important characteristic of confidence intervals is that this probability is controlled by the experimenter in the following way.

confidence coefficient   A probabilistic "figure of merit," called a **confidence coefficient**, is chosen in advance of the experiment. This coefficient specifies the probability that the confidence interval will cover or contain the true parameter value, so we want it to be as close to 1 as possible. Values typically chosen for the confidence coefficient are .90, .95, and .99. If, for example, the quantity .95 is selected, the interval would be called a 95% *confidence interval* for the parameter.

More formally, *the quantities L and U are said to determine a 95% confidence interval for a parameter if*

1. $P(L \leq \text{true parameter value} \leq U) \geq .95$ *and*
2. the values of L and U are completely determined (that is, have numerical values) once the sample has been taken and used in the computation of these limits.

Confidence intervals with other confidence coefficients are defined by replacing .95 with the desired coefficient in property (1).

It follows that once the experiment has been performed and the confidence limits have been computed, nothing probabilistic remains. For example, a 95% confidence interval for the mean mobility in the arthritis medication example might be from $L = 45$ to $U = 60$ degrees. The true value of the parameter is also a number; it is simply unknown. Consequently, it makes no sense to claim that $45 \leq \mu \leq 60$ with probability .95, since nothing is random. What we will claim instead is that the inequalities $45 \leq \mu \leq 60$ hold with 95% *confidence*. Hence the term "confidence interval."

Because of the relative frequency interpretation of probability (chapter 5), the qualification "with 95% confidence" can be viewed as follows. If the experiment of taking a random sample (of given size) from the population and calculating the values of L and U is repeated a large number of times, the interval will contain the true parameter value ($L \leq \mu \leq U$) for about 95% of the samples. Consequently, the conclusion that the interval contains the true parameter value will be correct unless we have been so unfortunate as to have gotten one of the 5% of "bad" samples; we will be "95% confident" of the conclusion.

This frequency interpretation of confidence intervals is demonstrated in display 8.1. One hundred random samples of size $n = 50$ were drawn (by a computer) from the student population represented by display 1.18. To simulate sampling from a large population, the samples were drawn with replacement. For each sample, a 95% confidence interval was computed for the mean of the weight distribution by the method of the next subsection. The confidence limits are shown in the display. Those confidence intervals that do not contain the true population mean weight ($\mu = 140.86$ pounds) are marked with asterisks. Since these are 95% confidence intervals, only 5 out of every 100 intervals should miss the true mean, on the average. For the 100 intervals shown in the display, there were 4 misses.

In chapter 7 it was claimed that the sampling distributions of estimators provide the vital ingredient in statistical inference. In the next subsection, the distribution of Student's t statistic will be used to obtain a confidence interval for the population mean.

Confidence Interval for the Population Mean

We begin with an example.

EXAMPLE 8.1

The steps for finding a confidence interval for a population mean were shown in example 7.5 (p. 214). These steps will be repeated here. The students of display 1.1 (p. 9) are assumed to represent a random sample of size $n = 36$ from a large student population. A 95% confidence interval will be constructed for the mean population age, μ.

The normal approximation for the distribution of Student's t statistic, $t = (\overline{X} - \mu)/(s/\sqrt{n})$, was used to find the number z such that

$$P(|t| \leq z) = .95$$

The solution $z = 1.960$ was obtained from table 1B.

Display 8.1

95% upper and lower confidence limits for the mean weight of the student population represented by display 1.18. Asterisks mark the confidence intervals that did not contain the true mean, $\mu = 140.86$ pounds.

SAMPLE NUMBER	L	U	SAMPLE NUMBER	L	U
1	127.0	144.2	51	128.7	144.5
2	136.8	152.1	52	135.5	150.6
3	133.1	147.0	53	131.2	148.3
4	129.0	142.5	54	128.8	141.9
5	134.1	152.0	55	130.6	143.2
6	134.5	148.0	56	134.0	147.6
7	128.9	143.8	57	131.2	145.9
8	133.3	145.9	58	136.4	150.7
9	132.4	147.9	59	130.5	145.4
10	130.5	146.4	60	130.9	145.7
11	126.0	141.1	61	136.7	152.9
12	130.4	147.6	62	134.7	149.6
13	139.9	156.5	63	135.3	149.9
14	133.1	147.9	64	135.8	152.1
15	140.0	155.8	65	139.6	155.7
16	131.0	146.4	66	128.9	145.8
17	136.3	152.6	67	129.2	144.1
18	135.1	150.7	68	128.7	145.1
19	133.8	148.9	69	130.7	145.0
20	137.7	153.8	70	130.9	145.2
21	128.5	141.1	71	140.2	157.2
22	130.4	145.2	72	131.1	145.9
23	133.6	150.4	73	135.7	154.3
24	130.6	145.8	74	141.7	155.8*
25	131.1	147.3	75	132.4	146.9
26	135.0	152.3	76	139.8	153.9
27	138.0	154.9	77	132.6	150.7
28	129.6	143.8	78	133.5	148.4
29	131.9	147.0	79	134.9	150.5
30	133.4	149.9	80	129.2	145.5
31	131.7	147.7	81	131.6	147.2
32	133.3	149.4	82	127.0	142.5
33	130.6	146.5	83	136.2	154.1
34	130.3	144.2	84	131.1	147.1
35	131.3	145.8	85	132.7	146.9
36	136.1	149.1	86	125.9	138.2*
37	137.7	155.1	87	132.6	145.5
38	139.2	154.7	88	132.6	149.1
39	129.1	143.7	89	130.3	144.4
40	133.0	148.6	90	135.1	151.4
41	134.9	151.2	91	140.4	154.5
42	132.2	147.6	92	125.0	138.8*
43	130.8	144.3	93	134.0	148.0
44	126.5	140.9	94	132.0	148.2
45	133.2	149.1	95	132.8	147.7
46	136.7	152.3	96	134.8	147.3
47	126.1	139.3*	97	131.9	147.0
48	132.8	148.3	98	131.4	147.4
49	128.5	143.3	99	136.4	153.1
50	126.0	142.2	100	138.3	154.0

Now, by the definition of Student's t statistic, the inequality $|t| \leq 1.960$ is the same as the inequality

$$\left| \frac{\overline{X} - \mu}{s/\sqrt{n}} \right| \leq 1.960$$

With some algebra, this inequality becomes the two inequalities

$$\overline{X} - 1.960 \frac{s}{\sqrt{n}} \leq \mu \leq \overline{X} + 1.960 \frac{s}{\sqrt{n}}$$

Because we have not changed the event by doing this algebra, these inequalities also hold with probability .95. Note that everything in the two outside expressions is either given or can be calculated from the sample: n and z are given numbers, and values of the statistics $\overline{X}$ and s can be computed by the methods of chapter 2. It follows from the definition of a confidence interval that the quantities

$$L = \overline{X} - 1.960 \frac{s}{\sqrt{n}} \quad \text{and} \quad U = \overline{X} + 1.960 \frac{s}{\sqrt{n}}$$

qualify as the limits of a 95% confidence interval for the population mean, μ.

To complete the calculation for the age data, the values $n = 36$, $\overline{X} = 22.03$ years, and $s = 7.648$ years are entered in the expressions for the confidence limits, yielding

$$L = 22.03 - 1.960 \times \frac{7.648}{\sqrt{36}} = 19.53 \text{ years}$$

and

$$U = 22.03 + 1.960 \times \frac{7.648}{\sqrt{36}} = 24.53 \text{ years}$$

We would conclude *with 95% confidence* that the mean age for the population of statistics students from which this sample was drawn is between 19.53 and 24.53 years.

Confidence Limits for the Mean

For any given confidence coefficient, the expressions

$$L = \overline{X} - z \frac{s}{\sqrt{n}}$$

and

$$U = \overline{X} + z \frac{s}{\sqrt{n}}$$

critical value

will be the limits of a (large sample) confidence interval for μ if the number z, called the **critical value**, is adjusted to satisfy the equation

$$P(|t| \leq z) = \text{confidence coefficient}$$

Table 1B is in the appropriate form for obtaining critical values quickly and easily for a variety of confidence coefficients. However, for convenience, a table of critical values for the more commonly used confidence coefficients is given in display 8.2.

Display 8.2

Some commonly occurring critical values for large-sample confidence intervals

CONFIDENCE COEFFICIENT (%)	TWO-SIDED INTERVAL CRITICAL VALUE z	ONE-SIDED INTERVAL CRITICAL VALUE z
80	1.282	.842
90	1.645	1.282
95	1.960	1.645
99	2.576	2.326

Interpretation of the Length of a Confidence Interval

The length of a confidence interval,

$$l = U - L$$

is a measure of the *precision* with which the interval pinpoints the parameter value. Since all candidates for the true parameter value lie in the interval, no two can differ by more than the interval length. For example, since

$$l = 24.53 - 19.53 = 5 \text{ years}$$

for the age confidence interval computed earlier, there was sufficient information in the sample to pinpoint the true mean age to within 5 years.

A quantity that is often used as a measure of precision is the *half-length* of the confidence interval,

$$h = l/2$$

This quantity is the distance from either confidence limit to the parameter estimate at the center of the interval. We will use the half-length for sample size design in section 8.3.

EXERCISES

8.1   Use the data from the calculation in example 8.1 to

(a)   calculate a 90% confidence interval for mean age, μ

(b)   calculate a 99% confidence interval for μ

8.2   In a random sample of size 55 from a population, the sample mean and standard deviation of a measurement variable are $\bar{X} = 24.66$ and $s = 1.454$, respectively.

(a)   Find a 95% confidence interval for the population mean, μ, of this variable.

(b)   Compute the length of this interval.

(c)   If the same sample mean and standard deviation given above were obtained for a sample of size 150 from the population, find a 95% confidence interval for μ. Compute the length of the interval.

(d)   How does the interval length found in part (c) compare to that of part (b)? What appears to happen to the length of the interval as sample size increases?

8.3   The mean and standard deviation of students' heights given in display 1.1 are $\overline{X} = 66.11$ inches and $s = 4.076$ inches. Assuming these students to be a random sample of size 36 from a large student population, find a 95% confidence interval for the population mean height.

8.4   In a random sample of 60 dealers from the national dealer registry for a particular brand of automobile, the average price of the economy model was found to be $7830, with a standard deviation of $320. Find a 95% confidence interval for the national economy model mean price.

8.5   The data given in display 1.10 (p. 20) represent the capacitances of a random sample of size 40 from a large lot of mylar capacitors delivered to a government defense laboratory for use in missile guidance circuits. The sample mean and standard deviation are $\overline{X} = .63767$ and $s = .01081$ microfarads (μf). Find a 99% confidence interval for the mean capacitance of the lot.

8.6   A survey to determine the mean yardage per play gained by college football running backs for 1982 was made by randomly selecting the records of 45 players from a national sports listing. (The yardage per play is the total yardage gained by the player for the year divided by the number of his attempted runs.) The average yardage per play for the sample was 3.7 yards, with a standard deviation of 1.6 yards. Find a 90% confidence interval for the mean yardage per play for all college running backs.

8.7   Use the algebraic expressions for the confidence limits for a population mean (that is, the expressions involving z, n, $\overline{X}$, and s) and the definition of interval length, $l = U - L$, to show that the length of the confidence interval is

$$l = 2z\,\frac{s}{\sqrt{n}}$$

(a)   Use this expression to draw the conclusions of exercise 8.2 directly.

(b)   How does the length of the confidence interval depend on the sample variation as measured by s? (That is, if s is larger for one sample than another, all other quantities remaining the same, what would be the relationship between the length of confidence intervals for μ computed from the two samples?)

(c)   How does interval length depend on the confidence coefficient? (See display 8.2.)

The Pivotal Method for Obtaining Confidence Intervals

Note the following features of the derivation of the confidence interval for the population mean. An estimator EST, in this case the sample mean $\overline{X}$, was given for a parameter PARM, here the population mean μ. The estimated version of the standard error, SE, was used in the Z-score transform to form the ratio (EST $-$ PARM)/SE. In the case of the mean, this ratio is simply the one-sample Student t statistic. The normal approximation to the sampling distribution was used to furnish the critical value z, for which the probability of the inequality

$$\left| \frac{\text{EST} - \text{PARM}}{\text{SE}} \right| \leq z$$

is equal to the confidence coefficient specified for the problem. Algebra was then applied to this inequality to produce the pair of inequalities

$$\text{EST} - z\text{SE} \leq \text{PARM} \leq \text{EST} + z\text{SE}$$

Because the limits in these inequalities satisfied the required properties, we concluded that

$$L = \text{EST} - z\text{SE} \quad \text{and} \quad U = \text{EST} + z\text{SE}$$

are the limits of a confidence interval for PARM with the prescribed confidence coefficient.

pivotal method

This is an outline of a general method for constructing confidence intervals called the **pivotal method**. The pivotal method will be used to construct *all* of the confidence intervals we will use in this text, so its importance as a unifying idea is considerable. It will be used in the next subsection to obtain confidence intervals for a population proportion.

Confidence Intervals for a Population Proportion

Confidence intervals for a proportion, p, follow immediately from an application of the pivotal method to the statistic

$$W = \frac{\hat{p} - p}{\text{SE}_{\hat{p}}} = \frac{\hat{p} - p}{\sqrt{\hat{p}(1 - \hat{p})/n}}$$

introduced in chapter 7. The distribution of W was seen to be approximately standard normal. (A rough judgment of the adequacy of the normal approximation can be based on a version of the rule given in chapter 7: The smaller of the numbers $n\hat{p}/(1 - \hat{p})$ and $n(1 - \hat{p})/\hat{p}$ should be at least 9.) Thus, if z is the standard normal critical value corresponding to the prescribed confidence coefficient, the limits of a confidence interval for p are

$$L = \hat{p} - z \sqrt{\frac{\hat{p}(1 - \hat{p})}{n}}$$

and

$$U = \hat{p} + z \sqrt{\frac{\hat{p}(1 - \hat{p})}{n}}$$

EXAMPLE 8.2

We will view the $n = 5800$ coin tosses pictured in display 5.1 (p. 123) as tosses of a coin with unknown *heads* probability, p. The experimental evidence suggests that p is near the fair-coin probability of .5. We will confirm this observation by computing a 95% confidence interval for p.

The number of heads seen in the 5800 tosses was 2917. Thus, the sample proportion of *heads* is

$$\hat{p} = \frac{2917}{5800}$$

$$= .5029$$

For confidence coefficient 95%, the critical value is $z = 1.960$. It follows that the confidence limits are

$$L = \hat{p} - z\sqrt{\frac{\hat{p}(1-\hat{p})}{n}} = .5029 - 1.960 \times \sqrt{\frac{.5029(1-.5029)}{5800}}$$

$$= .5029 - .0129$$

$$= .4900$$

and

$$U = \hat{p} + z\sqrt{\frac{\hat{p}(1-\hat{p})}{n}} = .5029 + 1.960 \times \sqrt{\frac{.5029(1-.5029)}{5800}}$$

$$= .5029 + .0129$$

$$= .5158$$

Rounding to three decimal places, we would conclude with 95% confidence that the true *heads* probability for this coin is between .490 and .516.

Since the fair coin probability, .5, falls in this interval, the data support the contention that the coin is fair.

The length of the interval is

$$l = U - L = .516 - .490$$

$$= .026$$

The shortness of the interval indicates that the true *heads* probability is being pinpointed quite accurately. Since large samples should produce informative (that is, short) confidence intervals, this is, perhaps, to be expected in this example.

EXAMPLE 8.3

A random sample of $n = 150$ students from a large university contains 55 women. Find a 90% confidence interval for the proportion of women students at the university.

The critical value for 90% confidence is $z = 1.645$. Since the sample proportion of women is

$$\hat{p} = \frac{55}{150} = .367,$$

the confidence limits are

$$L = \hat{p} - z\sqrt{\frac{\hat{p}(1-\hat{p})}{n}} = .367 - 1.645\sqrt{\frac{.367(1-.367)}{150}}$$

$$= .302$$

and

$$U = \hat{p} + z\sqrt{\frac{\hat{p}(1-\hat{p})}{n}} = .367 + 1.645\sqrt{\frac{.367(1-.367)}{150}}$$

$$= .432$$

We would conclude with 90% confidence that the proportion of women students at the university is between 30.2% and 43.1%.

EXERCISES

8.8   In a random sample of size $n = 300$, if the number of individuals with property A is found to be 120, find a 90% confidence interval for the proportion of the population possessing property A.

8.9   In a random sample of size $n = 200$ from a large city, the number of people with at least one year of college was 44. Find a 95% confidence interval for the proportion of individuals in the city with at least one year of college.

8.10   In a study to determine whether alcoholism has (in part) a genetic basis, genetic markers were observed for a group of 50 Caucasian alcoholics. For 5 alcoholics the antigen (marker) B15 was present. Estimate with a 95% confidence interval the proportion of all Caucasian alcoholics having this antigen.

8.11   In a sample of 174 University of New Mexico students, 35 claimed to spend at least 26 hours per week studying for all of their classes. Find a 90% confidence interval for the proportion of all UNM students who study at least 26 hours per week.

8.12   A newspaper sports reporter found that by the end of the football season he had successfully predicted the winners of 117 games and had guessed wrong on 79. Find a 95% confidence interval for his correct-choice probability.

8.13   An Associated Press release during a recent election reported the results of a political poll (in part) as follows. (Names have been omitted.)

"The poll, commissioned by XXX-TV, asked registered voters in three congressional districts who they would vote for if the election were held today.

"In the First Congressional District, 144 voters were polled, and between 60 percent and 76 percent said they would vote for Republican candidate A; between 13 and 29 percent said they favored Democrat candidate B; as many as 9 percent said they would vote for Libertarian candidate C, and between 2 percent and 18 percent said they were undecided."

(a)   What, if anything, is wrong with this news report?

(b)   If you found fault with the report in part (a), rewrite it so that it makes sense.

More Accurate Confidence Limits When p Is Near 0 or 1*

When the true value of p is near 0 or 1, the confidence interval given in the last subsection is somewhat inaccurate; the true probability with which the given limits contain p is smaller than the quoted confidence coefficient. An interval with this property is said to be *invalid.* (See section 8.3 for a discussion of validity.) More accurate limits in this case, given by Blyth and Still [39], are as follows. The upper limit is

$$U = \frac{(x + .5) + z^2/2 + z\sqrt{(x + .5) - (x + .5)^2/n + z^2/4}}{n + z^2}$$

where x is the number of "successes" (for example, the number of *heads* in example 8.2) and n is the number of trials. Again, z is the critical value from the normal tables for the given confidence coefficient.

To find the lower limit L, replace x by $x - 1$ and z by $-z$ in the expression for U:

$$L = \frac{(x - .5) + z^2/2 - z\sqrt{(x - .5) - (x - .5)^2/n + z^2/4}}{n + z^2}$$

For the data of example 8.2, these limits are $L = .4900$ and $U = .5159$, whereas the limits computed in the example were $L = .4900$ and $U = .5158$. Thus, when n is large and p is near .5, these and the limits computed in the last subsection are almost identical. That this is not always the case will be illustrated in exercise 8.14.

In what follows, we will use the more easily calculated limits given in the last subsection. However, when more accuracy is required, the limits given here should be used when the parameter p is near 0 or 1. The survey of confidence intervals for p by Blyth and Still [39] contains further comparisons and details for the confidence intervals of this and the last section, among others.

EXERCISE

8.14   A random sample of 2000 individuals is taken in order to estimate the incidence (proportion) of the disease AIDS in a large city. The disease was detected in 17 of the sample members. Find a 95% confidence interval for the proportion of people in the city with AIDS,

(a)   using the standard confidence limits of the last subsection

(b)   using the more accurate limits of this subsection.

(c)   Describe the difference in the limits found in (a) and (b).

---

* Optional section.

One-Sided Confidence Intervals

The topic of one-sided confidence intervals is linked with one-sided hypothesis tests, which will be taken up in the next chapter. For problems in which one-sided tests answer the question "Is there a difference (in distributions)?" in the affirmative, the natural confidence interval for addressing the question "How large is the difference?" is often a one-sided interval.

The study of one-sided intervals can also be justified on its own merits. In many problems, only one limit of a confidence interval is of interest. For example, a manufacturer interested in ensuring the quality of his product would be interested in the maximum proportion of defective items in an outgoing shipment. This information would be provided by an upper confidence limit for this proportion.

An educator concerned about poor student performance on an achievement test might want to estimate how low the mean score is for all students in a given year. He could do this by finding a lower confidence bound for the mean score.

The procedure for calculating the limit of a one-sided confidence interval is quite straightforward. For a given confidence coefficient, take the appropriate limit for the corresponding two-sided interval and replace the two-sided critical value, z, with the one-sided value. This critical value is found as follows. Locate the row in the ONE-SIDED TESTS column of table 1B in which 1 minus the confidence coefficient appears. The critical value is then read from this row in the last column of the table.

Thus, for example, to obtain a 95% upper one-sided confidence limit for p, the critical value $z = 1.645$ is found in the last column opposite the number $1 - .95 = .05$ in the ONE-SIDED TESTS column. Now, use this number in the expression for the upper limit of the two-sided confidence interval for p. The resulting limit is

$$U = \hat{p} + 1.645 \sqrt{\frac{\hat{p}(1 - \hat{p})}{n}}$$

We would conclude with 95% confidence that the true population proportion p is no larger than this limit.

Similarly, a 95% lower one-sided confidence limit for μ would use the one-sided critical value in the lower limit of the two-sided confidence interval:

$$L = \overline{X} - 1.645 \frac{s}{\sqrt{n}}$$

We would conclude with 95% confidence that the true value of μ is no smaller than this quantity.

Some of the more commonly used one-sided critical values are shown in display 8.2.

Intuitively, by giving up one of the limits in a two-sided confidence interval, we should be able to coax more information from the sample about the other. In fact, the one-sided limit is tighter than the corresponding two-sided limit, where "tightness" is measured by the half-interval length h. This quantity can be written in the form $h = z\mathrm{SE}$ for both the mean and proportion, where SE is the estimated standard error. For 95% confidence, the half-length is $1.645 \cdot \mathrm{SE}$ for a one-sided interval, which is smaller than the value $1.960 \cdot \mathrm{SE}$ for the two-sided interval. This comparison will hold for any confidence coefficient, so the one-sided bounds will always be tighter than the corresponding two-sided bounds.

EXAMPLE 8.4

A computer program simulates accidents for a given nuclear reactor under a particular set of operating conditions. The program is run 500 times and an accident is experienced in 7 of the runs. Find an upper 99% confidence limit on the probability of an accident under these operating conditions.

The sample proportion of accidents is

$$\hat{p} = \frac{7}{500} = .0140$$

From table 1B, or display 8.2, the one-sided critical value for 99% confidence is seen to be $z = 2.326$. Thus, the one-sided confidence limit is

$$\hat{p} + 2.326 \sqrt{\frac{\hat{p}(1 - \hat{p})}{n}} = .0140 + 2.326 \sqrt{\frac{.0140(1 - .0140)}{500}}$$

$$= .0262$$

We would conclude with 99% confidence that the probability of a nuclear accident under the given operating conditions is no greater than .0262.

EXAMPLE 8.5

Because of a concern that the graduating students at a large university were not meeting acceptable performance standards, 100 seniors were chosen at random to take an achievement test. The average score on the test was 67, with a standard deviation of 8.3. Find a 90% lower confidence limit on the mean score for the entire senior class.

The one-sided critical value for 90% confidence is 1.282. The confidence limit is then

$$\overline{X} - 1.282 \times \frac{s}{\sqrt{n}} = 67 - 1.282 \times \frac{8.3}{\sqrt{100}}$$

$$= 65.94$$

We would conclude with 90% confidence that the mean achievement score for the senior class is at least 65.94.

EXERCISES

8.15   Suppose that it is of interest to determine whether the mean capacitance of the lot of mylar capacitors described in exercise 8.5 (p. 231) is large enough to satisfy a certain government specification. Use the fact that the sample size is $n = 40$ and the mean and standard deviation are $\overline{X} = .63767$ and $s = .01081$ μf to find a lower 99% confidence bound for the mean capacitance of the lot. Do the data support a mean lot capacitance above the rated value of $.6\mu f$? Explain.

8.16   In the survey to determine UNM student study habits, considered in exercise 8.11 (p. 234), 35 students in a sample of 174 claimed to study at least 26 hours a week. The dean of the college of arts and sciences is interested in determining how small a proportion of students in the entire university study at least 26 hours per week. Assuming the survey data to constitute a random sample from the student population, find a lower 95% confidence limit for this proportion.

8.17   Find a 95% lower confidence limit for the correct-choice probability of the sports reporter of exercise 8.12 (p. 234). Recall that he predicted the outcomes of 117 games correctly and was wrong for 79 games. Is it reasonable to believe that his performance is better than simply predicting the outcomes of games with a (fair) coin toss? Explain.

◪ SECTION 8.3

Experimental Design and the Choice of Sample Size

A strategy for a statistical investigation consists of four stages:

1. A *design stage*, in which the experimental units are chosen and assigned "treatments," the statistical procedures are (tentatively) decided upon, the sample size is selected, and so forth.

2. An *experimental stage*, in which the experiment is carried out and data are acquired.

3. An *analysis stage*, in which an exploratory analysis of the data is made and the inferential analyses are carried out.

4. An *evaluation and report writing stage*, in which the inferences and exploratory analyses are interpreted and evaluated, conclusions are drawn, and reports of the experimental results are prepared for publication.

We will concentrate primarily on stages 3 and 4 in this text. However, statistics also plays an important role in the design of experiments, stage 1. A brief introduction to this topic will be given in chapter 17. One aspect of experimental design, the selection of sample size, is tied to properties of confidence intervals and hypothesis tests. Some details will be given here for confidence intervals and in chapter 9 for hypothesis tests.

The idea behind sample size design is that by specifying goals for a confidence interval or hypothesis test, it is possible to compute the size of the sample required to achieve these goals. The goals will be stated in terms of two criteria for judging or evaluating statistical procedures: *validity* and *sensitivity*. We consider these criteria for confidence intervals in the next subsection.

The Criteria of Validity and Sensitivity for Confidence Intervals

The term *validity* has the sense of reliability or dependability—of something doing what it is supposed to do. A confidence interval is supposed to contain the true value of the parameter of interest. Validity for confidence intervals is defined in terms of this idea.

The specification of a confidence coefficient, and the requirement that the confidence interval cover the true parameter value with probability equal to the confidence coefficient, is called the *validity requirement*. This requirement is an attempt to *design* validity into the inference procedure. However, what is planned by design may not hold in practice, due to unknown features of the underlying distribution of the observations. Thus, we will say that a confidence interval is **valid** if the *actual* probability with which it covers the true parameter value is equal to, or exceeds, the prespecified confidence coefficient. An interval for which this is not true is said to be *invalid*.

valid

In the confidence interval context, the term *sensitivity* has the sense of precision or accuracy. Since confidence intervals are used to pinpoint the true value of a parameter, a reasonable measure of sensitivity would be the length of the interval—or the half length, h. Although this is not the original sensitivity measure used for confidence intervals by their inventor, J. Neyman, it is intuitively attractive and is easily applied to the sample size design problem.

Sample Size Design for Confidence Intervals for a Population Mean

The expression for the half-length of a confidence interval for μ, $h = zs/\sqrt{n}$, can be solved for n to give

$$n = \frac{z^2 s^2}{h^2}$$

This is the expression we will use to compute the sample size for a confidence interval for μ. The specification of a confidence coefficient (validity requirement) will determine the critical value, z, and the sensitivity of the interval will be prescribed by giving a value for h. Thus, the only unknown in this equation is the standard deviation s.

Since the calculation of sample size is made at the design stage, before the sample is taken, a value of s to use in the equation cannot be obtained from the sample; it is necessary to obtain a preliminary value of s from another source.

Occasionally, a similar experiment will have been run, and the sample standard deviation from this experiment can be used as the preliminary value of s. This is not always a reliable source, however, since it is often difficult to verify the comparability of the conditions under which the experiment was run with those of the experiment under design. A popular alternative method is to run a "small" preliminary experiment, called a *pilot study*. The standard deviation estimated from this study is then used as the preliminary value of s to determine the sample size for the full experiment.

Once a value of s is available, the sample size can be computed as shown in the next example.

EXAMPLE 8.6

The mean height of statistics students at the University of New Mexico is to be estimated with a confidence interval. A high level of confidence is desired, so a confidence coefficient of 99% is chosen. The mean is to be estimated to within an interval length of 1 inch.

The 99% critical value for a two-sided confidence interval is $z = 2.576$. The half-length of the interval is to be $h = .5$ inches.

We will use the data of display 1.1 (p. 9) to provide a value of the standard deviation. The standard deviation of heights for these data was computed to be $s = 4.076$ inches.

Substituting these quantities into the above expression for sample size yields

$$n = z^2 s^2/h^2 = 2.576^2 4.076^2/.5^2$$
$$= 440.98$$

To obtain an integer sample size, this value is rounded to the next larger whole number. Thus, we would take the sample size to be $n = 441$.

With this sample size, the experiment will produce a 99% confidence interval of *approximately* 1 inch. The reason for the qualification is that the standard deviation computed from the experimental data, not the preliminary value of s, will be used to calculate the confidence limits. This strategy will ensure the proper confidence level (validity). However, since the computed standard deviation is unlikely to be exactly equal to the preliminary value used for experimental design, the actual confidence interval length will be only approximately equal to the prescribed length.

If, in the last example, budgetary limitations or other constraints made a sample size of 441 prohibitively large, a compromise in either the confidence level or interval length would be necessary in order to reduce it. An advantage of experimental design is that these issues can be resolved before the experiment is performed. In some cases, a satisfactory compromise may not be possible within the limitations of available resources. The wise decision will then be to abandon the project until adequate resources are available rather than carry out an experiment that is unlikely to achieve the desired goals.

If it is acceptable to reduce the confidence level to 95% in the last example, then $z = 1.960$, and the experimental goals can be met with a sample size of

$$n = \frac{1.960^2 4.076^2}{.5^2} = 255.30$$

or $n = 256$.

If, in addition, the length restriction can be relaxed from 1 inch to 1.5 inches, say, then the sample size would be

$$n = \frac{1.960^2 4.076^2}{(1.5/2)^2} = 113.5$$

or $n = 114$.

Note that modest changes in requirements can lead to substantial reductions in sample size.

EXERCISES

8.18   Find the sample size required to construct a 95% confidence interval of length 1.2 units for a population mean if the preliminary value of the standard deviation is $s = 12.34$.

8.19   (a)   Use the sample of .6 μf mylar capacitors in exercise 8.5, page 231, ($s = .01081$ μf) as a pilot study to design a 99% confidence interval of length .001 μf for the mean capacitance of a future lot.

   (b)   Recompute the sample size of part (a) for a confidence coefficient of 95%.

   (c)   Recompute the sample size of part (b) if the interval length is to be .002 μf.

8.20   Use the sample standard deviation $s = \$320$ from exercise 8.4 (p. 231) as a preliminary value of s to find the sample size required to estimate the mean price of economy cars to within an interval length of \$40, with a 95% confidence interval.

8.21   How many players should be sampled from the roster described in exercise 8.6 (p. 231) in order to estimate the mean number of yards gained with a 90% confidence interval to within .60 yards? Use the preliminary value $s = 1.6$ in the computation.

Sample Size Design for a Population Proportion

The equation for computing sample sizes of confidence intervals for a population proportion is

$$n = \frac{z^2 p(1 - p)}{h^2}$$

The number z is the critical value for the given confidence coefficient, h is the prescribed half-length of the interval, and p is a preliminary or trial value of the proportion.

If it is known that the true proportion is near .5—say, between .3 and .7—then a slightly conservative sample size (that is, one somewhat larger than needed) can be computed by taking $p = .5$ in the sample size expression. This is because the product $p(1 - p)$ takes on its largest value, .25, when $p = .5$ and is never smaller than .21 for p in the range from .3 to .7. Consequently, the conservative sample size will not be too far wrong in this case.

However, when p is outside of this range, the product $p(1 - p)$ becomes small rather rapidly; using $p = .5$ will produce sample sizes that are much larger than needed to achieve the given goals. Another method should then be used to obtain a preliminary value of the proportion. For example, a pilot study or information from a smaller experiment might be used.

EXAMPLE 8.7

How many tosses would it take to estimate the probability of *heads* for a coin with a 95% confidence interval of length .01?

The critical value for 95% confidence is $z = 1.960$, and the required half-interval length is $h = .01/2 = .005$. Assuming that the coin is nearly unbiased, it is reasonable to use the preliminary value $p = .5$. The sample size can then be calculated from the above expression:

$$n = z^2 p(1 - p)/h^2 = \frac{1.960^2 .5(1 - .5)}{.005^2}$$

$$= 38,416$$

The coin tosser is clearly in for a very sore thumb. He or she is likely to seek a compromise between the validity and sensitivity goals of the problem on one hand and the comfort of his or her thumb on the other. Note that by relaxing the interval length to .1, the sample size is reduced to $n = 385$. Again, reasonably modest changes in goals can lead to substantial differences in sample sizes.

The next example demonstrates a sample size design when the value of p is too near 1 to use $p = .5$ as the preliminary value.

EXAMPLE 8.8

Monte Carlo experiments of the type that produced display 8.1 can be used to investigate the influence that "unusual" features of the underlying distribution, such as extreme skewness or long tails, have on the actual probability with which a confidence interval covers the true value of a parameter. A computer can be programmed to take random samples from such a distribution. For the purposes of the Monte Carlo experiment, the true value of the parameter for this distribution will be known. The coverage probability can then be approximated by drawing many random samples and computing the confidence interval for each sample. The approximation to the actual coverage probability would then be the proportion of times the computed confidence intervals contains or covers the true parameter value.

To design such a Monte Carlo experiment, it is necessary to tell the computer how many confidence intervals to compute. As we now see, this is simply a sample size design problem for a population proportion.

The reason for this fact is that each confidence interval calculation results in a value of a two-valued variable, X: either the confidence interval covers the true parameter value ($X = 1$), or it doesn't ($X = 0$). The probability, $p = P(X = 1)$, that the interval covers the true parameter value is the unknown coverage probability we are after. Consequently, we can determine the sample size in terms of goals for a confidence interval for p.

Suppose, for example, that we require a 90% confidence interval for the coverage probability to have length .009. With this length, the confidence interval will distinguish between coverage probabilities that differ by as much as 1%. That is, we will not mistake a true coverage probability of 94% as having the value 95%, for example. The half-length of the interval is then $h = .0045$.

This design lacks only a preliminary value of p in order to determine the number of intervals to compute. A sensible choice for p would be the design value for the intervals. Thus, if we are using intervals designed to have confidence coefficient 95%, then we would expect (or hope) that the true coverage probability is close to this value. It follows that $p = .95$ would be a reasonable preliminary choice. We will use it in this example.

To meet the specified goals, the number of confidence interval calculations to make in the Monte Carlo experiment is

$$n = \frac{1.645^2.95(1 - .95)}{.0045^2}$$

$$= 6347.446$$

or

$$n = 6348$$

With this number of repetitions of the experiment, we would expect with 90% confidence to estimate the true coverage probability of the intervals (which are designed to have confidence coefficient 95%) to within .9%.

Relative versus Absolute Precision

The product $p(1 - p)$ in the expression for sample size seems to suggest something contrary to intuition. It should be more difficult, thus require a larger sample, to estimate the population proportion of a rare event (one for which p is near 0) than an event for which p is near .5; we would expect that a large sample would be required to see enough occurrences of the event in order to get a reliable estimate of the proportion. However, for a given confidence coefficient and interval length, the expression produces a *smaller* sample size for values of p near 0 than for p near .5.

The resolution of this paradox is that it is not sensible to take the same interval length for values of p near 0 as for values near .5. Intervals should be shorter for smaller values of p. In fact, reasonable interval lengths would be proportional to p. Put another way, the proper measure of sensitivity when estimating a population proportion is not absolute interval length but *relative length*:

$$\text{relative length} = \frac{\text{absolute length}}{p}$$

For example, a population proportion of $p = .5$ that is estimated with an interval of absolute length .05 is pinpointed to within a relative length of $.05/.5 = .1$ or 10%. On the other hand, an interval length of .05 would be of no value at all for pinpointing a population proportion of .005. In this case, the relative length would be $.05/.005 = 10$ or 1000% of the true proportion.

In order to achieve the same relative precision for $p = .005$ as for $p = .5$, the interval length would have to be $.1 \times .005 = .0005$. To calculate the sample size required to achieve this precision, the half-interval length $h = .0005/2 = .00025$ would be used in the sample size expression.

More generally, the sample size required to achieve a specified relative interval length can be computed by using the half-length

$$h = p \times (\text{relative length})/2$$

in the sample size expression.

EXAMPLE 8.9

The Pima Indians of Arizona have an abnormally high incidence of diabetes and thus represent a particularly good population for the study of the relationship between the diabetic status of mothers and the occurrence of birth defects in their children. In this example we will consider the design of an experiment to estimate the incidence (proportion) of birth defects among children of nondiabetic mothers. The comparison of this proportion with the incidence of birth defects for diabetic mothers will be made in a later chapter.

A preliminary value of the birth defect proportion will be obtained from a study of Comess and colleagues [8], made during the years 1964–1967. In 785 live births, 31 children were observed to have at least one birth defect. Thus, the estimated incidence of birth defects in this population is

$$\hat{p} = \frac{31}{785} = .0395$$

or 3.95%.

Assuming that these 785 births represent a random sample from the population of all live births (past, present, and future) for Pima mothers, a 95% confidence interval for the overall birth defect incidence is

$$.0259 \le p \le .0531$$

or from 2.59% to 5.31%.

The length of this confidence interval is .0272. This is very close to the length, .026, of the confidence interval for the *heads* probability in the coin toss experiment of example 8.2. However, the relative interval length for the coin toss experiment is $.026/.503 = .052$ or 5.2% (we have used the estimated value of p in the expression for relative length), whereas for the birth defect proportion it is $.0272/.0395 = .689$ or 68.9%.

The relative precision of this interval is rather poor and we might ask how large a sample would be required to achieve the same relative precision for the birth defect study as for the coin toss experiment. Using the proportion $\hat{p} = .0395$ from the Comess study as the preliminary value of p, the half-interval length for 5.2% relative precision is

$$h = .0395 \times .052/2 = .001027$$

Entering this number and the value $p = .0395$ in the sample size expression for 95% confidence yields

$$n = \frac{(1.960^2).0395(1 - .0395)}{.001027^2}$$

$$= 138186.53$$

or $n = 138,187$.

Based on an optimistic estimate of 300 births per year to nondiabetic Pima women, this study would take over 460 years to complete—not a very appealing prospect.

The last example illustrates that it is very expensive, in terms of sample size, to achieve good relative precision for rare events. For this reason it is seldom even attempted in practice, unless it is very important and the effort must be made—for example to estimate the probability of a catastrophic accident involving a nuclear power reactor. In general, good relative precision can only be achieved if observations can be made quickly and cheaply, as in a Monte Carlo study carried out on a computer.

EXERCISES

8.22   The preliminary value of a population proportion to be used for sample size calculations is $p = .25$.

(a)   Find the sample size required to estimate the true proportion with a 95% confidence interval of length .02.

(b)   Repeat the calculation of part (a) for a confidence coefficient of 90%.

(c)   How large a sample is needed to achieve a 90% confidence interval of length .05?

8.23   How many repetitions of the Monte Carlo experiment of example 8.8 (p. 242) would be required to estimate the true coverage probability for 90% confidence intervals to within an interval of length .001 with 99% confidence? (Use the device for obtaining a preliminary value of p given in the example.)

8.24   It is expected that a candidate for election to the council of a large city will capture about 55% of the votes. His election staff decides to take a telephone poll of voters to get a better idea of his prospects. How large a sample should they take to get within 1% (i.e., an interval length of .01) of the actual proportion of city voters who favor this candidate with a 95% confidence interval, if the expected vote proportion is used as the preliminary value?

8.25   A preliminary study has established that about 3% of a large population possess a certain genetic trait. A study is to be carried out to pinpoint this population proportion more precisely with a 95% confidence interval.

(a)   Find the sample size required to estimate the proportion to within an interval length of .01.

(b)   Find the sample size required to estimate the proportion to within a relative interval length of 10%.

◢ SECTION 8.4

Finite Population Corrections*

Let n denote sample size and N population size. When the sampling fraction n/N is not negligible, the lengths of the confidence intervals for μ and p can be made smaller by multiplying the expression for the standard error SE in the endpoints of the intervals by the finite population correction factor. The corrected confidence intervals are

$$\bar{X} - z \sqrt{\frac{N - n}{N - 1}} \frac{s}{\sqrt{n}} \leq \mu \leq \bar{X} + z \sqrt{\frac{N - n}{N - 1}} \frac{s}{\sqrt{n}}$$

and

$$\hat{p} - z \sqrt{\frac{N - n}{N - 1}} \sqrt{\frac{\hat{p}(1 - \hat{p})}{n}} \leq p \leq \hat{p} + z \sqrt{\frac{N - n}{N - 1}} \sqrt{\frac{\hat{p}(1 - \hat{p})}{n}}$$

The lengths of the intervals are now of the form

$$l = 2z \sqrt{\frac{N - n}{N - 1}} \text{ SE}$$

where SE represents the infinite population standard error. Since the finite population correction factor is less than 1 for $n > 1$, the precision of the intervals would be improved by its use. On the other hand, there will be little loss in precision incurred by using the infinite population confidence intervals of section 8.2 when the sampling fraction is small.

One-sided confidence limits with the finite population correction are constructed as before by choosing the corresponding (corrected) two-sided limit and replacing the two-sided critical value by the one-sided value.

EXAMPLE 8.10   Suppose the size of the student population in example 8.1 (p. 227) is N = 300. What are the confidence limits for mean age after the finite population correction?

---

* Optional section.

The sample size is $n = 36$, so the finite population correction factor is

$$\sqrt{\frac{N - n}{N - 1}} = \sqrt{\frac{300 - 36}{300 - 1}} = .940$$

The corrected confidence limits are found by multiplying the infinite population standard error, $7.648/\sqrt{36}$, in the original confidence limits by this value:

$$L = 22.03 - \frac{1.960(.940)7.648}{\sqrt{36}} = 19.68 \text{ years}$$

and

$$U = 22.03 + \frac{1.960(.940)7.648}{\sqrt{36}} = 24.38 \text{ years}$$

The length of the interval has been reduced from 5 to 4.7 years by accounting for the finiteness of the population—not a great gain in precision.

The Finite Population Correction in Sample Size Design

Suppose the population of statistics students at the University of New Mexico totaled only 400 students in the year the survey of example 8.6 (p. 240) was to be run. It is rather unlikely that the (infinite population) sample size of $n = 441$ would meet with much enthusiasm from the survey designer. The problem here is that the expression for sample size based on the infinite population model cannot be used effectively when the constraints of the design are likely to require a sizable sampling fraction. It is at this point that the finite population correction factor can play its most important role.

The expression for the confidence interval half-length when the finite population correction factor is included is

$$h = z \sqrt{\frac{N - n}{N - 1}} \frac{s}{\sqrt{n}}$$

This expression can be solved for n to obtain

$$n = \frac{Nz^2s^2}{h^2(N - 1) + z^2s^2}$$

The critical value, z, and interval half-length, h, are prescribed just as for infinite populations. Moreover, the strategy for obtaining a preliminary value of the standard deviation, s, is also the same.

The finite population sample size for a population proportion is

$$n = \frac{Nz^2p(1 - p)}{h^2(N - 1) + z^2p(1 - p)}$$

Again, the quantities z, h, and p to enter in this expression are determined just as they were for infinite populations.

   Note that in order for these sample sizes to achieve the desired goals, the corresponding finite population corrected confidence intervals must be used.

EXAMPLE 8.11

The sample size will be computed for example 8.6, assuming a population of size $N = 400$. The critical value for 99% confidence is $z = 2.576$, and the prescribed interval half-length is $h = .5$. Again, take $s = 4.076$. Substituting these values into the sample size expression for a population mean yields

$$n = \frac{400 \times 2.576^2 \times 4.076^2}{.5^2 \times (400 - 1) + 2.576^2 \times 4.076^2}$$

$$= 209.996$$

or

$$n = 210$$

Thus, the sampling fraction required to yield a 99% confidence interval for mean height of length 1 inch from this population is $n/N = 210/400 = .525$ or 52.5%. Over half of the population must be sampled in order to achieve the design goals. When the sample is taken, the confidence interval *with* the finite population correction will be calculated.

EXERCISES

8.26   (a)   Suppose the size of the lot of mylar capacitors for which the 99% confidence interval was calculated in exercise 8.5 (p. 231) is $N = 100$. Recalculate the confidence interval for the mean capacitance, using the finite population correction factor. (Recall that $n = 40$, $\bar{X} = .63767$, and $s = .01081$ μf.)

(b)   How large a sample would be needed in part (a) to estimate the mean capacitance to within .001 μf? (Use the same confidence coefficient and take the computed standard deviation as the initial value of s.)

8.27   Suppose that the study data for the 174 University of New Mexico students of exercise 8.11 (p. 234) came from questionnaires given to randomly selected individuals from among 350 students who visited the office of the Dean of Arts and Sciences during a given month. Because students who visit the dean's office are frequently having scholastic problems, it was decided that the data could not be extrapolated to the student population in general, and the "target" population was taken to be the 350 visitors for that month. Recall that 35 of the students sampled claimed to spend at least 26 hours per week studying for all of their classes. Find a 90% confidence interval for the proportion of students in the target population who study at least 26 hours per week.

Chapter 8 Quiz

1. What is meant by the true value of a parameter?

2. What is the purpose of a confidence interval in relation to the true value of a parameter?

3. What is the confidence coefficient for a confidence interval?

4. What properties are the endpoints L and U of a confidence interval required to satisfy?

5. If the limits of a 95% confidence interval are computed to be $L = 12$ and $U = 15$ for a given sample, what is wrong with the interpretation "the probability that the true parameter value is between 12 and 15 is .95"?

6. What is the relative frequency interpretation of the phrase "with 95% confidence"?

7. What information is required to calculate the limits of a confidence interval for a population mean? Give a numerical example of a confidence interval computation for this parameter.

8. What does the length of a confidence interval measure?

9. What is the pivotal method for constructing confidence intervals?

10. What information is required to calculate the limits of a confidence interval for a population proportion? Give a numerical example of a confidence interval calculation for this parameter.

11. What is a one-sided confidence limit? Under what circumstances would one use a one-sided limit instead of the two-sided version? What is gained in doing so?

12. How is a two-sided confidence limit for a population mean or proportion modified in order to obtain the appropriate one-sided version?

13. What are the stages in a statistical investigation?

14. What is the validity requirement for confidence intervals? When is a confidence interval valid?

15. What is the sensitivity measure we use for confidence intervals?

16. How are validity and sensitivity requirements used to design sample sizes for a population mean confidence interval? Give an example.

17. How is the sample size for an experiment to determine a confidence interval for a population proportion computed if the population proportion is (a) assumed to be between .3 and .7? (b) assumed to be outside of this interval?

18. What is the difference between absolute and relative precision? Why is relative precision a more sensible measure when the true proportion is near 0?

19. Why is it difficult to achieve good relative precision in confidence intervals for p when p is near 0?

These questions are for the optional section 8.4

20. How are the confidence limits for population means and proportions for infinite populations modified in order to account for finite populations when sampling fractions are large?

21. How are sample sizes calculated to obtain confidence intervals for μ and p with prescribed length and confidence when the population is finite?

SUPPLEMENTARY EXERCISES

8.28   The placement scores of exercises 1.23 (p. 33) represent those for a sample of a large group of students being placed in mathematics courses at the University of New Mexico for a given year.

(a) Find a 95% confidence interval for the mean arithmetic score for these students. (Note: $n = 44$, $\bar{X} = 14.523$, $s = 4.369$.)

(b) Find a 95% confidence interval for the mean algebra score. (Note: $n = 44$, $\bar{X} = 3.955$, $s = 3.530$.)

(c) Using this sample as a pilot study to sample placement scores the following year, determine how large a sample should be taken to estimate the average algebra score to within one unit, with a 95% interval.

8.29   The population of interest in a study of post-secondary educational needs for a large western city was comprised of people in the age range 18–35 years. In a sample of 2250 people from the city, 50.1% were in this age range.[*] Find a 99% confidence interval for the actual proportion of people in this city between 18 and 35 years of age.

8.30   It is suspected that oral contraceptives reduce the hemoglobin levels of their users. In a study in which several blood chemistry measurements were made on a random sample of people, the mean hemoglobin level for the 37 women reporting that they were taking oral contraceptives was 14.491, with a standard deviation of 1.079.[†] Find a 95% upper confidence limit for the (population) mean hemoglobin level of women on oral contraceptives.

8.31   In a study of energy consumption for heating nonresidential buildings in a large southern city, each building in a random sample of 43 was monitored for one heating season, yielding mean heating rate $\bar{X} = 27.472$ and standard deviation $s = 22.925$. Find a 95% confidence interval for the citywide mean heating rate for nonresidential buildings.

8.32   Of the 61 children (families) sampled for exercise 4.17 (p. 115), 33 spoke their native language at home. Estimate, with a 95% confidence interval, the proportion of Native American families (in the population from which the data were sampled) for which the native language is the predominant language used in the home.

---

[*] Data courtesy of G. Mallory, Albuquerque Urban Observatory.
[†] Data courtesy of P. Garry, School of Medicine, University of New Mexico, Albuquerque.

8.33   The table of exercise 4.19 (p. 116) is as follows:

	NPR AWARENESS	
SEX	AWARE	NOT AWARE
Male	81	8
Female	43	4

Estimate, with a 95% confidence interval, the proportion of the adult Albuquerque population aware of National Public Radio (NPR) or the Albuquerque affiliate.

8.34   The table of exercise 4.22 (p. 117) is as follows:

HOUSEHOLD SIZE (NO. OF PEOPLE)	NPR AWARENESS	
	AWARE	NOT AWARE
1	26	3
2	43	5
3	22	3
4	24	1
5 or more	9	0

Estimate, with a 90% confidence interval, the proportion of Albuquerque households containing five or more people.

8.35   The mean and standard deviation of winter travel distances for the $n = 34$ captured mice of exercise 4.10 (p. 107) are $\overline{X} = 15.029$ and $s = 17.133$ meters. Estimate, with a 95% confidence interval, the mean distance traveled by field mice during the winter months. If the more active mice are more likely to be trapped at least twice, discuss the possible bias in this estimate.

8.36   The average and standard deviation of the $n = 36$ items purchased during 1980 by the shopper of exercise 3.28 (p. 92) are $\overline{X} = \$1.551$ and $s = \$1.426$. By means of a 95% confidence interval, estimate the average item price of items purchased by the grocery shopper during 1980.

8.37   (Data coding) Experiments of historical importance to both science and statistics were performed beginning in the eighteenth century to determine such physical constants as the mean density of the earth, the distance from the earth to the sun (the so-called astronomical unit), and the velocity of light. An interesting series of experiments to determine the velocity of light was begun in 1875. The first method used, and reused with refinements several times thereafter, was the rotating-mirror method. In this method a beam of light is reflected off a rapidly rotating mirror to a fixed mirror at a carefully measured distance from the source. The returning light is rereflected from the rotating mirror at a different angle, because the mirror has turned slightly during the passage of the corresponding light pulses. From the speed of rotation of the mirror and from careful measurements of the angular difference between the outward-bound and returning light beams, the passage time of light can be calculated for the given distance. After averaging several calculations and apply-ing various corrections, the experimenter can combine mean passage time and distance for a determination of the velocity of light. Simon Newcomb, a distin-guished American scientist, used this method during the year 1882 to generate the passage time measurements given below. The travel path for this experiment was 3721 meters in length, extending from Fort Myer, on the west bank of the Potomac River in Washington, D.C., to a fixed mirror at the base of the Washington Mon-ument.

28	27	24	31	36	37	36	27	26	39	29
26	16	21	19	28	25	26	27	33	28	27
33	40	25	24	25	28	30	28	26	24	28
24	−2	30	20	21	26	22	27	32	25	29
34	29	23	36	28	30	36	31	32	32	16
−44	22	29	32	29	32	23	27	24	25	23

Newcomb's (coded) measurements of passage time of light[a]

Source: Stigler [31].
[a] These measurements were made in the period from July 24, 1882, to September 5, 1882. The given values times .001 plus 24.8 are Newcomb's measurements, recorded in millionths of a second or microseconds (μs).

The problem is to determine a 95% confidence interval for the "true" passage time, which is taken to be the mean of the population of measurements that were or could have been obtained by this experiment.

In order to save space in the table, the data have been *coded* by the linear transformation

$$Y = 1000(X − 24.8)$$

Thus, for example, the first observation $X = 24.828$ has become $Y = 28$ by the coding process. Coding has preserved all of the information in the observations while re-moving the redundant first three digits.

In order to find the confidence interval for the mean, it is unnecessary to "decode" the observations before calculating the sample mean and standard devia-tion. The reason for this is that the sample mean and sample standard deviation obey the same rules of expectation and variance for linear transformations as do their population counterparts. Thus, referring to chapter 5, we find that if variables X and Y are related by the expression

$$Y = aX + b$$

then

$$\bar{Y} = a\bar{X} + b$$

and

$$s_Y = |a| s_X$$

(a) Apply these expressions to the coding transformation above, after finding the sample mean $\bar{Y}$ and s_Y for the coded data, and solve for $\bar{X}$ and s_X.

(b) Use the results of part (a) to find a 95% confidence interval for the (uncoded) mean passage time of light.

(c) Find a 95% confidence interval for the *coded* mean passage time—that is, use the coded data for the interval computation. Now, decode the *endpoints* of this interval (using $X = .001Y + 24.8$) and show that the decoded limits are the same as those found in part (b). It follows that confidence interval calculations done by hand can often be simplified by coding. The advent of computers, however, has made the coding process (essentially) obsolete.

8.38   Taking Simon Newcomb's experiment, exercise 8.37, as a pilot study, determine how large a sample would be required, using his experimental setup, to estimate the mean passage time of light with a 95% confidence interval to within .001 millionths of a second. Newcomb's experiment required exceptionally good atmospheric conditions, something not always available in Washington, D.C. He was able to carry out 66 determinations over a period of 44 days, an average of 1.5 observations per day (assuming he worked seven days a week). Also, if it is assumed that the atmospheric conditions in Washington are suitable for experimentation 100 days per year, one could hope to make about 150 determinations a year. How many years would it take to complete the experiment?

Hypothesis Testing—Large-Sample Theory

Introduction

In this chapter we begin the study of the oldest and most often used form of statistical inference, the hypothesis test. Hypothesis testing is an outgrowth of experimental science, and both the terminology and philosophy of the subject reflect this history.

As we have seen in previous chapters, questions of statistical interest can be rephrased as questions concerning parameters of population frequency distributions. In this chapter, we will be interested in questions that can be posed in terms of the mean μ of a measurement variable or a population proportion p.

statistical hypothesis A **statistical hypothesis** is a statement concerning the values of a population parameter. One way of posing a question about such a parameter is to ask whether or not a particular statistical hypothesis is valid. A hypothesis test is a procedure for challanging the hypothesis on the basis of a random sample from the population.

We will study two methods of hypothesis testing, the so-called *inductive inference method*, whose chief architect was the famous geneticist-statistician R. A. Fisher, and the *decision theory method*, which was developed by two of the founders of modern statistics, J. Neyman and E. S. Pearson. The inductive inference method, which is the topic of section 9.2, will provide us with an intuitively appealing quantity for reporting the results of hypothesis tests: the P-value. The Neyman-Pearson theory, to be studied in sections 9.3–9.5, overcomes inadequacies of the inductive inference method by providing a framework for the important dimension of experimental design. The two methods are tied together in section 9.6 by showing that P-values can be used to carry out hypothesis tests designed using the Neyman-Pearson theory.

Supplementary topics that will not be explicitly used in later chapters are given in the optional section 9.7.

Measuring the Plausibility of Statistical Hypotheses: *P*-Values

In this section, the statistical hypotheses to be considered will specify a single number as a candidate for the true value of a parameter. A measure of the plausibility of this hypothesis, called the P-value, will be defined and illustrated.*

P-Values for Hypotheses Concerning a Population Proportion

The parameter of interest in this subsection is a probability or population proportion, p. For example, p might be the probability of *heads* in a coin toss, or the proportion of women in a population.

A statistical hypothesis will specify a candidate for the true value of p. Thus, the premise that a coin is fair, for example, could be expressed as the statistical hypothesis that the *heads* probability is .5. We will use the symbolic notation

$$H_0\colon p = .5$$

as shorthand for the statement: the hypothesis to be tested (H_0) is that the true value of p equals .5.

The true *heads* probability is unknown and must be estimated from a sample. That is, after tossing the coin several times and observing the number of *heads*, the true *heads* probability will then be estimated by the value of the sample proportion $\hat{p}$.

A sensible test of the fair-coin hypothesis can be based on a comparison of the estimated value of p and the hypothetical value, .5. From our knowledge of the sampling distribution of $\hat{p}$, the probability distribution of this estimator is concentrated around the true value of p. Consequently, the hypothesis will be plausible if the estimated value is near .5. The farther the estimate is from .5, the less plausible the hypothesis will be. The **P-value** is a quantitative measure of this plausibility, which is constructed from the sampling distribution of $\hat{p}$. As will be seen, the P-value will be used to assess the plausibility of many types of hypotheses—not just hypotheses about sample proportions.

P-value

The construction of the P-value will be illustrated for the fair-coin hypothesis. Suppose, for example, that 60 *heads* were seen in 100 tosses of the coin. In the light of this information, how plausible is the hypothetical value $p = .5$?

Reformulate this question as follows. Suppose that the true value of p were equal to the hypothetical value $p = .5$. Now, if the random experiment

---

* The P-value is sometimes called the *observed significance level*.

of tossing the coin 100 times were to be repeated, how likely would it be to see an outcome "at least as extreme" (relative to the hypothesis) as the one observed (60 *heads*)?

Because this reformulated question deals with a future sample, the answer will be a probability. *This probability is the P-value.*

The P-value for the example is calculated as follows. The value of $\hat{p}$ from the sample is 60/100, or .6. A value of $\hat{p}$ from another sample will be "at least as extreme" as the one observed if it is no closer to the hypothetical value $p = .5$:

$$|\hat{p} - .5| \geq |.6 - .5| \qquad (1)$$

Because, for the purposes of the calculation, the true value of p is equal to the hypothetical value $p = .5$, the standard error of $\hat{p}$ is

$$\sigma_{\hat{p}} = \sqrt{\frac{p(1 - p)}{n}}$$

$$= \sqrt{\frac{.5(1 - .5)}{100}}$$

$$= .05$$

Dividing both sides of the inequality (1) by this standard error, the left-hand side becomes the absolute value of the variable $V = (\hat{p} - p)/\sqrt{p(1 - p)/n}$ for $p = .5$.

The right-hand side of the inequality becomes the absolute value of V_{obs}, the (observed) value of V computed from the sample:

$$V_{\text{obs}} = \frac{.6 - .5}{.05}$$

$$= 2$$

Thus, the P-value can be written as

$$P\text{-value} = P(|V| \geq |V_{\text{obs}}|)$$

$$= P(|V| \geq 2)$$

Now, the variable V was seen in chapter 7 to have approximately a standard normal distribution. (The rule for the adequacy of the normal approximation that can be used here and in later applications is that the smaller of $np_0/(1 - p_0)$ and $n(1 - p_0)/p_0$ should be no smaller than 9, where p_0 is the hypothetical value of p. Thus, $p_0 = .5$ in this example.) Ignoring the fact that an approximation is being used, we calculate the P-value as

$$P\text{-value} = P(|Z| \geq |V_{\text{obs}}|)$$

where Z is a standard normal variable. Thus, table 1A yields

$$P\text{-value} = 2(.5 - .4772)$$

$$= .046$$

It is important to note that although the *P*-value is a probability, it is not a probability for the hypothesis. That is, it doesn't make sense to say the probability that the hypothesis is true is equal to .046; the hypothesis is either true or it is false.

The sense in which it is a probability, and the interpretation one would give to the quantity .046, is as follows: In a repeated sample of size 100, if $p = .5$ is the true *heads* probability for the coin, then an outcome as extreme as the one observed would occur with a probability of only .046. That is, in only about 5 out of every 100 subsequent experiments consisting of 100 tosses of the coin would we see a proportion of *heads* this far from .5. This suggests that the hypothetical probability of .5 is not supported by the data.

We would conclude, on the basis of the observed data, that it is rather implausible that the coin is fair. The reason for this conclusion is that small *P*-values correspond to implausible hypotheses.

A *P*-value provides a basis for deciding whether or not to believe in the truth of the given hypothesis; it summarizes the evidence against the hypothesis provided by the sample. We would be led to believe that the hypothesis is false if the *P*-value is small enough. It thus constitutes a *hypothesis test* of the type to be considered in the next section—a method for using the sample to choose between belief and disbelief in the hypothesis.

The test is said to be *significant* if the *P*-value is small enough to cause disbelief in the hypothesis. R. A. Fisher suggested the value .05 as the dividing line between significant and insignificant test outcomes in the 1920s, and this has been the basis of choice ever since.

An advantage of the *P*-value is that it supports a broader spectrum of choices than simply believing or not believing in the hypothesis. We will find useful the following classification of test outcomes that better utilizes the shadings of the *P*-value. (*P*-values are denoted by cap *P* in this text.) A test will be said to be

not significant if $P > .10$

mildly significant if $P \le .10$ but $P > .05$

significant if $P \le .05$ but $P > .01$

highly significant if $P \le .01$

Each succeeding classification indicates a decreased degree of belief in the hypothesis. For example, we would have no reason to disbelieve the hypothesis if the test outcome is not significant, but a mildly significant outcome would cause some doubt, and so forth.

We will use this classification in the conclusions to problems. However, a conclusion concerning the significance of a test will also be supported by giving the *P*-value or bounds for the *P*-value in parentheses within the statement. Thus, for the coin-toss experiment we would conclude that the test is significant ($P = .046$).

Because the observed sample proportion, $\hat{p} = .6$, is expected to be near the true *heads* probability, we could augment the conclusion by stating that the evidence favors a *heads* probability larger than .5—that is, a bias towards *heads*.

EXAMPLE 9.1

A few years ago at the University of New Mexico, the proportion of upperclassmen and graduates among students taking beginning statistics was 48.2%. The administration is interested in whether this proportion has changed since then. The data of display 1.1, viewed as a sample from the current population, will be used to address this question.

The appropriate hypothesis is that the current proportion of upperclassmen and graduates, p, is the same as the proportion for the earlier year:

$$H_0: p = .482$$

How plausible is this hypothesis in light of the current data?

The sample proportion of juniors, seniors and graduates from display 1.1 (for which $n = 36$) is

$$\hat{p} = \tfrac{10}{36} = .278$$

Under the given hypothesis, the observed value of the V statistic is

$$V_{obs} = \frac{.278 - .482}{\sqrt{\dfrac{.482(1 - .482)}{36}}}$$
$$= -2.450$$

Using the normal approximation, the P-value can be calculated from table 1A:

$$P = P(|Z| \geq |V_{obs}|)$$
$$= P(|Z| \geq 2.450)$$
$$= 2 \times (.5 - .4929)$$
$$= .0142$$

A quick pair of bounds for the P-value that requires no calculation can be obtained from table 1B. Search down the last column for the two numbers bounding the quantity $|V_{obs}| = 2.450$. These numbers are seen to be 2.326 and 2.576. The bounds for the P-value are then the corresponding two numbers in the TWO-SIDED TESTS column, in reverse order:

$$.01 < P < .02$$

Consequently, we would conclude that the test based on the given sample is significant ($P = .014$); it is implausible that the proportion of upperclassmen and graduates in elementary statistics is the same as it was before. It would be acceptable to substitute the bounds $.01 < P < .02$ for the value $P = .014$ in this statement, since the two expressions convey essentially the same information.

Since $\hat{p} = .278$, the conclusion could be augmented by the statement that the current proportion of upperclassmen and graduates in elementary statistics appears to be smaller than before. Apparently, students are now taking statistics earlier in their college careers.

P-Values for Hypotheses Concerning a Population Mean

P-values for hypotheses of the form

$$H_0 : \mu = \mu_0$$

can be obtained for large samples by applying the normal approximation to the distribution of Student's t statistic. If t_{obs} denotes the value of $t = (\bar{X} - \mu)/(s/\sqrt{n})$ obtained by replacing μ by the hypothetical value μ_0 and $\bar{X}$ and s by the values computed from the sample, then the (approximate) P-value is

$$P = P(|Z| \geq |t_{obs}|)$$

EXAMPLE 9.2

A quality check is to be run on a random sample of $n = 57$ vials of frozen blood taken from a large number of vials that have been stored for one year. The purpose of the test is to determine whether the levels of high-density lipoprotein (HDL) have changed due to storage. The mean HDL level μ of the stored blood will be compared with a published value of 47 milligrams per 100 milliliters (abbreviated as mg%) for a comparable group of people. The mean and the standard deviation for the sample of frozen blood are $\bar{X} = 27.84$ and $s = 10.61$ mg%. How plausible is it that the mean HDL value of the population of stored blood vials is 47 mg%?

The stem and leaf plot and box plot for the data are given in display 9.1. From the box plot it seems rather implausible that the center of the distribution is 47. We will be able to assign a quantitative value to this implausibility with the hypothesis test.

Display 9.1
Stem and leaf plot and box plot of 57 HDL cholesterol determinations (mg%)

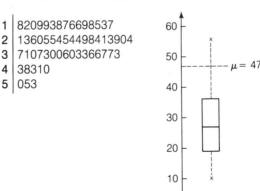

Data courtesy of R. Philip Eaton, M.D., Department of Endocrinology, School of Medicine, University of New Mexico, Albuquerque.

The hypothesis under test is H_0: $\mu = 47$. Thus, the observed value of the t statistic is

$$t_{obs} = \frac{27.84 - 47}{10.61/\sqrt{57}} = -13.634$$

Thus, $|t_{obs}| = 13.634$, and the P-value is

$$P = P(|Z| \geq 13.634)$$

This probability lies well outside of the range of both tables 1A and 1B, and the most we can say is that

$$P < .000001$$

We would report that the test is highly significant ($P < .000001$); it is quite implausible that the true mean level is 47 mg%. There is strong evidence of a decrease in mean HDL level.

EXERCISES

9.1   A coin is tossed 100 times, resulting in 43 heads. Find the P-value for the hypothesis that the coin is fair ($H_0:p = .5$).

9.2   The proportion of students taking business courses at a large college was 28% in 1980. In a random sample of 300 currently enrolled students, 110 are taking business courses. How plausible is it that the proportion of students in the college taking business courses has not changed since 1980?

9.3   The mayor of a large city is standing for reelection. In the election to his first term, he won over his opponent by earning 58% of the vote. His reelection staff has taken a telephone poll of 200 voters and found that 102 of them plan to vote for the mayor. How plausible is it that he will win over his current opponent by taking the same proportion of the vote as in the last election? Does he appear to be more popular or less popular than in the first race?

9.4   In example 8.9 (p. 244), it was seen that in 785 births to nondiabetic Pima women, 31 babies had at least one birth defect. How plausible is it that the true proportion of birth defects for children born to nondiabetic Pima mothers is 5%?

9.5   The mean of a certain measurement variable X in a population is hypothesized to be $\mu = 3.8$. For a random sample of 120 individuals from the population, the sample mean and standard deviation were $\bar{X} = 4.8$ and $s = 2.1$. How plausible is the hypothetical value of the mean in the light of this information?

9.6   A manufacturer of soft drinks is required to make periodic checks of the accuracy of the machine that automatically fills the cans the drinks are sold in. How plausible is it that the machine is filling cans with 12 ounces of fluid, on the average, if in a random sample of 150 cans the sample mean of the contents is 11.79 ounces, with a standard deviation of .354 ounces?

9.7   The average weight of a variety of fish caught in a river at the site of a nuclear power plant was 6.6 pounds before the plant's construction. River water is used to cool the reactor, and the water is recycled to the river after use. In a recent sample of 53 fish caught at the same site, the mean weight was 6.3 pounds, with a standard deviation of 1.7 pounds. How plausible is it than the mean weight of the fish is the same as before? If the test result showed a significantly smaller mean weight, could the power plant be blamed for the change? Why?

The Neyman-Pearson Method of Hypothesis Testing

Examples 9.1 and 9.2 illustrated situations in which the sample information provided strong evidence against the proposed hypotheses. This will not always be the case, of course.

When the outcome of a test is judged plausible or not significant ($P >$ .10), problems of interpretation arise. The difference between the true parameter value and the hypothetical one may actually be small enough to be of no practical importance. In this case, we would certainly want H_0 to be judged plausible. However, as was seen in chapter 7, the amount of information provided by a sample depends on the *size* of the sample. Thus, it is possible that a difference of practical importance exists, but the sample is simply not large enough to detect this difference. (Another possibility is that the information is there but is not being utilized properly. We will confront this problem in chapter 12.)

Unfortunately, by the time the P-value is computed, it is no longer possible to deal with the problem of inadequate sample size—the sample has already been taken. The selection of sample size is a preexperiment design problem—one the P-value methodology does not address. This deficiency (among others) in the inductive inference methodology led J. Neyman and E. S. Pearson to consider a different approach—one that made possible an experimental design component.

As was true for confidence intervals, experimental design for hypothesis tests depends on concepts of validity and sensitivity for statistical procedures. These ideas are discussed in this section. Their use in sample size design is considered in the optional section 9.7.

Statistical Hypotheses and the Hypothesis-Testing Decision Problem

A statistical hypothesis is a statement about the true value of a population parameter—the value of the parameter actually governing the frequency distribution of the observations.

hypothesis under test
alternative hypothesis

A key feature of the Neyman-Pearson theory is that in addition to the **hypothesis under test**, H_0, introduced in the last section, an **alternative hypothesis**, H_1, is always specified—sometimes implicitly. The two hypotheses H_0 and H_1 are alternatives to one another in the sense that they never specify the same parameter values, and together they specify all parameter values considered to be possible or of interest in the given problem.

For example, if p is the *heads* probability of a coin and the problem is to determine whether or not the coin is fair, the fair-coin hypothesis would be formulated as the statistical hypothesis H_0: $p = .5$, while all possible alternative conditions of bias would be represented by the hypothesis H_1: $p \neq .5$. Note that H_1 encompasses both the range of values $p < .5$, corresponding to a bias towards *tails*, and $p > .5$, representing a bias towards *heads*. The two hypotheses together specify all values of p. (The restriction of p to the range from 0 to 1 is understood and need not be explicitly stated.)

hypothesis testing (decision) problem

Now, the **hypothesis testing (decision) problem** is the problem of deciding between H_0 and H_1 on the basis of a random sample. Thus, after n tosses of the coin, it would be necessary to make one of the two decisions: that the coin is fair (decide in favor of H_0) or that it is biased (decide in favor of H_1). In the vernacular of hypothesis testing, the two decisions are called **accept H_0** and **reject H_0** (thus, accept H_1).

accept H_0
reject H_0

Since only these two decisions are possible, it is necessary to specify only the circumstances under which one would reject H_0. It is understood that in all other circumstances one would accept H_0. This fact is used to define a **hypothesis test.**

hypothesis test
test rule

A hypothesis test is determined by a **test rule** that specifies the sample outcomes for which the decision to reject H_0 will be made. For example, a rule to test the coin hypotheses is

Reject H_0 if $|V| \geq 1.960$

where $V = (\hat{p} - .5)/\sqrt{.5(1 - .5)/n}$ and $\hat{p}$ is the proportion of heads in a sample of n tosses of the coin. Every sample of n coin tosses will lead to a value of $\hat{p}$, thus to a value of V. Then if $|V| \geq 1.960$, H_0 will be rejected, whereas if $|V| < 1.960$, H_0 will be accepted. For example, if 60 *heads* are observed in $n = 100$ coin tosses, then $\hat{p} = 60/100 = .6$ and $V = (.6 - .5)/\sqrt{.5(1 - .5)/100} = 2$. Since $|V| = 2 > 1.960$, H_0 would be rejected.

An advantage of specifying a hypothesis test in this way is that it is possible to argue the merits of the test *before carrying out the experiment*. This is what makes experimental design possible.

The Two Types of Error

The performance criteria established by Neyman and Pearson are based on the possible errors that can be made in using a test. For example, the test rule may lead us to reject H_0 when the true parameter value is the one (or among the ones) specified by H_0—that is, when H_0 is *true*. This is called a **type I error.** A **type II error** is made if H_0 is accepted when H_1 is true, that is, when the true parameter value is one of those specified by H_1.

type I error
type II error

For example, a type I error in the coin-toss experiment would be to decide that the coin is biased (reject H_0) when it is actually unbiased ($H_0: p = .5$ is true). A type II error would be to decide the coin is unbiased (accept H_0) when it is not ($H_1: p \neq .5$ is true).

Since the decisions to accept or reject H_0 are made on the basis of a random experiment, the decisions are random events. Moreover, the probabilities of these events will depend on the true value of the parameter. Although we will not know this value, we can calculate the probabilities of the events for various possibilities. For example, we can calculate the probability that the above coin-toss rule leads us to reject H_0 when $p = .5$. This would be the probability of a type I error, since $p = .5$ when H_0 is true. As we will see in the next subsection, the criteria of validity and sensitivity for hypothesis tests are specified in terms of the probabilities of the two types of errors.

Validity, Sensitivity, and Power

An aspect of the decision theory approach is that decisions lead to *actions*. If H_0 is rejected, one action will be taken, while if it is accepted, a different action will be taken. A particularly important "action" is no action at all. In this case, the hypothesis H_0 is designated as the "no action" or *null hypothesis*, in that no action will be taken if H_0 is accepted. The hypothesis H_1 will then be the "action" or "research" hypothesis; whatever action is to be taken will be taken only if H_0 is rejected and H_1 is accepted.

In this special case, the type I error is to take action (reject H_0) when this action is unwarranted (H_0 true), while the type II error is to take no action (accept H_0) when action should have been taken (H_1 true.)

The viewpoint taken in the Neyman-Pearson theory of hypothesis testing is that the two types of errors are not of equal importance. A type I error is considered to be more serious than a type II error. The concept of test validity is based on an attempt to control the probability of a type I error.

significance level

The control is exerted as follows: a probability, α, called the **significance level** of the test, is specified (typically, before the experiment is carried out). The test is then designed to have its type I error probability no larger than α:

$$P(\text{Reject } H_0 \text{ when } H_0 \text{ true}) \leq \alpha$$

In designing a test, we will attempt to make the type I error probability *equal* to α. Such a test will then be called a *level α test*.

Ideally, the choice of a significance level would be based on an evaluation of the seriousness of the type I error. A small significance level would be assigned to an error with potentially serious consequences, while an error with indifferent consequences would be assigned a large level. In practice, it is seldom possible to make such an assessment, and the standard levels of .10 (10%), .05 (5%), and .01 (1%) are used to represent mild, ordinary, and extreme concern for making a type I error, respectively. Unless a different significance level is specified, the 5% value should be used in exercises.

Because the assumptions underlying the construction of the test may not hold in a given problem, the actual probability of a type I error may

valid

differ from the designed probability. The test is said to be **valid** if the true probability of a type I error does not exceed α. That is, an invalid test will be a test designed to have level α, for which the true type I error probability exceeds α.

power

The most commonly used measure of sensitivity in hypothesis testing is the **power** of the test. Power is the probability of rejecting H_0 when the true parameter value is one specified by H_1. The type II error probability is also used as a measure of sensitivity in practice. However, since accepting and rejecting H_0 are complementary events, it follows that the type II error probability is simply one minus the power of the test. Thus, the two measures are closely related. A more complete discussion of power is given in section 9.7.

Because high power means high probability of rejecting H_0 when H_0 is false, this measure of sensitivity can be interpreted as the ability of the test to distinguish the true value of the parameter (specified by H_1) from the one specified by H_0.

The concepts of validity and sensitivity play an important role in experimental design. As in the case of confidence intervals, sample size can be calculated to achieve prescribed validity and sensitivity levels for a test. How this is done will be shown in section 9.7. Methods for constructing hypothesis tests will be illustrated in the next two sections.

◥ SECTION 9.4

Two-Sided Hypothesis Tests

There are two principal applications of two-sided tests, illustrated in the next two subsections.

The Two-Sided, Two-Decision Problem for a Population Proportion

Consider hypotheses of the form

$$H_0: p = p_0$$

and

$$H_1: p \neq p_0$$

two-sided hypothesis

Because H_1 specifies values of p both above and below the hypothetical proportion p_0, it is called a **two-sided** alternative hypothesis.

One context in which two-sided alternative hypotheses are appropriate is in problems for which one action (or no action) would be taken if $p = p_0$ and another if $p \neq p_0$.

For example, suppose a casino game consists of a coin toss, using a coin with *heads* probability p. A player wins on a toss if *heads* appears; otherwise the house wins. The fairness of the game depends on the fairness of the coin. Periodically, the coin is checked by an official of the state gambling commission. The viewpoint of the commission is that it is bad for the reputation of the state if the customer is being cheated—that is, if $p <$ .5—while it is hard on the state's gambling revenues if too much money goes to gamblers because of a coin biased in their favor ($p >$ .5). Thus, the inspector will take action—change the coin and possibly fine the casino—if the coin is found to be biased in either direction, and no action will be taken if the coin is fair. Since the inspector is interested in detecting a bias in either direction, the above hypotheses with $p_0 = .5$ would be adopted.

Given a significance level α, the appropriate test for the above hypotheses is given by the rule

Reject H_0 if $|V| \geq c$

cutoff value
test statistic

two-sided test

where $V = (\hat{p} - p_0)/\sqrt{p_0(1 - p_0)/n}$. The number c is called the **cutoff value** or *critical value* for the test, and V is called a **test statistic**. Because the test rejects H_0 both for values of V above c and below $-c$, it is called a *two-tailed* or **two-sided test**.

Note that, from the viewpoint of detecting alternative values of p both above and below p_0, the form of the two-sided test is sensible. Since we expect the value of p to be near the true value of p, "large" values of V (that is, values sufficiently greater than 0) are indicative of a true value of p larger than p_0. Similarly, values of V that are negative but "large" in absolute value favor $p < p_0$. Both sets of V values are specified by the single inequality $|V| \geq c$.

The cutoff value quantifies what is meant by "large." It is chosen to satisfy the above-mentioned criterion that the type I error probability is to equal the (preassigned) significance level α. However, the test rule prescribes that H_0 is to be rejected only if $|V| \geq c$. Thus, the type I error probability is $P(|V| \geq c)$, where this probability is to be computed under the assumption that H_0 is true. But, as was seen in chapter 7, V has approximately a standard normal distribution in this case. Thus, if Z is a standard normal variable, the appropriate cutoff value for the above test is, to good approximation, the solution of the equation

$$P(|Z| \geq c) = \alpha$$

As was seen in chapter 6, table 1B is expressly designed to find solutions to equations of this type. For example, if a 5% significance level is chosen for the test, read down the column in the table labeled TWO-SIDED TESTS to the row containing the number .05. In this row, read the critical value $c = 1.960$ from the last column of the table.

For example, the rule for the two-sided 5% test the gambling inspector would use is

Reject H_0 if $|V| \geq 1.960$

where $V = (\hat{p} - .5)/\sqrt{.5(1 - .5)/n}$.

The Information Assessment Problem

A second kind of problem that leads naturally to two-sided alternative hypotheses is the problem of testing whether a sample contains enough information to proceed with a statistical investigation. In the last subsection, the emphasis on actions and their consequences suggested the potential for determining sample sizes through experimental design. In this way, the ability to detect "interesting" deviations of the true value of p from the hypothetical value would be designed into the experiment. The decision to accept H_0 would then be an indication that the actual difference between true and hypothetical value is too small to be of interest.

In too many problems dealt with in practice, the sample size is not chosen to achieve prescribed validity and sensitivity requirements. Sometimes, the actions to be taken or their consequences are not sufficiently well understood to carry out a design. In other instances, the availability of data is a primary motivation for the study; because data are often hard to come by, they may be used in studies other than the ones for which they were initially gathered. Then, sample sizes in the later studies are fixed by the size of the data collection.

In this kind of problem, a critical issue is whether the data contain enough information to draw any useful conclusions. This can be checked by a hypothesis test, which is used to "screen" the data for the existence of relevant information.

The null hypothesis H_0 usually represents the state of no (new) information (or no difference, no change, etc.)—the status quo. This state will be represented by a single parameter value. The alternative hypothesis, H_1, which represents the state of "some" information, will then specify all other possible values of the parameter. In the present context, this hypothesis will be two-sided.

In comparison problems of the type introduced in chapter 3, the null hypothesis H_0 is the hypothesis of no difference in the distributions being compared. This is actually the context that led to the terminology "null hypothesis." The sample is screened to determine whether sufficient information exists to "see" any difference in distributions. If there is (that is, if H_0 is rejected), then the inference process continues to determine the nature of the differences and, if desired, to quantify their magnitudes by means of confidence intervals. If not, the process stops. An illustration of such a comparison problem, called a *one-sample problem* because a random sample is taken from only one of the populations, is given in the following example.

EXAMPLE 9.3

The problem of example 9.1 (p. 258) is actually a one-sample comparison problem. We are asked to compare the proportion, p, of upperclassmen and graduate students in the population of current statistics students with the proportion, p_0, of such students from a previous year. It is assumed that we know the population proportion from the previous year to be $p_0 = .482$, so it is only necessary to sample from the current population.

The question to be answered is whether there has been a change in the proportion over the years. More precisely, since the data of display 1.1 are available and will be used in the study, the question is whether there is sufficient information in the data to detect any change in the proportion. Since either an increase or decrease would be of interest, a two-sided alternative hypothesis is used.

The hypotheses to test are then

H_0: $p = .482$ (the hypothesis of no change)

versus

H_1: $p \neq .482$ (the hypothesis of some change)

We will test these hypotheses at the 5% significance level. As before, the test rule is

Reject H_0 if $|V| \geq 1.960$

where $V = (\hat{p} - .482)/\sqrt{.482(1 - .482)/36}$.

In example 9.1 the observed value of the test statistic was shown to be $V = -2.450$. Thus, $|V| = 2.450 > 1.960$ and, following the test rule, our decision would be to reject H_0.

We would conclude *at the 5% significance level* that there is evidence that the current proportion of upper-division and graduate students differs from the previous value of .482. Because $\hat{p} = .278$, we could augment this conclusion by stating that the evidence supports a current proportion smaller than .482.

Had the data in the last example been such that the decision was to accept H_0, the nature of the conclusion would have been rather different. We would have concluded at the 5% significance level there is *insufficient evidence* to infer that a difference exists between the current proportion and the previous value of .482. That is, we could *not* conclude that the proportion is equal to .482 or even that the difference is too small to be of interest.

Because the experiment has not been designed to guarantee that interesting differences will be seen with high probability, a rather large difference may actually exist, but the experiment simply does not contain enough information to detect it. This flaw is the reason for the less satisfactory form of conclusion.

Two-Sided Tests for a Population Mean

Large-sample tests for a population mean follow the ideas and details of tests for a population proportion, with the difference that the hypotheses involve μ rather than p and the test statistic $t = (\overline{X} - \mu)/(s/\sqrt{n})$ is used in place of V. This is Student's t statistic introduced in chapter 7, and the tests are *Student's t tests*   called (**Student's**) t **tests.**

The hypotheses leading to a two-sided t test are of the form

H_0: $\mu = \mu_0$

versus

H_1: $\mu \neq \mu_0$

The rule for a level α test is

Reject H_0 if $|t| \geq c$

where $t = (\overline{X} - \mu_0)/(s/\sqrt{n})$. For large samples ($n \geq 30$) an approximate cutoff value c can be found from tables of the standard normal distribution as the solution of the equation

$P(|Z| \geq c) = \alpha$

An illustration of a t test is given in the following example.

EXAMPLE 9.4   The sample mean age and standard deviation for the 36 students of display 1.1 are $\overline{X} = 22.03$ and $s = 7.648$. These data are to be used to test at the 10% significance level whether the mean age μ of current statistics students differs from 19.5 years. The phrase "differs from" suggests that the problem poser would be interested in a shift in either direction from the hypothetical value of 19.5 years. It follows that the hypotheses to test are

H_0: $\mu = 19.5$

and

H_1: $\mu \neq 19.5$

The 10% test rule based on the large sample approximation is

Reject H_0 if $|t| \geq 1.645$

where $t = (\overline{X} - 19.5)/(s/\sqrt{36})$ and $c = 1.645$ is the two-sided 10% critical value from table 1B.

The value of t for the sample is

$$t = \frac{22.03 - 19.5}{7.648/\sqrt{36}}$$
$$= 1.985$$

Since $|t| = 1.985$ exceeds the cutoff value of 1.645, H_0 is rejected. We conclude at the 10% significance level that the current mean age differs from 19.5 years. Because $\overline{X} = 22.03$, the data support a mean larger than 19.5.

We will carry this example one step further and answer the question, "How large is the difference (between μ and 19.5)?" with a confidence interval for μ (chapter 8). It is sensible to take the confidence coefficient to be one minus the significance level of the test, namely 90%. This selection can be viewed as matching, with the confidence interval, the degree of concern for validity designed into the hypothesis test. (See also the discussion of confidence intervals and hypothesis tests in section 9.7.)

The bounds for the confidence interval are then

$$\overline{X} \pm 1.645 \frac{s}{\sqrt{n}}$$

Inserting the data yields

$$22.03 \pm 1.645 \frac{7.648}{\sqrt{36}}$$

or

$$19.93 \leq \mu \leq 24.13$$

Thus, the data support a true mean age of between 19.93 and 24.13 years. Equivalently, the mean age exceeds the hypothetical age of 19.5 years by an amount between $19.93 - 19.5 = .43$ and $24.13 - 19.5 = 4.63$ years. We would report with 90% confidence that current statistics students are between .43 and 4.63 years older than the hypothetical mean age, on the average.

EXERCISES

9.8   A coin with *heads* probability p is tossed 100 times, resulting in 43 *heads*. Test at the 5% significance level the hypothesis that the coin is fair.

9.9   The proportion of students from a university taking at least one business course in 1980 was 28%. The university administration wants to determine whether this proportion has changed for the current year in order to adjust the business school's budget allocation. A sample of 300 students will be taken from school records and a test will be carried out at the 1% significance level.

(a) Set up the hypotheses and give the rule for the test.

(b) If 110 students in the sample are taking at least one business course, what decision would be made on the basis of the test in part (a)?

(c) If the null hypothesis is rejected, do there appear to be more or fewer business students now than in 1980?

9.10 The proportion of the U.S. population with a particular genetic marker has been established to be 3%. An experiment to determine whether alcoholism is (in part) genetically determined will consist in measuring this marker for a sample of 350 individuals currently undergoing therapy in alcoholism treatment centers. A deviation in the proportion of alcoholics with the marker from that for the population at large would support the hypothesis of a genetic predisposition to alcoholism.

(a) Set up the hypotheses to be tested and give the rule for a 1% test.

(b) If 12 people in the sample of 350 alcoholics are found to possess the genetic marker, what conclusion would be drawn from the test of part (a)?

(c) If the null hypothesis is rejected in part (b), find a 99% confidence interval for the proportion of alcoholics who possess the genetic marker. By how much does the true proportion differ from 3%?

9.11 The mean of a certain measurement variable X in a population is hypothesized to be $\mu = 3.8$. Test this hypothesis at the 10% significance level, using the outcome of a random sample of 120 individuals from the population, for which the sample mean and standard deviation are $\overline{X} = 4.5$ and $s = 2.1$.

9.12 A manufacturer of soft drinks is required to make periodic checks of the accuracy of the machine that automatically fills the cans the drinks are sold in. Test at the 5% level the hypothesis that the machine is filling cans with 12 ounces of fluid, on the average, if in a random sample of 150 cans the sample mean of the contents is 11.79 ounces, with a standard deviation of .354 ounces.

SECTION 9.5

One-Sided Hypothesis Tests

The hypotheses that are to be tested in a given experiment depend critically on the viewpoint or perspective of the investigator. The two-sided alternative hypothesis established for the coin-toss game of the last section depended on the perspective of the gambling inspector. We will see that other perspectives can lead to *one-sided alternative hypotheses*. When this occurs, new tests, called *one-sided tests*, should be used.

Other Viewpoints and One-Sided Hypotheses

Let's adopt the viewpoint of a (slightly unscrupulous) gambling casino owner in the coin-toss game of the last section. The owner is in the business to make money and wants to avoid paying excessive winnings to the gambler, who wins when the coin comes up *heads*. Consequently, the owner will take action (replace the coin) if the coin is biased in the gambler's favor, namely for $p > .5$. No action will be taken if $p = .5$. Being slightly unscrupulous, the owner will "look the other way" and also take no action for a coin with *heads* probability $p < .5$. Thus, if we continue to interpret H_1 as the "action" hypothesis, the hypotheses the casino owner would want to test are

$$H_0: p \leq .5$$

versus

$$H_1: p > .5$$

one-sided hypothesis

The alternative hypothesis (and the null hypothesis) is called **one-sided** in this case, because it involves parameter values on only one side of the hypothetical value .5.

If we now shift the viewpoint to that of the gambler, the same problem context produces yet another form for the hypotheses. The gambler would be concerned about being cheated by a coin that is biased in favor of the house ($p < .5$) and would want to take action in such a case—for example, quit playing, accuse the casino management of cheating, etc. Presumably, the gambler would not stop playing if the coin is fair or is biased in his or her favor ($p \geq .5$). Consequently, the hypotheses to test are

$$H_0: p \geq .5$$

versus

$$H_1: p < .5$$

The alternative hypothesis is again one-sided; to distinguish it from the other form, we could call it a *lower* one-sided hypothesis. The casino owner's alternative would then be an *upper* one-sided hypothesis.

One-Sided Tests for a Population Proportion

At a given significance level, the two-sided test constructed for a two-sided hypothesis in the last subsection is (at least in design) a valid test for either of the one-sided hypotheses. That is, if the two-sided test is designed to have a type I error probability no larger than 5%, say, then the probability of rejecting either one-sided H_0 when true will also be no larger than 5%. Consequently, it is "legal" from the validity point of view to use a two-sided test in a one-sided problem.

However, from the sensitivity viewpoint, we should be able to do better. Intuitively, if we give up the possibility of detecting a value of p on the uninteresting side of $p = .5$ (namely for the range of values specified by H_0), we should be able to concentrate the information in the sample more effectively to detect a value on the other side (specified by H_1). This is the

one-sided tests

motivation for introducing new tests, called **one-tailed** or **one-sided tests** for these hypotheses.

Consider first one-sided hypotheses of the form

$H_0: p \leq p_0$

versus

$H_1: p > p_0$

This is the upper one-sided alternative hypothesis. The appropriate test statistic is again

$$V = \frac{\hat{p} - p_0}{\sqrt{p_0(1 - p_0)/n}}$$

but now we are interested in rejecting H_0 only for values of $\hat{p}$ "sufficiently far" above p_0, or equivalently, for "large" values of V. The test rule is

Reject H_0 if $V \geq c$

Note that this differs from the rule for the two-sided test in that the absolute value of V is not used. The cutoff value is again chosen to hold the probability of a type I error at the given significance level α. However, since H_0 is rejected only for values of V at or above c, using the normal large-sample approximation to the distribution of V, the critical value is now the solution of the equation

$P(Z \geq c) = \alpha$

For example, to construct a 5% one-sided test for the gambling casino owner, who wishes to test $H_0: p \leq .5$ versus $H_1: p > .5$, the test statistic would be $V = (\hat{p} - .5)/\sqrt{.5(1 - .5)/n}$ as before, but the test rule would be

Reject H_0 if $V \geq 1.645$

Recall from chapter 7 that the critical value $c = 1.645$ is the solution to the equation

$P(Z \geq c) = .05$

found from table 1B using the ONE-SIDED TESTS column.

Similarly, the test for

$H_0: p \geq p_0$

versus

$H_1: p < p_0$

uses the rule

Reject H_0 if $V \leq -c$

where the cutoff value $-c$ is the solution to the equation

$P(Z \leq -c) = \alpha$

and V is the test statistic given above.

Thus, a 5% test of the gambler's hypotheses $H_0: p \geq .5$ versus $H_1: p < .5$ would use the rule

Reject H_0 if $V \leq -1.645$

where the test statistic $V = (\hat{p} - .5)/\sqrt{.5(1 - .5)/n}$ is the same one used by the casino owner.

One-Sided Tests for a Population Mean

Large sample tests for a population mean follow the ideas and details of tests for a population proportion, with the difference that the hypotheses involve μ rather than p and the test statistic $t = (\overline{X} - \mu)/(s/\sqrt{n})$ is used in place of V. Applications are given in the following examples.

EXAMPLE 9.5

A consumer protection agency plans to investigate complaints that a certain retail store is overcharging customers relative to regional price norms. To test this hypothesis, the regional mean price for a population list of 500 items is computed from a current trade journal to be $35.40. A buyer is sent to the retail store to randomly sample prices for 100 of the items on the list. The mean and standard deviation of the prices for the sampled items are $37.25 and $15.45, respectively. Test the hypothesis that the store is overcharging customers, on the average, at the 5% significance level.

If μ denotes the true mean price of the 500 items for the retail store, since action will be taken if the accusation of overcharging is substantiated, the appropriate hypotheses are

$H_0: \mu \leq 35.40$

and

$H_1: \mu > 35.40$

Using the normal approximation to the distribution of the t statistic, the rule for a 5% test of these hypotheses is found to be

Reject H_0 if $t \geq 1.645$

where $t = (\overline{X} - 35.40)/(s/\sqrt{100})$. The value of the t statistic is

$$t = \frac{37.25 - 35.40}{15.45/\sqrt{100}}$$
$$= 1.197$$

Since this number does not exceed the cutoff value of 1.645, the decision would be to accept H_0. We would conclude at the 5% significance level that there is insufficient evidence to accuse the management of the retail store of overcharging its customers (on the average).

EXAMPLE 9.6

Refer to example 9.2 (p. 259). The question of interest is whether the HDL cholesterol levels for the stored blood have deteriorated—that is, have decreased. If so, the blood samples would have to be discarded. This question will be addressed by performing a one-sided hypothesis test at the 1% significance level.

We will take the viewpoint that the more serious error is to discard the samples if they are still good. This determines H_1 to be the hypothesis for which the action to discard the samples would be taken, if it is accepted. The hypothetical "fresh blood" level is 47 mg%, so if μ represents the current mean HDL level for the stored blood, the hypotheses to test are

$$H_0: \mu \geq 47$$

and

$$H_1: \mu < 47$$

Since the sample size of $n = 57$ is in the large-sample range, the 1% test rule is taken to be

Reject H_0 if $t \leq -2.326$

where $t = (\overline{X} - 47)/(s/\sqrt{57})$.

The sample mean and standard deviation for the stored blood HDL values were seen to be $\overline{X} = 27.84$ and $s = 10.61$ mg%. The value of the test statistic for this sample is

$$t = \frac{27.84 - 47}{10.61/\sqrt{57}} = -13.634$$

Since $-13.634 < -2.326$, H_0 would be rejected. We would conclude at the 1% significance level that the stored blood has deteriorated and is no longer suitable for studies involving HDL cholesterol levels.

How much has the blood deteriorated? That is, how much smaller than 47 mg% is the mean HDL level of the stored blood? This question can be answered by using a one-sided confidence interval that provides an upper bound for the mean level. A 99% confidence coefficient (equal to 1 minus the significance level) will be used.

The 99% upper confidence limit for μ is

$$\overline{X} + 2.326s/\sqrt{n} = 27.84 + 2.326\frac{10.61}{\sqrt{57}}$$

$$= 31.11 \text{ mg}\%$$

Thus, with 99% confidence, the true mean HDL level of the stored blood has fallen below the level of 47 mg% by at least $47 - 31.11 = 15.89$ mg%.

The Stages of a Hypothesis-Testing Problem Solution

Even in textbook solutions to inference problems, it is useful to follow a strategy that breaks the solution into the four stages given in section 8.3: (1) a design stage, (2) a data acquisition stage, (3) an analysis stage, and (4) an interpretation and reporting stage.

1. The design stage becomes the problem setup stage in the textbook context. The kind of variable being considered, measurement or categorical, determines which population parameter will be of interest in the problem. The design stage continues with the identification of the hypotheses from the problem statement. From the viewpoint presented, what is the "action" hypothesis? If each hypothesis leads to an action, which is the more important action to avoid if unwarranted (that is, if the other hypothesis is true)? This will determine the alternative hypothesis H_1. The null hypothesis H_0 will specify the collection of parameter values complementary to those specified by H_1. Note that the null hypothesis always contains the equals sign (for example, $\mu \geq \mu_0$, $p \leq p_0$, $\mu = \mu_0$, and so forth). If a one-sided alternative hypothesis is not clearly indicated, assume a two-sided alternative. Now, based on these hypotheses, the given significance level for the test, and the sample size, specify the test rule, complete with the cutoff value, and give the form of the test statistic with only those numbers missing that will come from the sample.

2. The data acquisition stage will be missing in textbook problems, but its inclusion in the above list of stages is to remind you that the data given in the problem (if any) should be set aside until the design stage is completed. The data should play no role in the selection of hypotheses or significance level for the test.

3. Once the data are available, the appropriate summary statistics are calculated and the test statistic is evaluated. The test rule then leads to the decision to accept or reject H_0 on the basis of the test statistic.

    Upon deciding that a difference exists between the true and hypothetical parameter values, if the problem statement requires an evaluation of how large the difference is, a confidence interval would also be calculated here.

4. The problem solution is not complete until the information provided by the last step is interpreted and put into an understandable form. In solving the problem, you play the role of a consulting statistician, working for the person or agency that posed the problem. Your conclusions should be written in terms this individual or agency can understand, namely the terms of the original problem statement. The appropriate significance level and/or confidence coefficient should be included in the conclusions.

The examples, starting with example 9.3, illustrate the steps of this solution strategy.

EXERCISES

9.13 A county agricultural agency is responsible for the protection of crops against dangerous insects. A variety of fruit fly is suspected of being responsible for damage seen on some locally grown peaches. If the infestation occurs in more than 10% of the peach trees in the county, the potential loss to growers would make it economically justifiable to undertake a massive spraying of the peach groves with a pesticide. However, because of the expense and the possible side effects of the pesticide on humans and animals, the agency does not want to use it unless necessary. Fly traps are set out in 500 randomly selected trees. A 1% hypothesis test will be used to decide whether or not to use the spray.

(a) Give the parameter of interest and its interpretation.

(b) Specify the hypotheses using the agency's viewpoint.

(c) Give the test statistic and test rule.

(d) If fruit flies are found in 74 of the 500 traps, what decision would the test of part (c) lead to?

(e) What is the appropriate bound on the proportion of trees infested by the fly that is supported by the data at the 99% confidence level?

(f) Based on parts (d) and (e), formulate conclusions suitable for reporting to the agricultural agency.

Note: Parts (a)–(f) of the solution to exercise 9.13 correspond to the steps in the solution strategy suggested in Stages of a Hypothesis-Testing Problem Solution, above. Include these steps in the solutions to the following exercises.

9.14   Repeat exercise 9.13 from the fruit growers' perspective.

9.15   Refer to exercise 9.6 (p. 260). An inspector from a consumer protection agency is sent to the soft-drink plant to check on a complaint that the 12-ounce cans are being filled with less than the advertised amount of soft drink. Using the data from the spot check of 150 cans ($\overline{X}$ = 11.79 ounces and s = .354 ounces), devise a 10% hypothesis test to investigate the complaint. If it is substantiated, use a 90% confidence interval to determine by how much the cans are being underfilled, on the average.

9.16   Refer to exercise 9.7 (p. 260). Construct a 5% hypothesis test from the viewpoint that action will be taken if the average weight of fish caught is smaller than it was before the construction of the power plant. Use a 95% confidence limit to give an upper bound for the current mean weight of fish.

9.17   Advertisers claim an average weight loss of at least 15 pounds in the first month of use for a new diet control pill. To test this claim, a consumer organization enlists the services of 35 individuals who use the medication for one month. The average weight decrease in the test group is 8.2 pounds, with a standard deviation of 3.6 pounds. What should the consumer organization conclude about the pills on the basis of a 10% hypothesis test? Use a two-sided confidence interval to determine bounds on the actual average weight loss supported by the data at the 90% confidence level.

9.18   The lot of mylar capacitors from which the data given in chapter 2 represent a random sample (n = 40, $\overline{X}$ = .63767 μf, s = .01081 μf) will be accepted for use in missile guidance circuitry only if the mean capacitance for the lot is at least .6 μf.

Take the viewpoint of the buyer for whom the important action is to reject the lot if it is below standard.

(a) Formulate the appropriate hypotheses for this problem.

(b) Test these hypotheses at the 1% significance level.

◢ SECTION 9.6

P-Values and Neyman-Pearson Hypothesis Tests

The Neyman-Pearson theory of hypothesis testing has the advantages of a well-defined solution strategy that includes the possibility of sample size design, but the decision to accept or reject the null hypothesis at a given significance level lacks the impact of a P-value. Moreover, because P-values are provided by the commonly available statistical computing packages, it has become standard practice to report the outcomes of statistical studies in terms of P-values. For these reasons, it is important to understand the relationship between P-values and Neyman-Pearson tests.

equivalent tests
P-values and Neyman-Pearson tests are linked through the concept of **equivalent tests.** At a given significance level, two rather different-looking test rules can actually lead to tests that are equivalent in the sense that they lead to the same decisions for all outcomes of sampling—that is, whenever test 1 rejects H_0, so does test 2, and conversely. The details of these equivalences will be given in the following subsections.

Hypothesis Tests and P-Values

Suppose a large sample P-value

$$P = P(|Z| \geq |t_{obs}|)$$

has been computed for the measurement variable hypothesis $H_0: \mu = \mu_0$. Recall that t_{obs} is the numerical value of the one-sample t statistic $(\overline{X} - \mu_0)/(s/\sqrt{n})$ computed from the sample.

Then, for any significance level α, the test rule

Reject H_0 if $P \leq \alpha$

produces a level α test of the given hypothesis H_0 against the two-sided alternative $H_1: \mu \neq \mu_0$.

Thus, to obtain a test with a 5% significance level, the rule to reject H_0 if $P \leq .05$ would be used; H_0 would be accepted if $P > .05$. For a 1% test, the rule would be to reject H_0 if $P \leq .01$, and so forth. An interpretation of the P-value in this context is that *P is the smallest significance level for which H_0 would be rejected.* For example, if $P = .02$, then H_0 would be rejected by these tests at significance levels $\alpha \geq .02$, but would be accepted at significance levels smaller than .02.

Now, for matching significance levels, *this* P-*value test is equivalent to the two-sided* t *test given in section 9.4.* That is, both test rules lead to exactly the same decision for all sampling outcomes; from a practical stand-point, it does not matter which of the test rules is used. But this means that once the P-value is given, we have the potential of performing a two-sided *t* test at any significance level whatever, without going through the steps of constructing the test rule and carrying out the calculations and decision procedure.

Consequently, by quoting the P-value (or P-value bounds) in a conclusion, we have not only the interpretation and impact of the P-value but the full scope of the Neyman-Pearson tests as well. For this reason, in subsequent chapters results of hypothesis tests will frequently be reported by computing and quoting P-values instead of quoting significance levels and decisions.

It is important to remember, however, that the design stage (1) of the solution process cannot be forgotten or overlooked. It is still important that the hypotheses be set up *before the data are acquired.* The P-value is then calculated at stage (3) of the solution process.

The *P*-Value Test for a Population Proportion

Let

$$P = P(|Z| \geq |V_{\text{obs}}|)$$

where V_{obs} is the value of $V = (\hat{p} - p_0)/\sqrt{p_0(1 - p_0)/n}$ when $\hat{p}$ has been computed from the sample. Then the P-value decision rule to reject H_0 if $P \leq \alpha$ yields a test for the hypotheses

$$H_0: p = p_0$$

versus

$$H_1: p \neq p_0$$

equivalent to the Neyman-Pearson two-sided test given earlier.

Thus, in order to extend the P-value reporting method to all tests so far considered, we lack only P-values for one-sided tests. They are given in the next subsection.

One-Sided *P*-Values

The large-sample P-value for the upper one-sided hypothesis problem

$$H_0: \mu \leq \mu_0$$

versus

$$H_1: \mu > \mu_0$$

is

$$P = P(Z \geq t_{\text{obs}})$$

where, as before, t_{obs} is the observed value of the one-sample t statistic. Similarly, the P-value for the lower one-sided case

$$H_0: \mu \geq \mu_0$$

versus

$$H_1: \mu < \mu_0$$

is

$$P = P(Z \leq t_{obs})$$
$$= P(Z \geq -t_{obs})$$

Both P-values are easily computed by the methods of chapter 6. An "exact" value can be obtained from table 1A. Alternatively, bounds can be quickly found using table 1B as follows.

For the upper one-sided test, if t_{obs} is negative, quote $P > .5$. If t_{obs} is positive, then find the numbers in the last column bounding t_{obs}. The P-value bounds are then given in the corresponding rows of the ONE-SIDED TESTS column. For example, if $t_{obs} = 1.467$, then the bounding values in the last column are 1.282 and 1.645. The corresponding bounds from the ONE-SIDED TESTS column are

$$.05 < P < .10$$

For the lower one-sided test, follow this same process with $-t_{obs}$. Thus, for example, if $t_{obs} = -2.05$, find the numbers bounding $-(-2.05) = +2.05$ in the last column. The P-value bounds are then

$$.01 < P < .025$$

The one-sided P-values for a population proportion are defined and computed in exactly the same fashion with t_{obs} replaced by V_{obs}. For the hypotheses $H_0: p \leq p_0$ versus $H_1: p > p_0$,

$$P = P(Z \geq V_{obs})$$

whereas for $H_0: p \geq p_0$ versus $H_1: p < p_0$, it is

$$P = P(Z \leq V_{obs})$$
$$= P(Z \geq -V_{obs})$$

Some applications are given in the following examples.

EXAMPLE 9.7

In example 9.5 (p. 272), a consumer protection agency tested the hypothesis that a retail store was overcharging customers at the 5% significance level. The appropriate hypotheses were $H_0: \mu \leq \$35.40$ versus $H_1: \mu > \$35.40$. The value of t_{obs} was found to be 1.197. Thus, the P-value for the test, computed from table 1A, is (to two decimal places)

$$P = P(Z \geq 1.197)$$
$$= .5 - .38$$
$$= .12$$

We would report that the test is insignificant $(P = .12)$; there is insufficient evidence to conclude that the store is overcharging customers, on the average. Since $P > .05$, this conclusion agrees with the results of the Neyman-Pearson test for which the hypothesis of no overcharge was accepted at the 5% level.

EXAMPLE 9.8

Suppose that a gambler who wins on *heads* in a coin-toss game suspects that the casino is using a coin biased in favor of *tails*. If p is the *heads* probability, the gambler's hypotheses are $H_0: p \geq .5$ and $H_1: p < .5$.

On the basis of a significance level of 1% and other design considerations (section 9.7), the gambler decides to play 100 times and take action against the house if a 1% hypothesis test leads to the decision that the casino is cheating.

Although this is a problem that uses the design aspect of the Neyman-Pearson theory, it can be solved with a P-value calculation.

Suppose that in the 100 plays the gambler wins 38 times. The proportion of wins is then $\hat{p} = .38$, and

$$V_{obs} = \frac{.38 - .5}{\sqrt{.5(1 - .5)/100}}$$
$$= -2.400$$

Bounds for the P-value from table 1B are $.005 < P < .01$.

The outcome of the test would be reported as highly significant $(P < .01)$. It follows from the P-value decision rule that H_0 would be rejected at the 1% significance level, and the gambler would proceed to take action against the casino.

One-Sided *P*-Values from Two-Sided *P*-Values

Statistical computer packages typically report only two-sided P-values, even when a one-sided test is to be performed. For tests involving population means and proportions, like those we have considered and will consider in later chapters, one-sided P-values can be easily derived from the two-sided versions.

For an upper one-sided hypothesis, if the value of the test statistic (t_{obs} or V_{obs}) is positive, then the one-sided P-value is one-half of the two-sided P-value. If the test statistic is negative, then report $P > .5$.

Thus, for a t test of $H_0: \mu \leq \mu_0$ versus $H_1: \mu > \mu_0$, if $t_{obs} > 0$ and the two-sided P-value is $P = .04$, for example, then the P-value for the one-sided test would be $P = .02$. If $t_{obs} < 0$, we would report $P > .5$.

For a lower one-sided hypothesis, if the value of the test statistic is negative, then the one-sided P-value is one-half of the two-sided P-value. If the statistic is positive, we would report $P > .5$.

To test $H_0: p \geq p_0$ versus $H_1: p < p_0$, if $V_{obs} < 0$ and the P-value for the corresponding two-sided test has bounds $.005 < P < .01$, for example, then the one-sided P-value is obtained by dividing these bounds by 2: $.0025 < P < .005$. If $V_{obs} > 0$, report $P > .5$.

EXERCISES

9.19   Calculate the P-value and formulate an appropriate conclusion based on the P-value for the hypotheses of exercise 9.9 (p. 268). Verify that the outcome of the P-value test agrees with that of the Neyman-Pearson test.

9.20   Calculate the P-value and use it to formulate an appropriate conclusion for the hypotheses of exercise 9.10 (p. 269). Verify that the outcome of the P-value test agrees with that of the Neyman-Pearson test.

9.21   Calculate the P-value and use it to formulate an appropriate conclusion for the hypotheses of exercise 9.12 (p. 269). Verify that the outcome of the P-value test agrees with that of the Neyman-Pearson test.

9.22   Calculate the P-value and use it to formulate an appropriate conclusion for the hypotheses of exercise 9.13 (p. 274). Verify that the outcome of the P-value test agrees with that of the Neymann-Pearson test.

9.23   Calculate the P-value and use it to formulate an appropriate conclusion for the hypotheses of exercise 9.17 (p. 275). Verify that the outcome of the P-value test agrees with that of the Neyman-Pearson test.

9.24   Calculate the P-value and use it to formulate an appropriate conclusion for the hypotheses of exercise 9.16 (p. 275). Verify that the outcome of the P-value test agrees with that of the Neyman-Pearson test.

SECTION 9.7

Power, Sample Size, and Other Topics*

Topics of interest in hypothesis testing but not pursued in later chapters are collected here.

Hypothesis Tests from Confidence Intervals

The intuitive idea of a confidence interval is that it specifies those values that the data support as possible candidates for the true parameter value. This suggests that one would accept a hypothetical value of the parameter as plausible only if it were contained in the confidence interval. To formalize this idea into a test of a null hypothesis H_0, we could adapt the rule:

> Reject H_0 if none of the parameter values specified by H_0 are contained in the confidence interval

---

* The material in this section is not essential for the methods of subsequent chapters.

This is a legitimate test rule, because it leads to a decision to accept or reject H_0 for every possible outcome of sampling. Moreover, the significance level of the test is related to the confidence coefficient for the confidence interval by the simple expression

significance level = 1 − confidence coefficient

Thus, a 95% confidence interval will produce a 5% test, a 90% interval a 10% test, and so forth.

Two-sided confidence intervals, the kind first introduced in chapter 8, will produce two-sided tests of two-sided hypotheses. Thus, for example, the standard confidence interval for a population mean, which has limits $\overline{X} \pm cs/\sqrt{n}$, can be used to test the hypotheses

$$H_0: \mu = \mu_0$$

versus

$$H_1: \mu \neq \mu_0$$

The rule would be to reject H_0 if μ_0 does not lie in the confidence interval. If the critical value c determines a 95% interval, for example, then the test would have significance level 5%. In fact, this test is equivalent to the two-sided t test given in section 9.4. Thus, the two steps—carrying out the hypothesis test and calculating the confidence interval at confidence coefficient = 1 − significance level—could have been compressed into the calculation of the confidence interval alone.

The confidence interval contains the information to answer all three questions: 1) Is there a difference? 2) What is the nature (direction) of the difference? 3) How large is the difference?

One-sided hypothesis tests can be constructed from one-sided confidence intervals. Thus, for example, the one-sided confidence interval with lower bound $\overline{X} - cs/\sqrt{n}$ can be used to test the one-sided hypotheses

$$H_0: \mu \leq \mu_0$$

and

$$H_1: \mu > \mu_0$$

(Note that the direction of the confidence interval, $\mu \geq$ confidence limit, is the same as the direction of the alternative hypothesis.) Now, the rule is to reject H_0 if none of the values of μ specified by H_0 are contained in the confidence interval. This rule will produce a one-sided test, with significance level = 1 − confidence coefficient that is equivalent to the one-sided t test.

Similarly, a one-sided test of the lower one-sided hypotheses $H_0: \mu \geq \mu_0$ versus $H_1: \mu < \mu_0$ can be constructed from the one-sided confidence interval with upper bound $\overline{X} + cs/\sqrt{n}$. This test is also equivalent to the corresponding one-sided t test.

When these ideas are applied to the population proportion, the large sample confidence intervals again produce one- and two-sided tests of the corresponding one- and two-sided hypotheses considered in sections 9.4 and 9.5, but these tests are not quite equivalent to the tests constructed there. The reason for this is that different versions of the standard error for $\hat{p}$ are used to construct the statistic W used for confidence intervals and the statistic V that appears in hypothesis tests. This means that there will be some sampling outcomes for which the two tests will arrive at different decisions. However, as the sample size increases, these discrepancies become increasingly rare, and the tests are said to be *asymptotically equivalent*. For our purposes, since we will deal exclusively with large samples for categorical variable problems, the difference between the two tests will be ignored.

EXAMPLE 9.9

A 95% confidence interval for the proportion of University of New Mexico upper-division and graduate students taking a statistics course, based on the data of example 9.1, has limits

$$\hat{p} \pm 1.960 \sqrt{\frac{\hat{p}(1-\hat{p})}{n}} = .278 \pm 1.960 \sqrt{\frac{.278(1-.278)}{36}}$$

or

$$.132 \le p \le .424$$

The hypotheses under test in example 9.1 were $H_0: p = .482$ versus $H_1: p \ne .482$. The 5% test based on the confidence interval would reject H_0 if the hypothetical value $p_0 = .482$ does not lie in the interval. Since this is the case, the decision would be to reject H_0, as before.

Note that the confidence interval covers values of p strictly smaller than .482. Thus, the conclusion would be that the current proportion of juniors and above is smaller than the value 48.2% observed in the previous year.

Moreover, we could add that the amount by which the proportion has decreased is between $.482 - .424 = .058$ and $.482 - .132 = .350$, or between 5.8% and 35%. Since all of this information comes from the confidence interval alone, the 95% confidence coefficient covers these statements and becomes a global confidence level for the two-stage procedure—screening for sufficient information to conclude that a difference exists, and then determining the nature and size of the difference.

EXAMPLE 9.10

In example 8.4 (p. 237), a 99% upper confidence limit for the probability of a nuclear accident, computed on the basis of a simulation, was seen to be .0262. A nuclear safety committee will certify the safety of the reactor design if the accident probability is smaller than 2%. What action should the committee take on the basis of a 1% hypothesis test?

If p denotes the true accident probability, the appropriate hypotheses are $H_0: p \ge .02$ and $H_1: p < .02$. A 1% test can be based on the confidence interval rule:

Reject H_0 if none of the values of p specified by H_0 fall in the confidence interval.

Now, since the values of p from .02 to .0262 are specified by H_0 and lie in the confidence interval, this rule would lead to the decision to accept H_0. At the 1% significance level, we would conclude that the sample evidence does not support the desired safety criterion for the reactor, and we would advise that it not be certified.

EXAMPLE 9.11

In example 8.5 (p. 237), a 90% lower confidence bound on the mean achievement score for seniors from a given high school was computed to be 65.94%. The school administration has decided that no remedial action will be taken if the true mean score exceeds 65%. Note that the important action to avoid if unwarranted in this case, the type I error, is to do nothing. Consequently, the term "action" has to be interpreted with care. Here, we assume that the basic goal of the school administration is to provide an adequate education. Thus, the importance of the error of deciding not to provide remedial aid when it is needed overrides the cost of the error of instituting a remedial program when it is not. If μ denotes the true mean score, this argument determines the hypotheses to be

$$H_0: \mu \le 65$$

and

$$H_1: \mu > 65$$

A 10% test of these hypotheses can be based on the 90% lower confidence bound. Since none of the values of μ specified by H_0 exceed 66.10%, the decision would be to reject H_0. At the 10% significance level, it would be decided that remedial action was not needed.

EXERCISES

9.25   A 90% confidence interval for the mean height, μ, of the population of statistics students from which the sample of display 1.1 (p. 9) was drawn is $64.99 \le \mu \le 67.23$ inches. Use this interval to test the hypotheses $H_0: \mu = 69$ inches versus $H_1: \mu \ne 69$ inches. What is the significance level for the test?

9.26   A 95% confidence interval for the true *heads* probability, p, of the coin tossed 5800 times in the experiment reported in display 5.1 (p. 123) is $.490 \le p \le .516$. This interval will be used to test whether the coin is fair.

(a)   Set up the appropriate hypotheses to test using this interval.

(b)   Give the significance level of the test and the test rule based on the confidence interval.

(c)   Give your decision and conclusions for the test of part (b).

9.27   A 95% confidence interval for the proportion, p, of babies with birth defects born to nondiabetic Pima Indian mothers, calculated in example 8.9 (p. 244), was $.0259 \le$

$p \leq .0531$. Use this interval to test at the 5% significance level that this proportion is no different from the incidence of birth defects in babies born to all mothers in the United States, which is 5%.

9.28   The lot of mylar capacitors described in exercise 8.5 (p. 231), for which the sample summary values are $n = 40$, $\bar{X} = .63767$, and $s = .01081$ μf, will be accepted for use in missile guidance circuitry only if the mean capacitance for the lot exceeds .6 μf. Use a 99% lower confidence bound to test the null hypothesis that the mean capacitance is no larger than .6 μf.

Power Curves and Sample Sizes

A coin-toss game is to be repeated to determine whether or not the coin being used is fair. The hypotheses to be tested are

$$H_0: p = .5$$

and

$$H_1: p \neq .5$$

where p is the true probability of *heads*. Consider first the two-sided 5% test of this hypothesis with the rule constructed in section 9.4: Reject H_0 if $|V| \geq 1.960$.

Denote by $\pi(p')$ the probability of rejecting H_0 when the true *heads* probability p has the value p'. Thus, for example, the probability of rejecting H_0 when it is "true," thus $p = .5$, is denoted by $\pi(.5)$. Since this probability is the type I error probability, we have $\pi(.5) = .05$.

There is a probability π of rejecting H_0 for every value of p between 0 and 1. For $p \neq .5$, this is what we have called the *power* of the test. A graph of $\pi(p')$, called a **power curve**, for the test with $n = 50$ is given in display 9.2. It is seen, for example, that if the true *heads* probability is .6, the probability

power curve

Display 9.2

Power curve for the test of $H_0: p = .5$ versus $H_1: p \neq .5$ based on $n = 50$ trials; significance level is $\alpha = .05$

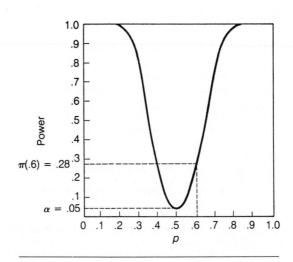

that the test correctly rejects H_0 is .28. If it were important to detect a coin bias as large as $p = .6$, this would not be a very useful test; in only about 28 out of every 100 experiments in which the coin is tossed 50 times would we be led to reject the hypothesis that the coin is fair. However, power depends on sample size, and the test can be improved by increasing the number of coin tosses, n.

How power depends on sample size is demonstrated by display 9.3. The 5% test for H_0: $p = .5$ versus H_1: $p \neq .5$ has a different power curve for each n. This graph contains the curves for $n = 10, 50, 100,$ and 500. From this pattern of curves, we can see how such graphs could be used to determine sample size.

As in the case of sample size selection for confidence intervals, by establishing certain performance *goals* for a hypothesis test, we can select the sample size needed to achieve the goals. For example, suppose that it is important to detect a bias in the coin represented by $p = .6$ with a probability of at least .90. This criterion determines the circled point in display 9.3 at

Display 9.3

Power curves for coin toss hypothesis test ($\alpha = .05$) for varying sample sizes; point determined by requirement that power be at least .90 at $p = .6$ [$\pi(.6) \geq .90$] is circled

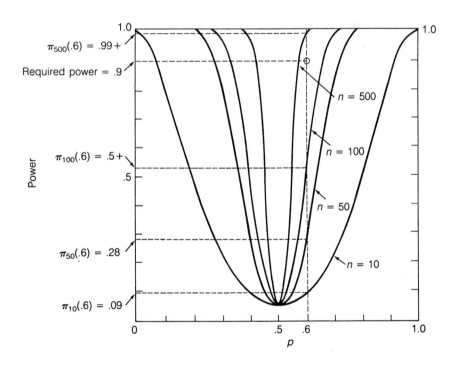

the intersection of the vertical line at $p = .6$ and the horizontal line at $\pi = .90$. Any sample size for which the corresponding curve passes over this point will satisfy the performance goal. For example, the test for sample size $n = 500$ has a power in excess of .99 at $p = .6$. On the other hand, $n = 100$ is not large enough. The cost-effective thing to do is to use the smallest sample size that achieves the goal. Either by a more extensive graph of power curves or by an explicit formula that is available for this problem, it can be shown that the required sample size for this example is $n = 320$.*

Sample size design is an important activity that is often ignored or overlooked, with the result that many experiments are doomed to fail even before they are carried out. For example, a special article in the *New England Journal of Medicine* by Freiman and colleagues [13] reported on a study of 71 clinical trials. The purpose of the trials was to evaluate whether certain drugs or treatments improved patients' condition. However, in these studies no improvement was detected. The authors demonstrated that in 57 of these trials a potential 25% improvement in response was actually indicated (an improvement of considerable economic importance) but went undetected because of inadequate sample size. In 34 of the experiments a 50% improvement was indicated but went undetected. Since clinical trials are performed only after a substantial number of laboratory tests have been carried out and a large investment of time and money has been made, incorrectly rejecting a potentially beneficial drug or treatment at this stage not only deprives the public of its use but also adds substantially to the financial liability of the company. These costs are passed on to the consumer in the increased prices of the drugs that are marketed. Consequently, in this one activity alone, the costs to the public of inadequate sample size design have, quite probably, been astronomical. This is just one area of possible statistical application in which the financial stakes and the stakes in potential human benefit are high. Experimental design, including sample size design, is too important to be ignored. This is one of the activities professional statisticians are trained to carry out and one for which their counsel can be uniquely important.

Power Computations for Student's *t*-Tests

Display 9.4 (p. 287) is designed to calculate the power for the level $\alpha = .05$ *t*-tests for a population mean constructed in sections 9.4 and 9.5. Power curves for tests of both the two-sided hypotheses $H_0: \mu = \mu_0$ versus $H_1: \mu \neq \mu_0$ and the upper one-sided hypotheses $H_0: \mu \leq \mu_0$ and $H_1: \mu > \mu_0$ are given in the display. To read the power for a given value of μ, it is first necessary to calculate the corresponding value of δ:

$$\delta = \frac{\mu - \mu_0}{\sigma/\sqrt{n}}$$

where σ is the population standard deviation. The power is then read from the appropriate curve for the value of δ rounded to one decimal place. (The power can be read to two decimal places for δ values rounded to the nearest one-tenth.) As in the case of confidence intervals, the population standard

---

* Both methods are used in practice. Explicit sample size formulas exist in some simple hypothesis-testing situations, but this is the exception rather than the rule. In more complex problems tables of power curves or charts based on power curves are used. A reference devoted to this topic is the book by Odeh and Fox [26].

Display 9.4

Power curves for 5% large-sample ($n \geq 30$) test of $\mu = \mu_0$ against one-sided hypothesis $\mu > \mu_0$ and two-sided hypothesis $\mu \neq \mu_0$ based on Student's t statistic. Horizontal scale in units of $\delta = (\mu - \mu_0)/(\sigma/\sqrt{n})$. Thus, $\delta = 0$ for $\mu = \mu_0$.

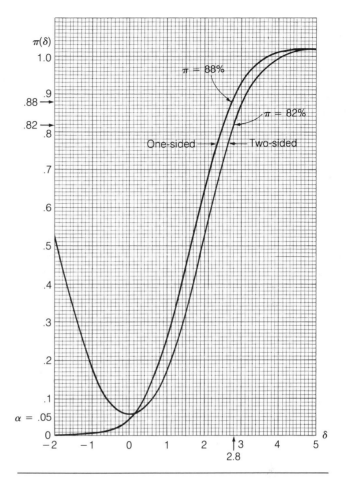

deviation, σ, must be estimated, either from historical data or from a pilot study.

The power curve for tests of two-sided hypotheses is symmetric around $\delta = 0$, and the power curve for tests of lower one-sided hypotheses is the mirror image (around the vertical line through $\delta = 0$) of the one for tests of upper one-sided hypotheses. Thus, power can be calculated for negative values of δ in both cases by the simple expedient of taking the absolute value of δ before using the display.

For example, suppose a two-sided test of the hypothesis that $\mu = 30$ is to be performed, and the power at $\mu = 29$ is desired. The sample size is $n = 50$, and historical data produce an estimate of 0 equal to 2.5. The value of δ is

$$\delta = \frac{29 - 30}{2.5/\sqrt{50}}$$

$$= -2.828$$

Take the absolute value and round δ to $+2.8$. Now, enter this number on the horizontal axis and read the power $\pi = .82$ on the vertical axis opposite the intersection of $\delta = 2.8$ with the two-sided test curve of display 9.4.

Note that a lower one-sided test of the hypothesis that $\mu \geq 30$ versus $\mu < 30$ for $n = 50$ and $\sigma = 2.5$ would have a power of about .88 at $\mu = 29$. The calculation of δ is the same as for the two-sided test, but the power is now read from the one-sided test curve. It follows that, by using a one-sided test instead of a two-sided one, the probability of detecting values of the mean equal to or less than 29 (by rejecting H_0) increases from 82% to 88%.

EXAMPLE 9.12

Assume that the number of items sold by the retail store being investigated for overcharging customers in example 9.5 is large. The hypotheses being tested are $H_0: \mu \leq \$35.40$ and $H_1: \mu > \$35.40$, where μ is the true mean price of items being sold in the store. Suppose that a \$5 average overcharge—yielding a true mean price of $\$35.40 + \$5 = \$40.40$—is serious enough to cause concern for the consumer protection agency. How likely is it that the t-test based on the current sample size of $n = 100$ will detect an average overcharge of this magnitude?

To calculate the power at $\mu = \$40.40$, we must first calculate the value of δ. Since $\mu_0 = \$35.40$, the numerator of the expression for δ is simply the overcharge amount, \$5. If we use example 9.5 as a pilot study and estimate the population standard deviation by the sample standard deviation $s = \$15.45$, then the value of δ for samples of size $n = 100$ is

$$\delta = \frac{5}{15.45/\sqrt{100}}$$

$$= 3.236$$

The power for this value of δ can be seen from the one-sided power curve to be approximately .94.

It follows that we are almost certain to detect an average overcharge as large as \$5. In fact, if we were satisfied with a more modest power of, say, .90 at $\mu = \$40.40$, a sample of size 100 is larger than needed. A more realistic sample size design will be obtained in the next subsection.

Sample Size Design for Student's *t*-Tests

To determine sample size using display 9.4, the power calculation process is reversed. Along with the significance level of 5% and a preliminary value of σ, a minimum power must be specified for the range of values of μ we are interested in detecting. In example 9.12 we might specify that the power be at least .90 for values of $\mu \geq \$40.40$. This could be guaranteed by taking the power to be equal to .90 at $\mu = \$40.40$, since the shape of the power curve then shows that the power will be larger than .90 for values of μ greater than \$40.40.

A design criterion often used in practice specifies for a 5% test a minimum power of 90% over the "interesting" parameter values. However, this would be subject to negotiation at the design stage of the experiment. If this criterion led to an unacceptably large sample size, the power and/or significance level of the test would have to be modified.

The minimum power, for example $\pi = .90$, is entered on the vertical axis of display 9.4, and δ is read from the appropriate curve on the horizontal axis. Thus, for a one-sided test with a minimum power of .90, we read $\delta = 2.9$.

To calculate the sample size, the expression for δ given in the last subsection is solved for n:

$$n = \left(\frac{\sigma\delta}{\mu - \mu_0} \right)^2$$

where μ is the parameter value at the smallest distance from μ_0 for which the minimum power is to hold. Thus, in example 9.12, μ would be $40.40, and the smallest distance from $\mu_0 = \$35.50$ at which the power of .90 is desired is $\mu - \mu_0 = \$5$. The sample size required to achieve the design goals would then be

$$n = \left(\frac{15.45 \cdot 2.9}{5} \right)^2$$

$$= 80.3$$

It would thus require a sample size of approximately 81 to detect a difference of $5 or more between the true and hypothetical means with a probability of at least 90% using a one-sided 5% t test.

EXERCISES

9.29   A test to see whether motorists are exceeding the 55-mile-an-hour speed limit is to be designed to detect an average speed of 56 miles per hour or over with a probability of at least .90. A pilot study consisting of a sample of 300 motorists yielded a standard deviation of $s = 6.89$ miles per hour. Use this value in place of σ in the design. What sample size will be required in order for a 5% test to achieve the design goal?

9.30   A 5% test to determine whether a new tool improves assembly time will be designed to detect an improvement of one-half minute or more, on the average, with a probability of at least .99. The standard deviation of $s = 1.2$ minutes, obtained from a pilot study, will be used in place of σ in the design. Find the number of assemblies to carry out with the new tool in order to meet this goal.

9.31   What sample size will be required to test, at the 5% significance level, whether the advertisers of the diet control pills in exercise 9.17 (p. 275) are making a valid claim if a weight loss of 1 pound or more below the advertised loss of 15 pounds is to be detected with a probability of .90? Use the given standard deviation of 3.6 pounds in the experimental design.

9.32 The experiment of exercise 9.6 (p. 260), in which the accuracy of a soft-drink filling machine was assessed, will be used as a pilot study to design a new 5% test that will detect deviations in mean accuracy as small as .05 ounces with a probability of at least .95. Recall that the estimated standard deviation was $s = .354$ ounces. What sample size would be required for a two-sided test to meet this goal?

The Finite Population Correction for *P*-Values

When the sample is large relative to the size of the population, it is desirable to include a finite population correction in the calculation of *P*-values. This correction involves the *division* of the observed value of the test statistic based on an infinite population (as computed in section 9.2) by the finite population correction factor. Thus, *P*-values for hypotheses concerning a population proportion can be computed by replacing V_{obs} in expressions for the *P*-values by

$$V_{obs\ finite\ pop} = \frac{V_{obs\ infinite\ pop}}{\sqrt{\dfrac{N-n}{N-1}}}$$

Similarly, *P*-values for hypotheses concerning a population mean can be obtained for large samples from small populations by dividing the observed value of the Student's *t* statistic, t_{obs}, by the finite population correction factor before calculating the probability.

EXAMPLE 9.13

In example 9.1, the *P*-value for testing the hypothesis that the current proportion of upperclassmen and graduates in statistics courses at UNM is the *same* as the proportion for an earlier year was calculated to be

$$P = P(|Z| \geq |V_{obs}|)$$
$$= P(|Z| \geq 2.450)$$
$$= .0142$$

Suppose that the population of current statistics students is rather small compared to the sample size of $n = 36$. For example, suppose that $N = 100$. The only modification needed in the *P*-value calculation to account for the large sampling fraction of 36/100 or 36% is to divide $V_{obs} = -2.450$ by

$$\sqrt{\frac{N-n}{N-1}} = \sqrt{\frac{100-36}{100-1}}$$
$$= .804$$

The corrected value of V_{obs} is $-2.450/.804 = -3.047$. The corrected *P*-value is then

$$P = P(|Z| \geq |-3.047|)$$
$$= 2(.5 - .4989)$$
$$= .002$$

The result of applying the finite population correction has been to make the hypothesis appear more implausible than it is relative to the same data from an infinite population.

EXAMPLE 9.14

Refer to example 9.2 (p. 259). The one-sided P-value for testing whether there was a decrease in HDL levels for vials of stored blood can be calculated to be

$$P = P(Z \leq t_{obs})$$
$$= P(Z \leq -13.634)$$
$$< .0000005$$

Suppose there were $N = 150$ vials in the batch from which the $n = 57$ vials were selected. To correct the P-value for this fact, divide the infinite population value $t_{obs} = -13.634$ by the finite population correction factor

$$\sqrt{\frac{N - n}{N - 1}} = \sqrt{\frac{150 - 57}{150 - 1}}$$
$$= .790$$

The corrected value of t_{obs} is then $-13.634/.790 = -17.257$, and the P-value would be $P(Z \leq -17.257)$. This is even farther off the scale of table 1A than was the infinite population value, and we are not able to improve on the bound $P < .0000005$.

Finite Population Corrections for Neyman-Pearson Tests

The finite population correction can be applied in any testing problem of sections 9.4 and 9.5 by *dividing* the value of the infinite population test statistic by the finite population correction factor before applying the decision rule.

EXAMPLE 9.15

Suppose that the number of items sold by the retail store in example 9.5 is $N = 500$. How does this fact alter the 5% test of the hypothesis that the store is overcharging customers?

It was seen that the rule for a 5% test of the appropriate one-sided hypotheses is

Reject H_0 if $t \geq 1.645$

where t is Student's t statistic.

Since the sampling fraction for this problem is $n/N = 100/500 = .2$, the finite population correction should be included in the computations.

The value of the infinite population t statistic was computed to be $t = 1.197$. The finite population correction factor is $\sqrt{(500 - 100)/(500 - 1)} = .895$. Thus, the corrected value of t is

$$t = \frac{1.197}{.895} = 1.337$$

Since this number does not exceed the cutoff value of 1.645, the decision would be to accept H_0. The finite population correction has made no difference in the decision; we would again conclude at the 5% significance level that there is insufficient evidence to accuse the management of overcharging its customers.

The Finite Population Correction for Power Calculations

A nonnegligible sampling fraction indicates the need for a finite population correction in the calculation of power for a t-test from display 9.4. The correction can be accomplished by the simple expedient of *dividing* the infinite population value of δ, given earlier, by the finite population correction factor before reading the power from the graph:

$$\delta_{\text{finite pop}} = \frac{\delta_{\text{infinite pop}}}{\sqrt{\dfrac{N-n}{N-1}}}$$

The following is an example of a power calculation involving the finite population correction.

EXAMPLE 9.16

Suppose that the number of items sold by the retail store being investigated for overcharging customers in example 9.5 is $N = 500$. How likely is it that the t-test based on a sample size of $n = 100$ will detect an average overcharge of $5 or more?

In example 9.12, the infinite population value of δ was computed to be

$\delta = 3.236$

The power for this value of δ was found to be approximately .94.

The sampling fraction of $n/N = 100/500 = .2$ is large enough to influence the power calculation.

To correct for this fact, the infinite population δ value will be divided by the finite population correction factor, which was computed in example 9.15 to be .895. The corrected value of δ is $3.236/.895 = 3.616$. It is seen from the one-sided curve of display 9.4 that the power corresponding to this value of δ exceeds .97. It follows that we are almost certain to detect an average overcharge as large as $5.

The Finite Population Correction for Computing Sample Sizes for t-Tests

For finite populations of size N, the sample size expression is

$$n = \frac{NB}{N+B-1}$$

where $B = (\sigma\delta/(\mu - \mu_0))^2$ is the infinite population expression for sample size. Thus, for the problem of example 9.12, if the total number of items is

$N = 500$, then the finite population sample size required to achieve a power of .90 at $\mu = \$40.40$ would be

$$n = \frac{500 \cdot 80.3}{500 + 80.3 - 1}$$

$$= 69.3$$

That is, the finiteness of the population reduces the required sample size from 81 to $n = 70$.

EXERCISE

9.33   The experiment of exercise 9.6, in which the accuracy of a soft-drink filling machine was assessed, will be used as a pilot study to design a new 5% test that will detect deviations in mean accuracy as small as .05 ounces with a probability of at least .95. Recall that the estimated standard deviation was $s = .354$ ounces.

Suppose that adjustments are to be carried out daily and a two-sided test is to be applied to the 5000 cans produced by the machine every day. What daily sample size would guarantee the above goals? How does this sample size compare with the infinite population value?

Chapter 9 Quiz

1. What is a statistical hypothesis? Give an example.

2. How is the *P*-value for a hypothesis of the form $p = p_0$ (for a population proportion) calculated? Give an example.

3. What is the interpretation of a *P*-value?

4. How are *P*-values used to quantify the plausibility (significance) of a hypothesis? Give the ranges of *P*-values associated with insignificant, mildly significant, significant, and highly significant hypothesis tests.

5. What is the format for reporting a *P*-value in the conclusion portion of a statistical investigation?

6. How is the *P*-value for a hypothesis of the form $\mu = \mu_0$ (for a population mean) calculated? Give an example.

7. What deficiency in the *P*-value method of hypothesis testing motivates the use of the Neyman-Pearson methodology?

8. What is the alternative hypothesis relative to a given hypothesis under test?

9. What is the hypothesis-testing decision problem?

10. How is a hypothesis test specified?

11. What are the two types of errors in a hypothesis testing problem?

12. What is the significance level of a test? How is it used in the test design? What is meant by a valid test?

13. What is the measure of sensitivity used in hypothesis testing? How is it interpreted?

14. What is meant by a two-sided (alternative) hypothesis for a parameter?

15. What is the test for a two-sided hypothesis for a proportion, p? Give an example.

16. What is the form of the two-sided, one-sample Student's t test? Give an example of its use.

17. What is a one-sided hypothesis? Give an example.

18. What determines whether a one- or two-sided hypothesis will be used in a problem solution? How can the form of the alternative hypothesis be determined from a problem statement?

19. What are one-sided hypothesis tests? What advantage do they have over two-sided tests for one-sided hypotheses?

20. What are the forms of the one-sided tests for a proportion? Give examples.

21. What are the forms for the one-sided Student's t tests for a population mean? Give examples.

22. What are the stages of a hypothesis-testing problem solution? Illustrate each step by an example.

23. What is meant by the term "equivalent tests"?

24. What is the P-value test rule for a two-sided population mean hypothesis that is equivalent to a level α two-sided Student's t test?

25. What is the P-value test rule for a two-sided population proportion hypothesis that is equivalent to the Neyman-Pearson test for the same hypothesis?

26. What are the one-sided P-values for population mean hypotheses? How are they calculated? Give examples.

27. What are the one-sided P-values for population proportion hypotheses? How are they calculated? Give examples.

28. How can one-sided P-values be obtained from two-sided P-values for tests concerning population means and proportions?

The following questions are for the optional section 9.7

29. How are level α hypothesis tests obtained from confidence intervals? Give the test rule and the relationship between the confidence coefficient and the significance level of the test.

30. For what kind of confidence intervals will the test be (a) two-sided? (b) one-sided?

31. What tests are equivalent to the tests based on the standard confidence intervals for the mean?

32. What is a power curve?

33. How can power curves be used to design sample sizes?

34. Why is sample size design an important part of the design of an experiment?

35. What is the process for using display 9.4 to determine the power of a (large sample) one-sample t test? Give an example.

36. How can display 9.4 be used to design sample sizes for (large sample) one-sample t tests? Give an example.

37. How are P-value calculations modified to account for finite populations?

38. How are Neyman-Pearson hypothesis tests modified to account for finite populations?

39. How are power and sample size calculations for t tests modified to account for finite populations?

SUPPLEMENTARY EXERCISES

9.34 How plausible is it that a sports reporter who successfully predicts the winners in 117 football games and guesses wrong in 79 is simply "acting like" a fair coin in making the choices? (Find the P-value and draw conclusions concerning the significance of the test.)

9.35 In a Monte Carlo experiment to demonstrate the meaning of the confidence coefficient for a confidence interval (see chapter 8), 96 out of the 100 randomly generated intervals covered the parameter of interest.

(a) Find the P-value for the hypothesis that the true confidence coefficient is 95%.

(b) Find the P-value for the hypothesis that the true confidence coefficient is 99%. Is 90%.

(c) Based on the P-values of parts (a) and (b), which of the three confidence coefficients is best supported by the data?

9.36 The posted speed limit on a certain section of road is 55 miles per hour. The average speed of a random sample of 300 motorists on this section was 57.1 miles per hour, with a standard deviation of 6.89 miles per hour. How plausible is it that the actual mean speed of motorists is equal to the posted speed limit?

9.37 The sample mean and standard deviation of the hemoglobin levels for $n = 37$ women on oral contraceptives are $\overline{X} = 14.491$ and $s = 1.079$. Find the P-value for the hypothesis that the population mean hemoglobin level for women taking oral contraceptives is 15.34.

9.38 The (population) mean algebra placement scores for students at a certain university who took the placement exam last year was 5.15. The summary data for a sample of students who took the exam this year are $n = 44$, $\overline{X} = 3.955$, and $s = 3.530$. Find the P-value for the hypothesis that there has been no change in the mean placement score from the previous year.

9.39 Test at the 5% significance level the hypothesis that the sports reporter who successfully predicted the winners in 117 football games and guessed wrong in 79 is simply "acting like" a fair coin in making the choices. If the null hypothesis is rejected, is the reporter's performance better or worse than tossing a fair coin?

9.40   A theory concerning the cause of the Sudden Infant Death Syndrome (SIDS) is that the lungs of SIDS children do not develop at the same rate as those of normal children. In particular, the muscle thicknesses of lung arterioles differ. In a random sample of 128 arterioles taken from a SIDS victim, the mean muscle thickness (as a percentage of total arteriole diameter) was 9.10, with a standard deviation of 2.150.

(a)   Test at the 5% significance level whether the mean arteriole muscle thickness for this SIDS victim is the same as the mean of 6.04 found for normal children of the same age.

(b)   If the null hypothesis of part (a) is rejected, does it appear that the arterioles of this SIDS victim are more or less heavily muscled than normal?

9.41   The following determinations of the parallax of the sun (the angle spanned by the earth's radius as if it were viewed and measured from the sun's surface), made in 1761 and analyzed by James Short, a noted telescope manufacturer, are given by Stigler [31]. Units are in seconds of a degree (1/360 degree).

8.50	8.06	8.65	9.71	8.80	7.99
8.50	8.43	8.35	8.50	8.40	8.58
7.33	8.44	8.71	8.28	8.82	8.34
8.64	8.14	8.31	9.87	9.02	9.64
9.27	7.68	8.36	8.86	10.57	8.34
9.06	10.34	8.58	5.76	9.11	8.55
9.25	8.07	7.80	8.44	8.66	9.54
9.09	8.36	7.71	8.23	8.34	9.07
8.50	9.71	8.30	8.50	8.60	

With a careful determination of the radius of the earth and a good average value of parallax, the basic measurement unit of astronomy, the astronomical unit (which is the average distance from the earth to the sun), can be obtained. The currently accepted value of the parallax of the sun is 8.798.

(a)   Find the summary statistics n, $\overline{X}$, and s for the data.

(b)   Test at the 5% significance level whether the mean of the (potential) population of Short's parallax determinations is equal to the currently accepted value. If not, is it greater or less than the accepted value?

9.42   It has been observed that sample surveys made by telephone tend to contact a higher proportion of women than is representative of the population—especially if too many of the interviews are done during weekdays. In a city known to be 52.2% female, 2567 of the 4286 telephone calls in a recent survey were answered by women. If p denotes the proportion of potential telephone interviews that would be completed by women if the entire population were surveyed,

(a)   test at the 1% level the null hypothesis that p is no larger than the population proportion of 52.2%.

(b)   The survey bias would be the amount by which the proportion p exceeds the population proportion of 52.2%. If the hypothesis of part (a) is rejected, how small a bias is supported by the given data?

9.43   A new tool is to be tested to determine whether it reduces the average time required to assemble a particular part in an aircraft plant. If it does, the assembly lines that produce this part will, at some expense, be reequipped with it. The assembly time using the current tools averages 13.9 minutes. If the average assembly time for 50 experimental assemblies with the new tool is 13.1 minutes, with a standard deviation of 1.2 minutes, what recommendation should be made to the plant management based on a test at the 5% significance level? Using a 95% confidence limit, how great an improvement in assembly time is supported by the data?

9.44   The traffic safety division of a city plans to install road barriers in a school zone if the average speed of motorists is found to exceed the posted speed of 15 miles per hour. The average speed of a random sample of 150 cars, clocked in the school zone, is observed to be 15.3 miles per hour, with a standard deviation of 2.5 miles per hour. Based on a 5% test, is there sufficient evidence to warrant installing the barriers? Give a 95% lower confidence bound for the true mean speed of cars through the school zone.

9.45   Calculate the P-value and use it to formulate an appropriate conclusion for the hypotheses of exercise 9.40. Verify that the outcome of the P-value test agrees with that of the Neyman-Pearson test.

9.46   Calculate the P-value and use it to formulate an appropriate conclusion for the hypotheses of exercise 9.42. Verify that the outcome of the P-value test agrees with that of the Neyman-Pearson test.

9.47   Calculate the P-value and use it to formulate an appropriate conclusion for the hypotheses of exercise 9.44. Verify that the outcome of the P-value test agrees with that of the Neyman-Pearson test.

PART III

Statistical Inference Methodology

In this third part of the book we return to the problems dealt with in the first section to confirm, by methods of statistical inference, the effects indicated by exploration. By *confirm* we do not mean that we will prove the effects to be real; that cannot be done statistically. However, we will be able to extend the idea of the weight of evidence in a sample about a hypothesis, introduced in chapter 9, to a large class of problems in a variety of settings through the use of hypothesis tests. When an effect is "confirmed," we will quantify its magnitude by means of a confidence interval.

Over the years, many statistical methods have been devised to deal with these problems. Many of the methods introduced rather early in the history of statistical inference, such as the chi-square methods of Karl Pearson, the Student t tests of W. S. Gosset and R. A. Fisher, and Fisher's analysis of variance, have withstood the test of time and now form the core of statistical methodology. They will hold center stage in this part of the book.

However, as the understanding of these methods has grown in the process of solving real problems, some difficulties have been uncovered. The first response was to create a new and seemingly distinct collection of methods, known as *nonparametric statistics*, which is still being actively developed. Although theoretically attractive, nonparametric methods often lack the classical methods' unity of conception and application. Moreover, it was subsequently shown that the problem originally addressed by nonparametric methods—that of lack of validity—is not the main problem with the classical methods. In an important publication, J. W. Tukey [33] established that in commonly occurring practical situations—those in which outliers are present in the data—it is the *sensitivity* of the classical procedures that can be diminished. This observation sparked the development of new statistical techniques, known as *robust methods*.

Briefly, a robust method is one whose properties of sensitivity and validity do not change much if the underlying distribution of observations deviates from that assumed in the construction of the method. An appealing feature of robust methods is that they modify the classical methods in intuitively reasonable ways. Moreover, these modifications are sometimes based on easily computable estimators. For example, the trimmed mean, introduced in chapter 2, is simply an ordinary mean with some of the observations (the outliers among them) removed. In other cases the estimators on which the methods are based, although intuitively reasonable, cannot be readily evaluated without a specialized computer program.

Since our goal in this text is to give methods that are easily implemented either by hand calculation or by readily available computer packages, many robust methods will be beyond our reach. However, much of their flavor can be gained from applications of those robust procedures we will present.

It has been established that many of the now classical nonparametric procedures, originally introduced for other reasons, are robust. R. L. Iman and others [17, 18] have also shown that for even modest sample sizes, excellent approximations to these nonparametric procedures can be obtained by using the same computations and the same tables used for the classical methods. Consequently, important robust alternatives to the classical methods will be made available with only a modest investment in effort to learn a special data transformation: the rank transform.

The robust and nonparametric methods for comparison problems will be given in chapter 12, after a classical introduction to these problems in chapters 10 and 11. Inference methods for categorical data problems will be given in chapter 13. Chapters 14 and 15 deal with simple linear regression, the description of the relationship between two measurement variables by the straight-line regression of one on the other. Some curvilinear regression models are also presented in chapter 15, and the important topic of regression diagnostics is considered in some detail.

An introduction to multiple linear regression is given in chapter 16. Some important diagnostic methods are described and the flexibility of the multiple regression model is illustrated with several examples. Finally, some ideas of experimental design are introduced in the context of randomized block experiments and the analysis of variance in chapter 17. The two-way analysis of variance is given in that chapter, and a graphical method is stressed for the analysis of interactions.

One- and Two-Sample Problems—Small-Sample Theory

Introduction

The tests and confidence intervals for the mean of a measurement variable in the small-sample case differ in one basic respect from those given in chapters 8 and 9—namely, in the sampling distribution for Student's t ratio. The publication of the small-sample distribution of this statistic in 1908 by W. S. Gosset, who published under the pseudonym of Student, marked an important point in the development of modern statistics. We will learn to use Student's t distribution for this and the two-sample t statistic in this chapter.

Recall that the two-sample problem involves the comparison of the distributions of a given variable for two populations based on samples from each. When the distributions are normal and samples are independently drawn from each population, the solution of the problem relies on the two-sample t test, to be covered in section 10.4. However, another frequently occurring mode of sampling is to draw observations in pairs, one member from each population. For example, to test whether a city is contributing a significant amount of sodium ion pollution to a river, one might look at the concentration of sodium ions in paired samples of water, one taken upriver from the city and the other downriver, on several days throughout the year. The problem would then be to test for the equality of mean upriver and downriver ion concentrations.

A difficulty arises in the use of the two-sample t test in such a problem. If the concentration from other cities upriver is large, it is also likely to be large both upriver and downriver from the city in question that day. Thus large concentrations are likely to occur together, as are small ones. This result indicates that the upriver and downriver concentrations have a positive association or, to use another term, they are positively *correlated*. Thus the assumption that the two samples are independent of one another, which is made in the two-sample problem, will be violated. In this situation, another form of test is needed, called the *paired difference* t *test*. This test will be presented in section 10.5.

◢ SECTION 10.2

Student's *t* Distribution

We will be on familiar ground for the hypothesis tests and the confidence intervals for a population mean μ given in this chapter because, except for the use of a different table to determine cutoff values, they are identical to the measurement variable procedures given in chapters 8 and 9. The key to these procedures is again Student's *t* statistic,

$$t = \frac{\overline{X} - \mu}{s/\sqrt{n}}$$

and its sampling distribution. However, for small to moderate sample sizes, the normal approximation for the sampling distribution is no longer adequate.

Two problems arise with small sample sizes. The first is that the standard error of the mean,

$$\sigma_{\overline{X}} = \frac{\sigma}{\sqrt{n}}$$

is larger simply because the factor $1/\sqrt{n}$ is larger for small n. The increased natural variability of $\overline{X}$ decreases the sensitivity of tests and confidence intervals. The effect of sample size on sensitivity was seen in the two previous chapters.

If the population standard deviation were known, this disadvantage would be the only penalty against sensitivity for small sample size. However, since the population standard deviation is almost never known, a second problem arises from the use of the estimator s for σ in the expression for the t statistic. This estimator depends on the particular observations obtained in the sample and thus will vary from sample to sample. For large samples the variation could be justifiably ignored, which made it possible for us to use the normal distribution to approximate the sampling distribution of t. For small samples the normal approximation is no longer adequate.

The effect of the variation of s is to cause the t statistic to take on values large in magnitude with greater frequency than predicted by the normal distribution. This tendency is most pronounced for very small sample sizes and diminishes as n increases. In order for the quoted confidence coefficients and significance levels to be correct, the critical values must be increased beyond those provided by the table of the standard normal distribution. Different values will be needed for each sample size to account for the changing variability of the t statistic. The tables needed to extend the

Student's t distribution

tests and confidence intervals to small samples are based on **Student's *t* distribution,** which is the sampling distribution of the *t* statistic for normally distributed observations.

degrees of freedom

Student's *t* distribution is actually a family of distributions, each member of which is associated with a value of a parameter ν (Greek letter nu) called the **degrees of freedom.** This parameter takes on integer values 1, 2, ... Display 10.1 shows the probability curves for Student's *t* distributions with degrees of freedom $\nu = 1$ and $\nu = 10$. As ν increases, these curves tend rather rapidly to the probability curve of the standard normal distribution, which is also shown in display 10.1. Because it is the limiting curve for arbitrarily large ν, it is viewed as the member of the Student *t* family corresponding to infinite degrees of freedom ($\nu = \infty$).

Display 10.1

Probability curves of Student's *t* distribution for $\nu = 1, 10, \infty$

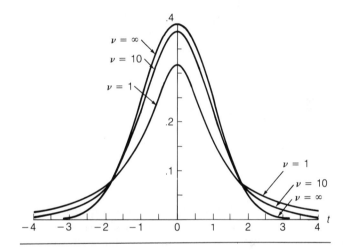

For the one-sample *t* statistic the degrees of freedom and sample size are related by the expression

$$\nu = n - 1$$

Consequently, large samples imply large degrees of freedom and thus a nearly normal sampling distribution for *t*. We have used the fact that the Student *t* distribution is very close to its limiting normal form for large samples in chapters 8 and 9.

Large degrees of freedom will imply accurate estimation of the population standard deviation σ for measurement variable problems and thus improved sensitivity for the statistical procedures. This result should not be surprising since ν is linked to sample size, and we have seen that increasing sample size improves sensitivity. However, as will be seen in later discussions, the degrees of freedom are computed differently for different procedures, and it is the value of ν rather than the sample size that more accurately reflects sensitivity.

Table 3 at the back of the book contains critical values for confidence intervals and hypothesis tests based on Student's t distribution. The body of the table contains the critical values corresponding to the degrees of freedom listed in the first column and to the confidence coefficients or significance levels appearing along the top. The various uses of the table will be illustrated in the following sections.

◢ SECTION 10.3

One-Sample t Tests and Confidence Intervals

The confidence intervals and tests for a population mean given in chapters 8 and 9 extend immediately to the small-sample case by replacing the use of table 1 for obtaining critical values with the use of table 3. This procedure will be demonstrated in a series of examples.

Two-Sided Tests, *P*-Values and Confidence Intervals

The following example illustrates a small sample application of the two-sided t test introduced in section 9.3. The P-value will be calculated, and an answer to the question "how much of a difference?" will be provided by a confidence interval.

EXAMPLE 10.1

Suppose that a hypothesis of solar evolution predicted a mean cooling rate of $\mu = .54$ degrees per million years for the Tocopilla meteorite. (See exercise 3.26 p. 90.) Do the observed cooling rates, given in the stem and leaf plot below, support this hypothesis? Test at the 5% significance level.

Note that the box plot shows good symmetry and no outliers. Thus, the assumption of a normal distribution of observations, basic to the t test, appears to be realistic.

To regain original units, move decimal 1 place to the left.

```
1|5
2|794
3|068
4|03
5|6
6|27
```

Since deviations from the hypothetical value in either direction would tend to discredit the hypothesis, a two-sided alternative is indicated:

H_0: $\mu = .54$ and H_1: $\mu \neq .54$.

The sample statistics for the Tocopilla meteorite are

$n = 12$ $\bar{X} = 3.892$ $s = 1.583$

The appropriate test rule is to reject H_0 if $|t| \geq c$, where the critical value, c, is to be obtained from table 3. Read the value $c = 2.201$ from the intersection of the row labeled degrees of freedom $\nu = n - 1 = 11$ and the TWO-SIDED TESTS column labeled with the significance level $\alpha = .05$.

The value of the t statistic is

$$t = \frac{\overline{X} - .54}{s/\sqrt{n}}$$
$$= \frac{3.892 - .54}{1.583/\sqrt{12}}$$
$$= 7.335$$

The decision would be to reject the null hypothesis. We would conclude at the 5% significance level that the true mean cooling rate for this meteorite is greater than .54 degrees per million years.

A conclusion with more impact would be obtained from the use of the P-value

$$P = P(|t| \geq |t_{obs}|)$$

Bounds for this P-value can be obtained from table 3 by searching along the row corresponding to degrees of freedom $\nu = 11$ to find values bounding $|t_{obs}| = 7.335$. The two-sided P-value bounds would then be read, in reverse order, from the same columns of the TWO-SIDED TESTS row. Since 7.335 falls to the right of the last number in the $\nu = 11$ row, the table provides only the upper bound

$$P < .001$$

We would conclude that the hypothetical value $\mu = .54$ is highly implausible on the basis of this sample.

A confidence interval for μ has the familiar form

$$\overline{X} - t\frac{s}{\sqrt{n}} \leq \mu \leq \overline{X} + t\frac{s}{\sqrt{n}}$$

seen in chapter 8. The only difference is that the critical value, t, will now be obtained from table 3 at the intersection of the row labeled $\nu = n - 1$ and TWO-SIDED TESTS column for which the probability is equal to $1 -$ confidence coefficient.

Thus, to obtain a 95% confidence interval for the mean cooling rate of the Tocopilla meteorite, find the critical value $t = 2.201$ in the row labeled $\nu = 11$ and TWO-SIDED TESTS column labeled $1 - .95 = .05$. The limits for the confidence interval are then

$$3.892 \pm 2.201 \frac{1.583}{\sqrt{12}}$$

$2.886 \leq \mu \leq 4.898$ degrees per million years

With 95% confidence, we are now able to conclude that the true mean cooling rate exceeds the hypothetical value by an amount between $2.886 - .54 = 2.346$ and $4.898 - .54 = 4.358$ degrees per million years.

One-Sided Tests, *P*-Values and Confidence Intervals

The following examples show how to extend the ideas of one-sided *t* tests and confidence intervals to small samples.

EXAMPLE 10.2

We will test the hypothesis that cholesterol levels for women increase with age. This will be done by comparing the levels given in the stem and leaf plot below, which provides cholesterol levels for a sample of women 40 years of age or older with the mean level for women under 30. The mean level for women under 30 has been established to be 200 mg%. The test will be carried out at the 10% significance level.

```
1*
 · | 62, 67
2* | 01, 14, 20, 29, 39
 · | 53, 71, 91
```

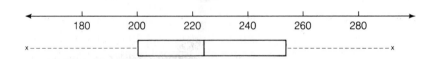

As seen from the box plot, the assumption of normality for this sample is reasonable.

If μ denotes the true mean level for women 40 years of age and older, then the hypotheses to test are

$$H_0: \mu \le 200 \quad \text{and} \quad H_1: \mu > 200$$

The test rule is to reject H_0 if $t \ge c$, where

$$t = \frac{\bar{X} - 200}{s/\sqrt{10}}$$

The degrees of freedom are $\nu = 10 - 1 = 9$ for this problem and, entering table 3 at the ONE-SIDED TESTS column corresponding to $\alpha = .10$, the critical value $c = 1.383$ is found.

The sample statistics are

$$n = 10 \quad \bar{X} = 224.7 \quad s = 41.53$$

Thus, the value of the *t* statistic is

$$t = \frac{224.7 - 200}{41.53/\sqrt{10}} = 1.881$$

Because $t = 1.881$ exceeds the cutoff value $c = 1.383$, the null hypothesis would be rejected and we conclude at the 10% significance level that the mean cholesterol level for women 40 years of age or older exceeds that for women under 30. Thus, there is evidence of an increase in cholesterol level with age.

The P-value for this problem,

$$P = P(t \geq t_{\text{obs}}) = P(t \geq 1.881)$$

can be bounded by finding the numbers in the $\nu = 9$ row of table 3 that bound 1.881. They are 1.833 and 2.262. The corresponding columns in the ONE-SIDED TESTS row yield the bounds

$$.025 < P < .05$$

Thus, the null hypothesis would be rejected by a one-sided t test at the 5% significance level, but not at the 2.5% level.

A 90% lower confidence limit for μ is of the form

$$\overline{X} - t\frac{s}{\sqrt{n}},$$

where the critical value t is now to be obtained from table 3. Locate $t = 1.383$ in the $\nu = 9$ degrees of freedom row and ONE-SIDED TESTS column for significance level $= 1 - .90 = .10$. The one-sided confidence interval is then

$$\mu \geq 224.7 - 1.383\frac{41.53}{\sqrt{10}}$$

or

$$\mu \geq 206.54 \text{ mg\%}$$

Thus, with 90% confidence, the mean cholesterol for women aged 40 and over exceeds that for women under 30 by at least $206.54 - 200 = 6.54$ mg%.

EXAMPLE 10.3

A consumer group suspects that the average weight of canned tomatoes being produced by a large cannery is less than the advertised weight of 20 ounces. To check their conjecture, the group purchased 14 cans of the cannery's tomatoes from various grocery stores. The weights of the contents of the cans, to the nearest half ounce, were as follows.

20.5, 18.5, 20.0, 19.5, 19.5, 21.0, 17.5,
22.5, 20.0, 19.5, 18.5, 20.0, 18.0, 20.5

Do the data confirm the group's suspicion? Test at the 5% significance level.

The stem and leaf plot and box plot of the data indicate no problems with the normality assumption:

To regain original units, move decimal one place to the left.

```
17 | 5
18 | 055
19 | 555
20 | 00055
21 | 0
22 | 5
```

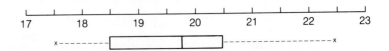

The 5% one-sided cutoff value for $14 - 1 = 13$ degrees of freedom from table 3 is 1.771. Thus, the test rule for the hypotheses

$$H_0: \mu \geq 20 \quad \text{and} \quad H_1: \mu < 20$$

is to reject H_0 if $t \leq -1.771$, where

$$t = \frac{\bar{X} - 20}{s/\sqrt{14}}$$

The summary statistics are

$$n = 14 \qquad \bar{X} = 19.68 \qquad s = 1.295$$

Thus,

$$t = \frac{19.68 - 20}{1.295/\sqrt{14}}$$
$$= -0.925$$

Since this value is larger than -1.771, the decision would be to accept H_0. We would conclude at the 5% significance level that there is insufficient evidence to support a mean weight of less than 20 ounces for the canned tomatoes.

To compute the P-value

$$P = P(t \leq t_{\text{obs}}) = P(t \geq -t_{\text{obs}})$$

enter table 3 with the negative of the observed value of t, $-(-0.925) = +0.925$. The P-value bounds

$$.10 < P < .25$$

are read from the columns in the ONE-SIDED TESTS row corresponding to the first two entries of the row for $\nu = 13$. It follows that the hypothesis H_0 would have been accepted even with a 10% test. Consequently, there is little support for the consumer group's suspicion.

Robust procedures for the one-sample problem to deal with deviations from the normality assumption will be given in chapter 12. We next take up one of the most important and widely used procedures in all of statistics—Student's t test for the two-sample problem.

EXERCISES

10.1   The summary statistics for the parachute fill times given in exercise 1.12 (p. 31) are $n = 27$, $\bar{X} = .453$ seconds, and $s = .141$ seconds.

(a) test at the 5% level that the mean fill time is .5 seconds for this type of parachute.

(b) Find a 95% confidence interval for the mean fill time.

10.2 The following data are drop rates in percents for students in a sample of sections of an elementary mathematics course over a period of years: 13, 10, 13, 16, 3, 14, 17, 10, 5.

(a) Test at the 10% significance level the null hypothesis that the mean drop rate for this course is no larger than 10% against the alternative that it exceeds 10%.

(b) If the hypothesis of part (a) is rejected, find a 90% lower confidence bound for the mean drop rate.

10.3 Ages at first transplant of a sample of heart transplant patients are as follows: 54, 42, 51, 54, 49, 56, 33, 58, 54, 64, 49.

(a) Find bounds for the P-value for the hypothesis that the mean age at first transplant is 50 years.

(b) Find a 95% confidence interval for the mean age at first transplant.

10.4 Because of the need to dispose of their older cars through conventional used car outlets, car rental firms pay close attention to the used car market. In a recent survey of 578 purchasers of used cars, a large car rental corporation determined that the average price paid for used cars purchased from franchised new car dealers was $4196, with a standard deviation of $1928.

(a) The average price (population mean) paid for cars purchased from independent used car dealers was reported in a trade magazine to be $3678. Test, at the 5% level, whether the average prices for used cars bought from the two types of dealers (new and used car dealers) are the same. If they are not the same, for which type of dealer is the average price greater?

(b) Find a 95% confidence interval for the nationwide mean purchase price μ of used cars bought from new car dealers.

10.5 A random sample of $n = 17$ bicyclists who rode 100 miles in the 10th Annual Tour of the Rio Grande Valley had a mean completion time of $\overline{X} = 7.993$ hours, with a standard deviation of $s = 1.281$ hours.

(a) The (population) mean completion time for the 50-mile riders, extrapolated to 100 miles, is 8.74 hours. Recalling that speed varies inversely with completion time, set up the appropriate statistical hypotheses to test whether the 100-milers are speedier than the 50-milers, on the average.

(b) Test the hypotheses of part (a) at the 10% significance level.

(c) Find the appropriate 90% confidence bound for a one-sided confidence interval on the mean completion time for the cyclists who rode 100 miles.

10.6 The cooling rates for the Coahuila meteorite of exercise 3.26 (p. 90) are: 5.6, 4.9, 3.3, 1.8, 4.8, 3.6, 3.2, 1.4, 1.6, 3.2, 8.2.

(a) Make a box plot of these observations and comment on any indications of nonnormality.

(b)   Find *P*-value bounds for the *t* test of the hypothesis that the true mean cooling rate is .54 degrees per million years. Is 4.0 degrees per million years.

(c)   Find a 90% confidence interval for the mean cooling rate of this meteorite.

10.7   Glucagon levels have been obtained for ten children being given a course of treatments for dwarfism using a human growth factor. The data are to be used to evaluate the treatments for future use. If the treatments lead to a mean glucagon level below 200, they will be considered satisfactory and no further treatments would be prescribed. If not, the course of treatments would be repeated. The levels for the children are as follows: 60, 215, 270, 40, 183, 117, 179, 42, 125, 120.

(a)   Make a box plot of the data and comment on any indications of nonnormality.

(b)   Formulate the statistical hypotheses from the viewpoint that the most serious error is to discontinue treatments if the patients have not yet been cured.

(c)   Find the *P*-value for the hypotheses of part (b) and use it to test these hypotheses at the 5% significance level.

(d)   If the null hypothesis of part (c) is rejected, construct a 95% one-sided confidence interval to evaluate the degree to which the true mean glucagon level for patients under treatment differs from the normal level of 200.

◤ SECTION 10.4

The Two-Sample Problem

The two-sample problem is the problem of comparing the frequency distributions of a variable for two populations based on samples from *both* populations. Here, the variable will be of the measurement type. Some of the problems studied exploratively in chapter 3 were of this type. For example, the problem given in example 3.1 (p. 71), of assessing a common genetic origin for modern Englishmen and Celts (the two populations), based on the distributions of maximum head breadths, was of this variety.

As before, the comparison will be based on a location measure for the frequency distributions. When the measure is the population mean and when the frequency distributions are normal, the appropriate statistical test is the two-sample Student *t* test. The popularity of the test is well deserved. Not only is it used as is for a wide variety of problems, but it can also be extended to a much broader class of problems through simple modifications. Two such modifications will be given in chapter 12.

Two-Sample Student's *t* Test

The basic assumptions underlying this test are as follows. *First, it is assumed that the variable of interest, X, is normally distributed for both populations.* The means and standard deviations of X are subscripted to designate the population, as in the following table:

POPULATION PARAMETERS	POPULATION	
	1	2
Mean	μ_1	μ_2
Standard deviation	σ_1	σ_2
Common value of standard deviation	σ	

Second, it is assumed that the population standard deviations are equal. The common value of the unknown standard deviation is denoted by σ. Because normal distributions are determined by their means and standard deviations, these two distributions will differ only in the values of their means μ_1 and μ_2.

The null hypothesis of no difference in distributions corresponds to the statistical hypothesis

$$H_0: \mu_1 = \mu_2$$

This can be tested against the two-sided alternative,

$$H_1: \mu_1 \neq \mu_2$$

corresponding to a difference of some kind—that is, a difference in either direction.

Similarly, the hypothesis $H_0: \mu_1 \leq \mu_2$ can be tested against $H_1: \mu_1 > \mu_2$, corresponding to a larger average value of X for population 1, or $H_0: \mu_1 \geq \mu_2$ can be tested against $H_1: \mu_1 < \mu_2$, corresponding to a smaller value.

Finally, it is assumed that random samples are independently taken from the two populations. That is, sampling is carried out in such a way that the values obtained in one sample do not influence the probabilities with which the values in the other are selected.

The observations from each sample are now summarized by the sample sizes, sample means, and standard deviations:

	POPULATION	
	1	2
Sample size	n_1	n_2
Sample mean	$\overline{X}_1$	$\overline{X}_2$
Standard deviation	s_1	s_2

These statistics will be used to construct tests for the pairs of hypotheses stated above.

Until now, statistical hypotheses have involved values for a single parameter. If we are given an estimator for this parameter, an expression for its standard error and the sampling distribution of the Z-score transform, we can easily obtain hypothesis tests and confidence intervals. It is also possible to use this method here. What is needed is a single parameter that combines μ_1 and μ_2 and allows us to formulate H_0 and H_1 in more familiar terms. The appropriate parameter is

$$\Delta = \mu_1 - \mu_2$$

the *difference* between the two population means (Δ is the capital Greek letter delta). In terms of Δ the hypotheses $H_0: \mu_1 = \mu_2$ and $H_1: \mu_1 \neq \mu_2$ can be expressed in the equivalent form

$$H_0: \Delta = 0 \text{ and } H_1: \Delta \neq 0$$

The equivalent forms for the one-sided hypotheses listed above are, respectively,

$$H_0: \Delta \leq 0 \text{ and } H_1: \Delta > 0$$
$$H_0: \Delta \geq 0 \text{ and } H_1: \Delta < 0$$

The natural estimator for Δ is the difference in sample means

$$\hat{\Delta} = \overline{X}_1 - \overline{X}_2$$

The expression for the standard error of $\hat{\Delta}$ is

$$\sigma_{\hat{\Delta}} = \sqrt{\frac{\sigma_1^2}{n_1} + \frac{\sigma_2^2}{n_2}}$$

This expression is seen to be the square root of the sum of squares of the individual standard errors of $\overline{X}_1$ and $\overline{X}_2$. The independence of the samples from the two populations plays its role in the form of this expression. (Note that it can be derived using the ideas in chapter 5. See exercise 5.39, p. 177.)

Because of the assumption $\sigma_1 = \sigma_2 = \sigma$, the expression simplifies to

$$\sigma_{\hat{\Delta}} = \sigma \sqrt{\frac{1}{n_1} + \frac{1}{n_2}}$$

If a good estimator of the common standard deviation σ can be found, it can be substituted for σ to obtain an estimated standard error $\mathrm{SE}_{\hat{\Delta}}$. Then if the sampling distribution of $(\hat{\Delta} - \Delta)/\mathrm{SE}_{\hat{\Delta}}$ can be determined, the steps for finding tests and confidence intervals given in chapters 8 and 9 can be followed to round out the inference for this problem.

Since σ is assumed to be the standard deviation for *both* populations, it is intuitively clear that a good procedure will use information from both samples to estimate σ. The estimator of choice is the so-called **pooled standard deviation,**

pooled standard deviation

$$s_P = \sqrt{\frac{(n_1 - 1)s_1^2 + (n_2 - 1)s_2^2}{n_1 + n_2 - 2}} \tag{1}$$

Now if we write

$$\mathrm{SE}_{\hat{\Delta}} = s_P \sqrt{\frac{1}{n_1} + \frac{1}{n_2}}$$

then theory asserts that

$$t = \frac{\hat{\Delta} - \Delta}{SE_{\hat{\Delta}}}$$

has Student's t distribution with $\nu = n_1 + n_2 - 2$ degrees of freedom.
Expanded out in all its grandeur, this rather simple expression has the somewhat more awesome form

$$t = \frac{\overline{X}_1 - \overline{X}_2 - (\mu_1 - \mu_2)}{s_P \sqrt{\dfrac{1}{n_1} + \dfrac{1}{n_2}}} \tag{2}$$

Student's two-sample t statistic

where s_P is given by (1). This expression is called **Student's two-sample t statistic.** The form of the statistic required for testing the above hypotheses has $\Delta = \mu_1 - \mu_2 = 0$ in the numerator:

$$t = \frac{\overline{X}_1 - \overline{X}_2}{s_P \sqrt{\dfrac{1}{n_1} + \dfrac{1}{n_2}}} \tag{3}$$

Thus when $\mu_1 = \mu_2$, this statistic has Student's t distribution with $\nu = n_1 + n_2 - 2$ degrees of freedom.

We will use form (2) shortly to obtain confidence intervals for $\Delta = \mu_1 - \mu_2$. At this point examples of the use of form (3) for testing the above hypotheses will be given, along with a simple strategy for evaluating this statistic by hand.

EXAMPLE 10.4

Are the Celts really different from modern Englishmen in head breadth? Let μ_1 be the mean head breadth for the English population and let μ_2 be that for the Celts. With no preconceived notions of size direction (before sampling), it would be reasonable to test

$$H_0: \mu_1 = \mu_2$$

against the two-sided alternative

$$H_1: \mu_1 \neq \mu_2$$

The test will be carried out by using the data of example 3.1, (p. 71), at a significance level of $\alpha = .05$.

The test rule is to reject H_0 if $|t| \geq c$, where c is obtained from table 3 for a two-sided test, $\nu = n_1 + n_2 - 2$ degrees of freedom, and t is the two-sample t statistic (3). Here $n_1 = 18$ and $n_2 = 16$, so $\nu = 32$. Consequently, the cutoff value is

$$c = 2.021$$

We have taken the critical value corresponding to $\nu = 40$, since the value for 32 is not tabulated. All that remains in order to carry out the test is to compute the value of t for this sample. This statistic can be hand-calculated efficiently by the following steps.

Step 1.   Obtain the values of n, $\overline{X}$, and s for each sample. For the data of example 3.1 they are as follows.

ENGLISH (1)	CELTS (2)
$n_1 = 18$	$n_2 = 16$
$\overline{X}_1 = 146.500$	$\overline{X}_2 = 130.750$
$s_1 = 6.382$	$s_2 = 5.434$

A rough check on the assumption $\sigma_1 = \sigma_2 = \sigma$ and the assumption of normality can be obtained by comparing s_1 and s_2 and by inspecting the schematic diagram of the data given in display 10.2. The boxes are nearly identical in length. The slightly longer tail of the English head breadth distribution is reflected in the slightly larger value of s_1. There are no outliers. The t test should perform beautifully.

Display 10.2

Box plots for Celtic and English head breadths

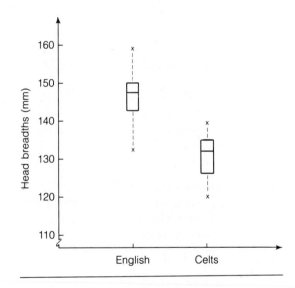

Step 2.   The denominator of the t statistic is calculated first. In the first stage s_P is calculated:

$$s_P^2 = \frac{(n_1 - 1)s_1^2 + (n_2 - 1)s_2^2}{n_1 + n_2 - 2}$$

$$= \frac{17 \times 6.382^2 + 15 \times 5.434^2}{32}$$

$$= 35.4792$$

thus

$$s_P = \sqrt{35.4792} = 5.9564$$

A useful check on this computation depends on the fact that s_P must be between s_1 and s_2. If it isn't, recompute it.

Step 3. Next compute $\sqrt{(1/n_1) + (1/n_2)}$:

$$\sqrt{\frac{1}{16} + \frac{1}{18}} = .34359$$

Now, multiply s_P by this number. The result is the denominator of the t statistic:

$$s_P \sqrt{\frac{1}{n_1} + \frac{1}{n_2}} = 2.0466$$

Step 4. Next, compute $\overline{X}_1 - \overline{X}_2 = 146.500 - 130.750 = 15.750$

Step 5. Divide this number by the result of step 3 to obtain the value of t:

$$t = \frac{15.750}{2.0466} = 7.696$$

Note that although the intermediate steps have been written down, for those using hand calculators all results can be stored in the memory to full accuracy until this last step. Round the value of t to 3 decimal places.

To complete the test, since $|t| = 7.696$ exceeds the cutoff value 2.021, the decision is to reject H_0. Thus we would conclude at the 5% significance level what was already obvious from the box plots—namely, that there is a difference in mean head breadths and, in fact, the English head breadth is larger than the Celts', on the average.

The strength of evidence in the sample against the equality of mean head breadths can be reported as bounds for the P-value

$$P = P(|t| \geq |t_{obs}|)$$

The observed (absolute) t value, 7.696, is well beyond the largest value in the $\nu = 40$ row of table 3. Consequently, from the two-sided test row we find

$$P < .001$$

This result is highly significant.

A Two-Sided Confidence Interval for $\mu_1 - \mu_2$

"By how much does one population mean exceed the other?" The question can be answered with a confidence interval for $\Delta = \mu_1 - \mu_2$. If the null hypothesis of no difference in means is rejected by a two-sided, two-sample t test, then the confidence interval with confidence coefficient $= 1 -$ significance level will contain either all positive values or all negative values of Δ.

If positive, the limits of the interval will encompass the amounts by which μ_1 exceeds μ_2. For example, the interval $3 \leq \mu_1 - \mu_2 \leq 5$ is interpreted to mean that μ_1 exceeds μ_2 by an amount between 3 and 5 units.

Negative values, on the other hand, imply that μ_2 exceeds μ_1, and *the negatives* of the confidence limits, *in reverse order,* are the bounds on the excess of μ_2 over μ_1. Thus the interval $-5 \le \mu_1 - \mu_2 \le -3$ would be interpreted to mean that μ_2 exceeds μ_1 by an amount between 3 and 5 units.

The derivation of the confidence interval is based on a now-familiar idea from chapter 8. Since

$$t = \frac{\hat{\Delta} - \Delta}{\mathrm{SE}_{\hat{\Delta}}}$$

has Student's t distribution, the appropriate confidence interval, based on the pivotal method, is

$$\hat{\Delta} - t\mathrm{SE}_{\hat{\Delta}} \le \Delta \le \hat{\Delta} + t\mathrm{SE}_{\hat{\Delta}}$$

The critical value t would be found from table 3 for $\nu = n_1 + n_2 - 2$ degrees of freedom and the appropriate confidence coefficient. Expanded in full detail, the interval becomes

$$\bar{X}_1 - \bar{X}_2 - ts_P \sqrt{\frac{1}{n_1} + \frac{1}{n_2}} \le \mu_1 - \mu_2 \le \bar{X}_1 - \bar{X}_2 + ts_P \sqrt{\frac{1}{n_1} + \frac{1}{n_2}}$$

EXAMPLE 10.5

By how much does the mean head breadth of modern Englishmen exceed that of the Celts? The computations for the test statistic in example 10.4 (with only one change) already contain the answer. The 5% cutoff value, $t = 2.021$, is also the critical value for a 95% (two-sided) confidence interval for $\mu_1 - \mu_2$. Multiply the denominator of the t statistic obtained in step 3 of the computation by this critical value:

$$ts_P \sqrt{(1/n_1) + (1/n_2)} = 2.021 \times 2.0466$$
$$= 4.136$$

Add and subtract this number from the value of $\bar{X}_1 - \bar{X}_2$ obtained in step 5 to find the limits of the confidence interval:

$$L = 15.750 - 4.136 = 11.614$$

and

$$U = 15.750 + 4.136 = 19.886$$

Thus, the 95% confidence interval is

$$11.614 \le \mu_1 - \mu_2 \le 19.886$$

We conclude with 95% confidence that the mean head breadth of Englishmen exceeds that of the Celts by an amount between 11.61 and 19.89 millimeters.

Suppose that instead of the given labeling, the Celts were population 1 and the English population 2. The only difference in the results of the computation would be that the difference in sample means is -15.750 instead of +15.750. The confidence interval would then be

$$-19.886 \leq \mu_1 - \mu_2 \leq -11.614$$

From this we would conclude that μ_2 exceeds μ_1 and, reversing the negatives of the limits, we would come to exactly the same conclusion as given above.

One-Sided Tests and Confidence Intervals

The extension of the theory from two-sided tests and confidence intervals to the one-sided versions will be illustrated by an example.

EXAMPLE 10.6

We are now in a position to address quantitatively the burning question—is the average height of men (μ_1) really greater than the average height (μ_2) of women? Bounds for the P-value will be obtained for the hypotheses

$$H_0: \mu_1 \leq \mu_2$$

versus

$$H_1: \mu_1 > \mu_2$$

using the data of display 1.1 (p. 9). The data are as follows. (The height for observation 27 has been corrected from 54 to 64 inches.)

MEN	WOMEN
72	66
71	65
66	62
71	69
72	63
67	63
73	67
65	67
67	66
68	60
65	68
70	65
66	64
72	64
	63
	70
	64
	64
	66
	65
	67
	57

The following schematic diagram shows an uncharacteristically short-tailed distribution of men's heights and a somewhat skewed distribution of women's heights. While these effects should not be unduly troublesome, it would be wise to back up the analysis with a procedure that "corrects" for tail length. Such a procedure will be given in chapter 12.

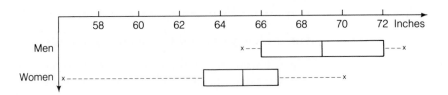

Briefly, the results of the computational steps are as follows.

Summary statistics

	MEN	WOMEN
n	14	22
$\overline{X}$	68.929	64.773
s	2.9211	2.9103

$$s_P = \sqrt{\frac{13 \times 2.9211^2 + 21 \times 2.9103^2}{34}}$$

$$= 2.9144$$

$$\sqrt{\tfrac{1}{14} + \tfrac{1}{22}} \times s_P = .3419 \times 2.9144$$

$$= .9964$$

$$\overline{X}_1 - \overline{X}_2 = 4.1560$$

Thus the observed value of t is

$$t = \frac{4.1560}{.9964}$$

$$= 4.1711$$

Now $\nu = 34$. (Note that the value of ν is the number in the denominator of the expression for s_P.) The observed value of t is seen to lie to the right of the largest entry in the $\nu = 40$ row of table 3. Consequently, from the probabilities for a one-sided test, we find

$$P < .0005$$

Thus it is very implausible, on the basis of this sample, that men and women have the same mean height. We conclude that men are, on the average, taller than women.

How much taller? A one-sided confidence interval for $\mu_1 - \mu_2$ will provide an answer. First note that the hypothesis $H_1: \mu_1 > \mu_2$ can be written in the alternate form $\mu_1 - \mu_2 > 0$. This suggests that the appropriate one-sided confidence interval will use the *lower limit* for the two-sided interval given above, with the two-sided critical value replaced by the appropriate one-sided value. The one-sided 95% critical value for 40 degrees of freedom, from table 3, is $t = 1.684$. Thus, the confidence limit is

$$\bar{X}_1 - \bar{X}_2 - 1.684 s_P \sqrt{\frac{1}{n_1} + \frac{1}{n_2}} = 2.4781$$

We would conclude with 95% confidence that the average height of men exceeds that of women by at least 2.48 inches.

EXERCISES

10.8   The following are androstenedione levels for samples of diabetic men and women.

MALES	FEMALES
217	84
123	87
80	77
140	84
115	73
135	66
59	70
126	35
70	77
63	73
147	56
122	112
108	56
70	84
	80
	101
	66
	84

Data courtesy of F. Purifoy, Ph.D., Department of Anthropology, University of New Mexico.

(a)   Make a schematic diagram of the data and describe any possible violations of the assumptions for a two-sample t test.

(b)   Find the P-value for the hypothesis that the mean androstenedione levels are the same for the populations of diabetic men and women.

(c)   Find a 95% confidence interval for the difference in mean levels for the populations of male and female diabetics.

10.9   Correctness scores on essays (defined as the number of words written divided by the number of errors) were used to study whether students for whom English is a second language would require special instruction in English composition. Scores were obtained for a sample of these "special" students and from a sample of students for whom English is the first or only language (control students). Summary data from the samples are as follows.

	SPECIAL STUDENTS	CONTROL STUDENTS
n	54	54
$\overline{X}$	11.546	14.894
s	6.142	7.714

Data courtesy of G. M. Gallant, Department of English, University of New Mexico.

(a) Test at the 1% significance level the null hypothesis that the mean score for special students is greater than or equal to the mean score for control students against the alternative hypothesis that it is lower. Find the P-value.

(b) If the null hypothesis of part (a) is rejected, find a 99% confidence bound on how much the mean score for controls exceeds that for the special students.

10.10   In a study to determine variables that would be useful for identifying archaeological sites in the Seep Ridge area of Utah, the viewspread (the angle in degrees for which a clear view of surrounding terrain is available) was measured for 34 known sites and for 68 randomly selected nonsites (fixed areas in which sites are known not to exist). The sample mean and standard deviation for viewspreads of sites are $\overline{X}$ = 243.09 and s = 86.11 degrees, whereas for nonsites they are $\overline{X}$ = 118.60 and s = 89.32 degrees.

(a) Formulate the appropriate statistical hypotheses for the conjecture that the mean viewspread for sites (μ_S) is greater than that (μ_N) for nonsites. Test the hypotheses at the 5% level. What is the P-value?

(b) Find the 95% one-sided confidence interval for the amount by which the site mean viewspread exceeds that for nonsites.

10.11   A current theory holds that people with pain in the lower back also have an increased electrical skin resistance in the area of the pain. To test this hypothesis, a physical therapy graduate student measured the skin resistances of 11 normal individuals and 9 people suffering from back pain. The data are as follows.

Normal group: 4.03, 3.83, 4.92, 4.00, 3.08, 8.26, 3.29, 4.24, 4.28, 4.48, 3.00

Pain group: 6.30, 3.08, 9.22, 3.08, 1.62, 4.10, 4.87, 4.44, 4.92

(a) Make a schematic diagram of the data and note any possible violations in the assumptions for a two-sample t test.

(b) Set up the statistical hypotheses to make the above theory the alternative (research) hypothesis.

(c) Test the hypotheses of part (b) at the 5% significance level.

10.12   The summary algebra scores for students making arithmetic scores of 14 or less and those scoring 15 or more on the placement exam described in exercise 1.23 (p. 33) are as follows.

| ALGEBRA | ARITHMETIC | |
SUMMARY	≤ 14	≥ 15
n	19	34
$\bar{X}$	1.474	5.840
s	2.038	3.262

(a) Test at the 5% level whether the mean algebra score for students with arithmetic grades 15 or larger exceeds the mean score for students scoring 14 or less in arithmetic.

(b) Use a 95% one-sided confidence interval to demonstrate the amount by which the mean algebra scores of students scoring high (≥ 15) on the arithmetic test exceeds that for the low-scoring (≤ 14) students.

10.13 The data are the breaking strengths in grams from tests of the durability of rice starch and canna starch film.

RICE STARCH	CANNA STARCH
556.7	791.7
552.5	610.0
397.5	710.0
532.3	940.7
587.8	990.0
520.9	916.2
574.3	835.0
505.0	724.3
604.6	611.1
522.5	621.7
555.0	735.4
561.1	990.0
	862.7

Data courtesy of the Statistical Laboratory, University of California, Berkeley.

(a) Make a schematic diagram of the data and comment on any deviations from the two-sample t test assumptions.

(b) Test for a difference in film strength at the 5% significance level.

(c) Give a 95% confidence interval for the difference in mean film strengths for the two starches.

An Extension to Non-Null Hypotheses*

Thus far in the two-sample problem we have considered hypotheses of the form $H_0: \mu_1 = \mu_2$—i.e., null hypotheses or hypotheses of no difference or no

---

* This discussion depends on material from the optional section 9.7. It will not be used in subsequent chapters.

change. Suppose it is of interest to test a hypothesis of the form

$$H_0: \mu_1 - \mu_2 = k \quad (\neq 0)$$

against either a one- or two-sided alternative hypothesis. The number k represents a hypothetical amount by which μ_1 exceeds μ_2. Thus, in example 10.4, an evolutionary theory might have hypothesized that head breadths for modern Englishmen should exceed those for the Celts by 15 millimeters. The hypotheses to test would then be

$$H_0: \mu_1 - \mu_2 = 15$$

versus

$$H_1: \mu_1 - \mu_2 \neq 15$$

These hypotheses can be tested using form (2) of the two-sample t statistic (p. 313), with $\mu_1 - \mu_2 = 15$ in the numerator:

$$t = \frac{\bar{X}_1 - \bar{X}_2 - 15}{s_P \sqrt{1/n_1 + 1/n_2}}$$

Once this t statistic has been computed, the actual test rule and cutoff value are exactly the same as those of the two-sided tests given earlier.

Alternatively, a test can be formulated from the equivalence of tests and confidence intervals shown in section 9.7. If a two-sided confidence interval for $\mu_1 - \mu_2$ is available, a test can be based on the rule to reject H_0 if the hypothetical value of this difference, $\mu_1 - \mu_2 = 15$, does not lie in the confidence interval. Thus, for example, a 5% test of the above hypotheses for head breadth can be obtained from the 95% confidence interval

$$11.614 \leq \mu_1 - \mu_2 \leq 19.886$$

derived in example 10.5. Since 15 does fall in this interval, H_0 would be accepted and we would conclude at the 5% level that the evolutionary theory of a 15 millimeter head breadth difference is plausible on the basis of our data.

One-sided tests can be treated in exactly the same way. For example, if it were conjectured that the average height of men exceeds that of women by at least 2 inches, in the notation of example 10.6 we would reduce this conjecture to the hypotheses

$$H_0: \mu_1 - \mu_2 \leq 2$$

versus

$$H_1: \mu_1 - \mu_2 > 2$$

Again, a modification of the t statistic to include the hypothetical value 2 in the numerator can be used with the usual one-sided test. Alternatively, since the 95% upper one-sided confidence interval

$$\mu_1 - \mu_2 \geq 2.4781$$

is available from example 10.6, we can test the hypotheses at the 5% level by using the rule to reject H_0 if none of the values specified by H_0 fall in the interval. Since these values are the numbers less than or equal to 2 and the confidence interval contains only values above 2.4781, we would reject H_0. Our conclusion at the 5% significance level would be that the mean height of men exceeds that for women by more than 2 inches.

EXERCISE

10.14   Refer to exercise 10.10 (p. 320). Use the confidence interval constructed in part (b) to test, at the 5% level, the hypothesis that the viewspread for sites exceeds that for nonsites by at least 90 degrees.

◼ SECTION 10.5

The Method of Paired Differences

In some problems, observations occur naturally in pairs. Each pair will consist of measurements on "matched individuals" from two populations. Thus, for example, measurements made at two locations on a river on the same days, one measurement above and one below a pollution source, would constitute measurements from "above source" and "below source" populations, paired by being taken on the same days.

Included in this problem category are experiments in which individuals are purposefully matched by pairs and two treatments are randomly assigned to the members of each pair. The measurements resulting from these experiments are then used to test for differences in the responses to the treatments. The rationale behind this kind of experimental design will be discussed in chapter 17.

For paired observations, the assumption of independence of samples is usually violated. Because the two-sample t test requires independent samples, it is then desirable (even necessary) to replace it by a test of another kind.

The test developed in this section utilizes the differences of the paired observations, hence the name *paired-difference method*.

The method will be illustrated by means of the following data, which are nitrogen levels (in milligrams per liter) in water taken from the Red River in northern New Mexico during the winter of 1979–1980. The first member of each pair was sampled at a station upriver from the outflow of a sewage treatment plant, and the second was obtained on the same day at a station several miles below the outflow. A difference in mean nitrogen levels would be evidence that the plant may be doing an inadequate job of treating the sewage before discharging it into the river.

X_1	X_2	d
.014	.012	.002
.044	.016	.028
.048	.020	.028
.066	.054	.012
.029	.050	−.021
.051	.069	−.018
.110	.155	−.045
.119	.028	.091

If X_1 represents the measurement made above the plant and X_2 the corresponding value made below it, then the analysis will be based on the differences

$$d = X_1 - X_2$$

which are given in the third column of the table.

The differences can be viewed as a sample from a population of differences with mean μ_d. From the rule of expectation given in chapter 5, if μ_1 is the population mean of the upriver measurements and μ_2 the mean of the downriver values, then

$$\mu_d = \mu_1 - \mu_2$$

It follows that the null hypothesis of equality of means,

$$H_0: \mu_1 = \mu_2$$

can be expressed as the hypothesis

$$H_0: \mu_d = 0$$

In fact, all of the interesting hypotheses concerning the means of the two variables can be written as hypotheses for μ_d. It follows that it will be possible to base tests of these hypotheses on the sample of differences.

As was seen in chapter 6, if X_1 and X_2 are normal variables, then d is also a normal variable. It follows that if $\bar{d}$ and s_d are the sample mean and standard deviation of the differences, then under the assumption that $\mu_d = 0$, the **paired-difference t statistic**

paired-difference
t statistic

$$t = \frac{\bar{d}}{s_d/\sqrt{n}}$$

has Student's t distribution, with $n - 1$ degrees of freedom. Note that n is the sample size for the differences, which is the number of *pairs*, not the total number of observations.

This is a one-sample t statistic for which the observations are differences and the population mean is 0. It follows that the paired-difference t test is simply a one-sample t test of the type studied in section 10.3.

Applying these ideas to the Red River data, we see that the rule for a two-sided 5% test of the null hypothesis of no difference in mean nitrogen levels is to reject H_0 if

$$|t| \geq 2.365$$

The cutoff value comes from table 3 for $n - 1 = 7$ degrees of freedom. The summary statistics are

$$n = 8, \bar{d} = .0096, \text{ and } s_d = .04148$$

Thus, the paired-difference t value is

$$t = \frac{.0096}{.04148/\sqrt{8}}$$

$$= .656$$

Since the cutoff value is not exceeded, the hypothesis of no difference in means is accepted. We would conclude at the 5% significance level that there is no evidence of a difference in mean nitrogen levels at the two stations. Any effect the sewage treatment plant might have on the river's nitrogen levels is not detectable at this downriver station.

The Paired-Difference t Test versus the Two-Sample t Test

It would be possible to ignore the association between samples and apply a two-sample t test to paired data. How would this test "perform" relative to the paired-difference test? Are there situations in which it would be preferred to the paired-difference method?

The following summary can be argued on the basis of the discussion of the law of variances for the difference of correlated variables given in section 5.6 (p. 175). If the correlation between variables X_1 and X_2 is positive, then the paired-difference t test is more powerful than the two-sample t test. This will be illustrated in example 10.7. If the correlation is negative, then the two-sample t test is not even a valid test. Consequently, only when the variables are uncorrelated would the two-sample test be preferred. However, when data are naturally paired (or paired by design), some form of correlation is almost inevitable. Consequently, the paired-difference test should always be used when data are paired.

As will be shown presently, when the data are not naturally paired but simply have the same sample sizes, there is a difficulty with the use of the paired-difference method. Here, the two-sample test is preferable.

Briefly, then, *the paired-difference test should be used rather than the two-sample test if and only if the data are paired (either naturally or by design).*

EXAMPLE 10.7

In families for which both parents are smokers, is the father or the mother likely to be the heavier smoker? Data collected by J. Yerushalmy [36] on a group of children born in the Kaiser Foundation Hospital in Oakland, California, and on their families makes it possible to address this question. Among the items of information collected for each family was the average number of cigarettes smoked per day by the father and by the mother of each child. The following data were randomly subsampled from among families in which both parents smoked.

SMOKER	1	2	3	4	5	6	7	8	9	10	11	12	13	14
						FAMILY								
Father (X_f)	11	20	20	20	10	20	20	30	9	20	5	20	20	20
Mother (X_m)	20	10	6	4	6	30	4	20	6	20	2	20	18	10
$d = X_f - X_m$	-9	10	14	16	4	-10	16	10	3	0	3	0	2	10

If μ_f and μ_m denote the population averages of X_f and X_m, we want to test

$$H_0: \mu_f = \mu_m \quad \text{versus} \quad H_1: \mu_f \neq \mu_m$$

The data are paired by family. Moreover, it is certainly reasonable to believe that there is a positive correlation between X_f and X_m. For example, if the father is a heavy smoker, it is likely that the mother will also be a heavy smoker. Consequently, the paired-difference method should be used to test this hypothesis. The differences are given in the bottom row of the table. The summary data required for the test are

$$n = 14 \quad \text{(the number of pairs)}$$
$$\bar{d} = 4.929$$
$$s_d = 8.260$$

Thus,

$$t = \frac{\bar{d}}{s_d/\sqrt{n}} = \frac{4.929}{8.260/\sqrt{14}} = 2.233$$

and bounds for the two-sided P-value, based on $v = 14 - 1 = 13$ degrees of freedom, are found from table 3 to be

$$.02 < P < .05$$

At the 5% significance level, we would conclude that a difference in the amount of smoking exists between the parents. In fact, since $\bar{d}$ represents the average excess in number of cigarettes smoked by fathers and $\bar{d}$ is positive, we would conclude that the fathers smoke more than the mothers, on the average.

Had the two-sample t test been applied to this problem, we would have found

$$t_{\text{2-sample}} = 1.720$$

For $14 + 14 - 2 = 26$ degrees of freedom, bounds for the P-value of the two-sided test based on this statistic are

$$.05 < P < .10$$

Thus H_0 would have been accepted at the 5% level. The paired-difference method provides a more sensitive test here.

Paired-Difference Confidence Intervals for the Difference in Means

If $d = X_1 - X_2$ is the difference in the variables used in the paired-difference method, then since $\bar{d}$ estimates $\mu_d = \mu_1 - \mu_2$, the interval

$$\bar{d} - t\frac{s_d}{\sqrt{n}} \le \mu_1 - \mu_2 \le \bar{d} + t\frac{s_d}{\sqrt{n}}$$

will be a two-sided confidence interval for the difference in means. As usual, t is the critical value from Student's t table, table 3, for degrees of freedom $\nu = n - 1$ and for the given confidence coefficient.

EXAMPLE 10.8

How much more do the fathers in the Kaiser Study smoke, in comparison to the mothers? We will answer this question by means of a 95% confidence interval for $\mu_f - \mu_m$, where μ_f is the average number of cigarettes smoked per day by fathers and μ_m is the corresponding quantity for the mothers. For $\nu = 14 - 1 = 13$ degrees of freedom, the 95% critical value from table 3 is

$$t = 2.160$$

Consequently, the endpoints of the confidence interval are

$$\bar{d} \pm t\frac{s_d}{\sqrt{n}} = 4.929 \pm 2.160\frac{8.260}{\sqrt{14}}$$

Thus

$$.161 \le \mu_f - \mu_m \le 9.697$$

That is, with 95% confidence (rounding the limits), we conclude that the men smoke between .2 and 9.7 more cigarettes per day than the women smoke, on the average.

As before, one-sided confidence intervals for $\mu_1 - \mu_2$ can be constructed from the limits of the two-sided interval by replacing the two-sided critical value by the corresponding one-sided value. For example, a 95% lower confidence bound for $\mu_f - \mu_m$ in the last example would be

$$\bar{d} - 1.771\frac{s_d}{\sqrt{n}} = 1.019$$

Thus, had we only been interested in how much more men smoke than women, we could conclude with 95% confidence that men smoke at least 1 more cigarette per day than women, on the average.

EXERCISES

10.15   The following figures show the amount of sleep gained (in hours) from two sleep remedies, A and B, applied to 10 individuals who have trouble falling to sleep.

REMEDY	INDIVIDUAL									
	1	2	3	4	5	6	7	8	9	10
A	0.7	−1.6	−0.2	−1.2	0.1	3.4	3.7	0.8	0.0	2.0
B	1.9	0.8	1.1	0.1	−0.1	4.4	5.5	1.6	4.6	3.0

(a)   Compute the differences and make a box plot of them. Comment on the normality of the distribution of differences.

(b)   Does there appear to be any difference in the effects of the two remedies? Test at the 5% significance level.

(c)   Find a 95% confidence interval for the mean difference in sleep gain for the two drugs.

10.16   Measurements on two teeth for a sample of 20 monkeys, used in an anthropological study, are as follows.

TOOTH 1	TOOTH 2
73	75
76	73
77	77
73	73
72	74
68	70
71	74
71	72
68	67
70	70
75	76
77	77
70	70
75	74
75	76
75	72
70	70
74	75
66	77
66	74

Data courtesy of C. Shuster, student in the Department of Anthropology, University of New Mexico, Albuquerque.

(a)   Make a box plot of the differences and comment on any deviations from normality.

(b) Test at the 5% significance level whether there is any difference in the mean measurements for the two teeth. Give bounds for the P-value for the hypothesis of no difference.

10.17 The rates of erosion (in meters per year) of a type of geological formation for two consecutive time periods were computed for a sample of 12 specimens. The data are as follows.

SPECIMEN	PERIOD 1	PERIOD 2
1	.24	.19
2	.08	.53
3	.51	.40
4	.48	.06
5	1.00	.28
6	.12	.61
7	.11	.17
8	.23	.61
9	.13	.14
10	.12	.18
11	.16	.23
12	.07	.14

Data courtesy of C. Condit, Department of Geology, University of New Mexico, Albuquerque.

(a) Make a box plot of the differences and comment on any deviations from normality.

(b) Test at the 10% significance level whether there is a difference in mean rates of erosion for the two time periods. Give bounds for the P-value for the hypothesis of no change.

10.18 The accompanying data are calcium carbonate ($CaCO_3$) readings (parts per million cubic centimeters) for ten wells in the Atrisco well field (one of the water sources for Albuquerque, New Mexico) for 1961 and 1966.

WELL NO.	YEAR 1961	1966
1	185	258
2	92	58
3	112	190
4	82	98
5	108	142
6	117	142
7	62	138
8	64	166
9	92	64
10	76	130

Data courtesy of C. Rail, Environmental Health Department, Albuquerque, New Mexico.

(a) There was a concern that the $CaCO_3$ levels in the water supply were rising during that period. Is this concern substantiated by the data? Test at 10% significance level.

(b) Obtain a one-sided 90% confidence interval for the mean difference $\mu_{1966} - \mu_{1961}$ in $CaCO_3$ concentrations for the two years and use it to provide a lower bound on the increase in concentrations.

10.19   The responses of ten subjects to a stimulus given at two times, once near the beginning and once near the end of a psychological conditioning experiment, are given in the accompanying table. Success in conditioning would correspond to a decrease in the measured response. Is there any evidence of success in conditioning? Give the P-value.

	TIME	
SUBJECT	1	2
1	34	24
2	39	45
3	30	33
4	41	44
5	35	38
6	25	35
7	39	35
8	48	44
9	30	31
10	37	35

10.20   In a classical study of hereditary influences [*The Effects of Cross- and Self-Fertilization in the Vegetable Kingdom* (1876), John Murray, London], Charles Darwin (with the aid of one of the originators of statistics, Sir Francis Galton) analyzed the accompanying data, which represents the heights (in inches) of certain plants with cross- and self-pollinated parents. The plants are paired (somewhat crudely) according to ancestry.

PAIR NO.	HEIGHTS (INCHES)	
	CROSS-POLLINATED	SELF-POLLINATED
1	23.5	17.4
2	12.0	20.4
3	21.0	20.0
4	22.0	20.0
5	19.1	18.4
6	21.5	18.6
7	22.1	18.6
8	20.4	15.3
9	18.2	16.5
10	21.6	18.0
11	23.2	16.2
12	21.0	18.0
13	22.1	12.8
14	23.0	15.5
15	12.0	18.0

(a) Using paired differences, obtain a 95% confidence interval for the difference in means for the two types of plants.

(b) Use the result of part (a) to determine whether or not the data support a difference between the two types of plants, at the 5% significance level.

A Danger of Using Paired Differences When the Data Are Not Naturally Paired

The arithmetic for computing the one-sample t statistic is easier than that for the two-sample t. Thus it is tempting to use the paired-difference method on two-sample problems even when the data are independently selected from the two populations and there is no *natural* pairing of observations. It is important that this temptation be resisted, because the paired-difference method is very sensitive to the way the pairs are put together. Different relative orderings of the data (before pairing) can lead to different values of t. On the other hand, the two-sample method provides the same solution regardless of the order of the data, and it is the procedure to use in such problems. The following example demonstrates this phenomenon.

EXAMPLE 10.9

In this example we will look at a comparison of the chlorine ion content for the east and west sides of the Rio Grande River. The following data were extracted from the chlorine ion data of exercise 3.13, (p. 84).

EAST SIDE (X_1)	WEST SIDE (X_2)
.29	.21
.30	.24
.39	.25
.40	.37
.45	.40
.46	.44
.65	.62

Unless stated otherwise, it should not be assumed that there is a relationship between pairs of values in the same row. Actually, the data here have been reordered (from smallest to largest) for convenience. This operation is easily carried out by a computer, and it is not unusual that one is presented with data that have been so rearranged—especially if no natural ordering existed in the first place.

The correlation between two data sets is maximized by arranging both of them in increasing order and then pairing them, as is done in this data set. A large positive correlation will lead to a small standard error for $\bar{d}$ and thus to an unnaturally sensitive paired-difference test. The paired differences for this arrangement are .08, .06, .14, .03, .05, .02, .03, for which $\bar{d} = .0586$ and $s_d = .0414$. These values yield $t = 3.745$, a highly significant value for $\nu = 6$ degrees of freedom. On the other hand, the two-sample method yields $t = .823$, which is insignificant.

The order in which the data were actually extracted from that of exercise 3.13 is as follows:

EAST SIDE (X_1)	WEST SIDE (X_2)
.40	.37
.45	.40
.30	.62
.46	.44
.29	.25
.39	.24
.65	.21

An application of the paired-difference method to this pairing yields $t = .693$. This result is more in keeping with the two-sample t value, but it is still dependent on the rather arbitrary data ordering. Many other paired-difference t values can be constructed from the data by forming other arrangements of the sets. The smallest t value will be obtained in the following exercise.

EXERCISE

10.21   The minimum correlation, and thus the minimum t value, occurs when one data set is listed in increasing order and the other in decreasing order. Arrange the data of example 10.9 in this manner and calculate the paired-difference t value. Compare the results with those of the other tests.

Chapter 10 Quiz

1. Why is the normal distribution no longer an adequate approximation to the sampling distribution of Student's (one-sample) t statistic when sample sizes are small?

2. What is Student's t distribution?

3. What are the degrees of freedom for the one-sample Student's t statistic?

4. How is table 3 of critical values for Student's t distribution used to carry out one-sample t tests? Give an example.

5. How is table 3 used to find P-values bounds for hypotheses concerning a population mean? Give an example.

6. How is table 3 used to find confidence intervals for a mean? Give an example.

7. What is the two-sample (comparison) problem?

8. What is the formulation (distribution assumptions) of the two-sample problem that leads to Student's two-sample t test?

9. What are the forms of hypotheses considered in the two-sample t tests?

10. What is the expression for the two-sample t (test) statistic? What are the steps to calculate its value from sample data? Give an example.

11. What are the degrees of freedom for the two-sample t statistic?

12. How are two-sample t tests carried out? Give examples.

13. How are P-value bounds obtained for the hypotheses considered in two-sample t tests? Give examples.

14. How (and when) are confidence intervals computed for the difference in population means? How are positive confidence limits interpreted? How are negative limits interpreted?

The following two questions are for the optional extension of the two-sided t *test to non-null hypotheses*

15. How is the two-sample t test modified to account for non-null hypotheses?

16. How can a confidence interval for the difference in means be used to test non-null hypotheses?

17. When should a paired-difference t test be used? Why is a two-sample t test inappropriate (or less desirable) in such circumstances?

18. How is a paired-difference t test carried out? Give an example.

19. How can confidence intervals for the difference in population means be obtained for paired data? Give an example.

20. Why is it inadvisable to use the paired-difference t test on arbitrarily paired data from two equal-sized independent samples?

SUPPLEMENTARY EXERCISES

10.22   The summary statistics for the nuclear reactor weld bead heights given in exercise 1.11 (p. 23) are $n = 20$, $\overline{X} = .9625$, and $s = 1.3418$. These heights are used to test whether the reactor has been tampered with by someone trying to steal the radioactive material. By careful machining, the original heights of all weld beads have a mean height (relative to the baseline) of zero. If tampering has occurred, the mean of the sample measurements will be different from zero. Test at the 10% significance level whether it is reasonable to believe that tampering has occurred for the reactor from which these summarized measurements were taken.

10.23   The swimming tension data of exercise 1.14 (p. 32) repeated here, are the tension levels measured before swimming minus the corresponding level taken after. It follows that a reduction in tension due to swimming would lead to values that are positive, on the average. Test the hypothesis that swimming reduces tension at the 5% significance level.

$$-1, 10, 2, -9, 8, 2, 2, 3, -20, -3, 11, 7, -11, 1, -9, -7, -3, -6$$

(Note that this is a problem of paired measurements in which the raw data given are already the differences. This is a commonly occurring situation in practice, since the difference is often the natural measure to use when looking for a treatment effect by comparing "before" treatment with "after" treatment measurements.)

10.24   The samples of sodium ion content of the water on the east and west sides of the Rio Grande river, first given in exercise 3.13 (p. 84), are repeated here:

WEST SIDE	EAST SIDE
6.10	1.43
4.50	1.22
4.31	2.50
6.10	1.45
5.00	.90
6.04	1.60
5.92	2.70
5.92	3.90
6.70	2.30
6.12	2.60
4.50	1.46
4.17	1.90
2.80	2.10
4.60	1.65
2.90	.90
4.51	.70
5.04	1.18
4.92	1.20
	1.40
	1.10
	1.00
	1.10
	1.65

Do the data support a difference in sodium ion levels for the two sides of the river? If so, how much of a difference? Use a 5% test and 95% confidence interval.

10.25   In an effort to measure the differences in intensity between summer storms (occurring June through October) and those occurring during the rest of the year (nonsummer storms) in Albuquerque, the percentage of rain falling during the first 5 minutes of the storm was computed for 12 summer and 14 nonsummer storms from U.S. Weather Service data gathered in the period from 1951 through 1957. The data are given in the accompanying table.

SUMMER		NONSUMMER	
18.2	21.2	23.8	3.4
80.0	60.7	10.5	8.3
8.6	34.6	14.3	66.7
45.0	40.7	40.0	15.0
28.6	6.7	16.1	11.8
20.7		11.1	18.2
50.0		5.0	26.3

(a) Make a schematic diagram of these data. What features of the schematic diagram suggest the data are not from normal distributions?

(b) Test at the 5% level whether there is a difference in average storm intensities for the two parts of the year.

(c) Find a 95% confidence interval for the difference in mean storm intensities for the two parts of the year.

10.26 It is important in sample surveys for the proportion of individuals completing the entire questionnaire (called the *response rate*) to be as high as possible in order to avoid bias. A survey method for telephone and mail surveys, called the total design method (TDM), has been devised by Dillman [9] to increase response rates. The accompanying data, in back-to-back stem and leaf form, constitute response rates (percentage responding) for telephone surveys in which the precepts and procedures of the total design method were used completely in the surveys and those in which the method was followed only in part.

TDM USED COMPLETELY		TDM USED IN PART
	5	6078
8585056	6	0555948
24841055	7	31415
51047922	8	4087
302	9	43

(a) Is there evidence that the total design method improves response rates? Use a one-sided test at the 5% significance level.

(b) Perform a two-sided t test of the hypothesis of no difference in response rates at the 5% level. Compare the two-sided test with the results of part (a). Comment on possible reasons for any differences seen in the outcomes of these tests.

(c) Quantify the effect of part (a) with a one-sided 95% confidence limit.

10.27 A problem from the field of communicative disorders is the following. Speech disorders are diagnosed by having the subject take a test consisting of reading aloud a list of words into a tape recorder. The various speech errors are then evaluated from the tape by a trained judge. It is important to design tests that can be as unambiguously judged as possible. Such tests can be evaluated by comparing the proportion of times two or more judges agree on their scorings of words (interjudge evaluation) or the proportion of times a given judge agrees with his own scoring after a period of time has elapsed (intrajudge evaluation). The following data constitute intrajudge data for two tests that were administered to the same six subjects. The proportions of times the judge agreed with his previous scoring after a six-month period are given as percentages.

	TEST	
SUBJECT	1	2
1	85	77
2	82	73
3	92	78
4	75	83
5	88	62
6	77	69

Data courtesy of L. P. Evans, Albuquerque Aphasia and Speech Consultants.

Is there any evidence of a difference between tests? If so, which test appears to be better? Use the 10% significance level for the hypothesis test.

10.28 The Levi Strauss run-up values (see exercise 3.15 p. 86) for mills C and D are as follows.

Mill C: 1.21, .97, .74, −.21, 1.01, .47, .46, .39, .36, .96, .98, .65, .57, .51, .34, −.08, −.39, .09, .15

Mill D: 1.15, 1.02, .38, .83, .66, 1.02, .88, .27, .51, 1.12, .59, 1.30, .68, 1.45, .52, .73, .71, .34, .07

(a) Do an exploratory analysis of these data. Comment on any diagnostic problems observed.

(b) Test at the 5% significance level whether there is a difference in wastage (run-up) for these two mills.

(c) Find a 95% confidence interval for the difference in mean run-up for the two mills.

10.29 The data on natural gas consumption from exercise 3.14 (p. 85) for homes with and without an occupant change are repeated here.

	NATURAL GAS CONSUMPTION	
HOME PAIR	FOR HOMES WITH CHANGE IN OCCUPANT	FOR HOMES WITH NO CHANGE IN OCCUPANT
1	4.4	−36.5
2	−19.4	3.8
3	−20.5	−7.6
4	−3.0	5.3
5	−30.7	−13.0
6	12.1	−13.5
7	−19.0	.5
8	−19.3	−27.7
9	−61.7	−8.1
10	6.2	25.3
11	−37.0	−36.5
12	−11.1	−36.5
13	12.7	−6.7
14	−6.7	−4.1
15	22.7	−14.0

(table continued on next page)

	NATURAL GAS CONSUMPTION	
HOME PAIR	FOR HOMES WITH CHANGE IN OCCUPANT	FOR HOMES WITH NO CHANGE IN OCCUPANT
16	−35.4	7.7
17	−42.2	−14.9
18	−7.7	−5.3
19	−8.8	−21.3
20	19.3	−11.7
21	−77.5	−10.0
22	12.1	−10.7
23	−16.3	−10.7
24	−41.9	−14.4
25	−9.0	11.8

Source: Mayer [22].

(a) Test at the 10% significance level whether there is a difference in mean consumption rates for the two occupant classifications.

(b) If the hypothesis of part (a) is rejected, use a 90% confidence interval to quantify the difference in mean consumption rates.

10.30 To study the motivational effects of different kinds of instructions in experimental games, a psychologist set up the following experiment. Two groups of subjects were combined into player pairs, 12 pairs in each group. The first group was given competitive instructions, and the second group individualistic instructions. Then each subject pair played a game, having a choice of strategies 40 times, and the number of mutually competitive choices were recorded for each pair. The results were as follows.

COMPETITIVE INSTRUCTIONS	INDIVIDUALISTIC INSTRUCTIONS
32	24
19	23
8	13
19	25
9	26
15	25
17	6
16	26
21	22
24	13
10	7
27	16

Data courtesy of Richard J. Harris, Department of Psychology, University of New Mexico.

(a) What is the appropriate test for deciding whether there is a difference in behavior due to the type of instructions?

(b) Carry out the test decided upon in part (a) using the 5% significance level.

(c) What range of differences in mean scores do the data support at the 95% confidence level?

10.31   In an experiment to determine the best way to cut sheets of insulating material to retain its impact strength, ten specimens were selected from a production lot and cut lengthwise, while a second ten were cut crosswise. The resulting strengths in foot-pounds are as follows.

LENGTHWISE SPECIMENS	CROSSWISE SPECIMENS
1.15	0.89
0.84	0.69
0.88	0.46
0.91	0.85
0.86	0.73
0.88	0.67
0.92	0.78
0.87	0.77
0.93	0.80
0.95	0.79

Data courtesy of the Statistical Laboratory, University of California, Berkeley.

(a) Make a schematic diagram of the data and comment on any features that could cause difficulties for the methods of this chapter.

(b) Test whether there is a difference in mean strengths for the two types of cuts at the 10% significance level.

(c) Find and interpret a 90% confidence interval for the mean strengths.

10.32   Lee 1 and Lee 2 are wells in the Ambrosia Lake Mining District, near Grants, New Mexico, used to monitor water quality in the neighborhood of a number of uranium mines. The two wells are located about a mile apart. The following data are measurements of total dissolved solids made from the two wells on the same days during the indicated months.

DATE	LEE 1	LEE 2
3/78	528	554
10/78	562	580
1/79	517	480
5/79	548	461
12/79	612	466
3/80	536	475
6/80	572	491
9/80	611	494

Data courtesy of the Environmental Protection Agency, State of New Mexico, Santa Fe.

Test at the 5% level whether there is a difference in the mean total dissolved solids levels for the two wells.

10.33 **The need for Student's t distribution—A Monte Carlo study.** Recall that a 95% confidence interval should cover the true value of the parameter in question about 95 times in every 100 samples. For example, the usual confidence interval for the mean μ of a measurement variable should have this property. The sampling distribution of the t ratio,

$$t = \frac{\bar{X} - \mu}{s/\sqrt{n}}$$

governs the critical value that enters into the interval. For large samples we used the value $z = 1.960$ for a 95% interval. However, if, for example, $n = 5$, we are instructed to use the critical value $t = 2.776$.

To see whether this refinement is really necessary, you will obtain 20 samples of size 5 from a distribution for which μ is known. Both the normal intervals with endpoints $\bar{X} \pm 1.96s/\sqrt{n}$ and Student's t intervals with endpoints $\bar{X} + 2.776s/\sqrt{n}$ will be computed. Then the number of coverages of μ (out of 20) will be tallied for each type of interval. If the refinement is needed, you should find the proportion of coverages for the t intervals to be closer to the nominal value of .95 than that for the normal intervals. By combining the results for all the members of your class, you can obtain even better estimates of the actual coverage probabilities for the two types of intervals.

The population to be sampled is the first 100 students of the complete class data set of display 1.18 (p. 28). The variable of interest is the weight variable. The population mean is $\mu = 138.4$ pounds.

(a) Select a random starting point in display 5.9 (pp. 137–139) and use the Monte Carlo sampling method given in chapter 5 to draw, with replacement, 20 samples of size 5 from this distribution. Obtain $\bar{X}$ and s for each of the 20 samples.

(b) Use the procedure given in chapters 8 and 10 to calculate both the large sample (normal) 95% confidence interval and the 95% Student's t interval for each sample.

(c) Determine whether each interval in part (b) covers $\mu = 138.4$ and calculate the proportions of coverage for each. Which interval has a coverage proportion closest to the nominal value of .95? Is the additional width provided by the t distribution needed?

The *k*-Sample Problem for Measurement Variables: One-Way Analysis of Variance

Introduction

In the measurement variable *k*-sample problem we attempt to detect differences in the means of a variable for $k = 3$ or more populations, based on random samples taken independently from each population. For example, we might study the variation in quality of a statistics professor's teaching methods by comparing the mean final exam scores of his students for a three-year period.

Student's two-sample *t* test was used in the last chapter to compare the means for $k = 2$ populations. For $k \geq 3$ populations, it would seem a very natural generalization simply to compare every possible pair of means. The populations could be classified into groups by the criterion that if no difference is detected between a pair of means, they are put in the same group, and if a difference is detected, they are put in different groups. For example, we might find that there is no difference in mean exam scores between years 1 and 2 but that the mean score for year 3 is significantly higher than that for year 1 or for year 2. This grouping, which could be represented symbolically by {1, 2} < {3}, would suggest the possibility that the professor's teaching techniques jelled after a two-year trial period. This kind of grouping would provide the information found to be most useful in practice, and it is our goal in this chapter to study an efficient method for obtaining such groupings quickly and easily.

multiplicity problem The direct comparison of all possible pairs of means involves a technical snag of some importance, called the **multiplicity problem.** The reason for concern is that it affects the *validity* of the method in an unknown and

340

uncontrolled way—something we have previously seen to be unacceptable in a statistical procedure. The multiplicity problem will be discussed more fully below. It will be circumvented by preceding the pairwise comparisons with a statistical procedure that tests the equality of all means simultaneously. This procedure, of substantial interest in its own right, is the so-called *one-way analysis of variance*. It is the prototype of a large and important class of statistical inference procedures known collectively as the analysis of variance. We will use this method in the context of observational experiments in this chapter. It was originally devised (by R. A. Fisher) for what we have called *designed experiments* in chapter 4. The interplay between experimental design and the method of analysis provided by the analysis of variance represents one of the most significant achievements in modern statistics. This topic will be treated briefly in chapter 17.

The one-way analysis will be followed by a grouping method virtually identical to the pairwise two-sample *t* comparisons described above, but organized in a simple and convenient way. The resulting two-step procedure is called *Fisher's least significant difference* (FSD) *method*. It is the

multiple-comparison methods

oldest of a collection of grouping techniques called **multiple-comparison methods.** Recent simulation studies (Carmer and Swanson [6]) have demonstrated that it is not only one of the simplest methods to use but also one of the best in terms of having good sensitivity for detecting real differences while retaining very closely the prespecified validity.

The FSD procedure can be used directly to calculate confidence intervals for the differences in group means when each population mean is put in a different group by the procedure. However, when groups contain two or more means, it is necessary to find another way to assess this difference. A useful method for constructing confidence intervals for differences in group means will be given at the end of the chapter.

◢ SECTION 11.2

The Multiplicity Problem and the Bonferroni Inequality

Suppose the assumptions necessary to carry out the two-sample *t* tests for all pairs of means are met. A complete statistical test of the equality-of-means null hypothesis $H_0: \mu_1 = \mu_2 = \cdots = \mu_k$ based on 2-sample *t* tests would be composed of several subexperiments or tests, one for every possible pair of means. For $k = 3$, three pairwise subtests would be required, one each for $H_0: \mu_1 = \mu_2$, $H_0: \mu_1 = \mu_3$, and $H_0: \mu_2 = \mu_3$. When $k = 4$, six tests would be required; and so on. A difference detected by *any* of the subtests would be regarded as grounds for rejecting $H_0: \mu_1 = \mu_2 = \cdots = \mu_k$. Consequently, either formally or informally, the rule being used for the complete test is to reject H_0 if at least one of the subtests rejects the corresponding paired null hypothesis.

As in any hypothesis test, we want to be able to guarantee validity for the complete test. That is, for a given significance level α, the probability

experimentwise error rate
comparisonwise error rates

with which H_0 is rejected by the procedure when H_0 is true should not exceed α. This probability is often called the **experimentwise error rate.** However, the error rates we know how to control from chapter 10 are the so-called **comparisonwise error rates,** the significance levels for each subtest. The problem is that the relationship between the comparisonwise error rates and the experimentwise rate is extremely complex and, in most problems of practical interest, will be unknown. Consequently, we will not know, in general, how to adjust the significance levels of the subtests in order to guarantee a prescribed significance level for the complete test.

Bonferroni inequality

The pessimism of this situation is lessened to some extent by a probability inequality of a very general nature called the **Bonferroni inequality.** It states that *if several events are of interest, then no matter how complex their interrelationships are, the probability that at least one of the events will occur cannot exceed the sum of the probabilities of the individual events.*

Because of the rule given above for the complete test, the Bonferroni inequality tells us that the *experimentwise error rate can never be larger than the sum of the comparisonwise error rates.* Unfortunately, this criterion is not always very helpful. For example, suppose there are $k = 4$ populations, which requires six two-sample subtests, and each subtest is carried out at a significance level of $\alpha = .05$. The experimentwise error rate, then, cannot exceed $6 \times .05 = .30$. (And it will be at least .05, since the probability of at least one event occurring can be no smaller than the probability of any one of the events.) However, we have no way of knowing where in the range from 5% to 30% the actual significance level is. In particular, the choice of a 5% significance level for the individual subtests by no means guarantees a 5% level for the complete test. The lack of control of validity makes this procedure unacceptable.

But why not make the Bonferroni inequality work for us? If, for example, an experimentwise error rate of $\alpha = .05$ is desired and the complete test involves, say, ten subtests (which would be the case for $k = 5$ populations), why not perform each subtest at significance level $\alpha = .05/10 = .005$? Then since the level for the complete test is no larger than 10 times this value, the 5% experimentwise error rate (or less) is guaranteed. In fact, this is a useful general procedure that is often used in practice when the multiplicity problem arises and no clear alternatives exist. Examples will be given in this and subsequent chapters.

However, in some instances the very generality of the Bonferroni inequality leads to a procedure that, while valid, can suffer from lack of sensitivity, relative to alternative tests. There are often competing statistical procedures that have greater sensitivity and also satisfactorily protect validity. Such is the case for the k-sample problem under study in this chapter. A large number of multiple comparison procedures have been devised for the k-sample problem. Several are available in various computer packages. The reader interested in learning more about these procedures would do well to begin with the paper of Carmer and Swanson [6] which provides references to the most commonly used ones. In this text only one multiple comparison procedure will be used, namely Fisher's least significant difference method (FSD). There are two reasons for this. First, since some readers will carry out the procedure by hand, it is desirable that it be easy to implement. Because of the computing method we will use to perform the analysis of variance, the FSD procedure will require only one simple additional step. Secondly, of the procedures compared by Carmer and Swanson, the FSD procedure is one of the top two in performance over a wide range of experimental situations and is often the best. Thus, it combines two of the most desirable properties of a statistical procedure: simplicity and high performance.

The FSD procedure is a two-step method, the first step of which is a one-way analysis of variance. Because of its independent interest, the analysis of variance will be developed separately in the next section.

One-Way Analysis of Variance

The assumptions underlying the one-way analysis of variance are the following. *First, the distributions of the measurement variable of interest for the* k *populations are assumed to be normal.* Consequently, the distributions are completely determined by the population means and standard deviations, which are assumed to be unknown. *Second, the standard deviations are assumed to be equal.* The common standard deviation is denoted by σ.

	POPULATION		
	1	2 · · ·	k
Population Means	μ_1	μ_2 · · ·	μ_k
Common Standard Deviation	σ	σ · · ·	σ

The null hypothesis of no difference among distributions then corresponds to the statistical hypothesis

$$H_0: \mu_1 = \mu_2 = \cdots = \mu_k$$

The test will be designed to detect any deviation whatsoever from the state specified by H_0. Consequently, the alternative hypothesis will be the universal hypothesis of "some difference." It is difficult to write this symbolically in terms of the μ_i's. We will settle for the notation

$$H_1: \text{Not } (\mu_1 = \mu_2 = \cdots = \mu_k)$$

The test of H_0 versus H_1 will be based on the same *summary data* that the two-sample test depended upon—the sample size, the sample mean, and the sample standard deviation for the random sample from each population:

	POPULATION 1	POPULATION 2	$\cdots$ POPULATION k
Sample Size	n_1	n_2	$\cdots n_k$
Mean	$\bar{X}_1$	$\bar{X}_2$	$\cdots \bar{X}_k$
Standard Deviation	s_1	s_2	$\cdots s_k$

Again, the assumption of sample-to-sample independence is made.

 Statistical inference for the analysis of variance is primarily a hypothesis-testing methodology. For the test of H_0 versus H_1 a test statistic F will be defined such that (a) large values of F favor H_1 and (b) the sampling distribution of F is known (and its critical values are available in a table) when H_0 is true. These properties will allow us to set up tests at given significance levels and calculate P-value bounds just as we did in chapter 9.

 The mathematical theory that makes all this possible depends on the use of a rather old measure of the variability of numbers—sums of squares—and some rather sophisticated calculations of more recent vintage, attributed primarily to R. A. Fisher. Although we will make no attempt to follow the mathematical analysis, it is worthwhile to look at a heuristic derivation of the test statistic to see why it is intuitively reasonable as well as mathematically convenient.

Sums of Squares

sum of squares

We have encountered a sum of squares of the type used in the analysis of variance in the sample standard deviation s. By omitting the square root and the divisor $n - 1$ in the expression for s, the principal constituent is seen to be the **sum of squares**

$$\text{SS} = \Sigma(X_i - \bar{X})^2$$

If $X_1, X_2, \ldots, X_n$ are any n numbers, then SS is a measure of how variable these numbers are around their arithmetic mean $\bar{X}$. SS will equal 0 only if all the X_i's are equal. In all other situations SS will be greater than 0, and if n is fixed, the size of SS will depend on how large the deviations of X_i are from $\bar{X}$. The larger the variability, the larger will be the value of SS.

 Since this property is a property of any collection of numbers, a reasonable parameter for summarizing the configurations of the μ_i's corresponding to H_0 and H_1 would be the sum of squares

$$\gamma = \Sigma(\mu_i - \bar{\mu})^2$$

(The symbol γ is the Greek letter gamma.) Since $\gamma = 0$ if and only if all the μ_i's are equal, in terms of γ the null hypothesis H_0: $\mu_1 = \mu_2 = \cdots = \mu_k$ would be equivalent to

$$H_0': \gamma = 0$$

while H_1 would correspond to

$$H_1': \gamma > 0$$

Note that by using a sum of squares, we are able to reduce hypotheses involving several parameters to hypotheses involving a single parameter, as we did in the two-sample problem.

To estimate γ we can use the device introduced for the two-sample t test whereby parameters in the expression for γ are replaced by their corresponding sample estimators. Thus to construct the natural estimator $\hat{\gamma}$ for γ, we would replace the population means μ_i by their sample estimators $\overline{X}_i$. This leads to the sample sum of squares

$$\hat{\gamma} = \Sigma(\overline{X}_i - \overline{X}.)^2$$

where $\overline{X}.$ denotes the arithmetic mean of the sample means $\overline{X}_i$. Since it is reasonable to expect that large values of $\hat{\gamma}$ would occur more frequently when $\gamma > 0$ than when $\gamma = 0$, it should be possible to devise a test statistic from $\hat{\gamma}$.

Because of the possibility that the sample sizes n_i are unequal, theory dictates a somewhat more complicated sum of squares as the basis of the test statistic:

$$SS_B = \Sigma n_i(\overline{X}_i - \overline{X}.)^2$$

where $\overline{X}.$ is now a weighted average of the $\overline{X}_i$'s rather than the straight arithmetic mean. The details are not important here since the correct formula for computing SS_B will be given later. The subscript B designates SS_B as the *between-population* sum of squares, a term whose explanation will also be momentarily deferred.

Each sum of squares we will consider has an associated **degrees of freedom.** The degrees of freedom associated with SS_B are $\nu_B = k - 1$, the number of populations minus one. The construction of the F statistic will require that the relevant sums of squares be normalized by dividing them by their degrees of freedom. The resulting quantities are called **mean squares.** The between-population mean square is then defined by the expression

degrees of freedom

mean squares

$$MS_B = \frac{SS_B}{\nu_B} = \frac{\Sigma n_i(\overline{X}_i - \overline{X}.)^2}{k - 1}$$

Now MS_B has the same virtue as a potential test statistic for testing H_0 versus H_1 as did SS_B—large values favor H_1 over H_0. However, it also has a serious fault: Its sampling distribution depends on the unknown standard deviation σ. However, the distribution of the ratio MS_B/σ^2 does not have this disadvantage. But now we have simply transferred the unknown parameter from the sampling distribution to the test statistic, where it is just as much of a nuisance, since we must be able to evaluate the test statistic from the sample in order to carry out a test. Fortunately, the ploy of replacing σ^2 by a sample estimator, used in designing the t statistic, also works here.

The best procedure is to take advantage of the assumption that all population distributions have the same σ by forming the *pooled estimate* of σ^2,

$$s_P^2 = \frac{\Sigma(n_i - 1)s_i^2}{N - k}$$

where $N = \Sigma\, n_i$ is the total sample size. (Compare this expression with the formula for the pooled estimate of σ used in the two-sample t statistic of the last chapter.)

Although in somewhat disguised form, s_P^2 is the ratio of a sum of squares,

$$SS_W = \Sigma(n_i - 1)s_i^2$$

and its degrees of freedom,

$$\nu_W = N - k$$

where W denotes *within-population*. Thus s_P^2 is simply the within-population mean square

$$MS_W = \frac{SS_W}{\nu_W}$$

Now the ratio

$$F = \frac{MS_B}{MS_W}$$

F statistic

F distribution

called an **F statistic,** is the appropriate test statistic for testing H_0 versus H_1. It has the desired properties of a test statistic. As argued, large values of F favor H_1. Moreover, *when H_0 is true, F has the so-called F distribution* (named in honor of R. A. Fisher), with *degrees of freedom $k - 1$ and $N - k$.*

Use of the *F* Distribution Tables

numerator degrees of freedom

denominator degrees of freedom

Table 4 at the back of the book is a table of critical values for the F distribution. The degrees of freedom $\nu_B = k - 1$ are called the **numerator degrees of freedom** since they are associated with the sum of squares in the numerator of the F statistic. Values of this quantity are given along the top row of table 4. Values of the **denominator degrees of freedom** are given in the first column. To design a hypothesis test at a given significance level, locate the column for the appropriate numerator degrees of freedom and the collection of rows corresponding to the denominator degrees of freedom. Within these rows locate the one corresponding to the given significance level. The cutoff (critical) value for the test is then the number at the intersection of the row and column so found.

EXAMPLE 11.1

The three years of final exam scores given in example 3.2 (p. 73) (with outliers removed), will be used to illustrate the methods to be given in this chapter. The first step will be to design a 5% test of the null hypothesis that there are no differences in average scores for the three years.

After deleting outliers, the sample sizes for the $k = 3$ samples are $n_1 = 20$, $n_2 = 25$, and $n_3 = 20$. Thus the total sample size is

$$N = 20 + 25 + 20 = 65$$

The numerator degrees of freedom are

$$\nu_B = 3 - 1 = 2$$

while the denominator degrees of freedom are

$$\nu_W = 65 - 3 = 62$$

Table 4 doesn't have an entry for 62 degrees of freedom. The liberal design would use the table value for 120 degrees of freedom, while the conservative design would use the value for 60. In the .05 row for the column for numerator degrees of freedom $= 2$ and the group of rows for denominator degrees of freedom $= 120$, we find the cutoff value $c = 3.07$. Thus the rule for the liberal test would be to reject H_0 if $F \geq 3.07$. For the conservative test H_0 would be rejected if $F \geq 3.15$. These two tests actually differ by very little.

The hand-computing procedure for evaluating F is given next.

The ANOVA Table

ANOVA table

A useful display for organizing and summarizing the results of an analysis of variance computation is the **ANOVA table.** Its entries are the sums of squares, degrees of freedom, and mean squares required for the calculation of the F statistics of interest. For more complicated analyses of variance there can be several hypotheses to test, requiring several F statistics and several table entries. For the one-way ANOVA there is only one hypothesis test to be made and thus only one F statistic. The entries in the table are the following:

SOURCE OF VARIATION	SUMS OF SQUARES (SS)	DEGREES OF FREEDOM (df)	MEAN SQUARES (MS)	F
Between populations	SS_B	$k - 1$	$MS_B = \dfrac{SS_B}{k - 1}$	$F = \dfrac{MS_B}{MS_W}$
Within populations	SS_W	$N - k$	$MS_W = \dfrac{SS_W}{N - k}$	
Total	SS_T	$N - 1$		

The last row will not, in general, enter into our calculations and its uses will be explained shortly. The computing formulas for SS_B and SS_W are

$$SS_B = \Sigma n_i \overline{X}_i^2 - \frac{(\Sigma n_i \overline{X}_i)^2}{N}$$

$$SS_W = \Sigma (n_i - 1) s_i^2$$

The steps in the calculations of the table entries will be given first for a simple numerical example. Suppose that random samples from three populations yield the following results:

POPULATION 1	POPULATION 2	POPULATION 3
1	2	3
3	4	5
	6	7
		9

Step 1. The summary values n_i, $\overline{X}_i$, and s_i are obtained for each sample. For the given samples the following values are obtained:

	POPULATION		
	1	2	3
n	2	3	4
$\overline{X}$	2	4	6
s	1.414	2	2.582

Step 2. The following statistics are calculated:

$$N = \Sigma n_i = 2 + 3 + 4 = 9$$
$$\Sigma n_i \overline{X}_i = (2 \times 2) + (3 \times 4) + (4 \times 6) = 40$$
$$\Sigma n_i \overline{X}_i^2 = (2 \times 2^2) + (3 \times 4^2) + (4 \times 6^2) = 200$$
$$SS_W = \Sigma(n_i - 1)s_i^2 = (1 \times 1.414^2) + (2 \times 2^2) + (3 \times 2.582^2) = 30.000$$

Step 3. Calculate SS_B and the other terms needed to fill out the ANOVA table.

$$SS_B = \Sigma n_i \overline{X}_i^2 - \frac{(\Sigma n_i \overline{X}_i)^2}{N} = 200 - \frac{(40)^2}{9} = 22.222$$

The mean squares are

$$MS_B = \frac{22.222}{2} = 11.111$$

$$MS_W = \frac{30.000}{6} = 5.000$$

and the value of F is then

$$F = \frac{11.111}{5.000} = 2.222$$

SOURCE OF VARIATION	SS	df	MS	F
Between populations	22.222	2	11.111	2.222
Within populations	30.000	6	5.000	
Total	52.222	8		

Note how the table aids in organizing the calculations. First, the SS column is computed and entered. Next, we enter the df column using the expressions for degrees of freedom. Then, the mean squares are formed by dividing each SS by the corresponding df. Finally, the F statistic is the ratio of mean squares.

The importance of the final row, labeled "Total" will now be explained.

Decomposition of the Total Sum of Squares

Suppose the $N = 9$ sample values in the preceding example had not been broken down by population but rather represented a single collection of numbers. The variation in this unstructured collection could then be represented as usual by the standard deviation

$$s = \sqrt{\frac{\Sigma(X_i - \bar{X})^2}{N - 1}}$$

total sum of squares

Except for the square root, this expression is a mean square, MS_T, in which the **total sum of squares**

$$SS_T = \Sigma(X_i - \bar{X})^2$$

is divided by its degrees of freedom

$$\nu_T = N - 1$$

Recall from the calculation of the sample standard deviation in chapter 2 that

$$SS_T = \Sigma X_i^2 - \frac{(\Sigma X_i)^2}{N}$$

where the sum and sum of squares are computed for all N observations. However, if an automatic calculation of the sample standard deviations can be made, an easier way to compute SS_T is to use the following expression in terms of s:

$$SS_T = (N - 1)s^2$$

Verify that for the 9 numbers above

$$SS_T = 52.222$$

Note that the total sum of squares in the ANOVA table above is equal to the sum of the between and within sums of squares. This result is not just happenstance but rather represents an important algebraic fact about these sums of squares:

$$SS_T = SS_B + SS_W$$

This result indicates that the total variation in the sample has been broken up into separate component sums of squares, each measuring a different source of variation. As seen above, SS_B measures the variation attributable to differences in means between the populations, while SS_W measures the inherent variability within populations.

additivity property

Note also that the degrees of freedom for each sum of squares add up to the total degrees of freedom. This **additivity property** of the sums of squares and degrees of freedom in an ANOVA table will be seen later to have implications for experimental design.

A numerical implication of the additivity property is that the value of any one sum of squares can be obtained from the other two. This fact is useful in more complex ANOVA situations. If the computations are done by hand, this property can also be used to check the correctness of your sum-of-squares calculations until you are comfortable with the computing method: Calculate all three sums of squares and check for additivity. If the additivity property holds (allowing for round-off errors), the calculations are almost certainly correct.

EXAMPLE 11.2

The data and summary statistics for the three years of exam data of example 3.2, purged of outliers, are given in the following tables:

YEAR 1		YEAR 2		YEAR 3	
	43	54	45	59	47
49	34	48	59	51	64
31	54	36	39	58	32
41	28	53	50	45	55
26	48	45	33	53	50
52	40	31	47	41	42
39	22	49	43	50	62
46	32	42	57	44	36
40	35	46	27	38	49
37	45	44	37	56	
58		41	40	68	
		51	46		
		63			

	YEAR		
	1	2	3
n	20	25	20
$\overline{X}$	40	45.04	50
s	9.6245	8.6626	9.6245

The computations for the second step of the calculations are as follows:

$$N = 20 + 25 + 20 = 65$$
$$\Sigma n_i \bar{X}_i = 2926$$
$$\Sigma n_i \bar{X}_i^2 = 132,715.04$$
$$SS_W = \Sigma(n_i - 1)s_i^2 = 5320.953$$

Then

$$SS_B = 132,715.04 - \frac{(2,926)^2}{65} = 1000.025$$

The total sum of squares is

$$SS_T = 6320.985$$

Now, rounding the sums of squares to one decimal place, the ANOVA table is as follows:

SOURCE VARIATION	SS	df	MS	F
Between years	1000.0	2	500.02	5.826
Within years	5321.0	62	85.82	
Total	6321.0	64		

From the test rule set up in example 11.1, our decision would be to reject H_0 at the 5% level, since 5.826 is larger than 3.07 (and 3.15). Thus we would conclude at the 5% significance level that the (population) mean exam scores are different.

The P-value bounds for this test can be obtained from table 4 by finding the critical values bounding 5.826 in the $\nu = 2$ column and $\nu = 120$ collection of rows. Since 5.826 falls between 4.79 and 7.32, which correspond to significance probabilities .01 and .001, respectively, we conclude that

$$.001 < P < .01$$

The next step is to determine the nature of the difference indicated by this test. This topic will be the subject of the next section.

◢

EXERCISES

11.1   The teaching load data from exercise 3.1 (p. 75) is as follows.

STATISTICS	APPLIED MATH	PURE MATH AND MATH EDUCATION	
89	118	197	203
135	144	139	92
168	108	104	90
70	48	174	90
100	110	182	121
190	110	181	101
145	85	50	98
195	107	44	
	135		
	147		
	99		

Test, at the 5% significance level, the hypothesis that there are no differences in average teaching loads for the three types of mathematics and statistics faculty at the University of New Mexico.

11.2   Determine bounds for the P-value of the test in exercise 11.1.

11.3   Verify that $SS_T = SS_B + SS_W$ for the data of exercise 11.1 as a check on your computations. (Note: Small differences may occur because of rounding errors.)

◢ SECTION 11.4

Fisher's Least Significant Difference Method

The Fisher least significant difference (FSD) method is a two-step procedure for comparing individual pairs of population means (a so-called *multiple-comparison* procedure). The first step consists of a standard one-way analysis of variance—usually carried out at the 5% significance level. If this test accepts H_0, the process stops and it is concluded that there is insufficient evidence to support differences among the population means. Only if the decision is to reject H_0 do we proceed to the second stage of the procedure.

The second step essentially consists of the application of two-sample t tests to every pair of means, as described in section 11.2. Now, however, the difficulties with experimentwise validity due to multiplicity are avoided because of the initial use of the one-way ANOVA. The ANOVA provides a simultaneous screening of all differences, which holds the experimentwise error rate for the entire FSD procedure very close to the significance level used for the ANOVA.

Through the use of t tests, the population means are arranged into groups in the second stage. Two means are put into the same group if the corresponding two-sided t test *accepts* the hypothesis H_0 of no difference in the means. (The test is made at the same significance level as the ANOVA.) Means are put into different groups if H_0 is rejected. This process results in the grouping of means described in section 11.1.

Two modifications are made in the two-sample procedure. First, maximum sensitivity is achieved by using a pooled estimate of σ based on *all* the samples. From the discussion in the last section, the within-population mean square, MS_W, was seen to be a pooled estimate of σ^2. Consequently,

$$s_P = \sqrt{MS_W}$$

will be used to estimate σ. This choice has the added advantage of simplifying the computation in the second modification.

Instead of computing the statistic

$$t = \frac{\overline{X}_i - \overline{X}_j}{s_P\sqrt{\dfrac{1}{n_i} + \dfrac{1}{n_j}}}$$

for each pair of samples, we form a yardstick for comparing the differences in means $\overline{X}_i - \overline{X}_j$ by combining the denominator of the t statistic and the test cutoff value c into the quantity

$$LSD = c s_P\sqrt{\frac{1}{n_i} + \frac{1}{n_j}}$$

least significant difference

called the **least significant difference.** The basis for this name, and the way the LSD is used, is discussed next.

The LSD When Sample Sizes Are the Same

Assume for the moment that all sample sizes are the same so that the value of the LSD is the same for all pairs of samples. Now if the sample means are arranged in increasing order, and if, for example, $\overline{X}_i$ is larger than $\overline{X}_j$, then the quantity $|t|$ will be larger than or equal to c if and only if $\overline{X}_i - \overline{X}_j \geq LSD$. In this case a significant difference would be claimed for the corresponding pair of population means, and they would be put into different groups. If $|t| < c$, or, equivalently, $\overline{X}_i - \overline{X}_j < LSD$, the means would be put into the same group. Consequently, LSD is the smallest difference between (ordered) pairs of $\overline{X}$'s for which a significant difference in population means would be claimed—that is, it is the least significant difference.

The two-sided cutoff value c is obtained from the table of critical values of Student's t distribution, table 3, for the significance level of the initial ANOVA—usually 5%—and the degrees of freedom associated with the within-population sum of squares,

$$\nu_W = N - k$$

An efficient way of using the LSD to group the population means will be demonstrated in the following example.

EXAMPLE 11.3

The (beta) islets of Langerhans are the insulin-producing cells in the pancreas. An observed relationship between obesity and diabetes in humans is thought to be linked with differences in the way these islets produce insulin in normal and obese individuals. While this difference cannot be directly observed for humans, studies on special strains of rats provide information that can be extrapolated to humans. The following data represent the insulin production over a three-week period for islets isolated from lean rats. (These data will be compared with the insulin production of islets from obese rats in exercise 11.4.)

The islets are collected into 24 cultures. At the end of each week in the three-week period, a different group of 8 cultures is chemically pulsed to induce insulin production, and the amount of insulin produced per islet is calculated. The results are the data given in the following table.

WEEK 1	WEEK 2	WEEK 3
2.02	1.49	.33
3.83	2.67	1.67
6.67	4.62	4.67
5.38	4.18	2.45
5.49	2.78	2.29
3.50	2.56	1.95
5.90	4.46	.49
4.89	3.79	1.81

Data courtesy of A. Hayek, M.D., School of Medicine, University of New Mexico, Albuquerque.

The problem is to determine whether any significant trend in insulin production exists over this time period, and if so, to describe it. A schematic diagram of the data is given in display 11.1. A definite downward trend is seen. If μ_1, μ_2, and μ_3 are the means of the distributions of insulin production for the three populations of islets—those pulsed at one, two, and three weeks—the significance of this trend would be reflected in significant differences among the μ_i's. These differences will be assessed by using the FSD procedure at a level of $\alpha = .05$.

The summary data and ANOVA table are as follows:

	WEEK		
	1	2	3
n	8	8	8
$\overline{X}$	4.710	3.319	1.958
s	1.503	1.108	1.343

SOURCE OF VARIATION	SS	df	MS	F
Between weeks	30.306	2	15.153	8.593
Within weeks	37.030	21	1.763	
Total	67.336	23		

Display 11.1

Insulin production per islet for lean rats

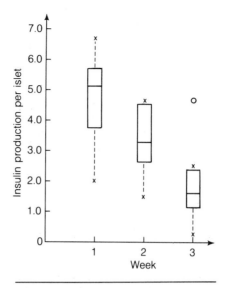

From table 4, with $\nu_B = 2$ and $\nu_W = 21$ (thus using rows corresponding to $\nu_W = 30$), we find the bounds on the P-value for this test to be

$.001 < P < .01$

In particular, since $P < .05$, the hypothesis H_0: $\mu_1 = \mu_2 = \mu_3$ would be rejected at the 5% level. Consequently, we proceed to the next step in the FSD method.

First, the LSD will be calculated. From table 3 for $\nu_W = 21$ degrees of freedom, the two-sided 5% cutoff value is found to be

$c = 2.080$

The pooled standard deviation is

$s_P = \sqrt{\text{MS}_W} = \sqrt{1.763} = 1.328$

Consequently,

$$\text{LSD} = c s_P \sqrt{\frac{1}{n_i} + \frac{1}{n_j}} = 2.080 \times 1.328 \times \sqrt{\frac{1}{8} + \frac{1}{8}} = 1.381$$

Now rewrite the $\overline{X}$'s in increasing order:

	WEEK			
	3	2	1	
$\overline{X}$	1.958	3.319	4.710	
	1.361			Difference < LSD
	2.752			Difference > LSD
		1.391		Difference > LSD

We will compare each mean in turn with those to its right.

Starting with the far left mean, we find the difference between it and the first mean to its right to be 1.361. This value is indicated in the display above. (The details in braces would not ordinarily appear in the figure. They are given here to explain the procedure.) Since this difference is *smaller* than the LSD, a line is drawn under these means, indicating that the corresponding population means will be in the same group.

Next we find the difference between the first and third $\overline{X}$'s to be 2.752. Since this is greater than the LSD, the line is not continued, indicating that μ_1 and μ_3 will be in different groups.

At this point we are finished with the first mean and go on to a comparison of the second mean with those to its right. The difference between the second and third means is 1.391. Since this value is greater than the LSD, no line is drawn. Whenever a difference larger than LSD is found, we go on to the next mean in sequence, here the third. But since there are no means to its right, we are finished.

The final grouping of means—or, equivalently, weeks—is

{week 2, week 3} < {week 1}

That is, the mean insulin production for week 1 is significantly greater than that for weeks 2 and 3, but the means for those two weeks are indistinguishable from one another. Thus the trend would seem to consist of a significant drop in production between weeks 1 and 2 followed by a leveling off thereafter. The reason for the inferred leveling off, which is contrary to the picture presented by the schematic diagram, can be clearly seen in the box plot for week 3. We will return to this example in the next chapter.

EXERCISES

11.4   Insulin production data for islets from the pancreases of obese rats are given in the following table.

WEEK 0	WEEK 1	WEEK 2	WEEK 3
31.2	18.4	55.2	69.2
72.0	37.2	70.4	52.0
31.2	24.0	40.0	42.8
28.2	20.0	42.8	40.6
26.4	20.6	26.8	31.6
40.2	32.2	80.4	66.8
27.2	23.0	60.4	62.0
33.4	22.2	65.6	59.2
17.6	7.8	15.8	22.4

Data courtesy of A. Hayek, M.D., School of Medicine, University of New Mexico, Albuquerque.

In addition to the production at the ends of the first three weeks, data are also available for the first day of the experiment (labeled week 0). The summary data are as follows:

	WEEK			
	0	1	2	3
n	9	9	9	9
$\bar{X}$	34.156	22.822	50.822	49.622
s	15.431	8.326	21.132	16.289

(a) Make a schematic diagram of the data and comment on the observed trend.

(b) Compare the trend in insulin production with that for the lean rats. Note any indications of possible violations of the assumption on which the ANOVA is based.

11.5 Apply the FSD procedure to the summary data given in exercise 11.4. Which of the distributional differences seen in the box plots does the analysis support as being significant?

The LSD for Different Sample Sizes

When the samples are of different sizes there is, in theory, a possibly different LSD to be computed for each pair of means. In practice, when calculations are being done by hand it is often possible to get away with calculating only two—a smallest (LSD_{min}) and a largest (LSD_{max}) value by using the two *largest* sample sizes and the two *smallest* sample sizes, respectively. These numbers will differ only in the factor $\sqrt{(1/n_i) + (1/n_j)}$, since c and s_P are determined once and for all from the test significance level and the ANOVA table values. Now, a difference in means larger than LSD_{max} would be larger than the LSD calculated for any pair of sample sizes, while a difference smaller than LSD_{min} would be smaller than all. This fact will usually serve to classify most means. Only in those instances in which a difference in means lies between LSD_{min} and LSD_{max} would the calculation of the LSD for that particular pair of sample sizes be required.

EXAMPLE 11.4

We are now in a position to complete the inferential analysis of the exam data begun in example 11.2 (p. 350). From the ANOVA table we find the pooled standard deviation estimate to be

$$s_P = \sqrt{85.82} = 9.230$$

and the within-population degrees of freedom to be

$$\nu_W = 62$$

The 5% two-sided critical value from table 3 is then

$$c = 1.980$$

The two largest sample sizes are 20 and 25. Consequently,

$$LSD_{min} = 1.980 \times 9.230 \times \sqrt{\frac{1}{20} + \frac{1}{25}} = 5.483$$

Since the smallest two sample sizes are 20 and 20, we have

$$\text{LSD}_{max} = 1.980 \times 9.230 \times \sqrt{\frac{1}{20} + \frac{1}{20}} = 5.779$$

The ordered means are as follows:

	YEAR 1	YEAR 2	YEAR 3
$\bar{X}$	40	45.04	50

The differences in means are seen to be $\bar{X}_2 - \bar{X}_1 = 5.04$, $\bar{X}_3 - \bar{X}_1 = 10$, and $\bar{X}_3 - \bar{X}_2 = 4.96$. Since the differences for years 1 and 2 and years 2 and 3 are both smaller than LSD_{min}, they are underlined. The difference for years 1 and 3 exceeds LSD_{max}.

The following rather curious grouping is a consequence of this analysis

{year 1, year 2} < {year 2, year 3}

The interpretation is that there was insufficient evidence in the sample to detect a difference over either one-year period, but the difference over the two-year period is significant. Thus a (possible) improvement in the professor's teaching is detectable over the three years of the study but not over a shorter time period.

Overlapping groupings like the one in the preceding example are clearly undesirable because they are difficult to interpret—but they are sometimes unavoidable. In some situations in which they occur, the added sensitivity provided by a robust procedure will eliminate the overlap and provide distinct groupings. In fact, it is often at this level of the refinement of groupings that an FSD procedure based on trimmed means or on ranks will pay off. This will be the topic of study in chapter 12.

EXERCISE

11.6   In an anthropological study of facial tissue thicknesses for different racial groups, observations were taken at several points on the faces of deceased individuals during autopsy. The Glabella measurements (taken at the bony ridge just above the nose) for people from three racial groups are as follows. (Values are in millimeters.)

CAUCASIAN	BLACK	NATIVE AMERICAN AND ORIENTAL
5.75	6.00	8.00
5.50	6.25	7.00
6.75	6.75	6.00
5.75	7.00	6.25
5.00	7.25	5.50
5.75	6.75	4.00

(table continued on next page)

CAUCASIAN	BLACK	NATIVE AMERICAN AND ORIENTAL
5.75	8.00	5.00
7.75	6.50	6.00
5.75	7.50	7.25
5.25	6.25	6.00
4.50	5.00	6.00
6.25	5.75	4.25
	5.00	4.75
		6.00

Data courtesy of Elliott C. Moore II, Ph.D, Department of Anthropology, University of New Mexico, Albuquerque.

(a)   Make a schematic diagram of these data.

(b)   Use the FSD procedure to determine whether groupings suggested by the schematic diagram are confirmed at the 5% significance level.

◪ SECTION 11.5

Confidence Intervals for Differences in Group Means*

The question of how much of a difference there is between population means becomes somewhat more complicated when the FSD procedure combines these means into groups. The parameter of interest would not be the individual means making up the group, but rather a mean for the group as a whole. If a group consists of a single mean, then the group mean would clearly be equal to that value. But how should we define the mean for a group consisting of three means, for example?

The fact that all three means occur in the same group suggests that we should be willing to believe they have the same value, and that value would be the group mean. However, in order to estimate the group mean, it is convenient to think of them as being distinct and to use an average to define the mean of the group. Thus, if the means in the group were μ_1, μ_2, and μ_3, then the group mean could be defined as

$$\mu_{\text{group}} = \frac{\mu_1 + \mu_2 + \mu_3}{3}$$

This parameter can be estimated by replacing each population mean in the definition by the corresponding sample mean:

$$\hat{\mu}_{\text{group}} = \frac{\bar{X}_1 + \bar{X}_2 + \bar{X}_3}{3}$$

---

* Optional material; not used in subsequent chapters.

The advantage of this definition is that, under the assumptions of the analysis of variance, given in section 11.3, the estimator has a normal distribution and the standard error of the estimator is easily computed to be

$$\sigma \sqrt{\frac{1}{9n_1} + \frac{1}{9n_2} + \frac{1}{9n_3}}$$

An estimate of this standard error, call it SE for short, can be obtained by replacing σ by the pooled standard deviation $s_P = \sqrt{MS_W}$, as before. At this point, if we wanted a confidence interval for the group mean, the pivotal method could be applied to the statistic

$$t = \frac{\hat{\mu}_{group} - \mu_{group}}{SE}$$

which has Student's t distribution with the degrees of freedom associated with s_P, namely, $N - k$.

However, our goal is to find a confidence interval for the difference of two group means, not for a single mean. The question we would like to ask is, "How different are the groups (or group means) for a given pair of groups?" Nevertheless, this construction contains the essence of the method for constructing confidence intervals for differences in group means and much more.

The more general notion we will need is that the parameter of interest is a linear combination of individual means of the form

$$\text{parameter} = a\mu_1 + b\mu_2 + c\mu_3$$

We will discuss the case of linear combinations of three means, but the ideas apply to any number. Note that the group mean considered above was of this form, with the three numbers a, b, and c all equal to $\frac{1}{3}$.

The estimator for this quantity is

$$\text{estimator} = a\bar{X}_1 + b\bar{X}_2 + c\bar{X}_3$$

Now, the estimated standard error of this estimator is

$$SE = s_P \sqrt{\frac{a^2}{n_1} + \frac{b^2}{n_2} + \frac{c^2}{n_3}}$$

This expression can be obtained by using the formula for the variance of a linear combination of independent random variables given in chapter 5. The common standard deviation σ is replaced by s_P. Then $t = (\text{estimator} - \text{parameter})/SE$ has Student's t distribution, with $N - k$ degrees of freedom. The pivotal method can now be employed to obtain the confidence interval

$$\text{estimator} - tSE \leq \text{parameter} \leq \text{estimator} + tSE$$

where the critical value, t, comes from table 3 for the desired confidence coefficient and degrees of freedom $= N - k$.

EXAMPLE 11.5

Let's apply these ideas to the insulin production data of example 11.3 (p. 354). We found that the insulin production groupings for the three weeks are {week 1} > {week 2, week 3}. The group mean for the first group is simply μ_1, the mean for week 1, and the mean for the second group is $(\mu_2 + \mu_3)/2$. Thus, the difference in group means, for which the confidence interval is to be constructed, is

$$\mu_{\text{group 1}} - \mu_{\text{group 2}} = \mu_1 - \frac{\mu_2 + \mu_3}{2}$$

$$= 1\mu_1 + (-\tfrac{1}{2})\mu_2 + (-\tfrac{1}{2})\mu_3$$

This parameter is a linear combination of means with $a = 1$ and $b = c = -\tfrac{1}{2}$. The estimator of the difference in group means is this same linear combination of sample means:

$$\text{estimator} = \overline{X}_1 - \frac{\overline{X}_2 + \overline{X}_3}{2}$$

From the summary data of example 11.3 we compute the value of this estimator to be

$$4.710 - \frac{3.319 + 1.958}{2} = 2.0715$$

Now, the standard error is

$$\text{SE} = s_P \sqrt{\frac{1^2}{n_1} + \frac{(-\tfrac{1}{2})^2}{n_2} + \frac{(-\tfrac{1}{2})^2}{n_3}}$$

$$= s_P \sqrt{\frac{1}{n_1} + \frac{1}{4n_2} + \frac{1}{4n_3}}$$

Using the results of example 11.3, the value of this standard error is

$$1.328 \sqrt{\frac{1}{8} + \frac{1}{4 \cdot 8} + \frac{1}{4 \cdot 8}} = .5750$$

To find a 95% confidence interval, we will need the critical value for 95% confidence and 21 degrees of freedom (from table 3). This number is 2.080. Thus, the limits for the confidence interval are

$$2.0715 \pm 2.080 \cdot .5750$$

or

$$0.876 \quad \text{and} \quad 3.268$$

With 95% confidence, we would conclude that the mean insulin level for week 1 exceeds the mean insulin level for weeks 2 and 3 by an amount between .88 and 3.27 mg%.

One question should be asked, even though there is no good answer. What happens to the validity of the FSD procedure if we go beyond the confidence intervals for pairs of means from which the groupings are made (see exercise 11.7) and compute confidence intervals for differences in group means as well? The conclusions based on these confidence intervals put an added burden on the experimentwise confidence level (one minus the experimentwise error rate), and it is no longer clear that the quoted level applies. The optimistic strategy, and the one often followed in practice, is simply to ignore the problem. A somewhat more conservative strategy would be to use the Bonferroni inequality *on the confidence intervals* but not on the original grouping method. Thus, if we were interested in bounds on k of the differences in means for groups determined by an FSD procedure carried out at the 5% significance level, each confidence interval could be computed using confidence coefficient $1 - .05/k$.

Another alternative that we will not pursue is to replace the FSD procedure with confidence intervals that have a specifiable global (i.e., experimentwise) confidence level for either all linear combinations of population means or, at least, for all pairwise differences of means. Classical procedures by H. Scheffé and J. W. Tukey, respectively, are available for these tasks. These procedures can be found in the text of Neter, Wasserman, and Kutner [24], for example. The price of guaranteed validity, provided by these methods, is usually a loss of sensitivity that shows up in the lengths of the confidence intervals. However, these procedures should be considered if an unqualified guarantee of validity is required.

EXERCISE

11.7   Find 95% confidence intervals (without Bonferroni adjustment) for the differences in insulin group means for the groupings found in exercise 11.5. Interpret your results.

Chapter 11 Quiz

1. What is the problem encountered in trying to test the equality of several means by carrying out two-sample t tests on each possible pair of means?

2. Why is the multiplicity problem of concern in a statistical investigation?

3. What is the Bonferroni inequality? What does it tell us about the relationship between the experimentwise and comparisonwise error rates in testing the equality of several means by using two-sample t tests on each pair of means?

4. How can the Bonferroni inequality be used to guarantee a prescribed experimentwise error rate for a procedure consisting of several subprocedures such as that described in question 3?

5. Why aren't procedures based on the Bonferroni inequality universally used in k-sample comparison problems?

6. What are the assumptions under which the one-way ANOVA is carried out?

7. What are the hypotheses being tested by a one-way ANOVA F test? Describe the alternative hypothesis in words.

8. What are the components of the ANOVA table for a one-way analysis of variance? How are they computed from sample data? Give an example.

9. How is table 4 of the F distribution used to carry out the test of the hypotheses of question 7?

10. What are the steps of the Fisher least significant difference method? Give an example of this method for data with equal sample sizes.

11. For hand calculation, what is a simple method for classifying most means using the FSD method when sample sizes are different?

12. What is the interpretation of an FSD grouping in which the same population mean occurs in two groups?

The following questions are for the optional section 11.5.

13. How is the mean defined for a group of means?

14. How is a confidence interval constructed for a group mean?

15. How is a confidence interval constructed for a difference in group means? Give an example.

SUPPLEMENTARY EXERCISES

11.8 The following table lists the weights in pounds of infants at the end of three months of feeding with three different milk formulas. Test for any differences in the effects of the formulas on weight at the 5% significance level.

FORMULA 1	FORMULA 2	FORMULA 3
12.0	12.5	12.3
12.2	12.6	12.5
12.4	12.9	13.0
12.5	12.7	12.7

11.9 The following data are summary indices derived from questionnaires designed to determine the political conservatism of college students. Questionnaires were given to a sample of students from the departments of Anthropology, Mathematics, and Psychology. Test for any difference in political philosophy. (Large scores indicate more conservative individuals.)

ANTHROPOLOGY		MATHEMATICS		PSYCHOLOGY	
58	55	56	56	59	57
58	56	57	54	60	60
59	56	58	55	61	63
57	57				

11.10 The summary data for the verbal ability scores analyzed in example 4.2 (p. 104) are given below. Determine, using an FSD procedure at the 1% level, which of the differences seen in the exploratory curves of displays 4.10 and 4.11 (pp. 105 and 106) are significant.

			CHILDREN'S AGES (YEARS)				
	3	4	5	6	7	8	9	10
n	10	11	11	10	11	11	12	11
$\bar{X}$	169.70	279.55	396.55	288.60	325.64	373.82	404.00	401.18
s	86.38	99.68	43.78	66.66	25.87	40.93	16.94	35.91

11.11   The field mouse activity data given in exercise 4.10 (p. 107) is repeated below.

(a)   Analyze at the 5% level the seasonal differences in activity with an FSD procedure.

(b)   Which, if any, of the ANOVA assumptions appear to be violated?

FALL		WINTER		SPRING	SUMMER	
0	15	0	30	15	60	0
0	8	34	15	0	21	0
21	29	0	15	15	15	21
0	15	15	8	18	15	17
15	46	15	21	109	15	0
0	39	87	15	15	33	15
15	30	15	15	0	24	106
15	15	0	15	15	33	17
0	11	5	22	47	42	21
8	0	0	0	30	54	21
0		0	15	15	11	
0		15	15	34	32	
15		8	33	47	8	
21		0	21	42	71	
0		15	15	0	150	
34		47	0	22	18	
0		0	0	34	12	

Data courtesy of J. Scheibe, Department of Biology, University of New Mexico, Albuquerque.

11.12   In an experiment to determine the effects of different chemicals on the deterioration of pottery, small pottery squares were stored in the chemicals for a period of time, and then their break strengths were measured on a crushing machine. The average break strengths for samples stored in the various chemicals, along with the sizes of the samples, are given in the accompanying table. An ANOVA yielded a significant difference in break strengths at the 5% level. The within-population mean square was computed to be $MS_W = 4.101$. Use the FSD procedure to determine the nature of the differences the five chemicals had on the break strengths of the pottery.

			CHEMICAL		
	A	B	C	D	E
n	12	18	14	15	10
$\bar{X}$	8.04	11.15	9.43	6.31	9.22

11.13   Twenty-five samples of a food product were assigned at random to five storage methods to test the effectiveness of these methods in preserving the water content (in percentages) of the foods for a given time period. Because of defective seals, several of the samples became unusable. The usable data after the storage period are as follows:

METHOD 1	METHOD 2	METHOD 3	METHOD 4	METHOD 5
7.8	5.4	8.1	7.9	7.1
8.3	7.4	6.4	9.5	
7.6	7.1		10.0	
8.4				
8.3				

What can be deduced from these data about the relative effectiveness of the five methods? (Use an FSD procedure at the 5% level.)

11.14   The free testosterone coefficients (FTC) for men over 50 years of age in three libido categories are as follows.

1	2	3
1.5	5.9	4.8
2.1	6.1	
.8	6.2	
2.3	3.6	
	2.2	
	1.2	
	2.0	
	2.3	
	1.4	
	7.8	
	4.2	
	1.2	

Test at the 10% level whether there is a difference in mean FTC levels for the three groups.

11.15   The free testosterone levels for women in three occupational categories from exercise 3.30 (p. 93) are repeated below.

(a)   Are the differences in levels indicated by the schematic diagram constructed in exercise 3.30 confirmed at the 5% significance level? (If you didn't do exercise 3.30, construct the schematic diagram here.)

(b)   What features of the data suggest the ANOVA assumptions are violated?

HOUSEWIVES		SECRETARIES AND OFFICE WORKERS		PROFESSIONAL CAREERS			
8	2.3	1.1	2.3	1.2	2.9	3.4	5.2
1.1	2.3	1.8	2.3	1.5	3.0	3.5	5.3
1.2	2.3	2.0	2.4	2.1	3.1	3.6	5.7
1.4	3.9	2.1	2.5	2.2	3.2	3.6	6.1
1.8	4.2	2.1	5.9	2.2	3.3	4.8	6.2
2.0		2.2		2.7	3.4	4.9	7.8

11.16   Octane ratings for a particular brand of gasoline were sampled from service stations selected randomly from four regions of the northeastern United States, which we will designate by A, B, C, and D. Analyze the data for regional differences in octane ratings at the 10% significance level.

REGION A		REGION B		REGION C		REGION D		
84.0	83.0	82.4	82.0	83.2	83.7	80.2	83.5	81.5
83.5	85.8	82.4	83.2	82.8	83.6	82.9	83.6	81.9
84.0	84.0	83.4	83.1	83.4	83.3	84.6	86.7	81.7
85.0	84.2	83.3	82.5	80.4	85.1	84.2	82.6	82.5
83.1	82.2	83.1		82.7	83.1	82.8	82.4	
83.5	83.6	83.3		83.0	84.2	83.0	83.4	
81.7	84.9	82.4		85.0	80.6	82.9	82.7	
85.4		83.3		83.0	82.3	83.4	82.9	
84.1		82.6		85.0		83.1	83.7	

Source: O. C. Blake, "National Motor-Gasoline Survey," Bureau of Mines Report of Investigation No. 5041, 1954; given by Brownlee [5].

11.17   The Tour of the Rio Grande Valley bicycling speeds for people choosing three different distances, given in exercise 3.19 (p. 88) are as follows.

50 MILES	75 MILES	100 MILES	
9.0	7.2	20.2	12.5
8.1	11.3	12.0	11.4
11.2	8.8	12.4	8.1
8.7	9.5	17.6	10.8
10.2	8.0	16.5	14.7
9.3		12.6	11.0
9.0		11.1	
9.0		12.3	
9.0		21.1	

(a)   Compare the mean bicycling speeds for the three groups at the 10% significance level.

(b)   Which of the ANOVA assumptions appear to be violated?

11.18   The complete data set from the study of the effect of different types of instructions on the competitive choices made by game players, introduced in exercise 10.30 (p. 337) is given below. In addition to the groups receiving competitive and individualistic instructions, two additional groups were given cooperative instructions and no (specific) instructions, respectively.

(a) Analyze the effects of the different types of instructions at the 5% significance level.

(b) Which ANOVA assumptions appear to be violated?

	INSTRUCTION TYPE			
Numbers of competitive choices (out of 40)	COMPETITIVE	INDIVIDUAL	NO INSTRUCTIONS	COOPERATIVE
	32	24	10	0
	19	23	30	5
	8	13	10	4
	19	25	15	2
	9	26	13	3
	15	25	15	0
	17	6	10	2
	16	26	19	0
	21	22	13	3
	24	13	8	18
	10	7	16	0
	27	16	2	8

Data courtesy of Richard J. Harris, Department of Psychology, University of New Mexico, Albuquerque.

11.19 The following data are starch film thicknesses in ten-thousandths of an inch from a study of the break strengths of seven types of starch films. Determine at the 5% significance level what differences in thicknesses are supported by the data.

	STARCH						
WHEAT	RICE	CANNA	CORN	POTATO	DASHEEN	SWEET POTATO
5.0	7.1	7.7	8.0	13.0	7.0	9.4
3.5	6.7	6.3	7.3	13.3	6.0	10.6
4.7	5.6	8.8	7.2	10.7	7.1	9.0
4.3	8.1	11.8	6.1	12.2	5.3	7.6
3.8	8.7	12.4	6.4	11.6	6.2	
3.0	8.3	12.0	6.4	9.7	5.8	
4.2	8.4	11.4	6.9	10.8	6.6	
	7.3	10.4	5.8	10.1	6.6	
	7.5	9.2		12.7		
	7.8	9.0		13.8		
	8.0	12.5				

Data courtesy of the Statistical Laboratory, University of California, Berkeley.

11.20 In a previous exercise (exercise 3.20, p. 88) it was hypothesized that the sudden infant death syndrome (SIDS) is a lung disorder. Physical evidence suggesting this hypothesis was obtained from autopsy data by comparing the lungs of children diagnosed as dying of SIDS with those of children dying of causes unrelated to diseases of the lungs (controls). When a normal child is born, the musculature of small arteries (arterioles) in the lung begins to change. Before birth this muscle layer (media) is relatively thick, because it must help the arteriole withstand the systemic blood pressure of the mother, the blood pressure that causes blood to flow throughout the body. After birth a small valve closes and the blood circulating to the baby's lung is put under the control of the baby's pulmonary system, for which the blood pressure is substantially less. Like any muscle that is no longer used, the medial layer begins to atrophy and become thinner, reaching the thickness typical of adult lungs within a year or so after birth.

These arterioles are an integral part of the system that carries blood to the lung for oxygenation. Any deviation from the normal development of the lung after birth could upset the balance of this system and lead to death. One indication of abnormal development would be a deviation in the trend of the medial muscle thicknesses from the rather regular pattern of decrease with time.

Medial muscle thicknesses were measured microscopically for arterioles from the lungs of five SIDS children and five controls. The arterioles measured for each child represent a random sample of arterioles from its lungs. The sample mean then estimates the mean medial thickness for the entire lung. The sample means $\bar{X}$ and their standard deviation, along with sample sizes and the ages of the children at death are given below. These data are extracted from Koopmans and Fullilove [21]. Medial thickness is expressed as a percentage of total arteriole thickness.

	I.D.	AGE AT DEATH (DAYS)	NUMBER OF ARTERIOLES (n)	MEAN THICKNESS ($\bar{X}$)	STANDARD DEVIATIONS (s)
	C1	4	23	13.35	.144
	C2	30	20	10.43	2.19
CONTROL	C3	36	98	10.55	1.98
	C4	76	31	6.04	1.69
	C5	198	48	5.60	1.52
	S1	20	9	8.00	1.29
	S2	51	28	8.70	1.43
SIDS	S3	66	128	9.10	2.15
	S4	110	34	8.40	2.92
	S5	190	16	9.20	1.32

(a) At the 5% significance level, determine LSD groupings for these ten children. Do the data suggest differences between SIDS and controls? What is the nature of the difference?

(b) Plot the sample means versus age at death for SIDS and for controls on the same graph. Describe the differences in trends of means for the two curves.

11.21 The following (contrived) data are mileages (in miles per gallon) for ten automobiles used to test a new gasoline additive purported to increase mileage.

GASOLINE WITH ADDITIVE	GASOLINE WITHOUT ADDITIVE
15.8	15.4
17.9	17.1
20.3	20.1
15.6	14.9
12.4	12.3

(a) Apply the one-way ANOVA to these $k = 2$ populations. At the 5% level, test whether there is a difference in mileages for the gasolines with and without the additive.

(b) Compute the value of the two-sample t statistic for the data. Show that it is related to the value of the F statistic computed in part (a) by the equation $F = t^2$.

(c) Show that the 5% cutoff value for the F test of part (a) is the square of the two-sided 5% cutoff value for the two-sample t test that could be constructed in part (b). Does this relationship also hold for the 1% cutoff values? The 10% cutoff values?

(d) Using the results of part (c) and the fact that the equation $F = t^2$ is true in general for $k = 2$ populations, argue that the F test of part (a) is equivalent to the two-sided, two-sample t test.

Robust and Nonparametric Procedures for Measurement Variable Comparison Problems

---

Introduction

The statistical performance of the methods of chapters 10 and 11 depends on the assumptions of normality and equivariability of the underlying distributions. Unfortunately, as we have seen frequently in our exploratory analyses, nature appears to be rather unconcerned about the assumptions statisticians make. Outliers are a commonplace feature of real data, and differences in dispersion are often seen as well. In these situations the performance of the procedures we have used in the last two chapters can be badly affected.

robust

In response to this problem, statisticians have developed what are called **robust** methods. A procedure is said to be robust if its performance characteristics remain reasonably constant when the actual distribution deviates from the assumed one (usually the normal distribution) by amounts commonly seen in practice.

It is desirable that the procedures we use be both *validity-robust* and *sensitivity-robust*. While problems with the sample mean have been known and dealt with (by the substitution of the trimmed mean and median) since the 1700s, efforts to produce robust procedures date roughly from the 1940s, with the introduction of nonparametric methods. The focus was principally on validity robustness, and most nonparametric methods have the remarkable property that validity remains constant for extremely broad families of distributions.

However, in the pioneering paper [33], J. W. Tukey pointed out that the most frequent victim of nonnormality is sensitivity. The usual proce-

Display 12.1

Box plot and 95% confidence intervals for population center of household income data based on the sample mean and 10% trimmed mean

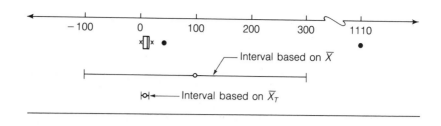

dures can be degraded, sometimes greatly so, by distributions with long tails—the kind that produce outliers in samples. An illustration is given in display 12.1, using the household income data introduced in example 2.9 (p. 62). The confidence interval for the mean is indicated by the line segment below the box plot. The horizontal lines at the ends of the segment are at the interval limits, and the dot at the center indicates the position of the sample mean. The confidence interval for the trimmed mean is shown, in the same format, below the one for the mean.

Note the extreme difference in interval lengths. Recall that for confidence intervals, good sensitivity means short intervals. In this regard, the confidence interval based on the sample mean makes a poor showing relative to the interval based on the trimmed mean in this example. Because the largest outlier is rather atypical of what one would ordinarily see in practice, this example is somewhat extreme. Yet the same phenomenon occurs to a lesser degree in many practical problems and often goes unnoticed or untended. When this happens, the time, effort and expense of the experiment under study and the analysis are frequently wasted.

A primary goal of this text is to provide tools for discovering when a traditional analysis will be in trouble and to provide remedial methods when it is. The exploratory methods of the first section of the book dealt with discovery. The schematic diagram pinpoints the major difficulties that can plague the classical comparison methodologies given in the last two chapters. Outliers and/or differences in dispersion were seen in a number of the schematic diagrams that were computed at the exploratory stage of the analyses. In this chapter, methods will be presented to counteract these difficulties.

The problems of outliers and differences in dispersion cannot be entirely separated, since they often occur together in the same data. When this happens, a single remedial procedure will frequently take care of both problems at once. Two such procedures are given in this chapter. Both involve transforming the data; the important rank methods of section 12.4 depend on the *rank transform,* and the family of *power transformations* is treated in section 12.3.

While the transform methods provide powerful hypothesis-testing procedures, the fact that measurement scales are distorted in the process of data transformation makes them poorly suited for quantifying differences in location by confidence intervals. Outlier correction based on trimmed means, on the other hand, preserves the measurement scale. Confidence intervals for differences in population trimmed means can thus be computed in the same manner as were confidence intervals for differences in means in chapter 10. Since a complete inference strategy can be based on trimmed means, they will be presented before the other methods in the next section.

While the method of trimmed means is effective in correcting for outliers, it is seldom useful in dealing with differences in dispersion. A further correction method, based on Satterthwaite's approximation, can be used either after a trimmed mean analysis or alone (if the data are already reasonably normal) to lessen the impact of unequal dispersions. This method will be applied to the two-sample problem in the next section.

It is of interest to note that all three methods to be given in this chapter depend on a common computing strategy. The data will be modified in some way and summary statistics will be computed. Then, the methods of chapters 10 and 11 will be applied *precisely as before*, but to the new summary data rather than to n, $\overline{X}$, and s. Because of this fact, with only modest effort required to learn one or two data transformations, we will acquire a large and important tool-kit of robust methods for dealing with problems that commonly occur with real data.

◢ SECTION 12.2

Methods Based on the Trimmed Mean

Hypothesis tests for the one-, two-, and k-sample problems and confidence intervals for the trimmed mean and difference in trimmed means will be illustrated in this section. Satterthwaite's approximation will be given to adjust two-sample tests for differences in dispersion.

Confidence Intervals for the Trimmed Mean

The calculation of confidence intervals based on the trimmed mean will be illustrated by going through the details of the computations for display 12.1. The stem and leaf plot for the household income data is as follows.

```
0 | 7758052 2 7
1 | 2
2 |
3 |
4 | 6
---------------
HI | 1110
```

As seen in chapter 10, the bounds for the usual confidence interval for the mean are

$$\overline{X} \pm t \frac{s}{\sqrt{n}}$$

where t is the appropriate critical value from Student's t table, table 3. The confidence interval based on the trimmed mean will be computed in exactly the same way, but with the sample size, mean, and standard deviation replaced by the trimmed sample size, trimmed mean, and trimmed standard deviation. Review section 2.5 for the details of the following computation.

We will adopt the strategy of selecting the trimming fraction to be the smallest multiple of 5% that just eliminates all outliers. This strategy determines a 10% trimming fraction for the household income data. The number of observations to be trimmed from each end of the sample is $.10 \cdot 12 = 1.2$, which rounds to 2. The observations to be trimmed are underlined in the above stem and leaf plot. The trimmed sample size is then the number of remaining observations after trimming, $n_T = 8$. The trimmed mean is

$$\overline{X}_T = 6.625$$

The Winsorized sample is obtained by replacing the underlined values in the stem and leaf plot by the numbers in the corresponding boxes. Thus 0 and 2 would be replaced by 2 on the low side and 46 and 1110 by 12 on the high side. The standard deviation of the Winsorized sample is

$$s_W = 3.793$$

It follows from the computation given in section 2.9 that the trimmed standard deviation is

$$s_T = \sqrt{\frac{(n-1)s_W^2}{n_T - 1}}$$

$$= \sqrt{\frac{(12-1) \cdot 3.793^2}{8-1}} = 4.755$$

The summary data n_T, $\overline{X}_T$, s_T now replace n, $\overline{X}$, and s in the confidence interval calculation. The degrees of freedom for the trimmed t are $n_T - 1 = 8 - 1 = 7$, so the 95% critical value from table 3 is 2.365. It follows that the endpoints of the 95% trimmed mean confidence interval are

$$L = \overline{X}_T - 2.365 \frac{s_T}{\sqrt{n_T}} = 6.625 - 2.365\frac{4.755}{\sqrt{8}} = 2.649$$

and

$$U = 6.625 + 2.365\frac{4.755}{\sqrt{8}} = 10.601$$

These limits are plotted in display 12.1. We would conclude with 95% confidence that the average income for the town is between $2,649 and $10,601.

For comparison purposes, verify that the confidence limits for the mean, also plotted in display 12.1, are as follows. The summary data are

$$n = 12, \qquad \overline{X} = 100.92, \qquad s = 318.01$$

For $n - 1 = 11$ degrees of freedom, the 95% critical value from table 3 is 2.201. Consequently, the endpoints of the confidence interval are

$$L = 100.92 - 2.201 \frac{318.01}{\sqrt{12}} = -101.14$$

and

$$U = 302.98$$

It is clear that the confidence interval based on the trimmed mean is substantially more useful for pinpointing the center of the household income distribution than is the usual confidence interval. The properties of the trimmed mean procedure were studied by Tukey and McLaughlin in [34].

The last example uses, implicitly, a property that trimmed mean confidence intervals share with those for the ordinary mean—a property that will not be preserved by the transformations to be considered later. This property is *invariance under linear transformation*. Thus, when data have been coded by a linear transformation, the confidence interval can first be calculated for the coded data. Then the confidence limits for the uncoded population trimmed mean can be obtained simply by applying the inverse of the transformation to the coded data limits. The household income data were coded in thousands of dollars. We calculated the limits for the coded data, then drew conclusions directly for the uncoded data.

It should be emphasized that the confidence intervals based on the sample mean and the sample trimmed mean are really for different measures of population center, namely the mean and the trimmed mean. When the population distribution is symmetric, these measures will be the same. However, for asymmetric distributions they can be quite different. It can be argued that for long-tailed distributions, a parameter such as the median or trimmed mean generally provides a more reasonable measure of center than does the mean. This observation will be our motivation for calculating confidence intervals for the trimmed mean.

One-Sample Trimmed *t* Tests

Hypothesis tests based on the trimmed mean can be constructed as in chapters 9 and 10, with the traditional *t* statistic replaced by the trimmed *t* statistic

$$t = \frac{\overline{X}_T - \mu_T}{s_T/\sqrt{n_T}}$$

Thus, for the above example, a two-sided test of the hypothesis that the average household income for the town is \$11,000 would reject H_0 if $|t| \geq 2.365$, where the critical value is the two-sided 5% value from table 3 for $n_T - 1 = 7$ degrees of freedom. The value of the trimmed *t* statistic is

$$t = \frac{6.625 - 11}{4.755/\sqrt{8}} = -2.602$$

At the 5% significance level, we would reject the hypothesis that the average income is \$11,000. Since the sample trimmed mean is smaller than 11, we would conclude that the average income for the town is less than $11,000. The confidence interval of the last subsection could now be used to say by how much.

EXERCISES

12.1 Tooth measurements for 20 monkeys, used in an anthropological study, are as follows:* 37, 38, 39, 38, 37, 50, 39, 37, 35, 37, 40, 39, 38, 38, 40, 36, 35, 36, 34, 37.

(a) Make a box plot of the data. Use this plot to determine a suitable trimming fraction for a trimmed mean procedure.

(b) Compute a 95% confidence interval for the mean of the tooth measurement population based on the methodology given in chapter 10.

(c) Compute a 95% confidence interval for the trimmed mean tooth measurement based on the trimming fraction found in part (a).

(d) Comment on the relative lengths of the intervals found in parts (b) and (c).

12.2 The weld bead height data of exercise 1.11 (p. 23) are repeated below.

3.45	−.37	−.85	.62
.94	.87	.24	4.11
.90	−.44	3.15	.98
−.43	1.35	1.23	.50
.52	−.34	2.11	.71

(a) Make a box plot of the weld data. Use this plot to determine a trimming fraction for a trimmed mean robust procedure.

(b) Use a trimmed t test with the trimming fraction found in part (a) to test at the 5% level whether or not the average weld bead height is equal to 0.

(c) Find a 95% confidence interval for the trimmed mean of the weld bead height distribution.

12.3 The Glabella measurement average of 4.75 millimeters (mm) has been used for many years in the reconstructions of facial features of Caucasians from skeletal remains. (The Glabella is the bony ridge just above the nose.) A recent study of nearly 300 individuals made during autopsy provided measurements from which the following 20 Glabella values were randomly sampled.

6.00	8.25	5.25	5.00
5.50	8.00	2.00	7.00
6.00	5.75	5.00	4.00
5.75	4.00	5.00	9.50
5.75	5.25	5.75	6.00

Data courtesy of E. S. Moore, Ph.D., Department of Anthropology, University of New Mexico.

---

* Data courtesy of Carl Shuster, student in the Department of Anthropology, University of New Mexico, Albuquerque.

(a) Make a box plot of the data. What features suggest the need for a robust procedure? What trimming fraction should be used in the analysis?

(b) Test at the 5% significance level whether the trimmed mean Glabella measurement of modern Caucasians differs from 4.75 mm.

(c) Find a 95% confidence interval for the trimmed mean and determine by how much this mean differs from the currently accepted value of 4.75 mm.

12.4 The mayor of a Southern city decided that if the average energy consumption for heating nonresidential buildings exceeds 20 units, he will initiate a campaign to familiarize the local business community with the advantages of solar heating. The following data represent the current consumption rates for a random sample of 43 buildings.

16.6	12.8	23.3	71.8	23.7
0.0	8.5	28.8	30.1	29.1
20.5	21.9	120.0	26.4	78.8
8.6	27.0	45.0	30.5	27.3
4.7	36.1	84.9	28.8	13.0
28.4	23.6	21.4	13.6	20.3
21.1	12.9	10.7	13.3	21.0
38.6	29.0	34.5	28.7	
0.0	13.4	8.8	23.8	

Data courtesy of D. Hadfield, R. & D. Associates, Albuquerque, New Mexico.

(a) Make a box plot of the data. What trimming fraction is required for these data?

(b) Test at the 10% significance level whether the population trimmed mean energy consumption exceeds 20 units. Should the mayor proceed with his campaign?

(c) Find the 90% lower confidence limit for the trimmed mean consumption.

(d) Use the summary data $n = 43$, $\overline{X} = 27.472$, and $s = 22.925$ to calculate a 90% lower confidence bound for the population mean consumption. How does the confidence interval determined by this limit compare with the interval for the trimmed mean in part (c)?

12.5 Refer to exercise 8.37 (p. 251) for a description of the problem leading to the following data. The data (repeated from this exercise) are coded passage times of light.

Newcomb's (coded) measurements of passage time of light[a]

28	27	24	31	36	37	36	27	26	39	29
26	16	21	19	28	25	26	27	33	28	27
33	40	25	24	25	28	30	28	26	24	28
24	-2	30	20	21	26	22	27	32	25	29
34	29	23	36	28	30	36	31	32	32	16
-44	22	29	32	29	32	23	27	24	25	23

Source: Stigler [31].
[a] These measurements were made in the period from July 24, 1882, to September 5, 1882. The given values times .001 plus 24.8 are Newcomb's measurements, recorded in millionths of a second.

(a) Make a box plot of these data. What features of the plot suggest the desirability of a robust procedure?

(b) The 10% trimmed sample size, mean, and standard deviation of the coded data are

$$n_T = 52, \qquad \overline{X}_T = 27.38 \qquad s_T = 4.686$$

Calculate a 95% confidence interval for the 10% coded trimmed mean passage time.

(c) The inverse of the coding transformation for the passage time data is $Y = 24.8 + .001X$. Use this transformation and the fact that trimmed means are invariant under linear transformations to find a 95% confidence interval for the uncoded trimmed mean passage time.

(d) Compare the length of the trimmed mean confidence interval of part (c) to the length of the 95% confidence interval for the mean,

$$24.8236 \le \mu \le 24.8288$$

What influence has the use of the trimmed mean had on the length of the interval? This exercise suggests that the use of robust procedures can be important even for relatively large sample sizes and relatively modest deviations from normality.

Two-Sample Tests and Confidence Intervals

Parts of two interesting data sets, the Levi Strauss mill run-up data of exercise 3.15 (p. 86), and the hydrocarbon automobile emission data of exercise 4.18 (p. 115), will be used to provide the following examples. We will use the data for mills A and C and the emission levels for the two groups of automobile manufacture years pre-1963 and 1968–1969.

To begin with, let's ignore exploratory analyses of the data and go straight to the hypothesis tests, first for the mill data. The two-sample method of chapter 10 requires the following summary. (The data are in units of run-up × 100.)

	MILL A	MILL C
n	22	19
$\overline{X}$	45.220	48.310
s	100.32	44.03

Now, the value of the two-sample t statistic is

$$t = -0.124$$

For 39 degrees of freedom, the P-value for the two-sided test of no difference in mean run-up exceeds .50. Thus, at any interesting significance level, we would accept the no-difference null hypothesis.

The summary statistics for the emission data are as follows.

	PRE-1963	1968–1969
n	10	16
$\bar{X}$	890.600	506.313
s	591.567	707.803

The value of the two-sample t statistic is

$$t = 1.430$$

Thus, for 24 degrees of freedom, bounds on the P-value of the two-sided test are

$$.10 < P < .20$$

Again, at significance levels at or below 10%, the decision would be to accept the null hypothesis of no difference in emission levels.

At this point, and studies too often end at this point, both analyses have been completely destroyed. If Levi Strauss accepts the run-up analysis and buys cloth from mill C instead of mill A, it will suffer significantly higher wastage than necessary. The unfortunate engineers who spent several months in the hot sun of a shopping center parking lot testing automobile emissions would be led to believe that the improvements in pollution controls initiated between 1963 and 1968 have had no detectable effect. As we will see, the failure to detect a difference in both cases is the result of using the standard t test when the assumptions on which it is based are violated. It is a failure of the analysis, not the data.

This problem can be avoided by carrying out an exploratory analysis before summarizing the data. Display 12.2 provides schematic diagrams of the two data sets under consideration. Note the outliers in both diagrams. The assumption of normality would not be reasonable in either case. On the basis of the box lengths (interquartile ranges), it would seem that the assumption of equal variation, especially for the emission data, is questionable as well.

The usual strategy for attempting to detect unequal population dispersions is to carry out a hypothesis test for the equality of standard deviations. Such tests would be carried out before performing the test for means. Unfortunately, as mentioned earlier, unequal dispersions are often linked with serious forms of nonnormality such as long-tailed distributions, and the standard tests for unequal dispersions are even more seriously affected by nonnormality than are the t tests. For example, the usual test for equality of standard deviations would be accepted at any reasonable level for the emission data ($P > .50$), but rejected for the mill data ($P < 0.002$). These conclusions are just the reverse of what we would come to from the schematic diagrams of display 12.2 and are almost entirely a function of the outliers!

The next step in a "standard" analysis for the emission data would be to use the two-sample t test, which would lead to the outcome shown above. The next step for the mill data would be to use the t-test with the modifica-

Display 12.2

Examples for which
usual two-sample *t* test
fails to detect
difference in means. (a)
Outliers present but
roughly comparable
IQRs (run-up data,
exercise 3.15). (b)
Outliers plus different
IQRs (automobile
emission data, exercise
4.18).

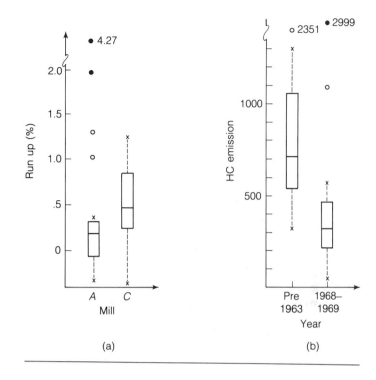

tion involving Satterthwaite's approximation, to be covered later in this section. In this case, the modified test would give almost exactly the same results as before.

Because of the potential for being misled—and because exploratory analyses based on schematic diagrams are, in any event, a far superior diagnostic tool—tests for equality of standard deviations will not be given in this text.

The two-sample *t* test based on trimmed means performs best when dispersions, as measured by box lengths in the schematic diagrams, are not too different. Since this is true of the mill data, the next example will illustrate the trimmed mean procedure for mills A and C.

EXAMPLE 12.1

We will test the hypothesis that the average run-ups for mills A and C are equal, where the location measure to be used is the population 20% trimmed mean. The choice of the trimming fraction is based on the observation that 20% is the smallest (multiple of 5) trimming fraction that just eliminates all of the outliers. Stem and leaf plots of the data are as follows.

MILL A		MILL C
2 0	-3	9
0	-2	1
7	-1	
8 7	-0	8
3	0	9
3 4 5 2	1	5
4 8 4 7	2	
0 5 2	3	9 6 4
	4	7 6
	5	7 1
	6	5
	7	4
	8	
	9	7 6 8
1	10	1
427,130,194	HI	121

For mill A, $n = 22$; thus $.20 \times 22 = 4.4$. This number rounds up to 5 observations to be trimmed from each side of the sample. The values to be trimmed are indicated by underlines in the stem and leaf plot. For mill C, $n = 19$, and the number trimmed from each end is 4. The summary data are the following (in units of percentage of run-up $\times$ 100):

	MILL	
	A	C
n_T	12	11
$\bar{X}_T$	17.917	50.909
s_T	22.343	42.427

The expression for the pooled (trimmed) standard deviation is exactly as before, but uses the trimmed summary data:

$$s_P = \sqrt{\frac{11 \cdot 22.343^2 + 10 \cdot 42.427^2}{21}}$$

$$= 33.446$$

The value of the two-sample trimmed t statistic is then

$$t = \frac{17.917 - 50.909}{33.446\sqrt{1/11 + 1/12}}$$

$$= -2.363$$

For 21 degrees of freedom, the two-sided P-value, $P = P(|t| > 2.363)$, is found from table 3 to satisfy the inequalities

$$.02 < P < .05$$

It follows that the hypothesis of no difference in average run-up would be rejected at the 5% significance level, and we would conclude that, in fact, the average run-up for mill C is greater than that for mill A.

How much greater? The expression for the confidence interval for the difference in means given in section 10.4 can be applied to the trimmed summary data to obtain a 95% confidence interval for the difference in 20% trimmed means. Moreover, since trimmed means preserve linear transformations, the confidence limits can be found using the units of run-up $\times$ 100 given in the above summary, then converting these limits to the original data units simply by dividing them by 100.

The order of the sample means is reversed so that the confidence interval will measure the amount by which the population trimmed mean for mill C exceeds that for mill A:

$$50.909 - 17.917 = 32.992$$

The critical value for 21 degrees of freedom from table 3 is 2.080. This number times the denominator of the t statistic is the value that is added and subtracted from the difference in sample means to obtain the confidence limits:

$$2.080 \cdot 33.446 \cdot \sqrt{\frac{1}{11} + \frac{1}{12}} = 29.039$$

Thus,

$$L = 32.992 - 29.039 = 3.953$$

and

$$U = 32.992 + 29.039 = 62.031$$

It follows with 95% confidence that the average run-up (as measured by the 20% trimmed mean) for mill C exceeds that for mill A by an amount between .0395% and .620%.

Satterthwaite's Correction for Unequal Dispersions

When a difference in dispersion exists for the two populations, it is important to modify the standard t test. The reason is that not only can the sensitivity of the test be affected, but in some instances its validity can be compromised. This effect is magnified by differences in sample sizes. For schematic diagrams showing reasonable normality but differences in dispersion, the method of this subsection can be applied to correct the standard two-sample t test and confidence intervals given in chapter 10. When outliers are present as well, the correction can be applied to the trimmed t test and confidence intervals.

The correction procedure consists of (1) replacing the expression for the standard error of the difference in sample means (the denominator of the t statistic) by the unpooled version

$$SE = \sqrt{\frac{s_1^2}{n_1} + \frac{s_2^2}{n_2}}$$

and (2) replacing the degrees of freedom by a different expression. This expression and the SE both can be conveniently calculated as follows. Define g_1 and g_2 by the expressions

$$g_1 = \frac{s_1^2}{n_1} \quad \text{and} \quad g_2 = \frac{s_2^2}{n_2}$$

The standard error is then

$$SE = \sqrt{g_1 + g_2}$$

and the expression for the degrees of freedom is

$$\nu = \frac{(g_1 + g_2)^2}{g_1^2/(n_1 - 1) + g_2^2/(n_2 - 1)}$$

The degrees of freedom are rounded to the nearest integer for use in table 3. The corrected t statistic

$$t = \frac{\overline{X}_1 - \overline{X}_2}{SE}$$

would now be used in hypothesis tests. Confidence intervals would be calculated using the corrected limits

$$\overline{X}_1 - \overline{X}_2 \pm t \cdot SE$$

where t is the critical value from table 3 corresponding to the given confidence coefficient and the corrected degrees of freedom.

EXAMPLE 12.2

The stem and leaf plot indicating the details of a 10% trimmed mean calculation* applied to the emission data for pre-1963 and 1968–1969 cars is as follows.

```
                           | 0* | 111, 071, 199, 188, 117
                    347    | t  | 388, 294, 211, 353, 241, 353
        570, 411, 541      | f  | 558, 460, 470
                    630    | s  |
              905, 800     | .  |
                    058    | 1* | 088
                    293    | t  |
                           | f  |
                           | s  |
                           | .  |
                           | 2* |
                    351    | t  |
                           | f  |
                           | s  |
                           | .  | 999
```

* Underlined values are those trimmed from the sample. They are replaced by the boxed values to form the Winsorized sample.

Verify that the trimmed mean summary statistics are

	PRE-1963	1968–1969
n_T	8	12
$\bar{X}_T$	776.0	319.33
s_T	380.683	189.036

Because of the difference in trimmed standard deviations, Satterthwaite's approximation will be applied to these data. We find

$$g_1 = \frac{380.683^2}{8} = 18114.943$$

and

$$g_2 = \frac{189.036^2}{12} = 2977.884$$

Thus the corrected standard error is

$$SE = \sqrt{18114.943 + 2977.884} = 145.234$$

and the corrected degrees of freedom are

$$\nu = (18114.943 + 2977.884)^2/(18114.943^2/7 + 2977.884^2/11)$$
$$= 9.330$$

The degrees of freedom would be rounded to 9. The corrected trimmed t statistic is then the difference in means divided by standard error:

$$t = \frac{776.00 - 319.33}{145.234}$$
$$= 3.144$$

For 9 degrees of freedom, the bounds on the P-value for the two-sided test of the hypothesis of no difference in means are

$$.01 < P < .02$$

Thus, we would reject the hypothesis at the 5% significance level, but not at the 1% level.

Had the uncorrected procedure been used, we would have obtained a computed t value of 3.578. For 18 degrees of freedom, this would have led to the P-value bounds $.002 < P < .005$. The uncorrected procedure seriously overstates the significance of the test in this case.

A 95% confidence interval for the difference in (10% trimmed) mean emission levels for 1963 versus 1968–1969 cars has limits

$$776.00 - 319.33 \pm 2.262 \cdot 145.234$$

where SE = 145.234 is the corrected standard error and $t = 2.262$ is the two-sided 5% critical value from table 3 for 9 degrees of freedom. These bounds are 128.151 and 785.189. Thus, with 95% confidence the average hydrocarbon emission level for 1963 automobiles exceeds that for 1968–1969 automobiles by an amount between 128.15 and 785.19 parts per million.

In general, when the larger sample size occurs with the smaller standard deviation, the uncorrected test will err against validity as it does in the last example. When the larger sample size weights the larger standard deviation, the uncorrected test will lose sensitivity. Both difficulties can be avoided by using either this correction procedure or a variation-equalizing transformation of the type to be given in section 12.3.

EXERCISES

12.6 The 10% trimmed summary data for the correctness scores reported in exercise 10.9 (p. 319) are as follows.

	SPECIAL STUDENTS	CONTROL STUDENTS
n_T	42	42
$\bar{X}_T$	10.763	14.107
s_T	4.886	5.585

(a) Find the P-value for the test of the hypothesis that there is no difference in (trimmed) mean correctness scores for the two categories of students.

(b) Find a 95% confidence interval for the difference in trimmed mean correctness scores.

12.7 The means and standard deviations of the ages at death by suicide of a sample of 90 Anglos and 15 Native Americans from the records of the Albuquerque Crisis Unit for 1978 are as follows.

	ANGLO	NATIVE AMERICAN
$\bar{X}$	42.21	25.07
s	18.30	8.51

Data courtesy of D. Bagley and J. Abbin, Albuquerque Suicide Crisis Unit.

(a) Find a 90% confidence interval for the difference in mean ages for Anglo and Native American suicide victims, using the method of chapter 10.

(b) Use Satterthwaite's approximation to find a 90% confidence interval for the difference in means of part (a) and compare the result with the confidence interval found there. What difference has the use of the approximation made?

12.8 The serum iron levels* of a random sample of 37 women on oral contraceptives was $\bar{X} = 113.34$ mg%, with a standard deviation of $s = 41.78$ mg%, and for 138 women

* Data courtesy of P. Garry, School of Medicine, University of New Mexico, Albuquerqué.

not on oral contraceptives, the mean was 96.05 mg%, with a standard deviation of 31.20 mg%.

(a) Find bounds on the P-value for the two-sided test of equal mean iron levels for women on and not on oral contraceptives using the assumption of equal variances (as in chapter 10).

(b) Repeat the computation requested in part (a) using Satterthwaite's approximation. What effect does the inequality of standard deviations appear to have had on the test of part (a)?

(c) Since the differences in standard deviations do not appear to be excessively large in this problem, why are the differences in the outcomes in parts (a) and (b) so large?

12.9 The mean speeds for the first 20 bikers electing the 100-mile option for the Tour of the Rio Grande Valley (display 1.17, p. 25) was $\overline{X} = 13.210$ miles per hour, with a standard deviation of $s = 3.450$ miles per hour. The mean and standard deviation of the speeds for the 9 individuals electing the 50-mile option were $\overline{X} = 9.233$ and $s = .805$ miles per hour.

Find a 95% confidence interval for the difference in speeds for 50 and 100 mile bike tourers

(a) using the equal variance theory of chapter 10

(b) using the correction of this subsection.

(c) Why does the correction improve sensitivity (interval length) in this problem?

12.10 The testosterone levels for samples of women in the two occupational categories (i) professional career women and (ii) secretaries and office workers (see exercise 3.30, p. 93) are repeated here.

PROFESSIONAL CAREERS				SECRETARIES AND OFFICE WORKERS	
1.2	2.9	3.4	5.2	1.1	2.3
1.5	3.0	3.5	5.3	1.8	2.3
2.1	3.1	3.6	5.7	2.0	2.4
2.2	3.2	3.6	6.1	2.1	2.5
2.2	3.3	4.8	6.2	2.1	5.9
2.7	3.4	4.9	7.8	2.2	

(a) Use a suitable trimmed mean to account for any outliers, then use Satterthwaite's correction to adjust the 5% test of whether professional career women have larger levels of testosterone in their blood, on the average, than do secretaries and office workers.

(b) Find a 95% lower confidence bound on the amount by which the average testosterone level for professional career women exceeds that for secretaries and office workers.

12.11 The androstenedione levels for male and female diabetics, given in exercise 10.8 (p. 319), are repeated here.

MALES	FEMALES
217	84
123	87
80	77
140	84
115	73
135	66
59	70
126	35
70	77
63	73
147	56
122	112
108	56
70	84
	80
	101
	66
	84

(a) The schematic diagram found in part (a) of exercise 10.8 shows a difference in dispersions for the two data sets, as well as two minor outliers in the data for females. Use this information to arrive at an appropriate trimming fraction for the data.

(b) Apply the trimming fraction of part (a) to construct a trimmed t test of the hypothesis of no difference between average levels of androstenedione for male and female diabetics.

(c) Apply Satterthwaite's approximation to the test of part (b). What difference does this approximation make in the tests?

The One-Way ANOVA and FSD Procedure Applied to Trimmed Means

In the following example, the steps of the FSD procedure given in chapter 11 are applied, step for step, to trimmed mean summary data. This procedure is an immediate extension of the two-sample method for which the extensive Monte Carlo study of Yuen and Dixon [37] showed excellent performance both in preservation of validity and improvement in sensitivity for distributions with long tails.

EXAMPLE 12.3   In example 11.3 (p. 354) it was observed that a single outlier in week 3 of insulin production for islets of Langerhans from lean rats caused an obvious downward trend in the second two weeks (display 11.1) to be missed by the analysis. In this example, a robust backup based on 10% trimmed means will be carried out on the same data. Fill in the computational steps by applying the methods of chapter 11 to the following trimmed mean summary.

	WEEK		
	1	2	3
n_T	6	6	6
$\overline{X}_T$	4.832	3.407	1.777
s_T	1.230	1.032	.946

The ANOVA table based on these data is as follows:

SOURCE OF VARIATION	SS	df	MS	F
Between weeks	28.041	2	14.021	12.112
Within weeks	17.364	15	1.158	

For 2 and 15 degrees of freedom (using 16 since 15 is not in the table), we find from table 4 that

$$P < .001$$

From table 3 the critical Student's t value for 15 degrees of freedom is

$$c = 2.131$$

Consequently,

$$\text{LSD} = 2.131 \times \sqrt{1.158} \times \sqrt{\frac{1}{6} + \frac{1}{6}} = 1.324$$

The ordered trimmed means are as follows:

	WEEK 3	WEEK 2	WEEK 1
$\overline{X}_T$	1.777	3.407	4.832
Differences:	├──── 1.63 ────┤		
		├──── 1.425 ────┤	

All differences exceed the LSD. Consequently, the grouping is now

$$\{\text{week 3}\} < \{\text{week 2}\} < \{\text{week 1}\}$$

Note that the use of the trimmed mean has broken the grouping of weeks 2 and 3 formed by the original analysis.

EXERCISES

12.12   Apply the FSD procedure with 20% trimmed means to the *original* exam scores of example 3.2 (p. 73)—the data with outliers included. The data are repeated below. How does the result of this analysis compare to the use of the standard FSD procedure applied to the outlier-deleted data in example 11.4?

YEAR 1		YEAR 2		YEAR 3	
88	43	54	45	59	47
49	90	48	59	51	64
31	34	36	39	0	32
86	54	53	50	58	55
41	28	45	33	45	0
26	48	31	47	0	50
52	40	49	43	53	42
39	89	42	57	41	62
46	22	46	27	50	0
40	32	44	37	44	0
37	35	41	40	38	36
58	45	51	46	56	49
97		63		68	

12.13   The complete Levi Strauss mill data set of exercise 3.15 (p. 86) is repeated here.

MILL A		MILL B		MILL C		MILL D		MILL E	
.12	.03	1.64	.63	1.21	.98	1.15	.59	2.40	2.23
1.01	.35	−.60	.90	.97	.65	1.02	1.30	−.37	.31
−.20	−.08	−1.16	.71	.74	.57	.38	.68	.82	1.68
.15	1.94	−.13	.43	−.21	.51	.83	1.45	.92	1.13
−.30	.28	.40	1.97	1.01	.34	.66	.52	−.93	1.23
−.07	1.30	1.70	.30	.47	−.08	1.02	.73	.80	1.69
.32	4.27	.38	.76	.46	−.39	.88	.71	1.58	
.27	.14	.43	7.02	.39	.09	.27	.34		
−.32	.30	1.04	.85	.36	.15	.51	.07		
−.17	.24	.42	.60	.96		1.12			
.24	.13	.85	.29						

Data courtesy of Levi Strauss Corp., Albuquerque, New Mexico.

(a)   Make a schematic diagram of the data. Note: you will have this diagram if you did exercise 3.15.

(b)   Use the diagram of part (a) to find the (multiple of 5%) trimming fraction that just removes the outliers for all mills.

(c)   Apply the classical FSD procedure (from chapter 10) to the data.

(d)   Apply a trimmed mean FSD procedure to the data using the trimming fraction found in part (b).

(e)   Describe the difference in the outcomes of the procedures of parts (c) and (d). How do the differences found in part (d) compare with the distributional differences seen in the schematic diagram?

A Paired-Difference Application

The following example will illustrate the application of trimmed means to paired differences.

EXAMPLE 12.4

June is the heavy air-conditioning month in New Mexico. However, because the dry climate makes evaporative air conditioning possible, the electrical consumption is not expected to be much different than it would be in an average-consumption month such as September. To check this expectation, the Public Service Company of New Mexico measured the average rate of electrical consumption (in kilowatts) for several households in the Albuquerque metropolitan area in September 1973 and in the following June. The data for 11 such households, along with the September-minus-June differences, are given in the following table:

| | MONTH | | DIFFERENCES |
HOUSEHOLD	SEPTEMBER	JUNE	(SEPTEMBER MINUS JUNE)
1	.60	1.00	−.40
2	2.56	1.12	1.44
3	.60	1.12	−.52
4	.28	2.56	−2.28
5	1.32	1.12	.20
6	.76	.44	.32
7	.28	.24	.04
8	.88	.92	−.04
9	1.92	1.76	.16
10	.00	.04	−.04
11	.20	.00	.20

Data courtesy of B. Waldman, Public Service Company of New Mexico, Albuquerque.

The box plot of these differences is given in display 12.3. The two outliers indicate the desirability of a robust analysis. The 95% confidence interval for the 10% population trimmed mean μ_{Td} of the differences in September and June measurements is

$$-.291 \leq \mu_{Td} \leq .325$$

This result was obtained by using the summary data

$$n_T = 7 \qquad \bar{d}_T = .0172 \qquad s_T = .3325$$

in place of n, $\bar{d}$, and s_d in the confidence interval formulation of section 10.5. This interval is plotted in display 12.3(c). The usual 95% confidence interval for the mean of the differences, μ_d based on the summary data

$$n = 11 \qquad \bar{d} = -.0836 \qquad s_d = .8853$$

is shown for comparison in display 12.3(b).

Both of these confidence intervals contain zero. Consequently, a value of zero for both the mean and the trimmed mean for the population of differences is consistent with the data. (See section 9.7 for a more formal, hypothesis-testing interpretation.) Thus in both cases we would conclude that the data do not support a difference in average electrical consumption for the two months.

An evaluation of the information in the sample concerning the "no

Display 12.3

Diagrams for electrical consumption rates of example 12.4. (a) Box plot. (b) 95% Student t confidence interval for μ_d. (c) 95% confidence interval for μ_{Td} based on trimmed-mean difference $\bar{d}_T$.

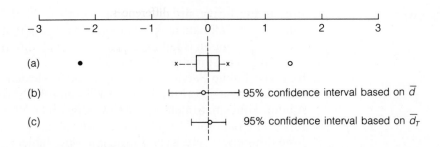

difference" hypotheses can be based on P-values for the tests. The t statistic for the usual test is

$$t = \frac{\bar{d}}{s_d/\sqrt{n}} = \frac{-.0836}{.8853/\sqrt{11}} = -.313$$

Replacing n, $\bar{d}$, and s_d, by h, $\bar{d}_T$, and s_T, the t statistic based on the 10% trimmed mean is

$$t = \frac{.0172}{.3325/\sqrt{7}} = .137$$

The degrees of freedom for the two tests are $\nu = 11 - 1 = 10$ and $7 - 1 = 6$, respectively. From table 3 we find that the P-values for both two-sided tests exceed .50. Thus the hypothesis of no difference in location measures is not in jeopardy from these data.

Because the test based on the trimmed means is more sensitive than the one that uses the mean, it lends more support to the possibility that the average of the differences is zero, since a smaller nonzero average difference would have been detected had it existed.

EXERCISES

12.14   The data from each row of the obese rat insulin data of exercise 11.4 (p. 356) are actually obtained from the same cell cultures. Thus, for example, the data for weeks 0 and 1, repeated below, are naturally paired.

WEEK 0	WEEK 1
31.2	18.4
72.0	37.2
31.2	24.0
28.2	20.0
26.4	20.6
40.2	32.2
27.2	23.0
33.4	22.2
17.6	7.8

(a) Box plot the paired differences.

(b) Test at the 5% significance level the hypothesis of zero difference in mean insulin levels for the two weeks, using the paired difference method of section 10.5.

(c) Test the hypothesis of part (b) replacing the mean difference with the 5% trimmed mean difference in insulin levels. Compare the result of this test with that of part (b).

12.15 The percentages of students who dropped the course are given below for the fall and spring semesters of 19 math courses taught at a particular college during a given year. The problem is to determine whether there is a significant difference in drop rates for the two semesters.

COURSE #	PERCENT DROPPING THE COURSE	
	FALL	SPRING
1	5	13
2	22	10
3	10	13
4	12	16
5	8	3
6	17	14
7	2	17
8	25	10
9	10	10
10	10	13
11	7	59
12	7	11
13	40	13
14	7	5
15	9	12
16	17	14
17	12	3
18	12	14
19	1	15

(a) Make a box plot of the differences and determine a suitable trimming fraction for a trimmed t paired-difference procedure.

(b) Use a classical paired-difference t test (chapter 10) to test whether there is a difference in drop rates for the two semesters.

(c) Repeat part (b) using a trimmed mean paired-difference test with the trimming fraction found in part (a).

(d) Discuss any differences found using the tests of parts (b) and (c).

The Use of Power Transformations to Equalize Variation

power transformation

When k-sample data are positive and dispersion increases as the group medians increase, then a **power transformation** is often an effective device for equalizing these dispersions. The family of power transformations is indexed by a real parameter q as follows.

If $q > 0$, the transformation converts y into y^q. Thus, for example, if $q = .3$, the data value 23.45 would become $23.45^{.3} = 2.577$ under this transformation. For $q = 0$, the power transformation is $\log(y)$, while if $q < 0$ the transformation is $-y^q$. Any base for the log is permissible, but base e is standard and will be used in this text.

Certain members of the power transformation family have been used frequently in practice to equalize dispersions and for other purposes. They are the square root transformation, $y = y^{1/2}$ ($q = .5$), the log transformation ($q = 0$) and the reciprocal transformation $-1/y = -y^{-1}$ ($q = -1$). These transformations are often arrived at by theoretical considerations. Our method for selecting transformations, however, will be purely empirical.

Two facts make the power transformations extremely useful. First, they are easy to implement both on computers and hand calculators. Most scientific calculators have both the log function and a y^x button that give the transformed values directly. Moreover, these are functions commonly implemented in computer languages. Second, a commonly observed feature of the frequency distributions of positive data is that variation tends to increase as the center of the distribution increases. This is an observation that is sometimes explained by the natural stacking up of data near the lower boundary at 0. A decrease in the stacking causes both the center and the variation to increase. This phenomenon can be seen in the schematic diagram of the hydrocarbon emission data given in display 12.4. Note how longer boxes are associated with larger medians.

Note also that rather unpleasant outliers occur for almost every car year in the emission data. An added attraction of the power transformations is that they will often tame outliers. The effect of the transformations for $q < 1$ (the range that applies when variation increases with medians) is to diminish large values differentially. The larger the value, the more it is decreased relative to others. Thus, outliers on the high side of the distribution will often be "brought into the fold" by a power transformation.

The ordering of the data by magnitude is preserved in the transformed data because the power transforms are what mathematicians call *monotone increasing transformations*. Thus, for example, the transform of the largest value in a data set will be the largest of the transformed values, and so on.

A somewhat unfortunate characteristic of power transformations is that they stretch the lower end of a distribution, sometimes creating outliers where none previously existed. If this happens, or outliers persist on the high side of the distribution, the transformation can be followed by a trimmed mean analysis. Since the variability will have been equalized, the

Display 12.4

Schematic diagrams of automobile emission data of exercise 4.18

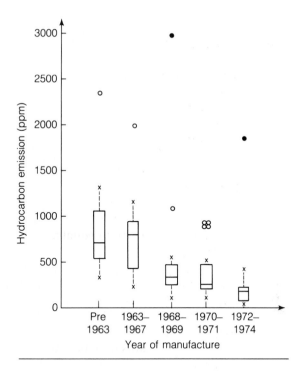

data will be in the best configuration for trimmed means. However, as we will show by transforming the emission data, considerable sensitivity will often be gained from the transformation alone.

Fitting Power Transformations to Data

The fitting method we will use is attributed to J. W. Tukey and is detailed in [32]. Another popular fitting method developed by G. E. P. Box and D. R. Cox is given in [3]. Tukey's method has the advantages of being easy to apply once an ordered data summary is available; moreover, the method is not influenced by outliers in the data.

Since our inference strategy will always begin with an exploratory analysis of the data, the raw materials of the fitting method, the medians and interquartile ranges for each group, will already be available. We will form a scatter diagram (chapter 4, p. 108) of logarithms of IQRs (the y's) versus the logarithms of the corresponding medians (the x's). A straight line will then be fitted to the data and the slope of the line, b, will be found. The index of the power transformation to use is then

$$q = 1 - b$$

Now, every number in the data set would be transformed by the chosen power transformation and the standard k-sample analysis would be carried out on the transformed data.

The least-squares fitting method, to be covered in chapter 14, which is implemented in most statistical computer programs and in many hand calculators, is perhaps the most convenient way to find the slope of the line. However, until that method has been covered, we will use a simple eye-fit method. The lack of precision in computing a slope by this method will not be of real concern, since transformations with indices in a small range of values generally work equally well.

The method consists of drawing a good-fitting line through the points. We will then locate two points (x_1, y_1) and (x_2, y_2) on the line near the two extremes of the scatter diagram. The slope of the line is then calculated by means of the expression

$$b = \frac{y_2 - y_1}{x_2 - x_1}$$

EXAMPLE 12.5   The medians and interquartile ranges of the automobile hydrocarbon emission data of exercise 4.18 (p. 115), and their logarithms are given in the following table.

YEARS OF MANUFACTURE	MEDIAN	IQR	LOG MEDIAN (x)	LOG IQR (y)
Pre-1963	715	517	6.57	6.25
1963–1967	780	517	6.66	6.25
1968–1969	323.5	271.5	5.78	5.60
1970–1971	244	269.5	5.50	5.60
1972–1974	160	156.5	5.08	5.05

The scatter diagram of log IQR versus log median and an eye-fitted line are given in display 12.5. Two points on the line are (5.00, 5.08) and

Display 12.5

Log IQR versus log median scatter diagram and linear fit for the hydrocarbon emission data of exercise 4.18

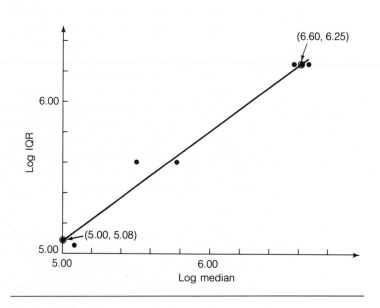

(6.60, 6.25), as shown on the plot. The slope of the line is then

$$b = \frac{6.25 - 5.08}{6.60 - 5.00} = .73$$

It follows that an appropriate index for the power transformation is

$$q = 1 - .73 = .27$$

In fact, indices close to this value will also be appropriate. The index $q = .25$ (one-fourth root) is a more natural choice; this is the value we will use.

To evaluate the effect of the transformation on the FSD procedure, the analysis will first be given for the raw emission data. The data summary, F table and FSD groupings for a 5% test are as follows.

Summary statistics

VARIABLE	SAMPLE SIZE	MEAN	STANDARD DEVIATION
Pre-1963	10	890.6000	591.5673
1963–1967	13	801.4615	454.9285
1968–1969	16	506.3125	707.8026
1970–1971	20	381.4500	287.8864
1972–1974	19	244.1053	410.7866

ANOVA table

SOURCE	SS	df	MS	F
Between	4226834.1537	4	1056708.53840	4.343
Within	17759967.8080	73	243287.23024	
Total	21986801.9610	77	P-value < 0.01	

Fisher's Least Significant Difference Test

Significance Level = 5.0%

LSD Groupings: {1968–1969, 1970–1971, 1972–1974} < {pre-1963, 1963–1967, 1968–1969}

The schematic diagram of display 12.4 suggests at least two distinct groupings—the first consisting of the pre–emission-control cars, built before 1968, and the second consisting of cars built after 1968. Another sharp decrease in emissions appears to have occurred between 1971 and 1972. Thus three groups appear possible. However, the analysis fails to separate any of these groups entirely.

The schematic diagram for the transformed data is given in display 12.6. Observe that the transformation has been rather successful in equalizing dispersions, as measured by box lengths. The three groupings suggested in display 12.4 are clearly visible here. Many of the original outliers are outliers no longer. Two of them were too tough for the transformation to handle completely, but, as we will see in the following analysis, they have been sufficiently tamed to allow the groupings in the data to show through.

Display 12.6

Summary and schematic diagram for the hydrocarbon emission data of exercise 4.18 (p. 115), rescaled by the power transformation with index $q = .25$

VAR. NAME	SAMPLE SIZE	MEDIAN	QUARTILES	ADJACENTS	IQR
Pre-1963	10	5.16413	4.82280	4.31601	0.88044
			5.70324	6.96327	
1963-1967	13	5.28474	4.53508	4.05360	1.00201
			5.53709	6.68740	
1968-1969	16	4.23769	3.72939	2.90278	0.91425
			4.64364	5.74324	
Major outliers:	7.4002				
1970-1971	20	3.95222	3.73665	3.16228	0.90404
			4.64068	5.53709	
1972-1974	19	3.55656	2.83783	2.11474	1.01998
			3.85782	4.47214	
Minor outliers:	6.5848				

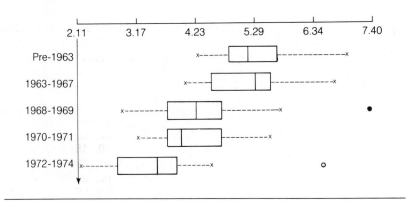

Summary statistics

VARIABLE	SAMPLE SIZE	MEAN	STANDARD DEVIATION
Pre-1963	10	5.3004	0.7843
1963–1967	13	5.1848	0.7160
1968–1969	16	4.3242	1.0767
1970–1971	20	4.2283	0.7449
1972–1974	19	3.4849	1.0374

ANOVA table

SOURCE	SS	df	MS	F
Between	32.6026	4	8.15065	10.086
Within	58.9897	73	0.80808	
Total	91.5923	77	P-value < 0.001	

Fisher's Least Significant Difference Test

Significance Level = 5.0%

LSD Groupings: {1972–1974} < {1968–1969, 1970–1971} < {pre-1963, 1963–1967}

As mentioned earlier, the leftover outliers in the last example could have been accounted for by doing a trimmed mean analysis instead of a standard one. Another alternative is the rank analysis of the next section.

EXERCISES

12.16   The meteorite cooling rate data of exercise 3.26 (p. 90) are repeated below.

(a)   Establish that $q = .25$ is a reasonable index for equalizing dispersions for the different meteorites.

(b)   Transform the data using the power transformation with $q = .25$ and determine the FSD groupings at the 5% significance level. How many distinct groups do there appear to be?

(c)   Many statisticians would use the logarithmic transformation to equalize variances for the meteorite data. Apply this transformation and repeat the FSD procedure on the results. How well does the logarithm separate groups in comparison to the fitted power transformation of part (b)?

METEORITE	COOLING RATES
Walker County	.69, .23, .10, .03, .56, .10, .01, .02, .04, .22
Lombard	.10, .15, .45, .50, .38, .45, .51, .61, 1.00, .84
Uwet	.21, .25, .16, .23, .47, 1.2, .29, 1.1, .16
Quillagua	.60, .23, .15, .33, .32, .58, .52, .42, 2.2, 1.4
Coahuila	5.6, 4.9, 3.3, 1.8, 4.8, 3.6, 3.2, 1.4, 1.6, 3.2, 8.2
Tocopilla	5.6, 2.7, 6.2, 2.9, 1.5, 4.0, 4.3, 3.0, 3.6, 2.4, 6.7, 3.8
Hex River	7.4, 5.6, 3.0, .64, 8.0, 7.4, 10.0

12.17   Apply a power transformation to the data of exercise 12.11 (p. 385) and follow the transformation with a classical two-sample t test. How do the results of this test compare with the outcome of the trimmed t test in exercise 12.11?

◢ SECTION 12.4

Methods Based on the Rank Transformation

rank transform

A variable transformation that is often able to account for several forms of nonnormality simultaneously is the **rank transform**. Because of this property, methods based on the rank transform tend to be used as general alternatives to the classical procedures. While this is a safe strategy from the viewpoint of validity robustness, it is not always the best in terms of sensitivity. Nevertheless, the rank transform methods are perhaps the most widely available and most useful robust methods.

Wilcoxon two-sample rank sum test

Kruskal-Wallis test

Our approach to rank transform methods will be that of Iman and others ([17], [18]) who studied and popularized the strategy of applying classical testing methods to ranked data. It is shown in [17] that when the two-sample *t* test is carried out for the ranked data, a test essentially equivalent to the popular nonparametric **Wilcoxon two-sample rank sum test** results. The ANOVA performed on ranks will provide a close alternative to the **Kruskal-Wallis test** (see Noether [25]). Technical details are given in [18]. Finally, the paired-difference *t* test will be applied to ranked data. Although this test has no close nonparametric relative, its performance is at least on a par with the *Wilcoxon signed rank test*, which is given for completeness in an optional subsection.

Ranks and a Method for Ranking Data

To rank a set of data simply means to arrange the data in increasing numerical order and to assign the *ranks* 1, 2, 3, . . . to them, from smallest to largest. For example, if the original numbers are 45, 89, 44, 74, 103, the number 44 would be assigned rank 1; 45 would be assigned rank 2; 74, rank 3; 89, rank 4; and 103, rank 5. Because of the ordering feature of the stem and leaf plot, the process of assigning ranks is easily implemented on a stem and leaf plot. Leave some space above each stem and write the rank of each number just above its leaf (preferably in a different color). The ranking of the five numbers given above is shown in the following stem and leaf plot.

```
        |  2 1
     4  |  5 4

     5  |

     6  |
        |  3
     7  |  4
        |  4
     8  |  9

     9  |
        |  5
    10  |  3
```

tied observations

When a data set contains the same number more than once, these **tied observations** are assigned the average of the ranks they would have received if they were different. For example, the two 44s in the data set 44, 89, 44, 89, 89 are each assigned the rank $(1 + 2)/2 = 1.5$ (abbreviated 1′ in the following stem and leaf plot), while the three 89s each receive rank $(3 + 4 + 5)/3 = 4$. (Note that if an odd number of digits are tied, the assigned rank is the middle value of the possible ranks; if an even number are tied, it is the average of the middle two.)

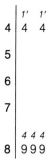

A Two-Sample Application

To carry out a two-sample t test on ranked data, the raw observations are first combined into a single group and ranks are assigned to the combined data. Then, the ranks are separated out into the same two groups the raw data occupied. For example, if the observation that received rank 1 came from group 2, then 1 would be assigned to rank group 2, and so on. Thus, two new groups would be formed that consist of the ranks of the raw data that formed the original groups. A convenient method for computing ranks by hand without losing track of the groupings is given in the next example.

EXAMPLE 12.6

A two-sample t test will be performed on the ranked cooling rates for the Walker County and Uwet meteorites of exercise 3.26 (p. 90). The following is a stem and leaf plot of the cooling rates ($\times 100$).

WALKER CO.		UWET
4213	0	
00	1	66
23	2	1539
	3	
	4	7
6	5	
9	6	
	7	
	8	
	9	
	10	
	11	0
	12	0

The ranking could be done directly on this stem and leaf plot. However, to illustrate a method useful in other multi-sample problems, we will put the observations into side-by-side stem and leaf plot forms, as shown below. Each sample is read into a column of the form, and the ranks are then assigned row by row as indicated. When this process is completed, the ranks already appear in the proper groups (columns) and the rank summary data can be computed directly from the stem and leaf plots.

STEM	WALKER CO.				UWET			
0	3³	1¹	2²	4⁴				
1	0⁵ʹ	0⁵ʹ			6⁷ʹ	6⁷ʹ		
2	2¹⁰	3¹¹ʹ			1⁹	5¹³	3¹¹ʹ	9¹⁴
3								
4					7¹⁵			
5	6¹⁶							
6	9¹⁷							
7								
8								
9								
10								
11					0¹⁸			
12					0¹⁹			

Check that the summary data are as follows. (Recall, $11' = 11.5$, for example.)

	WALKER CO.	UWET
n	10	9
$\overline{X}$	20.00	45.22
s	23.898	40.699

The value of the two-sample t statistic based on the rank summary is

$$t = -2.207$$

The P-value for the two-sided test of the hypothesis of no difference in cooling rates is obtained from table 3, as before. The degrees of freedom are those of the two-sample test, namely $n_1 + n_2 - 2 = 17$. The P-value bounds are

$$.02 < P < .05$$

Consequently, the hypothesis of no difference in cooling rates would be rejected at the 5% level, and we would conclude that the cooling rate for the Uwet meteorite is greater than that for the Walker County meteorite, on the average.

Display 12.7

Box plots of cooling rates for Walker County and Uwet meteorites. (a) Original data. (b) Ranks.

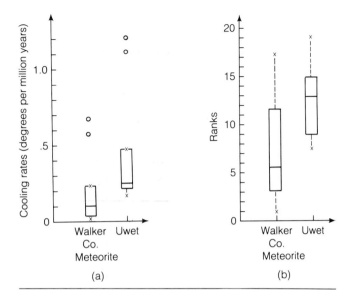

(a)   (b)

The motivation for using a robust test in this example would come from an exploratory analysis. The schematic diagram for the original data is given in display 12.7(a). After rank transformation, the schematic diagram has the more reasonable (normal) appearance seen in 12.7(b).

How different are the results of the t tests for the raw and ranked data? Had this test been carried out on the *original* data ($\times 100$), the following summary values would have been found:

	WALKER CO.	UWET
n	10	9
$\overline{X}$	20.00	45.22
s	23.898	40.699

The value of the two-sample t statistic would have been

$t = -1.669$

The bounds for the P-value of the two-sided test would then have been

$.10 < P < .20$

That is, had we applied the two-sample t-test to the original data, the hypothesis H_0 of no difference in cooling rates would have been *accepted* at the $\alpha = .05$ level. We would have decided that there was insufficient information in the sample to show a difference in distributions.

In fact, there *is* sufficient information to show a difference, but it must be measured with a little care. That is, we must use a procedure such as the one involving the rank transform to validate the assumptions underlying the t test.

The specific forms of the hypotheses for the ranked data in the last example were omitted for a reason. Until now, all of the measurement variable hypotheses have been for population means of either the raw data or, implicitly in the last two sections, of transformed data. Unfortunately, this convention does not carry over to ranks. There is no variable defined for all members of a population which, when sampled, yields the ranks observed in the sample. The rank assigned to a specific individual depends on the sample size and on that individual's raw measurement value as compared to those of his or her fellow sample members. Thus, there is no population mean of ranks. However, because we are used to thinking about population frequency shifts in terms of means, it is convenient to create a fictitious population rank mean μ_R to describe rank hypotheses. Thus, the hypothesis of no difference in cooling rates, above, could be written H_0: $\mu_{R1} = \mu_{R2}$. This convention will be a useful guide to intuition; however, it should not be reported as results. One can conclude, as was done in the above example, that the result found in the analysis holds on the average for the populations involved, without being specific about what is meant by "average." This strategy will generally provide an adequate intuitive picture of the nature of the differences in distributions.

EXERCISES

12.18 Refer to example 12.1. Do a two-sample rank test for the Levi Strauss mill data given in the example. Compare the P-value of the rank test with that of the test based on trimmed means.

12.19 Refer to exercise 12.10 (p. 385). Apply a one-sided, two-sample test based on ranks to the data and compare the outcome of this test to the outcome of the trimmed t test of that exercise.

12.20 Refer to exercise 12.11 (p. 385). Test the hypothesis of no difference in androstenedione levels for male and female diabetics, using a two-sample test based on ranks. Compare the results of this test with the outcome of the trimmed t test.

A k-Sample Example

The rank analysis method will be applied to the hydrocarbon emission data in the following example.

EXAMPLE 12.7

The stem and leaf plot form for ranking the automobile emission data of exercise 4.18 (p. 115) is given in display 12.8. The rank summary computed from this display is as follows.

	PRE-1963	1963–1967	1968–1969	1970–1971	1972–1974
n	10	13	16	20	19
$\overline{R}$	60.35	58.58	37.44	36.58	20.29
s_R	11.888	12.672	21.161	18.822	19.080

Display 12.8 **Stem and leaf plot form for ranking data from several samples, applied to automobile emission data of exercise 4.18. Ranks are in italics; prime (′) denotes a rank ending in .5.**

YEAR OF MANUFACTURE

STEM	PRE-1963	1963–1967	1968–1969	1970–1971	1972–1974
0			*7* 71		*2 2 5 2 9 6 4 8* 20, 20, 60, 20, 95, 70, 58, 85
1			*11 21 18′ 12* 11, 99, 88, 17	*15 10 20 13′ 18′* 41, 00, 90, 40, 88	*13′ 16 17* 40, 60, 75
2		*35* 70	*36 25 32′* 94, 11, 41	*34 23 23 29 32′ 29* 47, 00, 00, 23, 41, 23	*29 27 26 31 23* 23, 20, 17, 35, 00
3	*39* 47	*40* 50	*45 41′ 41′* 88, 53, 53	*43 38 37* 59, 06, 00	*44* 60
4	*48* 11	*49 47* 23, 05	*51 52* 60, 70	*53 50* 94, 35	*46* 00
5	*54 56* 41, 70		*55* 58		
6	*58* 30	*57* 20			
7		*59 60* 00, 80			
8	*61* 00	*62* 23		*64 63* 82, 80	
9	*66* 05	*68 65* 40, 00		*68 68* 40, 40	
10	*70′* 58	*70′* 58	*72* 88		
11		*73* 50			
12	*74* 93				
HI	*77* 2351	*76* 2000	*78* 2999		*75* 1880

The ANOVA table is as follows:

SOURCE	SS	df	MS	F	
Between years	16,329.711	4	4082.428	12.846	$P < .001$
Within years	23,199.615	73	317.508		

The ordered means with the underlines resulting from the FSD procedure are as follows:

	1972–1974	1970–1971	1968–1969	1963–1967	PRE-1963
$\bar{R}$	20.29	36.58	37.44	58.58	60.35

That is, the groupings are

$$\{1972\text{--}1974\} < \{1970\text{--}1971,\ 1968\text{--}1969\} < \{1963\text{--}1967,\ \text{pre-}1963\}$$

Thus the added sensitivity provided by the rank procedure verifies the significance, at the 5% level, of the three distinct groupings observed in the schematic diagram, display 12.4. Note also that the rank transform produces the same groupings as the power transform did.

EXERCISES

12.21   Apply a rank analysis to the complete meteorite data given in exercise 12.16 (p. 397). Compare the results of the rank analysis with the power transform analysis carried out in that exercise.

12.22   Apply a rank analysis to the complete Levi Strauss mill data set given in exercise 12.13 (p. 388). Compare the results of the rank analysis with the trimmed mean analysis carried out in that exercise.

A Paired-Difference Application

When an exploratory analysis *on the differences* in a paired-difference analysis reveals outliers, a rank alternative to the trimmed mean procedure given in section 12.2 can be applied to test for differences in distributions. The paired data are first combined into a single group and ranked. Then, a data set of paired ranks is formed by replacing each observation in the original paired configuration by its rank. Finally, the usual paired-difference analysis is performed on the paired ranked data.

This method will be applied to the electrical consumption data of example 12.4.

EXAMPLE 12.8   The data of example 12.4 (p. 389), in units × 100, are ranked in the following stem and leaf plots. Note that the pairing is destroyed in this process and must be reestablished before proceeding.

	SEPTEMBER	JUNE
0*	*1'* 00	*3 1'* 04, 00
0t	*6' 6' 4* 28, 28, 20	*5* 24
0f		*8* 44
0s	*9' 9' 11* 60, 60, 76	
0.	*12* 88	*13* 92
1*		*14 16 16 16* 00, 12, 12, 12
1t	*18* 32	
1f		
1s		*19* 76
1.	*20* 92	
2*		
2t		
2f	*21'* 56	*21'* 56

The re-paired ranks and their differences are given in the next table. Each pair consists of the ranks of the original data in that pair. Thus, for example, note that the ranks for household number 1, which has the September consumption value .60 and June value 1.00, are 9.5 and 14, respectively.

HOUSEHOLD	SEPTEMBER	JUNE	DIFFERENCES
1	9.5	14	−4.5
2	21.5	16	5.5
3	9.5	16	−6.5
4	6.5	21.5	−15.0
5	18	16	2.0
6	11	8	3.0
7	6.5	5	1.5
8	12	13	−1.0
9	20	19	1.0
10	1.5	3	−1.5
11	4	1.5	2.5

Now, the summary data for the differences are

$$n = 11, \qquad \bar{d} = -1.182, \qquad s = 5.728$$

The value of the paired-difference t statistic is then

$$t = \frac{\bar{d}}{s/\sqrt{n}} = \frac{-1.182}{5.728/\sqrt{11}} = -0.684$$

The two-sided P-value for $11 - 1 = 10$ degrees of freedom, from table 3, has the bound

$$P > .50$$

Consequently, as was concluded in example 12.4, there is no evidence of a difference in electricity consumption rates for the two months at any interesting significance level.

◤
EXERCISES

12.23 Apply the paired rank difference method to the insulin data given in exercise 12.14 (p. 390). Test the hypothesis of no difference in insulin levels for weeks 0 and 1 at the 5% significance level. How does the outcome of this test compare with the outcomes of the regular and trimmed mean tests obtained in exercise 12.14?

12.24 Apply the paired rank difference method to the drop percentage data of exercise 12.15 (p. 391). Comment on any differences in the outcome of this test and the tests carried out in exercise 12.15.

A Paired-Difference Method Based on Signed Ranks*

A more traditional way of dealing robustly with paired data is to form *signed ranks* of the differences. A t test applied to the signed ranks is then (roughly) equivalent to the **Wilcoxon signed rank test,** a "classical" nonparametric procedure.

Wilcoxon signed rank test

To form signed ranks, we deal directly with the differences of the original sample measurements, $d = X_1 - X_2$. The *absolute values* of these differences are ranked. Then, the signs of the original differences are applied to the ranks. Thus, for example, if $d = 12.5 - 13.4 = -.9$ and the absolute value, .9, has rank 3 among the absolute values of sample differences, then the original difference will be replaced by -3 in the sample of signed ranks.

Now, a standard one-sample t test for $\mu = 0$ is applied to the sample of signed ranks. The method is demonstrated in the following example.

EXAMPLE 12.9

The signed-rank method will be applied to the electrical consumption data of example 12.4 in order to compare this method to the one of the last subsection. The original sample differences, their absolute values, and signed ranks are shown in the following table.

---

* Optional material; not used in subsequent chapters.

SAMPLE DIFFERENCES	ABSOLUTE DIFFERENCES	RANKS OF ABSOLUTE DIFFERENCES	SIGNED RANKS
−.40	.40	8	−8
1.44	1.44	10	10
−.52	.52	9	−9
−2.28	2.28	11	−11
.20	.20	5.5	5.5
.32	.32	7	7
.04	.04	2	2
−.04	.04	2	−2
.16	.16	4	4
−.04	.04	2	−2
.20	.20	5.5	5.5

Note the difference between this method and the method of the last subsection, which was demonstrated in example 12.8. There, the ranks of the original observations were formed and the data set to be analyzed then consisted of the differences in these ranks. Here, the differences of the original data are formed first, then these differences are "sign-ranked." Compare the last column of the above table with that of the table in example 12.8. These are the data to which the t tests are applied in the two methods.

The signed-rank summary data (computed from the last column of the above table) are

$$n = 11, \quad \bar{R} = .1818, \quad \text{and} \quad s_R = 7.093$$

The value of the one-sample t statistic is then

$$t = \frac{\bar{R}}{s_R/\sqrt{n}}$$
$$= \frac{.1818}{7.093/\sqrt{11}}$$
$$= .0850$$

The P-value for a two-sided test based on this t value and $11 - 1 = 10$ degrees of freedom has the bound

$$P > .50$$

This is the same result as found in example 12.8.

EXERCISES

12.25 Repeat exercise 12.23 using the signed-rank method and comment on any differences in the two paired-difference methods based on ranks.

12.26 Repeat exercise 12.24 using the signed-rank method and comment on any differences in the two paired-difference methods based on ranks.

Chapter 12 Quiz

1. What is a robust statistical procedure?

2. What is the primary use of exploratory data analysis in relation to robust methods? What is the most useful exploratory plot from this viewpoint for measurement variable comparison problems?

3. How is the trimmed mean used to obtain robust alternatives to the classical procedures of chapters 10 and 11? Give an example for each problem type: one-sample, two-sample, k-sample, and paired-difference.

4. What is Satterthwaite's correction for unequal dispersions? How is it applied in the two-sample problem? Give an example.

5. What are the power transformations? How are they used to equalize dispersions?

6. What is the process for fitting a power transformation to equalize dispersions in a k-sample problem?

7. How is the stem and leaf plot used to rank (a) a single sample? (b) several samples together?

8. How is the rank transformation used to provide robust procedures? Give examples from the two-sample problem, the k-sample problem and the problem of paired data.

The following question is for the optional section on signed ranks

9. How is the method of signed ranks applied in a paired data problem?

◤

SUPPLEMENTARY EXERCISES

12.27   The storm intensity data of exercise 10.25 (p. 334) are repeated below. A schematic diagram of these data shows an outlier from one group overlapping the adjacent range for the other, probably due to a misclassification of storms for the two seasons. In this situation, especially when dispersions are otherwise comparable, a trimmed mean analysis is frequently more sensitive in detecting shifts in distributional centers than is a rank method. Demonstrate this effect for these data by finding and comparing the P-value bounds for two-sample tests based on the appropriate trimmed mean and the two-sample test based on ranks.

SUMMER		NONSUMMER	
18.2	21.2	23.8	3.4
80.0	60.7	10.5	8.3
8.6	34.6	14.3	66.7
45.0	40.7	40.0	15.0
28.6	6.7	16.1	11.8
20.7		11.1	18.2
50.0		5.0	26.3

12.28   In a study of the effect of a major pollution source on the Red River in northern New Mexico, readings of phosphorus levels were made at low flow conditions during the winter of 1979–1980 at six recording stations. The stations are located downstream from a reference point. Thus, for example, HRG 21 represents a recording site 21 miles downstream from the reference point. The pollution source is at 23.2 miles below the reference. The data are as follows.

HRG 21	HRG 22	HRG 23.1	HRG 23.3	HRG 24	HRG 25
0.000	0.005	0.005	0.040	0.030	0.040
0.005	0.005	0.020	0.040	0.020	0.010
0.000	0.000	0.040	0.030	0.010	0.020
0.010	0.005	0.020	0.160	0.050	0.030
0.005	0.000	0.024	0.370	0.050	0.030
0.005	0.030	0.030	0.190	0.040	0.070
0.008	0.020	0.064	0.240	0.080	0.060
0.010	0.047		0.092	0.067	0.194
0.008	0.059		0.060	0.045	0.024
	0.018		0.326	0.092	0.175
	0.028		0.269	0.035	0.062

Data courtesy of D. Tague and G. Jacobi, New Mexico Environmental Improvement Division, Santa Fe, New Mexico.

(a) Make a schematic diagram of the data. What indications of difficulties with the usual test for equal means are apparent?

(b) Apply the standard FSD method (chapter 11) to these data at the 5% significance level.

(c) Apply an FSD procedure based on ranks at the 5% level.

(d) Fit and carry out a power transformation and perform an FSD procedure at the 5% level on the transformed data.

(e) Describe the differences in outcomes of the procedures of parts (b)–(d). What trend in phosphorus levels with distance from the reference station is demonstrated by the more sensitive of these procedures?

12.29   The ages at death by suicide for three ethnic groups, obtained from randomly selected records of the Albuquerque Suicide Crisis Unit for 1978, are as follows.

Anglos: 21, 55, 42, 25, 48, 22, 42, 53, 21, 21, 31, 28, 24, 53, 66, 90, 27, 48, 47, 49, 53, 62, 31, 31, 32, 43, 57, 42, 34, 39, 24, 79, 46, 52, 27, 76, 44, 35, 32, 26, 51, 19, 27, 58

Hispanics: 27, 22, 20, 51, 60, 15, 19, 24, 24, 18, 43, 50, 31, 29, 21, 27, 34, 76, 35, 55, 24, 68, 45, 57, 22, 48, 48, 14, 52, 29, 21, 28, 17, 38

Native Americans: 26, 17, 24, 22, 16, 21, 36, 18, 48, 20, 35, 23, 25, 23, 22*

(a) Make a schematic diagram of the data.

(b) Test for a difference in average ages at suicide for the three groups using the standard FSD procedure of chapter 11 at the 5% significance level.

(c) Use the FSD procedure based on ranks at the 5% level to analyze the data.

---

* Data courtesy of D. Bagley and J. Abbin, Albuquerque Suicide Crisis Unit.

(d) Fit a power transformation to the data and use the FSD procedure at the 5% level on the transformed data.

(e) Describe any differences in the outcomes of the analyses made in parts (b)–(d).

12.30   The following determinations of the parallax of the sun (the angle spanned by the earth's radius as if it were viewed and measured from the sun's surface), made in 1761 and analyzed by James Short, a noted telescope manufacturer, are given by Stigler [31]. Units are in seconds of a degree (1/360 degree).

8.50	8.06	8.65	9.71	8.80	7.99
8.50	8.43	8.35	8.50	8.40	8.58
7.33	8.44	8.71	8.28	8.82	8.34
8.64	8.14	8.31	9.87	9.02	9.64
9.27	7.68	8.36	8.86	10.57	8.34
9.06	10.34	8.58	5.76	9.11	8.55
9.25	8.07	7.80	8.44	8.66	9.54
9.09	8.36	7.71	8.23	8.34	9.07
8.50	9.71	8.30	8.50	8.60	

With a careful determination of the radius of the earth and a good average value of parallax, the basic measurement unit of astronomy, the astronomical unit (which is the average distance from the earth to the sun), can be obtained.

(a) Make a box plot of the data and describe any deviations from normality.

(b) Demonstrate by calculating both the standard 95% confidence interval and the 95% confidence interval based on the appropriate trimmed mean that even for rather large data sets a robust analysis will often provide a nonnegligible increase in sensitivity.

(c) The currently accepted value of the parallax of the sun is 8.798. Test, using a trimmed mean test at the 5% level, the hypothesis that the center of the measurement population from which Short's data is taken is equal to this value.

12.31   Using a trimmed mean, test at the 10% significance level whether there was an increase in the average price of groceries from 1974 to 1980 for the family that gave rise to the data of exercise 3.28 (p. 92). The data are repeated here.

MAY 1974			APRIL 1980			
2.05	.46	.25	4.35	3.19	1.99	.46
1.32	.79	.50	4.80	1.66	2.99	.56
.50	.79	.56	.68	1.66	.99	.94
1.92	.79	.56	.68	2.85	.72	2.12
2.17	1.00	.25	1.93	1.63	.39	2.12
2.10	1.00	.56	.83	6.89	.39	1.65
2.23	1.00	.79	.79	.56	.46	
.80	1.00	2.12	.79	.89	.46	
.95	.25	2.12	.79	.89	.46	
.69	.25	5.55	1.19	1.63	.46	

12.32   The height data from example 10.6 (p. 317), used to test the (null) hypothesis that men are no taller than women, are repeated below. The P-value for this hypothesis, based on a classical two-sample t test, has the bound $P < .0005$. Back up the classical test with a test based on ranks. What, if any, effect on the P-value bound does the robust test have?

MEN	WOMEN
72	66
71	65
66	62
71	69
72	63
67	63
73	67
65	67
67	66
68	60
65	68
70	65
66	64
72	64
	63
	70
	64
	64
	66
	65
	67
	57

CHAPTER 13

Categorical Variable Methods—Chi-Square Tests

Introduction

In this chapter we return to the categorical variable comparison and association problems treated by exploratory methods in chapters 3 and 4. As with measurement variable problems, the treatment of categorical variable problems requires the dimension of statistical inference to answer two questions. "Is there a difference?" "How large is the difference?" Chi-square tests, which provide a preliminary screening for significance, will be introduced in this chapter to answer the first question. The confidence interval for a population proportion introduced in chapter 8, plus a new interval for differences in proportions, will be used to answer the second. Additionally, a method for spotting "interesting" differences will be introduced to aid in describing the nature of the differences between observed and hypothetical frequencies. As before, these procedures will fit together to provide a powerful, unified strategy for dealing with categorical variable problems.

The chi-square test for homogeneity, which treats the k-sample categorical variable comparison problem, and the chi-square test for contingency, which deals with the association problem for two categorical variables, are actually the same test. Consequently, the distinction between these two types of problems will be lost to a great extent. This test is given in section 13.3. To introduce the chi-square statistic and its distribution, which is basic to all of the methods of this chapter, we will first present the oldest significance test in the statistician's arsenal—Karl Pearson's chi-square test for goodness of fit, which first appeared in published form in 1900. Despite its age, this test is still important today and is used extensively in practice. It will be used to solve the one-sample problem for categorical variables in the case where the hypothesis H_0 completely specifies the distribution.

Finally, section 13.4 gives a two-sample test for proportions that allows one-sided as well as two-sided alternatives to be considered. This test and the associated one- and two-sided confidence intervals will provide tools for testing and quantifying differences in population proportions analogous to the two-sample t test for means given in chapter 10.

■ SECTION 13.2

The Chi-Square Goodness of Fit Test

To have a concrete example to work with in this discussion, we will consider a data set that was used as evidence in an actual court case. The data represent a random sample of 1336 individuals from the jury pool of a large municipal court district for the years 1975–1977. The fairness of the representation of various age groups on juries was being contested. The strategy for doing this was to challenge the representativeness of the pool of individuals from which the juries are drawn. This was done by comparing the age group distribution within the jury pool against the age distribution in the district as a whole, which was available from census figures. Since the jury pool was quite large, a random sample was taken from it for the study.

AGE GROUP (YEARS)	SAMPLE (OBSERVED) FREQUENCIES	POPULATION (CENSUS) PROPORTIONS
18–19	23	.061
20–24	96	.150
25–29	134	.135
30–39	293	.217
40–49	297	.153
50–64	380	.182
65 and over	113	.102
	1336 = sample size	

To formulate the problem in terms of a statistical hypothesis test, the null hypothesis of no age discrimination can be expressed as the statistical hypothesis that the true jury pool (population) proportions, denoted by p_1, $p_2, \ldots, p_7$, are equal to the corresponding census proportions:

$$H_0: \quad p_1 = .061, \ p_2 = .150, \ \ldots, \ p_7 = .102$$

The alternative hypothesis that is adopted here, and in all tests we will consider in this and the next section, is that *some* difference exists between the true and hypothetical proportions. That is, at least one of the equalities specified by H_0 is not true:

$$H_1: \text{not } H_0$$

Thus, the test of these hypotheses must be capable of detecting *any* deviation in the true population proportions from the hypothetical ones.

The construction of this global test relies on the idea of measuring the "distance" between actual and hypothetical parameter values by a sum of squares. Recall that this idea was also used in the analysis of variance in chapter 11. If the hypothetical proportions are represented symbolically by

p_i^0 for $i = 1, 2, . . . , k$, where k is the number of categories of the variable in question ($k = 7$ for the example), then a measure of how different the true and hypothetical parameters are is

$$\Sigma w_i(p_i - p_i^0)^2$$

where the w_i's are positive weights (to be given shortly) and the sum is over all k categories. This weighted sum of squares is 0 if and only if all of the true population proportions are equal to the corresponding hypothetical ones. Thus, the hypothesis H_0 corresponds to a zero value for the sum of squares and H_1 to any positive value.

Now, since the true proportions are not known, they would be estimated from the sample. If O_i represents the observed (absolute) frequency for the ith category of the variable (the numbers in the second column of the above table), and n is the sample size, then the natural estimate of the population proportion p_i is the sample proportion

$$\hat{p}_i = \frac{O_i}{n}$$

This quantity could then be substituted for p_i in the above sum of squares to form the basis for a test statistic to test H_0 against H_1. Since large values of the sum of squares favor the alternative hypothesis, an appropriate test rule would be to reject H_0 if the sum of squares exceeds some constant. To choose the constant, it will be necessary to know the sampling distribution of the test statistic under the null hypothesis, or a good approximation to it.

Both tradition and convenience dictate that the actual test statistic be written in terms of the absolute frequencies rather than the relative frequencies. This is accomplished by multiplying the terms in the sum of squares by the sample size n. The resulting numbers

$$E_i = np_i^0$$

expected frequencies

are called the **expected frequencies** and represent the number of observations we would expect to find in a sample of size n, on the average, if the hypothesis H_0 were true. The choice of the weights w_i that provides a test statistic whose sampling distribution can be approximated is then $1/E_i$, and

chi-square statistic

the test statistic itself, called the **chi-square statistic,** has the form

$$X^2 = \Sigma \frac{(O_i - E_i)^2}{E_i}$$

where the sum is over all categories of the variable.

A strategy for computing the chi-square statistic will soon be combined with a method for looking at the nature of the differences between observed and expected frequencies. First, the approximation to the sampling distribution of X^2 will be discussed.

The Chi-Square Distribution

The chi-square test will be based on the rule

Reject H_0 if $X^2 \geq c$

chi-square distribution

degrees of freedom

where the cut-off value c is to be determined to control the type I error probability to the value specified by the preassigned significance level α. The exact sampling distribution for X^2 is unknown, so c cannot be determined exactly. However, a very good approximation to this sampling distribution is given by the **chi-square distribution** for which critical values are given in table in the appendix.

Like the t distribution, the chi-square distribution is a family of distributions indexed by an integer-valued parameter ν called the **degrees of freedom**. For the goodness of fit test, considered in this section, the degrees of freedom are equal to the number of variable categories minus 1:

$$\nu = k - 1$$

For the above example, since $k = 7$ we would have $\nu = 6$.

For a given probability, denoted by Pr in the table, table 2 provides the number c for which

$$P(X^2 \geq c) = \text{Pr}$$

To find the cutoff value for a 5% test of the hypothesis for the above example, we would enter the table at row $\nu = 6$ and the column for which Pr is equal to the significance level .050. The resulting table entry, $c = 12.59$, is the cutoff value. The test rule would then be

Reject H_0 if $X^2 \geq 12.59$

We next consider the calculation of the test statistic itself.

Where Are the Interesting Differences?

The chi-square statistic is calculated at the same time that categories or *cells* with interesting differences between observed and hypothetical frequencies are pinpointed. The process involves calculating the quantities

$$z_i = \frac{O_i - E_i}{\sqrt{E_i}}$$

for all categories. When H_0 is true, these variables have, approximately, a standard normal distribution. Thus, deviations from 0 that signify interesting differences between observed and expected frequencies can be quantified by standard normal critical values.

A useful symbolic labeling of categories can be based on the 10%, 5%, and 1% standard normal critical values:

If $|z_i| \leq 1.645$ assign the symbol "·"
If $1.645 < |z_i| \leq 1.960$ assign "o"
If $1.960 < |z_i| \leq 2.576$ assign "O"
If $|z_i| > 2.576$ assign "@"

In the following table, the jury pool data table given above is augmented by columns showing the expected frequencies, the values of z_i, and these symbols.

Goodness of fit table

AGE	OBS. FREQ.	HYP. PROP.	EXP. FREQ.	z	SYMBOL
18–19	23	0.061	81.5	−6.480	@
20–24	96	0.150	200.4	−7.375	@
25–29	134	0.135	180.4	−3.452	@
30–39	293	0.217	289.9	0.181	.
40–49	297	0.153	204.4	6.476	@
50–64	380	0.182	243.2	8.776	@
65$^+$	113	0.102	136.3	−1.994	O
Sample Size =	1336				

$X^2 = 231.260$

This table can be used as a computing form to organize the calculations. First, the observed frequencies are entered and summed to obtain the sample size $n = 1336$. If this value is known in advance, this sum provides a check on the correctness of the given frequencies.

Next, the hypothetical proportions are entered, and the expected frequencies are computed by multiplying the hypothetical proportions by n. For example check that

$$E_1 = 1336 \times 0.061$$
$$= 81.496 \quad \text{(rounded to 81.5 in the table)}$$

Now, $z_i = (O_i - E_i)/\sqrt{E_i}$ is calculated for each row and is recorded in the z column of the table. The chi-square statistic is then the sum of the *squares* of these quantities:

$$X^2 = \Sigma z_i^2$$

If a hand calculator is used for this computation, the value of the chi-square statistic can be accumulated in the calculator memory at the same time the z's are computed. Square each z after recording it in the table and add it to the memory (making sure the memory is cleared at the beginning of the calculation). The memory will then contain the value of X^2 after the last z_i is computed.

Check that $X^2 = 231.260$ for the jury pool example. Since this value exceeds the cutoff value $c = 12.59$, we would reject the null hypothesis H_0. That is, at the 5% significance level we would conclude that the true proportions in the jury pool differ from those in the municipal court district population. An age bias in the formation of the jury pool is apparent. But what is the nature of this bias?

The symbols in the last column of the table are assigned according to the scheme given earlier. For example, because −6.480 exceeds 2.576 in absolute value, the category "18–19 years of age" is assigned the symbol "@". *The strategy we will adopt is to interpret only those categories with the most "extreme" symbols.* These symbols are "@" and "O" in the above table. Several of the categories should be interpreted in this example.

The signs of the z's play a role in the interpretation. A negative sign indicates a category for which the observed frequency is smaller than expected under the hypothesis H_0, and a positive sign indicates a greater than expected frequency. Thus, we see that the younger age groups (categories 18–29) occur less frequently than expected, thus are underrepresented in the jury pool, while the older age groups (40–64) are overrepresented. A somewhat milder indication of underrepresentation of people 65 years of age and older is also seen.

EXAMPLE 13.1

Do people know and report their weights accurately to the nearest pound? A test of this hypothesis (that the author has carried out on a generation or two of statistics students always with the same results) can be based on the following observation. Since weights vary over a wide range, we would expect that the last digit of weight (for example 3 for a weight of 193 pounds) would be equally likely to have any one of the possible values from 0 to 9. That is, each digit would have proportion .1 in the population. Now, if people accurately report their weights, then the same proportions should hold for the last digits of the reported weights. Here, the true proportions, p_i, of reported weights are of interest and the null hypothesis of no difference between actual and reported weights has the statistical form

$$H_0: \quad p_0 = .1, \ldots, p_9 = .1$$

We will use the data of display 1.1 (p. 9) to test this hypothesis. The data and the quantities derived for the test are given in the following table.

Goodness of fit table

LAST WT DG	OBS. FREQ.	HYP. PROP.	EXP. FREQ.	z	SYMBOL
0	14	0.100	3.6	5.481	@
1	0	0.100	3.6	−1.897	o
2	2	0.100	3.6	−0.843	.
3	4	0.100	3.6	0.211	.
4	1	0.100	3.6	−1.370	.
5	14	0.100	3.6	5.481	@
6	0	0.100	3.6	−1.897	o
7	1	0.100	3.6	−1.370	.
8	0	0.100	3.6	−1.897	o
9	0	0.100	3.6	−1.897	o
Sample Size =	36				

$X^2 = 79.000$

The cutoff value for a 5% test from table 2, corresponding to $\nu = 10 - 1 = 9$ degrees of freedom, is $c = 16.9$. The computed value of chi-square, $X^2 = 79.000$, exceeds this value. Consequently, the hypothesis that weights are reported accurately would be rejected at the 5% significance level.

From the symbol assignment it is seen that only the digits 0 and 5 appear to be interesting. Because the signs of the z's are positive, the observed frequencies exceed the expected frequencies; more than the expected number of people report weights ending in 0 or 5. Apparently, people tend to round their weights to the nearest multiple of 5 pounds when reporting them.

P-Values for Chi-Square Tests

More impact can be given to the outcome of a chi-square test by reporting bounds for the P-value

$$P = P(X^2 \geq X^2_{obs})$$

where X^2_{obs} is the computed chi-square value. To find these bounds, enter table 2 at the row corresponding to the degrees of freedom for the problem and find the consecutive values in that row that bound X^2_{obs}. The corresponding numbers in the Pr row, in reverse order, are the P-value bounds.

For example 13.1 we find that $X^2_{obs} = 79.000$ is larger than the largest value, 27.88, in the $\nu = 9$ row. Thus, we conclude that

$$P < .001$$

for this problem. As before, this indicates a highly significant test result. If, in fact, people accurately report their weights, an outcome as extreme as the one observed for this sample would be seen less frequently than once in every 1000 random samples of size 36, on the average.

Confidence Intervals to Quantify Interesting Differences

Once the chi-square test has shown support for a difference between hypothetical and true proportions and the nature of these differences has been described, it is natural to ask how large these differences are. This question can be answered by means of confidence intervals for the true proportions, p_i.

A multiplicity problem arises with the use of two or more confidence intervals for the p_i's. Since a multiple comparison method for estimating these parameters (like the Fisher least significance difference method of chapter 11) is not known, a strategy based on the Bonferroni inequality is adopted.

If the null hypothesis is rejected, confidence intervals will be computed for those proportions that received the most "extreme" symbols, for example the symbol "@," or the symbol "O" if no @'s appear in the table. On the other hand, if the null hypothesis is accepted, the inference process stops; no interpretations will be made nor will confidence intervals be computed. This conforms with the strategy of using the initial hypothesis test to screen the data for enough information to proceed with further analysis.

The joint confidence coefficient for the confidence intervals will be taken to be one minus the significance level of the test. Thus, if the significance level is α, and r intervals are to be computed, then the confidence coefficient to be used for each interval is

$$1 - \frac{\alpha}{r}$$

The confidence intervals are then computed exactly as described in chapter 8. For example, in the jury pool problem the significance level of the test was $\alpha = .05$, and $r = 5$ categories received the symbol "@." Thus, 5 confidence intervals will be computed, each with confidence coefficient $1 - .05/5 = .99$ or 99%.

From table 1B, the critical value for 99% confidence is seen to be $z = 2.576$. Thus, the confidence limits for the p_i's will be of the form

$$\hat{p}_i \pm 2.576\sqrt{\frac{\hat{p}_i(1 - \hat{p}_i)}{n}}$$

where the sample size is $n = 1336$ and $\hat{p}_i = O_i/n$ is the estimated proportion for the ith category.

The estimated proportions and confidence limits are given in the following table. The hypothetical (census) proportions are also given.

AGE CATEGORY	$\hat{p}_i$	LOWER LIMIT	UPPER LIMIT	HYPOTHETICAL p_i
18–19	.017	.008	.026	.061
20–24	.072	.054	.090	.150
25–29	.100	.079	.121	.134
40–49	.222	.193	.252	.153
50–64	.284	.253	.316	.182

As an example of the computation that was used to construct this table, note that the observed frequency for the 18–19-year category was 23 (from the original data table). Thus, the observed proportion is $\hat{p}_1 = 23/1336 = .017$. The lower confidence limit is then

$$.017 - 2.576\sqrt{.017(1 - .017)/1336} = .008,$$

and the upper limit is

$$.017 + 2.576\sqrt{.017(1 - .017)/1336} = .026.$$

That is, we would conclude that the true jury pool proportion of 18–19 year olds is between .8% and 2.6%. To stress the difference between hypothetical and true proportions, we could affirm that the true proportion lies at least $6.1 - 2.6 = 3.5\%$ below the census proportion of 6.1%.

Similar comparisons can be made for the other categories. The Bonferroni method then guarantees at least a $1 - \alpha = 1 - .05$ or 95% overall confidence level for all comparisons jointly.

EXAMPLE 13.2

As a continuation of example 13.1, we will compute confidence bounds for the true proportions of individuals who report their weights as ending with a 0 or a 5. From the table of example 13.1, we see that these are the only two categories for which the proportions differ significantly from the hypothetical value of 10%.

An overall significance level for the test will be taken to be 5%. For $r = 2$, the confidence coefficient for each interval will then be $1 - .05/2 = .975$.

Since a critical value is not given for this coefficient, the critical value $z = 2.326$ for the next larger coefficient, .98, will be used. This leads to slightly conservative intervals (that is, intervals somewhat longer than required to achieve the overall confidence level of 95%).

The estimated proportion for each category is $14/36 = .389$. Thus, the confidence limits are $.389 \pm 2.326\sqrt{.389(1 - .389)/36}$ or $L = .200$ and $U = .578$. Consequently, we would conclude with joint confidence of at least 95% that between 20% and 57.8% of all individuals (in the appropriate student population) record their weights as ending with 0 and a like proportion record their weights as ending with 5.

Guaranteeing the Adequacy of the Chi-Square Approximation

As mentioned earlier, the chi-square distribution of table 2 is only an approximate distribution for the chi-square statistic, X^2, of this section and the next. Guidelines for guaranteeing the adequacy of the approximation have been suggested by a number of investigators. A **rule of thumb** is that *all expected cell (category) frequencies should be larger than or equal to one.* (See Kempthorne [20].) A more conservative rule that is often used adds the proviso that at least 20% of the cells should have expected frequencies greater than or equal to 5. We will use the simpler and less conservative rule in this text.

rule of thumb

Since the hypothetical frequencies are known in advance for goodness of fit tests, sample sizes can be made large enough at the design stage of the experiment to guarantee the adequacy of the approximation. This is done simply by taking n larger than one divided by the smallest hypothetical proportion. Thus, in the jury pool problem, since the smallest hypothetical proportion is .061 (for the 18–19-year-old category), we would need a sample size of at least $n = 1/.061 = 16.39$ in order for the chi-square approximation to be adequate. The actual sample size of 1336 was taken substantially larger than this to guarantee good power for the test as well.

When a sample size is not large enough and the sample has already been drawn, a situation that will occur most frequently in the contingency tests of the next section, it will be necessary to resort to the strategy of combining categories in order to validate the approximation. An example of the use of this strategy is given next.

EXAMPLE 13.3

We will test whether the underlying distribution of the class variable for the data of display 1.1 (p. 9) is the same as the class variable distribution for the "population" of display 1.18 (p. 28). To do this, the class variable proportions from display 1.18 are taken as the hypothetical proportions in a chi-square goodness of fit test. A summary of the analysis is given in the following table. The observed frequencies are from display 1.1, and the hypothetical proportions were computed from display 1.18.

Note that the expected frequency for the "Other" category is smaller than 1. To correct this problem, the Grad. and Other categories will be

Goodness of fit table

CLASS	OBS. FREQ.	HYP. PROP.	EXP. FREQ.	z	SYMBOL
Fr.	11	0.220	7.9	1.094	.
Soph.	15	0.510	18.4	−0.784	.
Jun.	7	0.190	6.8	0.061	.
Sen.	1	0.040	1.4	−0.367	.
Grad.	2	0.030	1.1	0.885	.
Other	0	0.010	0.4	−0.600	.
Sample Size =	36				

$X^2 = 3.095$

combined. This is accomplished by adding together the observed frequencies for the two categories to obtain an observed frequency for the combined "Grad. & Other" category. Similarly, the hypothetical proportions are summed to obtain the hypothetical proportion for the combined category. The resulting table is the following.

Goodness of fit table for combined categories

CLASS	OBS. FREQ.	HYP. PROP.	EXP. FREQ.	z	SYMBOL
Fr.	11	0.220	7.9	1.094	.
Soph.	15	0.510	18.4	−0.784	.
Jun.	7	0.190	6.8	0.061	.
Sen.	1	0.040	1.4	−0.367	.
Grad. & Other	2	0.040	1.4	0.467	.
Sample Size =	36				

$X^2 = 2.169$

The expected frequencies are now all greater than 1, so the chi-square approximation will be valid for this table.

Note that the effect of combining cells has been to decrease the value of X^2. The degrees of freedom for the second table are $\nu = 4$, and we find from table 2 that the P-value has bound $P > .5$. That is, it is quite plausible that the underlying distribution of the class variable for the data of display 1.1 is the same as for display 1.18.

EXERCISES

13.1 A manufacturer's safety department wishes to determine whether particular days of the week are worse than others as far as accident frequency is concerned. The accompanying table lists the accident frequencies accumulated over a one-year period.

	MONDAY	TUESDAY	WEDNESDAY	THURSDAY	FRIDAY
Number of accidents	37	14	18	25	20

(a) Use an equally likely probability assignment to set up the probabilities for the "no difference between days" null hypothesis H_0.

(b) Find the bounds for the P-value for the chi-square goodness of fit test, and test H_0 at the 5% level.

(c) If H_0 is rejected, which days contribute most heavily to the value of X^2? Interpret these results and give possible reasons for them.

(d) If H_0 is rejected, give joint 95% confidence intervals for the "interesting" accident proportions.

13.2 The dean of arts and sciences at a certain university established the grading guidelines of 10% A's and F's, 20% B's and D's, and 40% C's for his faculty. In a statistics class consisting of 117 students, the numbers of individuals receiving the five letter grades were as given in the following table.

GRADE	NUMBER
A	16
B	50
C	31
D	11
F	9

(a) Test whether the professor for this course is following the dean's standards by testing whether this class can be reasonably considered to be a random sample from a population for which the dean's grading proportions are used. What is the null hypothesis for this problem? Use a 1% significance level for the test and find the P-value.

(b) If the null hypothesis is rejected by the test of part (a), describe the nature of the differences between the actual and hypothetical proportions receiving the five letter grades.

(c) Find joint 99% confidence intervals for the "interesting" letter grade proportions found in part (a).

13.3 In a study of suicide rates for New Mexico, the question of whether certain age groups were more or less prone to suicide was studied by comparing data from a random sample of suicide records for 1980 with the 1980 census age proportions for New Mexico. The data and census figures are given in the following table.

AGE	OBSERVED SUICIDE FREQUENCIES	1980 CENSUS PROPORTIONS
20–29	34	.253
30–39	26	.201
40–49	23	.148
50–59	22	.155
over 59	21	.243

(a) Test at the 5% significance level whether the age distribution of suicides is the same as the age distribution for the population. Find the P-value for the test.

(b)   If the test of part (a) is rejected, describe the nature of the differences between the true and hypothetical distribution of suicide ages.

(c)   If the test of part (a) is rejected, find joint 95% confidence intervals for the "interesting" proportions described in part (b).

13.4   Statisticians are increasingly being called upon to serve as expert witnesses in lawsuits. In this problem, a statistician was hired to discredit the testimony of an expert on man–machine interactions in a suit brought against the manufacturer of a commercial meat-processing machine. The suit was brought on behalf of a young woman whose hand was injured while using the machine in her job.

The expert proposed to demonstrate that the opening in the machine through which meat is fed is too large relative to the hand sizes of a large proportion of the U.S. female population. As a consequence, the machine is too dangerous for the "average" female to use.

To test this claim, the expert made hand measurements on several women entering and leaving his university's library and on a number of secretaries and office workers at the university. He was able to substantiate the claim on the basis of this "sample."

The statistician's challenge was based on showing that the expert's sample was not representative of the U.S. female population, as he stated in his claim. To do this, it was only necessary to demonstrate that the distribution of *some* variable measured by the expert could not be the same as that for the U.S. population at large; if the sample is "biased" for one measurement, it is quite likely that it will be biased for others as well, and little confidence should be placed in the expert's results.

The statistician chose to show that the age distribution for the expert's sample differed significantly from the age distribution for U.S. females aged 14 and over, which is readily available from census data. The observed frequencies from the expert's sample and the census proportions are given in the following table.

AGE GROUP	EXPERT'S OBSERVED FREQUENCIES	CENSUS PROPORTIONS FOR U.S. FEMALE POPULATION
14–24	47	.261
25–44	8	.315
45–64	4	.278
over 64	1	.146

(a)   Test the hypothesis that the expert's sample is a random sample from the U.S. female population at the 5% significance level. What are bounds for the P-value?

(b)   Describe the nature of the differences between the true proportions being estimated by the sample and the hypothetical proportions from the census. What does this description imply about the method used to draw the sample?

(c)   When a sample is drawn in such a manner that certain groups have probabilities of being selected that differ from their proportions of occurrence in the population, the sample is said to be *biased*. The sample proportions are then estimates of the probabilities with which the groups are drawn rather than of the true population proportions. Find joint 95% confidence intervals for the probabilities with which the expert is sampling the different age groups in the above table.

◢ SECTION 13.3
_____

Tests of Homogeneity and Contingency

The k-sample problem for a categorical variable is the problem of testing whether the distributions of the variable are the same for k populations, based on independent random samples from each population. The chi-square test for homogeneity is the most commonly used test for this purpose.

The two-variable association problem is to test whether two categorical variables are independent, in the sense defined in chapter 5. The most commonly used test for this purpose is the chi-square contingency test.

As we will see in this section, the tests for homogeneity and contingency are actually the same. For this reason, we will consider them together. The calculations follow closely those given for the goodness of fit test, so little additional computational detail will be needed.

As in the goodness of fit test, the chi-square test will be used to screen the sample or samples for adequate information to proceed with the analysis. The symbol method for pinpointing and interpreting cells will be the same as before. However, in order to quantify interesting differences, we will now require confidence intervals for *differences* in proportions. These intervals, and a strategy based on the Bonferroni inequality for guaranteeing prescribed joint confidence, will be given in section 13.4.

The Two-Way Table Data Representation

As seen in chapters 3 and 4, the observed frequencies that constitute the raw data for both the k-sample problem and the two-variable association problem can be conveniently displayed in a two-way table. The two-way table will also be exploited in the inference computations. An example that will be used to illustrate the methodology is as follows.

In a recent study of the Bernalillo County juvenile justice system, it was hypothesized that girls would receive more lenient treatment (case dispositions) than boys. The case records of 152 boys and 156 girls were randomly selected from the county files and the disposition of each case was classified in increasing order of severity as (1) counseled and released (C&R), (2) one intervention by the probation department (1 Int.), (3) two or more interventions (2+ Int.) and (4) referral to juvenile court (Court). The data are given in display 13.1.

The observed frequency in the ith row and jth column of the table will be denoted by O_{ij}. Thus, for example, $O_{11} = 63$ in display 13.1. The number of rows in the table will be denoted by r and the number of columns by c. The table is often referred to as *an r by c (r × c) table*. Thus, the table in display 13.1 is 4×2.

The row and column totals shown in display 13.1 will be important in the analysis. The total for the ith row will be denoted by R_i and the jth column total by C_j. For example, in display 13.1, $R_1 = 170$ and $C_1 = 152$. The grand total, which is the sum of either all row totals or all column totals, is denoted by T. For display 13.1, $T = 308$.

Display 13.1

Two-way table of observed frequencies and marginal totals for the juvenile justice system example

		SEX		
DISPOSITION	BOY	GIRL		ROW TOTALS
C&R	63	107		170
1 Int.	41	35		76
2+ Int.	18	7		25
Court	30	7		37
Column Totals	152	156		308

Data courtesy of S. Teaf, M.A., Department of Sociology, University of New Mexico.

Is the above juvenile justice problem a two-sample problem or an association problem? If it is a two-sample problem, then the variable of interest is the four-category variable Disposition, and the two populations are those of boys and girls. The statement of the problem suggests that records were separated into these two groups, and then random samples were taken: of size $C_1 = 152$ from the population of boys and $C_2 = 156$ from the population of girls. The question of equality of treatment of the two sexes then leads to the null hypothesis of no difference in the population frequency distributions of the disposition variable for the two sexes.

However, this problem could also be viewed as an association problem between the Disposition and Sex variables. An association would show up as a difference in the conditional distributions of Disposition given Sex for the two sex categories. Thus, both problems reduce to distributional comparisons. This is why there is no difference in the inference procedures for the two types of problems.

One distinction that is worth mentioning is that the two problems can lead to different experimental designs. For the comparison problem, the sample sizes $C_1, C_2, \ldots, C_c$ would ordinarily be fixed by the design of the sampling experiment. On the other hand, in an association problem, the sample size is the grand total T, which would be fixed by design. The C_i's would not be; the values of the C_i's would be random and would depend on the marginal distribution of the column variable—the Sex variable in the above example. However, the chi-square test for contingency is a test for equality of the conditional distributions, which treats the column sums as fixed. So this design difference is notable only because for the contingency problem it is more difficult to ensure the adequacy of the chi-square approximation; the expected cell frequencies depend on the row and column totals, and the latter are not under the designer's control. We look at the calculation of the expected frequencies next.

The Expected Frequencies

The (estimated) expected frequency for the ith row and jth column is denoted by E_{ij}. These frequencies represent the absolute frequencies to be expected if the null hypothesis under consideration were true. We will argue that for both the homogeneity and contingency tests, the expected frequencies have the form

$$E_{ij} = \frac{R_i \times C_j}{T} \tag{1}$$

That is, the expected frequency for a given row and column is the product of the row total and column total for the same row and column divided by the grand total.

The argument for the k-sample problem is as follows. The row total column of the table represents the observed frequencies for a sample from an (artificial) *pooled* population. Under the null hypothesis that the distributions are the same for all populations, each value of the variable would have the same population proportion for the pooled population as it does for each individual population. The expected frequency, E_{ij}, is defined to preserve this property for the estimated proportions. That is, under H_0 the expected proportion, E_{ij}/C_j, for the ith variable value of the jth population is equal to the pooled population estimate, R_i/T:

$$\frac{E_{ij}}{C_j} = \frac{R_i}{T}$$

Now, cross-multiply this expression by C_j to obtain expression (1).

The argument for the association problem is as follows. The expected frequency, E_{ij}, divided by the sample size, T, is designed to estimate the joint population proportion, p_{ij}, under the null hypothesis. However, the null hypothesis for the association problem is that the two variables are independent. It follows that the joint distribution will satisfy the product rule (chapter 5, p. 168) relative to the marginal distributions:

$$p_{ij} = p_{i \cdot} \times p_{\cdot j}$$

where $p_{i \cdot}$ is the marginal proportion for the ith category of the row variable and $p_{\cdot j}$ is the marginal proportion for the jth category of the column variable. But these proportions are estimated by the sample marginal proportions R_i/T and C_j/T respectively and the (estimated) expected frequency under the independence hypothesis is defined to preserve the product rule for the estimates:

$$\frac{E_{ij}}{T} = \frac{R_i}{T} \times \frac{C_j}{T}$$

This also reduces to expression (1) when both sides are multiplied by T.

As an example of the use of expression (1), the expected frequency for the cell in the first row and first column of display 13.1 is

$$E_{11} = \frac{R_1 \times C_1}{T} = \frac{170 \times 152}{308}$$

$$= 83.90$$

The remaining expected frequencies and the computations for the chi-square test will be given in the next subsection.

The Chi-Square Test

Once the expected frequencies are available, the rest of the chi-square test computations follow almost exactly the steps given for the goodness of fit test. One difference is that the degrees of freedom must be calculated in a different way. For the homogeneity and contingency test, the degrees of freedom are

$$\nu = (r - 1) \cdot (c - 1)$$

That is, the degrees of freedom equal the number of rows in the table (not counting the margins) minus one times the number of columns minus one. For the juvenile justice example,

$$\nu = (4 - 1) \cdot (2 - 1) = 3$$

The cutoff value for the test, or the P-value, are again found from table 2 (p. 606). Thus, for a 5% test of the hypothesis that the case disposition distributions for boys and girls are the same, the cutoff value would be $c = 7.815$. The rule would be to reject this hypothesis if $X^2 \geq 7.815$, where X^2 is the chi-square statistic

$$X^2 = \sum \frac{(O_{ij} - E_{ij})^2}{E_{ij}}$$

This sum is taken over all cells of the two-way table.

The computation of the chi-square statistic will again be carried out in a way that allows us to identify cells with "interesting" differences between observed and expected frequencies. We will compute

$$z_{ij} = \frac{O_{ij} - E_{ij}}{\sqrt{E_{ij}}} \tag{2}$$

for each cell and assign symbols to the cells according to the scheme given for the goodness of fit test in section 13.2 (p. 415). The chi-square statistic will then be computed by summing the squares of the z_{ij}'s:

$$X^2 = \Sigma z_{ij}^2$$

The results of the computations and the symbol assignments for the juvenile justice example (display 13.1) are given in display 13.2. The value of the computed chi-square statistic, 30.952, exceeds the cutoff value of 7.815. Consequently, the hypothesis of no difference in treatment of boys and girls would be rejected at the 5% significance level.

The most interesting cells (those receiving the symbol "@" in display 13.2) are seen to be the ones corresponding to court referrals. Significantly more boys were required to appear in court than were girls. To a somewhat lesser extent, the "counseling and release" category was also interesting. A significantly larger proportion of girls were counseled then released than

Display 13.2

Inference table for the juvenile justice example

DISPOSITION		SEX		ROW SUMS
		BOY	GIRL	
C&R	Obs.	63	107	170
	Exp.	83.9	86.1	
	z	−2.3	2.3	
	Sym.	O	O	
1 Int.	Obs.	41	35	76
	Exp.	37.5	38.5	
	z	0.6	−0.6	
	Sym.	.	.	
2+ Int.	Obs.	18	7	25
	Exp.	12.3	12.7	
	z	1.6	−1.6	
	Sym.	.	.	
Court	Obs.	30	7	37
	Exp.	18.3	18.7	
	z	2.7	−2.7	
	Sym.	@	@	
Column Sums		152	156	308

$X^2 = 30.952$

were boys. No interesting differences were observed in the other two disposition categories. We would conclude at the 5% significance level that the analysis supports the contention that girls are treated more leniently than are boys by the juvenile justice system of Bernalillo County.

Let us summarize the steps in the computation to this point.

1. Once the observed frequencies are placed in the two-way table, the first step is to calculate the row and column totals R_i and C_j for each row and column, and compute the grand total T.

2. Next, the expected frequencies are computed for each cell by means of expression (1).

3. The z_{ij}'s are computed and their squares are summed to find the value of the chi-square statistic X^2. Using these z's, symbols are assigned to cells according to the scheme given on page 415.

If the null hypothesis is accepted, the analysis stops at this point. If it is rejected, we continue to the interpretation of the interesting cells. We will often want to quantify the interesting differences by confidence intervals. The method for doing this will be given in the next subsection. First, some additional examples will be given.

EXAMPLE 13.4

In this example the baby food advertisement data introduced in section 3.3 (p. 77) will be analyzed. Recall that a producer of baby foods commissioned a survey of the reactions of mothers to a particular advertisement for the product in three living environments, urban, suburban, and rural. The mothers were asked if the advertisement persuaded them to use the product. The variable of interest was their response, which has values "yes" or "no."

The following table gives the results of the computations for the chi-square test.

RESPONSE		URBAN	SUBURB	RURAL	ROW SUMS
			POPULATION		
Yes	Obs.	30	19	21	70
	Exp.	31.1	16.5	22.4	
	z	−0.2	0.6	−0.3	
	Sym.	·	·	·	
No	Obs.	38	17	28	83
	Exp.	36.9	19.5	26.6	
	z	0.2	−0.6	0.3	
	Sym.	·	·	·	
Column Sums		68	36	49	153

$X^2 = 0.955$

For example, the computed expected frequency for the cell in the first row and first column of the table is

$$E_{11} = \frac{68 \cdot 70}{153}$$

$$= 31.11 \quad \text{(rounded to 31.1 in the table)}$$

The z value for this cell is then

$$z_{11} = \frac{30 - 31.11}{\sqrt{31.11}}$$

$$= -.199 \quad \text{(rounded to } -0.2 \text{ in the table)}$$

Since the absolute value of $-.199$ is smaller than 1.645, the symbol assigned to the cell is "·". The contribution of this cell to the value of X^2 is $-.199^2 = .040$.

In fact, the contributions for all cells are small, leading to the overall sum of $X^2 = .955$ for the chi-square statistic.

It is common practice to quote the P-value for a chi-square test. Here, the degrees of freedom are $(2 - 1) \cdot (3 - 1) = 2$, and the value of the chi-square statistic lies to the left of all entries in the 2 degrees of freedom row of table 2. Thus,

$$P > .5$$

The outcome of the test is highly insignificant and we would report that there is no evidence of a difference in reactions to the advertisement for the three regions.

EXAMPLE 13.5   A somewhat controversial yet not uncommon use of a hypothesis test is to apply the test to an entire population in order to quantify the magnitudes of observed differences in distributions. An example would be a test for the equality of voter preference distributions (at different voting locations) in the student election described in example 3.3 (p. 79). Since we focus attention on a single election, the students who voted in that election constitute the entire population of interest. However, under the null hypothesis of no difference in distributions for the various voting locations, we can imagine a large population of voters with a particular preference distribution from which we randomly select 1607 to vote at the Geology Building, 1567 to vote at the president's garage, and so forth. The overall magnitude of the differences in distributions can then be quantified by the P-value for a chi-square homogeneity test, while the more interesting individual differences can be pinpointed from the symbol table. The result of the analysis is as follows.

PARTY		GEOL. B	PRES. G	VOTING LOCATION ST. UNION	LAPOSADA	ENG. CTR.	ROW SUMS
USDA	Obs.	640	776	1628	537	436	4017
	Exp.	753.9	735.1	1663.9	538.5	325.6	
	z	−4.1	1.5	−0.9	−0.1	6.1	
	Sym.	@	.	.	.	@	
HowWhy	Obs.	78	47	170	51	23	369
	Exp.	69.2	67.5	152.8	49.5	29.9	
	z	1.1	−2.5	1.4	0.2	−1.3	
	Sym.	.	O	.	.	.	
St. Vce.	Obs.	390	348	776	232	107	1853
	Exp.	347.7	339.1	767.6	248.4	150.2	
	z	2.3	0.5	0.3	−1.0	−3.5	
	Sym.	O	.	.	.	@	
Progr.	Obs.	112	60	278	131	26	607
	Exp.	113.9	111.1	251.4	81.4	49.2	
	z	−0.2	−4.8	1.7	5.5	−3.3	
	Sym.	.	@	o	@	@	
Unaff.	Obs.	387	336	695	197	102	1717
	Exp.	322.2	314.2	711.2	230.2	139.2	
	z	3.6	1.2	−0.6	−2.2	−3.1	
	Sym.	@	.	.	O	@	
Column Sums		1607	1567	3547	1148	694	8563

$X^2 = 184.721$

With the degree of freedom equal to $(5 − 1) \cdot (5 − 1) = 16$, the P-value corresponding to $X^2 = 184.72$ is smaller than .001. The observed differences in distributions are highly significant. Moreover, the specific differences seen in the bar graph of display 3.7 (p. 81) can now be substantiated.

For example, it is seen that the major differences in voter preference from the "norm" occur at the Engineering Center. Engineers appear to favor USDA more than would be expected and the other parties less than expected relative to the hypothesis of equal distributions.

A feature of the chi-square test that is worth noting is that large contributions to X^2 can occur for cells lying in rows or columns with relatively small marginal sums. For example, several significant differences are seen for the Progressive Party in the last example, although that party received only 7% of the total vote. (See display 3.7.) This demonstrates that interesting differences shown by the test may not indicate interesting *real* differences in distributions. Although this phenomenon is not immediately apparent from the chi-square table, a comparison bar graph reveals it. For this reason, it is useful to look at both the two-way table analysis and a comparison bar graph when interpreting the results of a homogeneity or contingency analysis.

EXAMPLE 13.6

In this example, we demonstrate the method of combining categories to compensate for inadequate sample size—and the importance of doing so.

In a study of student's perceptions of their note-taking abilities, the Dean of the College of Arts and Sciences at the University of New Mexico asked a number of freshmen and sophomores the question, "How well do you take class notes?" The responses are given in the following table.

	CLASS	
RESPONSE	FRESH.	SOPH.
Extremely well	11	2
Well	56	15
Moderately well	76	6
Poorly	7	0
Not at all	0	0
No response	0	1

We see, after the fact, that too many response categories were included for this question. The "Not at all" category was chosen by no one. Since the row total will be 0, the computation of the expected frequencies for this row will require a division by zero. Since this is not allowed, the chi-square statistic cannot be calculated if this category remains. This category contains no useful information and we can remove the entire row from the table. The resulting table with all of the derived values is then as follows.

RESPONSE		CLASS		ROW SUMS
		FRESH.	SOPH.	
Extremely	Obs.	11	2	13
well	Exp.	11.21	1.79	
	z	−.062	.155	
	Sym.	.	.	
Well	Obs.	56	15	71
	Exp.	61.21	9.79	
	z	−.666	1.664	
	Sym.	.	o	

(table continued on next page)

| | | CLASS | | |
RESPONSE		FRESH.	SOPH.	ROW SUMS
Relatively	Obs.	76	6	82
well	Exp.	70.69	11.31	
	z	.632	−1.579	
	Sym.	.	.	
Poorly	Obs.	7	0	7
	Exp.	6.03	.97	
	z	.393	−.983	
	Sym.	.	.	
No response	Obs.	0	1	1
	Exp.	.86	.14	
	z	−.928	2.321	
	Sym.	.	O	
Column Sums		150	24	174

$X^2 = 13.501$

The *P*-value computed from this table is smaller than .01, and it appears that interesting differences exist in the "no response" category. However, note that the expected frequencies for the cells in that row are smaller than 1, as is the expected frequency in the fourth row and second column. The rule of thumb for the adequacy of the chi-square approximation is that all expected frequencies must be greater than 1. There is a good possibility that the results are not due to a difference in the response distributions but rather to the fact that the chi-square distribution does not provide an accurate *P*-value for this table.

To arrive at a table for which the rule of thumb is satisfied, the response categories "No response" and "Poorly" can be combined. The resulting table is as follows.

| | | CLASS | | |
RESPONSE		FRESH.	SOPH.	ROW SUMS
Extremely	Obs.	11	2	13
well	Exp.	11.21	1.79	
	z	−.062	.155	
	Sym.	.	.	
Well	Obs.	56	15	71
	Exp.	61.21	9.79	
	z	−.666	1.664	
	Sym.	.	o	
Moderately	Obs.	76	6	82
well	Exp.	70.69	11.31	
	z	.632	−1.579	
	Sym.	.	.	
Poorly and	Obs.	7	1	8
No response	Exp.	6.90	1.10	
	z	.039	−.098	
	Sym.	.	.	
Column Sums		150	24	174

$X^2 = 6.143$

The *P*-value for this table exceeds .100, and we would now conclude at significance levels 10% and smaller that there is no difference in the note-taking distributions for freshmen and sophomores.

It might be argued that we lost the important difference when the two categories were combined. This *is* always a risk when categories are combined. On the other hand, if the chi-square approximation is to be used, and there is no other real choice, the risk must be taken. A careful choice of the categories to combine can, at least, minimize this risk.

EXERCISES

13.5   In exercise 4.15 (p. 113), data were presented on the relationship between the incidence of birth defects among the Pima Indians and the diabetic status of the mother. The two-way table of observed frequencies is as follows.

	CHILD'S BIRTH DEFECT STATUS	
MOTHER'S DIABETIC STATUS	ONE OR MORE DEFECTS	NO DEFECTS
Nondiabetic	31	754
Prediabetic	13	362
Diabetic	9	38

(a)   Test the hypothesis of no association between number of birth defects and mother's diabetic status at the 5% significance level. What are bounds for the *P*-value?

(b)   If the hypothesis of part (a) is rejected, find the interesting cells by the symbol method and interpret what you find.

13.6   Are the *sex* and *class* variables independent for elementary statistics students? This question was addressed by exploratory methods in chapter 4, using the data of display 1.1. The observed frequencies from display 4.1 are duplicated below.

	CLASS				
SEX	FRESHMAN	SOPHOMORE	JUNIOR	SENIOR	GRADUATE
Male	3	6	4	0	1
Female	8	9	3	1	1

(a)   Find the expected frequencies for this table. What data grouping would you recommend in order to satisfy the rule of thumb for the adequacy of the chi-square approximation?

(b)   Make this grouping and carry out the test of the "no association" hypothesis at the 10% significance level. Give your conclusions.

13.7   A sample of death certificates of individuals over 20 years of age, known to have died by suicide in the Southwest during the period 1978 to 1981, revealed the following age and ethnicity information.

	ETHNICITY		
AGE	ANGLO	HISPANIC	NATIVE AMERICAN
20–29	34	25	11
30–39	26	9	2
40–49	23	6	1
50–59	22	6	0
over 59	21	4	0

(a) Find the bounds for P-value for the hypothesis that there is no difference in the distributions of age at suicide for the three ethnic groups. Use these bounds to test the hypothesis at the 5% significance level.

(b) If the hypothesis of part (a) is rejected, find the interesting cells in the table and interpret what you find.

13.8 In an arthritis clinic, patients are asked to evaluate the degree of activity of their disease on a scale from 1 to 3, with 1 indicating the least activity (severity) and 3 the greatest. At the same time, a clinical rating is made on the same scale by a physician. Data for a number of patients are given in the following table.

PHYSICIAN'S RATING	PATIENTS' RATING		
	1	2	3
1	31	14	4
2	7	5	5
3	2	9	12

(a) Does there appear to be an association between the physician's ratings and the ratings of the patients? Test at the 5% significance level.

(b) If the answer to part (a) is "yes," does it appear that the physician's and patients' rating schemes are in reasonable agreement? Give an explanation to support your answer.

13.9 In an experiment to determine what kind of vegetation is most likely to be associated with archaeological sites in the Southwest, 98 sites were randomly selected from a population of known sites, and 68 "non-sites" (locations known not to contain sites) were selected. The vegetation was classified for each location into one of the categories *juniper*, *salt-desert*, and *other*. The results of the classification are given in the following table.

VEGETATION	SITE TYPE	
	SITE	NON-SITE
Juniper	14	1
Salt-desert	64	52
Other	20	15

(a) Does there appear to be a difference in vegetation distribution for the two site types? What are bounds for the P-value?

(b) If the hypothesis is rejected at the 5% significance level in part (a), describe the nature of the difference in vegetation distributions. Which type of vegetation should an archaeologist look for to increase his or her chance of finding a site?

◤ SECTION 13.4

Tests and Confidence Intervals for Two Population Proportions

A test for the equality of two population proportions is presented in this section. The two-sided version of this test is equivalent to the chi-square test for a 2×2 table. However, our treatment will allow one-sided tests to be made easily as well. Confidence intervals for the difference in two proportions will make it possible to answer the question, "How large is the difference?" when interesting differences are seen in the two-way table analyses of the last section. Again, the Bonferroni inequality will be used to solve the multiplicity problem when more than one interval is calculated for a given data set.

Hypotheses for Two Proportions

In the comparison problem of this subsection, we will consider only two populations and the variable of interest will take on only two values. One of the categories will be singled out for special consideration. The population proportions for that category in the two populations will be denoted by p_1 and p_2. The hypothesis of no difference in distributions for the two populations then reduces to the null hypothesis $H_0: p_1 = p_2$. A test for this hypothesis against a two-sided alternative, based on random samples of sizes n_1 and n_2, respectively, will be constructed in this section.

One-sided hypotheses are also of interest in this context. An example that we will use to illustrate the application of the methods to comparison problems is the following.

EXAMPLE 13.7   This example is a continuation of the pneumatic wrench comparison which was described in exercise 3.27 (p. 91). A new wrench is compared with an old one to determine whether it produces significantly fewer out-of-tolerance measurements. If p_1 represents the proportion of out-of-tolerance measurements for the old wrench and p_2 the corresponding proportion for the new wrench, then this question leads to the hypotheses

$$H_0: \quad p_1 \leq p_2$$

and

$$H_1: \quad p_1 > p_2$$

From exercise 3.27, samples of size $n_1 = n_2 = 96$ are available to test these hypotheses. The test will be given shortly.

The association problem for two two-valued categorical variables will also be covered in this discussion. Now p_1 and p_2 will represent conditional population proportions or conditional probabilities. Symbolically, if X and Y are the two variables with values 1 and 2, then (in the probability context)

$$p_1 = P(X = 1|Y = 1)$$

and

$$p_2 = P(X = 1|Y = 2)$$

In words (for conditional proportions),

p_1 = conditional proportion of category 1 for the first variable among individuals with value 1 of the second variable

and

p_2 = conditional proportion of category 1 for the first variable among individuals with value 2 of the second variable

The hypothesis of no association, or independence, is then also the null hypothesis H_0: $p_1 = p_2$. The sample size will be $n = n_1 + n_2$, where n_1 and n_2 are the marginal frequencies for the two categories of variable 2. An application of the methods to be developed will be illustrated in the following example.

EXAMPLE 13.8

We will look at the association of the *birth defect status* of Pima Indian children (variable 1) with the *diabetic status* of their mothers (variable 2) using the data from the study reported in exercise 4.15 (p. 113). For this example, we will consider only the *nondiabetic* and *diabetic* categories for the diabetic status variable. The birth defect status variable has categories *one or more birth defects* (the category of primary interest) and *no defects*. The data are repeated in the following table.

BIRTH DEFECT STATUS	DIABETIC STATUS OF MOTHER	
	NONDIABETIC	DIABETIC
One or more defects	31	9
No defects	754	38
	785	47

The population parameters are

p_1 = proportion of children with one or more birth defects among children born to nondiabetic mothers

and

p_2 = proportion of children with one or more birth defects among children born to diabetic mothers

The hypothesis that diabetes in the mother increases the rate of birth defects in their children can be formulated as the alternative hypothesis that $p_1 < p_2$. Thus, the hypotheses to test are

$$H_0: \quad p_1 \geq p_2$$

and

$$H_1: \quad p_1 < p_2$$

A sample of $n = 832$ children, $n_1 = 785$ born to nondiabetic mothers and $n_2 = 47$ to diabetic mothers, is available for the test.

The Hypothesis Tests for Two Proportions

The construction of tests for the one- and two-sided hypotheses of the last subsection follows closely the method used to produce the two-sample t test. First, an estimator for the difference in proportions is selected. Then, the standard error of this estimator is found. Finally, the Z-score transform of the estimator will lead to a suitable statistic for testing the hypotheses.

The obvious choice for an estimator of the difference in population proportions is the difference in sample proportions, $\hat{p}_1 - \hat{p}_2$. The standard error of this estimator is

$$\sigma_{\hat{p}_1 - \hat{p}_2} = \sqrt{\frac{p_1(1 - p_1)}{n_1} + \frac{p_2(1 - p_2)}{n_2}}$$

Different estimates of the standard error will be used for hypothesis tests and confidence intervals. For hypothesis tests, a pooled estimate of the common value of p_1 and p_2 (under the assumption that $p_1 = p_2$) is formed:

$$\hat{p}_C = \frac{n_1 \hat{p}_1 + n_2 \hat{p}_2}{n_1 + n_2} = \frac{O_1 + O_2}{n_1 + n_2}$$

where O_1 and O_2 are the observed (absolute) frequencies of the category of interest from the two samples. For example, $O_1 = 31$ and $O_2 = 9$ in the diabetic status example. The estimated standard error of $\hat{p}_1 - \hat{p}_2$ for hypothesis tests is then

$$SE_{\hat{p}_1 - \hat{p}_2} = \sqrt{\hat{p}_C(1 - \hat{p}_C)(1/n_1 + 1/n_2)}$$

The test statistic is

$$Z = \frac{\hat{p}_1 - \hat{p}_2}{SE_{\hat{p}_1 - \hat{p}_2}}$$

When $p_1 = p_2$, this statistic has (approximately) a standard normal distribution. Thus, the cutoff values for the tests can be obtained from table 1B.

**EXAMPLE 13.7
(continued)**

For example, suppose that a 5% test is to be constructed for the one-sided hypotheses of the pneumatic wrench example. The one-sided 5% cutoff value from table 1B is $c = 1.645$. Thus the appropriate test rule would be

Reject H_0 if $Z \geq 1.645$

The calculation of the test statistic proceeds as follows. In exercise 3.27 it was seen that $O_1 = 15$ of the old wrench measurements were out of tolerance, whereas only $O_2 = 2$ of the new wrench measurements were out of tolerance. Recall that $n_1 = n_2 = 96$. Thus,

$$\hat{p}_1 = \tfrac{15}{96} = .1563$$
$$\hat{p}_2 = \tfrac{2}{96} = .0208$$

The pooled estimate of p is

$$\hat{p}_C = \frac{15 + 2}{96 + 96}$$
$$= \tfrac{17}{192} = .0885$$

The estimated standard error is then

$$SE_{\hat{p}_1 - \hat{p}_2} = \sqrt{.0885(1 - .0885)(\tfrac{1}{96} + \tfrac{1}{96})}$$
$$= .0410$$

Finally, the value of the test statistic is

$$Z = \frac{.1563 - .0208}{.0410}$$
$$= 3.303$$

Since this number exceeds the cutoff value 1.645, H_0 would be rejected. We would conclude at the 5% significance level that the proportion of out-of-tolerance measurements for the new wrench is smaller than that for the old one, and the manufacturer would be advised to buy the new wrench.

**EXAMPLE 13.8
(continued)**

As a second example, the rule for a 1% test of the hypothesis of no association between birth defects and mother's diabetic status is

Reject H_0 if $Z \leq -2.326$

The sample proportions are

$$\hat{p}_1 = \tfrac{31}{785} = .0395$$
$$\hat{p}_2 = \tfrac{9}{47} = .1915$$

The pooled estimate of p is

$$\hat{p}_C = \frac{31 + 9}{785 + 47}$$
$$= .0481$$

Consequently, the estimated standard error is

$$SE_{\hat{p}_1 - \hat{p}_2} = \sqrt{.0481(1 - .0481)(\tfrac{1}{785} + \tfrac{1}{47})}$$
$$= .0321$$

Finally, the value of the test statistic is

$$Z = \frac{.0395 - .1915}{.0321}$$
$$= -4.731$$

Since this number is smaller than the cutoff value of -2.326, the null hypothesis is rejected. We would conclude at the 1% significance level that diabetic mothers produce a significantly larger proportion of children with birth defects than do nondiabetic mothers.

How much larger is the proportion of children with birth defects born to diabetic mothers? How much smaller is the proportion of out-of-tolerance measurements produced by the new wrench? These questions can be answered by computing one-sided confidence intervals by the method of the next subsection.

Confidence Intervals for the Difference of Two Proportions

With a different estimator for standard error, the pivotal method can be used to find confidence limits for $p_1 - p_2$. The modification consists of using the unpooled estimates of the population proportions in the expression for the standard error of $\hat{p}_1 - \hat{p}_2$:

$$SE_{\hat{p}_1 - \hat{p}_2} = \sqrt{\frac{\hat{p}_1(1 - \hat{p}_1)}{n_1} + \frac{\hat{p}_2(1 - \hat{p}_2)}{n_2}}$$

If z is the (two-sided) critical value for the given confidence coefficient, then the limits for a two-sided interval are

$$L = \hat{p}_1 - \hat{p}_2 - z SE_{\hat{p}_1 - \hat{p}_2} \quad \text{and} \quad U = \hat{p}_1 - \hat{p}_2 + z SE_{\hat{p}_1 - \hat{p}_2}$$

One-sided confidence limits are of the same form as the corresponding two-sided limits, but with the two-sided critical value replaced by the one-sided value from table 1B.

The appropriate limit to use can be determined from the alternative hypothesis of the "screening" test. The hypothesis $H_1: p_1 > p_2$ is equivalent to $H_1: p_1 - p_2 > 0$, which suggests that a lower limit is desired for $p_1 - p_2$. Similarly, $H_1: p_1 < p_2$ will lead to an upper limit for $p_1 - p_2$.

EXAMPLE 13.7
(continued)

For example, in the pneumatic wrench problem for which the alternative hypothesis is $H_1: p_1 > p_2$, we will find a lower confidence bound for the old wrench minus new wrench difference $p_1 - p_2$. For 95% confidence, the ONE-SIDED TEST column of table 1B would be entered at the $1 - .95 =$

.05 row to find the critical value $z = 1.645$. The unpooled standard error estimate is

$$SE_{\hat{p}_1 - \hat{p}_2} = \sqrt{\frac{.1563(1 - .1563)}{96} + \frac{.0208(1 - .0208)}{96}}$$

$$= .0398$$

Thus, the lower confidence bound is

$$L = .1563 - .0208 - 1.645(.0398) = .0699$$

Rounding this to .07, we would conclude with 95% confidence that the old wrench produces at least 7% more out-of-tolerance measurements than does the new wrench. That is, the manufacturer can expect at least a 7% reduction in the number of improperly torqued screws in the meter assemblies.

**EXAMPLE 13.8
(continued)**

An upper confidence limit is indicated for the difference in birth defect proportions by the alternative hypothesis $H_1: p_1 < p_2$. For 99% confidence, the one-sided critical value is found from table 1B to be $z = 2.326$.

The estimated (unpooled) standard error is

$$SE_{\hat{p}_1 - \hat{p}_2} = \sqrt{\frac{.0395(1 - .0395)}{785} + \frac{.1915(1 - .1915)}{47}}$$

$$= .0578$$

Thus, the upper confidence bound is

$$U = .0395 - .1915 + 2.326(.0578) = -.0175$$

The inequality $p_1 - p_2 \leq -.0175$ is equivalent to $p_2 - p_1 \geq .0175$. Thus, with 99% confidence, we conclude that the proportion of children with at least one birth defect born to diabetic mothers exceeds that for nondiabetic mothers by at least 1.75%.

An example of a two-sided test and confidence interval is given next.

EXAMPLE 13.9

Suppose that in the student election problem treated in example 13.5 (p. 429) we were interested in comparing only the USDA party vote at the Geology Buildings and at the Engineering Center. The Party Preference variable could then be taken to have the two values *USDA* and *Other*, where "Other" pools all non-USDA categories. The data are summarized in the following table.

| | VOTING LOCATION | |
PARTY PREFERENCE	GEOLOGY BUILDING	ENGINEERING CENTER
USDA	640	436
Other	—	—
Sample sizes	1607	694

The question, "Is there a difference in preference for the USDA party at the two locations?" would be addressed by testing the hypotheses

$$H_0: p_1 = p_2$$

versus

$$H_1: p_1 \neq p_2$$

where p_1 is the proportion of the voters at the Geology Building who voted for USDA and p_2 is the proportion voting for USDA at the Engineering Center.

A 5% test will be performed. The two-sided 5% critical value from table 1B is 1.960. Thus, the test rule is

Reject H_0 if $|Z| \geq 1.960$

The sample proportions are

$$\hat{p}_1 = \tfrac{640}{1607} = .3983$$
$$\hat{p}_2 = \tfrac{436}{694} = .6282$$

The pooled estimate of p is

$$\hat{p}_C = \frac{640 + 436}{1607 + 694}$$
$$= .4676$$

and the pooled standard error for the test statistic is

$$SE_{\hat{p}_1 - \hat{p}_2} = \sqrt{.4676(1 - .4676)(\tfrac{1}{1607} + \tfrac{1}{694})}$$
$$= .02266$$

The value of the test statistic is then

$$Z = \frac{.3983 - .6282}{.02266}$$
$$= -10.148$$

Since the absolute value of the test statistic exceeds the cutoff 1.960, the null hypothesis is rejected. We conclude at the 5% level that there is a difference in voting proportions for USDA at these two locations. The sample proportions indicate a heavier vote for USDA at the Engineering Center.

Note that it would be more impressive in this problem to quote the P-value bound $P < .000001$.

A 95% confidence interval for $p_1 - p_2$ will have limits

$$\hat{p}_1 - \hat{p}_2 \pm 1.960 SE_{\hat{p}_1 - \hat{p}_2}$$

where the estimated standard error is now the unpooled version

$$SE_{\hat{p}_1 - \hat{p}_2} = \sqrt{\frac{.3983(1 - .3983)}{1607} + \frac{.6282(1 - .6282)}{694}}$$
$$= .02204$$

Check that these limits are

$$L = -.2732 \quad \text{and} \quad U = -.1868$$

We conclude from the negative range that p_2 exceeds p_1, and the amount of the difference is between 18.68% and 27.32%. That is, with 95% confidence, the proportion voting for USDA at the Engineering Center exceeds the USDA vote proportion at the Geology Building by an amount between 18.7% and 27.3%.

EXERCISES

13.10   In a study of the need for a community college in a Southwestern city, a random sample of potential students yielded 1257 who had at most a high school education and 754 who had at least one year of college. Of those with at most a high school education, 584 stated that their need for further education was low, while the number of people with some college indicating a low need for further education was 453.

   (a)   Is there a difference between the proportions of low-need individuals in the population for the two education levels? Test at the 5% significance level and find bounds for the P-value.

   (b)   Find and interpret a 95% confidence interval for the difference in proportions of low-need individuals for the two education levels.

13.11   In the bar exam example of exercise 3.17 (p. 87), of 168 candidates for the bar who took the exam, 30 were members of minority groups. Of the 127 who passed the exam, 9 were members of minority groups.

   (a)   Is it reasonable to believe that the probability of a minority group member passing the bar exam is smaller than the corresponding probability for a non-minority candidate? Test at the 5% significance level. Find bounds for the P-value.

   (b)   If the null hypothesis of part (a) is rejected, find a lower 95% confidence bound for the amount by which the pass probability for a non-minority candidate exceeds that for a minority candidate.

13.12   The management of the Albuquerque National Public Radio (NPR) affiliate wanted to know whether awareness of the existence of an NPR station among adults (18 years of age or older) in this community was greater than the awareness in other communities possessing NPR stations throughout the country. A telephone survey of 139 Albuquerque adults yielded 99 who were aware of the local NPR station. Compare this result with the results of a national telephone survey of 1067 adults of whom 224 were aware of their local NPR stations.

   (a)   Is the awareness proportion significantly greater for Albuquerque? Test at the 1% level and give bounds for the P-value.

   (b)   Find a 99% confidence bound for the amount by which the awareness proportion of Albuquerque exceeds that for the population of the national survey.

Quantifying Interesting Differences in Two-Way Tables

The confidence intervals of the last subsection, adjusted for multiplicity, can be used to answer the question, "How large are the differences?" in a two-way table. If r confidence intervals are to be computed and an overall confidence level of $1 - \alpha$ is desired, each interval will be calculated for confidence coefficient $1 - \alpha/r$.

For example, it was seen in example 13.5 (p. 429), that interesting differences (from the expected frequencies) occurred for the USDA party at the Geology Building and at the Engineering Center and for the Progressive Party at the president's garage, La Posada, and the Engineering Center. The USDA vote at the Geology Building was less than expected, while that at the Engineering Center was greater than expected. This suggests that it would be interesting to see by how much the proportions of individuals voting for USDA at the two locations differ. Similarly, the difference in voting proportions for the Progressive Party at the president's garage and La Posada and the difference in the Unaffiliated voting proportions at the Geology Building and the Engineering Center appear worth quantifying. We will calculate these $r = 3$ confidence intervals with an overall confidence level as close as possible to 95%.

In order to achieve a joint confidence level of $1 - \alpha = .95$ for 3 intervals, each should be calculated at confidence level $1 - .05/3 = .983$. Critical values from table 1B are available for confidence levels .98 (corresponding to a two-sided test significance level of .02) and .99, but not for .983. However, if we use the critical value for .98, the overall confidence level will be at least $1 - 3 \times (.02)$ or .94, not too far from the desired value of 95%. This is the value we will use. The critical value for the confidence intervals is then $z = 2.326$.

The confidence intervals are now calculated exactly as in the last subsection. For the difference in USDA voting proportions, note that $O_1 = 640$ of the $n_1 = 1607$ people voting at the Geology Building voted for USDA, while the number was $O_2 = 436$ of the $n_2 = 694$ people voting at the Engineering Center. Check that the sample proportions are (to three significant digits)

$$\hat{p}_1 = .398 \quad \text{and} \quad \hat{p}_2 = .628$$

The unpooled standard error is

$$\text{SE}_{\hat{p}_1 - \hat{p}_2} = \sqrt{\frac{.398(1 - .398)}{1607} + \frac{.628(1 - .628)}{694}} = .022$$

The confidence interval limits for $p_1 - p_2$ will then be

$$L = .398 - .628 - 2.326(.022) = -.281$$

and

$$U = .398 - .628 + 2.326(.022) = -.179$$

Similarly, check that the confidence limits for the difference in Progressive Party voting proportions at the president's garage (p_1) and La Posada (p_2) are

$$L = -.100 \quad \text{and} \quad U = -.051$$

while the confidence limits for the difference in the Unaffiliated voting proportions at the Geology Building (p_1) and the Engineering Center (p_2) are

$$L = .054 \quad \text{and} \quad U = .134$$

Thus, we can conclude with overall confidence of at least 94% that the USDA vote was between 17.9% and 28.1% larger at the Engineering Center than at the Geology Building, the vote for the Progressive party was between 5.1% and 10% larger at La Posada than at the president's garage, and the Unaffiliated vote was between 5.4% and 13.4% larger at the Geology Building than at the Engineering Center.

Confidence intervals for the individual proportions (see section 13.2) can also be included among the multiple comparisons. Each simply adds 1 to the value of r in figuring the individual confidence coefficient $1 - \alpha/r$. Clearly, trying to compute a large number of intervals adversely affects the sensitivity (interval length) of each, and one soon reaches a point where the individual intervals are no longer useful. For this reason, the fewer intervals calculated, the better each will be.

EXERCISE

13.13 Refer to the two-way table of display 13.2 (p. 428). Calculate and interpret confidence intervals for the difference of proportions of boys and girls who were counseled and released and for those who were sent to court. Choose the individual confidence coefficient for these two intervals so as to guarantee an overall confidence level of at least 95%.

Chapter 13 Quiz

1. What is the form of the hypotheses tested using the chi-square goodness of fit test?

2. How are the expected frequencies calculated for the goodness of fit test? What is the form of the chi-square statistic?

3. What is the chi-square distribution? How is it used in the goodness of fit test?

4. What is the purpose of the z-values and the symbols derived from them? How are they derived?

5. How is the value of the chi-square statistic obtained from the z-values?

6. What is the strategy for interpreting categories based on the symbols and z-value signs?

7. How are the goodness of fit computations organized and carried out in table form? Give an example.

8. When is it reasonable to calculate confidence intervals for the population proportions in a goodness of fit test?

9. How is the Bonferroni inequality used in the calculation of confidence intervals for population proportions for more than one category after a goodness of fit test?

10. What is the rule of thumb for guaranteeing the adequacy of the chi-square approximation used in the text?

11. What is a procedure for modifying a chi-square test to bring it into conformity with the chi-square approximation rule of thumb?

12. What is the statistical problem addressed by the chi-square test of homogeneity? By the test of contingency?

13. What are the two-way table data representations for homogeneity and contingency problems? What do the rows and columns represent for each problem? What do the marginal totals represent?

14. Why is it reasonable to use the same data representation for both homogeneity and contingency problems?

15. How are expected frequencies calculated for homogeneity and contingency problems? What is the form of the chi-square statistic?

16. What are the degrees of freedom for a homogeneity or contingency test?

17. What are the steps for carrying out a test of contingency or homogeneity? Give an example.

18. How can the rule of thumb for the adequacy of the chi-square approximation be "validated" in a test for homogeneity or contingency? Give an example.

19. What is the interpretation of the parameters p_1 and p_2 in a two-valued categorical variable comparison problem? In an association problem for two, two-valued categorical variables?

20. How are tests for one-sided hypotheses involving p_1 and p_2 constructed? Give an example.

21. How are one- and two-sided confidence intervals constructed for $p_1 - p_2$? Give examples.

22. How is the Bonferroni inequality used to guarantee a given overall confidence level for confidence intervals bounding differences in proportions for interesting cells found by a homogeneity or contingency test?

SUPPLEMENTARY EXERCISES

13.14 In exercise 3.7 (p. 82), the distributions of the *showup* variable for the bicycle Tour of the Rio Grande Valley (display 1.17) for those choosing the 50-mile, 100-mile, and other distance options was seen to be

VALUES OF *SHOWUP* VARIABLE	DISTANCE		
	100	50	OTHER
No-show	11	6	5
Show	62	10	6

(a)   Is there a difference in distributions for the three categories? Test at the 5% significance level. Give bounds for the *P*-value.

(b)   Find a 95% confidence interval for the difference in no-show proportions for the 100-mile and "other" categories.

13.15   The following data were reported in the study of the association of SIDS deaths and altitude, exercise 3.20 (p. 88).

SIDS VARIABLE	ALTITUDE		
	LOW	MEDIUM	HIGH
Deaths due to SIDS	27	38	12
Others	6601	10,512	2511
Total births (two-year period)	6628	10,550	2523

Analyze this table for differences in the rates (proportions) of SIDS deaths with altitude. Test for a difference at the 10% significance level and find bounds for the *P*-value.

13.16   In the study of possible discriminatory employment practices discussed in exercise 3.23 (p. 89), the following table of layoff levels by minority group status was given.

	LAYOFFS		
	LOW	MEDIUM	HIGH
Minority	95	13	19
Nonminority	120	8	0

(a)   Test for an association between layoff level and minority status at the 5% significance level and give bounds for the *P*-value.

(b)   Find and interpret a 95% confidence interval for the difference in minority layoff proportions for the low-layoff category and the combined medium- and high-layoff categories. (Pool the observed frequencies for these two categories.)

13.17   A more detailed breakdown of education level in the study of post-secondary education need, presented in exercise 13.10 (p. 442), is as follows.

EDUCATIONAL LEVEL	NEED TO UPGRADE EDUCATION	
	LOW	HIGH
Less than high school	198	192
High school diploma	386	481
Some college	803	924
Bachelor's degree	453	301
Graduate degree	383	126

Data courtesy of G. Mallory, Albuquerque, Urban Observatory, *Survey of Adult Needs for Postsecondary Education Programs in the Albuquerque Metropolitan Area*, 1980.

(a) Test for a difference in perceived need for the various education levels at the 5% significance level and give bounds for the *P*-value.

(b) What is the apparent trend in need proportions with education level? Discuss.

13.18 The accompanying table contains the numbers of individuals from an intermediate statistics class in various majors, by sex. Is there evidence that *major* and *sex* are dependent variables among students taking intermediate statistics? Test at the 5% significance level.

MAJOR	SEX	
	MEN	WOMEN
Life science	9	4
Engineering	8	1
Other	9	3

13.19 The famous geneticist Gregor Mendel was the first to develop a mathematical theory of trait inheritance. His theory was based on studies of plants. In an experiment to verify the theoretically derived ratios of $\frac{9}{16}$, $\frac{3}{16}$, $\frac{3}{16}$, and $\frac{1}{16}$ for round-yellow, wrinkled-yellow, round-green, and wrinkled-green peas, respectively, Mendel obtained 315, 101, 108, and 32 pea plants of these four types. Do the observed data support his theory? (Test at the 5% level.)

13.20 Using the complete class data set of display 1.18 (p. 28), test at the 1% significance level whether people favor particular second digits when recording their heights in inches. How does the outcome of this test compare with that for the last digit of weight given in example 13.2? What reasons can you give for the difference (if any)?

13.21 An article in a recent scientific publication reported the result of an experiment to test the effectiveness of the drug Noxoline in repairing traumatic spinal nerve damage. Twenty-two cats were administered identical nerve lesions; nine were then treated with Noxoline and thirteen with a saline solution. The conditions of the cats after a two-week period are given in the accompanying table. Is there evidence that Noxoline has had any effect on the condition of the cats? (Test at the 5% significance level.) If so, what is the nature of the effect?

CONDITION AFTER TWO WEEKS	FOR CATS TREATED WITH	
	NOXOLINE	SALINE SOLUTION
Dead	2	5
Normal	3	1
Mildly spastic	4	0
Very spastic	0	7

13.22 In the telephone survey to determine perceived educational needs in a southwestern metropolitan area of exercises 13.10 and 13.17, the numbers of individuals indicating high and low needs for basic educational skills (such as mathematics, writing, etc.), by ethnicity, are given in the accompanying table.

	SKILL NEED	
ETHNICITY	LOW	HIGH
Hispanic	515	532
Native American	46	37
Black	35	32
Anglo	2270	601
Other	75	63

Data courtesy of G. Mallory, Albuquerque, Urban Observatory, *Survey of Adult Needs for Postsecondary Education Programs in the Albuquerque Metropolitan Area*, 1980.

(a) Does there appear to be a difference in educational need for the various ethnic groups? Test at the 5% significance level and give bounds for the *P*-value.

(b) If a difference is found in part (a), construct joint 95% confidence intervals for the interesting differences in high-need proportions for the various ethnic groups. Interpret your results.

13.23 An archaeologist hypothesized that the quality of pots found at sites with a high density (HD) of artifacts should be different from that of pots found at low density (LD) sites because of the use of pots in transporting trade goods. To test this hypothesis, potsherds from the two types of sites were rated on a 3-point scale as being of low, medium, or high quality. The result of rating a number of pots from the two types of sites is given in the following table.

	SITE TYPE	
POT RATING	HD	LD
Low	5	8
Medium	12	8
High	7	5

(a) Find the *P*-value for the hypothesis of no difference in rating distributions for the two site types and test the hypothesis at the 10% significance level.

(b) If differences are seen in part (a), describe their nature.

13.24 In the jury pool selection discrimination study described in Section 13.2 (p. 413ff), the jury pool sample was also compared to census data for possible ethnicity biases. The data are given in the following table.

ETHNICITY	JURY POOL FREQUENCIES	COUNTY POPULATION PROPORTIONS
Anglo	961	.570
Hispanic	321	.382
Black	13	.021
Native American	17	.019
Other	16	.008

(a) Does there appear to be ethnic bias in the selection of jury pool members? Find bounds for the *P*-value and test the hypothesis of no discrimination at the 5% significance level.

(b) Find joint 95% confidence intervals for the true jury pool proportions for those ethnic groups significantly under- or over-represented.

13.25 In the study of a genetic basis for alcoholism and the association between alcoholism and cirrhosis of the liver, discussed in exercise 3.22 (p. 88), data on the presence of the genetic marker B12 from random samples of nonalcoholic Anglo patients (Controls) at a Veteran's Administration hospital and alcoholic patients with and without cirrhosis were as follows.

	ANGLO		
POSSESS B12	ALCOHOLIC, NONCIRRHOSIS	ALCOHOLIC, CIRRHOSIS	CONTROL
Yes	3	14	15
No	20	13	24

Data courtesy of R. T. Rada, M.D., Department of Psychiatry, University of New Mexico School of Medicine, Albuquerque.

(a) Is there a difference in B12 distributions for the three groups of patients? Test at the 5% significance level and find bounds for the *P*-value.

(b) If differences are indicated in part (a), describe the nature of the differences. What does the marker B12 appear to be distinguishing in Anglo patients?

13.26 The B12 data for Hispanic patients from the study described in exercise 13.25 are given in the following table.

	HISPANIC		
POSSESS B12	ALCOHOLIC, NONCIRRHOSIS	ALCOHOLIC, CIRRHOSIS	CONTROL
Yes	5	16	3
No	12	29	26

(a) Answer question (a) of exercise 13.25 for Hispanic patients.

(b) Answer question (b) of exercise 13.25 for Hispanic patients.

(c) If you solved exercise 13.25, does it appear that the marker B12 is distinguishing anything different for Anglos and Hispanics? Explain.

13.27 In a study of drug prescription prices in a large city, a physician was interested in determining whether a pharmacist's choice of filling a prescription with a brand-name drug or a generic drug depended on the number of different brands of that drug available in the pharmacy. The physician had the same prescription filled at a random sample of 39 pharmacies in the city. The following table summarizes his findings.

| | NUMBER OF AVAILABLE BRANDS | | |
TYPE USED	1	2	3 OR MORE
Brand name	6	17	4
Generic	2	2	8

Data courtesy of L. Chilton, M.D., Department of Pediatrics, University of New Mexico School of Medicine.

(a) Are there differences in the type of drug used by pharmacies having different numbers of brands of this drug available? Test at the 5% significance level.

(b) If the answer to part (a) is yes, describe the nature of the difference.

13.28   In the study of the association between the amount of testosterone in women's blood and their "intellectual drive" levels reported in exercise 3.30 (p. 93), the "drive" levels and the testosterone levels can be grouped in order to convert the measurement variable problem to a categorical data problem. The result of such a grouping is given in the following table.

| INTELLECTUAL DRIVE LEVEL | TESTOSTERONE LEVEL | | |
	LOW	MEDIUM	HIGH
Low	13	13	6
High	5	4	11

(a) Is there an association between testosterone and intellectual drive? Test at the 5% significance level and find bounds on the P-value.

(b) If the answer to part (a) is yes, describe the nature of the association.

13.29   A sample of bicycle riders from the data for the Tour of the Rio Grande Valley given in display 1.17 (p. 25) was taken to test whether there is an association between the age of the rider and the average speed with which he or she made the tour. The intent was to determine whether bicyclists slow down as they get older (that is, after they reach the age of 30). However, data for younger riders (less than 18 years of age) are also included to determine whether they differ in speed from older riders.

| AGE | SPEED (MPH) | | |
	<11	11–15	>15
≤17	14	9	0
18–29	11	11	5
≥30	10	9	7

(a) Is there any trend evident in speed with advancing age? Test at the 10% significance level and find bounds on the P-value.

(b) If the answer to part (a) is yes, describe the nature of the trend.

13.30   The University of New Mexico administration was concerned that the faculty was more lenient in grading students in elementary courses than other colleges and universities nationally. A national survey indicated that 23% of all elementary students receive A's and B's, 33% receive C's, and 44% receive D's and F's. A random sample of 175 UNM students taking elementary courses showed 65 receiving A's and B's, 47 C's, and 63 D's and F's.

(a) Is there evidence that UNM professors are deviating from national grading standards? Test the null hypothesis at the 5% significance level and find bounds for the P-value.

(b) If deviations are detected by the test of part (a), describe the nature of the differences. Is there evidence of grade inflation by the UNM faculty?

13.31 The following data were obtained in the study of the use of blood grouping to determine children at increased risk of dying of SIDS, discussed in exercise 3.24 (p. 89).

BLOOD GROUP	SIDS	CONTROLS
O	69	55
A	40	71
B	17	7
AB	9	2

(a) Test at the 5% significance level whether there is a difference in blood group distributions for SIDS children and controls. Find bounds on the P-value.

(b) If differences in distributions are indicated by part (a), describe the nature of the differences. Would this information be useful in predicting which children are at high risk of dying of SIDS? Why?

CHAPTER 14

Simple Linear Regression: Descriptive Methods

Introduction

This is the first of three chapters to consider the important topic of regression. In this and the next chapter we will discuss *simple linear regression*. The term "simple" means that a single independent variable, X, will appear in the regression model along with the dependent variable Y. Review chapter 4 for the meanings of independent and dependent variables in this context. *Multiple (linear) regression*, in which more than one independent variable is considered in the regression, will be the topic of chapter 16.

The term "linear" in both simple and multiple *linear* regression refers to the fact that the regression models are members of the important family of *linear models* that will occupy us for the rest of the book. Linear models will be introduced in section 14.3.

In chapter 4 we described the association between two measurement variables X and Y in terms of the flow of the conditional distributions of Y given X for changing values of X. An important characterization of this flow is the curve swept out by a location measure of the conditional distributions, called a *regression curve*. When the location measure is the (population) mean, the curve, $\mu(x)$, is the classical regression curve we will study in this and the next chapter.

The conditional distribution of Y for a given value x of X can be described in terms of random deviations from the regression curve. Thus, if in a sample we observe the value y of Y at x, we can write

$$y = \mu(x) + e \tag{1}$$

where the *error e* represents the amount by which y differs from the regression curve at x. The job of a regression analysis will then be to "extract" or estimate the regression function $\mu(x)$ from data assumed to arise according to the model (1) when the errors are random.

To compound the terminological confusion, in this chapter we will be interested in what is often referred to as the *linear regression model*, in which the regression function $\mu(x)$ is a linear function of x. The abbreviated terminology is an attempt to avoid such unwieldy, if more accurate, descriptions as "linear simple linear regression." The linear regression model and some justifications for its use begin our discussion in the next section.

◢ SECTION 14.2

The Linear Regression Model

linear regression model

Let $(x_1, y_1), (x_2, y_2), \ldots, (x_n, y_n)$ represent the paired values of X and Y for a sample of size n. The **linear regression model** relating these x and y values in terms of the representation (1) is

$$y_i = \beta_0 + \beta_1 x_i + e_i \tag{2}$$

where the random errors e_i are assumed to be independent, normally distributed variables with means and standard deviations given by

$$\mu_{e_i} = 0 \quad \text{and} \quad \sigma_{e_i} = \sigma$$

That is, the e_i's are all assumed to have the same normal distribution, with mean 0 and unknown standard deviation σ. The regression function is

$$\mu(x) = \beta_0 + \beta_1 x$$

linear function

where β_0 and β_1 are unknown parameters to be estimated from the sample. This expression is known as a **linear function** because it is a sum of terms and because the highest power of x occurring in the expression is the first power. We have seen linear functions before in the discussion of linear transformations in chapter 5, for example.

Regression Coefficients

regression coefficients
constant term

In statistical terms β_0 and β_1 are called **regression coefficients**, with β_0 representing the **constant term** in the regression. Graphically, a linear regression curve is a straight line, with its slope and location in space depending on the values of β_0 and β_1, as illustrated in display 14.1. In graphical representations the y coordinate of a curve corresponding to a given x coordinate is

Display 14.1

A linear regression function with intercept β_0 and slope β_1

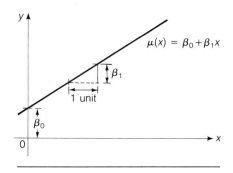

commonly labeled with the symbol *y*—leading, for example, to the expression

$$y = \beta_0 + \beta_1 x$$

for the linear regression curve. This representation has been carried over to statistical notation in many texts and articles. Because we have already adopted the symbol *y* to denote a value of the dependent variable Y, we will continue to use the more cumbersome but more explicit notation $\mu(x)$ to denote the value of the regression function, and thus the *y* coordinate of the curve, at $X = x$.

intercept

slope

The value of β_0, called the **intercept** of the line, is the *y* coordinate of the point at which the line passes through the *y* axis. That is, it is the value of $\mu(x)$ for $x = 0$. The more important parameter, β_1, is the **slope** of the line. In statistical terms it is *the average rate of increase (or decrease) of* X *with* X *in the following sense. At any value x of X, $\mu(x)$ is the average value of Y (i.e., the average of the conditional Y distribution given $X = x$). Now it is easy to see that if x is allowed to increase by one unit, $\mu(x)$ will increase by β_1 units. Thus β_1 is the increase in the average value of Y per unit increase in x. (Note that if β_1 is negative, this increase is actually a *decrease* of $|\beta_1|$ units.) Since a straight line has the same slope at every point, this single parameter represents the average increase in Y per unit increase in X over the entire range of X values if the regression is linear.

The Linear Component of a Regression Curve

From the discussion in chapter 4, it should be apparent that regression curves that are exactly linear are the exception rather than the rule. On the other hand, the most commonly used regression methodology is that of linear regression. How can one reconcile these two situations?

Unfortunately, perhaps the most commanding justification for the use of linear regression methodology is that it is the simplest such methodology. Since, as we will see, neither the computations nor the concepts of linear regression are entirely trivial, it takes a certain amount of courage to proceed to more complicated regression models.

general trend

A less often used but more reasonable justification is that the relevant feature of a regression curve for addressing questions often posed in a statistical study is a **general trend** representing the average increase or decrease of the curve over the (entire) observed range of the independent variable. For example, to determine whether the regressions of cholesterol levels with age are different for men and women, it is sufficient to demonstrate that the average rates of increase (over all ages) are different.

This trend can be viewed as the *linear component* of the more general regression curve about which the curvilinear features vary. See display 14.2 for an illustration. Then *the slope parameter of the linear component measures the average rate of increase of the regression curve over the entire (observed) range of the independent variable.* Since the slope parameter is estimated in a linear regression analysis of the data, this analysis will provide an important feature of the nonlinear regression curve.

Display 14.2

Linear trend component of general regression curve; first curve is sum, point by point, of second two curves

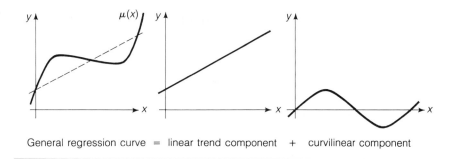

General regression curve = linear trend component + curvilinear component

Unfortunately, statistical inference based on linear regression methods for this general trend component will have good sensitivity only if the variation of the curvilinear component is not too large relative to that of the linear component—that is, unless the curve is almost linear. The linearity can usually be checked with sufficient accuracy by a scatter diagram, although an exploratory regression analysis of the type carried out in chapter 4 is usually more informative and often needed. A statistic for assessing the degree of linearity of a data set will be given in section 14.5. However, this statistic will not distinguish between deviations from linearity due to a curvilinear component and variations due to other sources such as random measurement errors. Consequently, it alone cannot be used as a diagnostic tool for the presence of a curvilinear regression component. A better diagnostic method, based on smoothing residuals, will be outlined in section 14.6. A more sophisticated type of inference, needed when the regression curve has a substantial curvilinear component, will be covered in chapter 16.

◪ SECTION 14.3

Linear Models and Least Squares

The linear regression model

$$y_i = \beta_0 + \beta_1 x_i + e_i$$

linear models

is a member of the family of **linear models**, so named because the parameters β_0 and β_1 in the regression term appear singly and to the first power. Here the term *linear* applies to the way the parameters appear in the model, not to the form of dependence on the independent variable. Thus, for example, the model

$$y_i = \beta_0 + \beta_1 x_i + \beta_2 x_i^2 + e_i$$

would be a linear model (when the assumptions about the e_i's are as given earlier) but would represent a quadratic (polynomial) regression model, since the highest power of x_i is two. Such models are considered in chapter 16.

On the other hand, the model

$$y_i = \beta_0\beta_1 + \beta_2^2 x_i + e_i$$

is not a linear model because β_0 and β_1 are multiplied together and β_2 is squared. It is, however, linear in the independent variable.

A rich and useful theory of parameter estimation and statistical inference is available for linear models. Without explicitly saying so, we have used the results of this theory in chapters 10, 11, and 12. The normal models for the one-, two-, and k-sample problems and for the paired-difference method can be written as linear models. The parameter estimates and F and t tests introduced for these problems are then a result of the general linear model theory. Because the notation is somewhat complicated and distracting, and because the parameter estimates and tests could be justified by appealing to intuition, it was convenient to ignore this fact. In this chapter we will compensate for previous neglect by outlining some of the details of the linear model theory in the context of the linear regression model. Additional details will be given in chapter 16.

Two Linear Models

In chapter 11 the use of sums of squares was introduced to measure the variability of observations from a mean. Sums of squares play an integral role in the statistical theory of linear models. They are used as distance or goodness of fit measures to assess how well given linear models fit the ob-

method of least squares   served data. The **method of least squares** for estimating parameters is based on this usage.

To illustrate these ideas, we will consider the following two linear models:

Model 1. $y_i = \beta_0 + e_i$, $i = 1, 2, \ldots, n$

where the e_i's are independently and normally distributed, with mean 0 and standard deviation σ.

Model 2. $y_i = \beta_0 + \beta_1 x_i + e_i$, $i = 1, 2, \ldots, n$

with the same assumption for the e_i's as for model 1.

Model 1 will play the role of the statistical model specified by a null hypothesis H_0. Model 2 is a global or *full* model assumed to contain the true distribution of the data. H_0 will be tested (in chapter 15) by comparing the fits of the two models to the given data.

Analysis of Model 1

How well does model 1 fit the observations? To answer this question, we will look at the *sum of squares of the errors* $e_i = y_i - \beta_0$, which also represent the deviations of the observations from the regression term of the model:

$$\sum_i (y_i - \beta_0)^2$$

least-squares criterion

Since β_0 is an unknown parameter and since this sum is to be a measure of how well the model fits the data, it is reasonable to take as an estimate of β_0 that number that makes the model best fit the observations at hand. Since small values of the sum of squares correspond to a good fit (the value zero would correspond to a perfect fit), *this criterion leads us to seek the value of β_0 for which the sum of squares above is least (or a minimum)*. Stated in this manner, this is the **least-squares criterion** for determining an estimator of β_0.

At this point the problem of obtaining the estimator becomes a purely mathematical problem, which, unfortunately, requires the calculus (or mathematical ideas of comparable complexity) to solve. Let it suffice to say that the solution leads to a simple equation involving β_0 and $y_1, y_2, \ldots, y_n$, which can be solved for the desired estimate $\hat{\beta}_0$. The solution is

$$\hat{\beta}_0 = \bar{y}$$

the sample mean.

least-squares estimator

Thus the sample mean is the **least-squares estimator** of the constant term β_0 in the simplest possible model in which each y_i is represented as a random deviation e_i from this constant. This result is, perhaps, not too surprising, since model 1 is just another way of stating that $y_1, y_2, \ldots, y_n$ are the values of a random sample of n observations from a population in which the variable Y is normally distributed, with mean β_0 and standard deviation σ. Thus the sample mean is the least-squares estimator of the population mean. This conclusion supports the heuristic derivation of the sample mean as a good estimator of this parameter, given in chapter 2.

The remaining parameter of the model, which must be estimated for inference purposes, is the standard deviation σ. The linear model theory dictates the following procedure. Form the sum of squares of the y_i's minus the estimated regression function, here $\hat{\beta}_0 = \bar{y}$:

$$SS_1 = \sum_i (y_i - \bar{y})^2$$

degrees of freedom

Associated with this sum of squares is a **degrees of freedom** ν. For regression models of the type we are considering, ν is equal to the sample size n minus the number of parameters being estimated in the regression term of the model. (For applications of the linear model theory in the analysis of variance, this definition must be modified slightly.) Since the regression term in model 1 contains only one parameter, β_0, the degrees of freedom are

$$\nu_1 = n - 1$$

The estimator of σ is then the square root of the mean square obtained by dividing SS_1 by ν_1:

$$s = \sqrt{\frac{\sum(y_i - \bar{y})^2}{n - 1}}$$

We recognize this expression as the sample standard deviation, which is the estimator of the population standard deviation introduced in chapter 2.

The statistical inference theory for linear models that is associated with model 1 is now precisely the inference theory for the one-sample problem presented in chapter 10. Hypothesis tests and confidence intervals for the population mean β_0 are based on the t statistic

$$t = \frac{\hat{\beta}_0 - \beta_0}{SE_{\hat{\beta}_0}} = \frac{\bar{y} - \beta_0}{s/\sqrt{n}}$$

and Student's t distribution, as before.

For model 2 both the least-squares estimation process and the statistical inference are more varied and complex. However, the linear model theory makes it possible to follow through these same steps to arrive at the appropriate results. This analysis is carried out next.

■ SECTION 14.4

Analysis of Model 2: Fitting the Regression Line by Least Squares

The sum of squares to be minimized to obtain the least-squares estimates of β_0 and β_1 is

$$\sum_i (y_i - \beta_0 - \beta_1 x_i)^2$$

where $(x_1, y_1), \ldots, (x_n, y_n)$ are the paired values of the independent and dependent variables. The mathematical treatment of this minimization problem leads to two equations to be solved for the estimators $\hat{\beta}_0$ and $\hat{\beta}_1$. The solution of these equations yields

$$\hat{\beta}_1 = \frac{S_{xy}}{S_{xx}}$$

and

$$\hat{\beta}_0 = \bar{y} - \hat{\beta}_1 \bar{x}$$

where

$$S_{xy} = \sum_i (x_i - \bar{x})(y_i - \bar{y})$$

and

$$S_{xx} = \sum_i (x_i - \bar{x})^2$$

We will go into the details of computing these quantities using more convenient computing formulas shortly. However, note that S_{xx} is the familiar sum of squares from the expression for the standard deviation of the x's. The estimator of σ for this model is based on the sum of squares

$$SS_2 = \sum (y_i - \hat{\beta}_0 - \hat{\beta}_1 x_i)^2$$

which again involves the differences of the y_i's and the estimated regression function

$$\hat{\mu}(x) = \hat{\beta}_0 + \hat{\beta}_1 x$$

The degrees of freedom are now

$$\nu_2 = n - 2$$

since two parameters are estimated in the regression term of the model. The estimator of σ is then

$$s = \sqrt{\frac{SS_2}{\nu_2}} = \sqrt{\frac{\Sigma (y_i - \hat{\beta}_0 - \hat{\beta}_1 x_i)^2}{n - 2}}$$

Again, a much simpler computing expression will be used for this quantity.

The computational strategy for obtaining $\hat{\beta}_0$, $\hat{\beta}_1$, and s best suited to hand calculation will be given next.

Computing Parameter Estimates for the Least-Squares Regression Line

The computations will be illustrated for a data set of developmental scores obtained for a group of children of preschool and elementary school ages. This set is a subsample of the data set used in example 4.2 (p. 104), except that here the developmental score, or D score as we will call it, is an average of scores for many characteristics measuring a child's development, such as verbal ability, graphical perception, and general communication ability. The D score will be the dependent variable (Y), and we will study the association of this variable with age, which is taken as the independent variable (X) in the linear regression model, model 2. The data are given in display 14.3.

The strategy for computing the parameter estimates of the model breaks the computation into three steps. In step 1 the basic summary statis-

Display 14.3

Ages (years) and developmental scores (D scores) for 12 children

CHILD I.D.	AGE (X)	D SCORE (Y)
1	3.33	8.61
2	3.25	9.40
3	3.92	9.86
4	3.50	9.91
5	4.33	10.53
6	4.92	10.61
7	6.08	10.59
8	7.42	13.28
9	8.33	12.76
10	8.00	13.44
11	9.25	14.27
12	10.75	14.13

Data courtesy of M. Kartas, M.S., Department of Communicative Disorders, University of New Mexico, Albuquerque.

tics, n, Σx_i, Σy_i, Σx_i^2, Σy_i^2, $\Sigma x_i y_i$, are computed. In step 2 the auxiliary statistics $\bar{x}$, $\bar{y}$, S_{xx}, S_{xy}, and S_{yy} will be calculated. Finally, in step 3 the parameter estimates $\hat{\beta}_0$, $\hat{\beta}_1$, and s will be produced from the results of step 2. The specific computing formulas for steps 2 and 3 are as follows:

Computing formulas for step 2

$$\bar{x} = \frac{\Sigma x_i}{n} \qquad \bar{y} = \frac{\Sigma y_i}{n}$$

$$S_{xx} = \Sigma x_i^2 - \frac{(\Sigma x_i)^2}{n}$$

$$S_{xy} = \Sigma x_i y_i - \frac{\Sigma x_i \cdot \Sigma y_i}{n}$$

$$S_{yy} = \Sigma y_i^2 - \frac{(\Sigma y_i)^2}{n}$$

Computing formulas for step 3

$$\hat{\beta}_1 = \frac{S_{xy}}{S_{xx}}$$

$$\hat{\beta}_0 = \bar{y} - \hat{\beta}_1 \bar{x}$$

$$s = \sqrt{\frac{S_{yy} - \hat{\beta}_1 S_{xy}}{n - 2}}$$

We now apply these steps to the data of display 14.3.

Step 1.   The sample size is

$$n = 12$$

and the sum and sum of squares for the age variable are

$$\Sigma x_i = 73.0800 \qquad \Sigma x_i^2 = 518.7598$$

while for the D scores they are

$$\Sigma y_i = 137.3900 \qquad \Sigma y_i^2 = 1616.6203$$

The sum of cross-products is

$$\Sigma x_i y_i = 3.25 \times 9.40 + 3.92 \times 9.86 + \cdots$$
$$+ 10.75 \times 14.13$$
$$= 890.9842$$

Step 2.

$$\bar{x} = \frac{73.0800}{12} = 6.090$$

$$\bar{y} = \frac{137.39}{12} = 11.449 \qquad \text{(to three decimal places)}$$

$$S_{xx} = \Sigma x_i^2 - \frac{(\Sigma x_i)^2}{n}$$

$$= 518.7598 - \frac{(73.0800)^2}{12}$$

$$= 73.703$$

$$S_{yy} = 1616.6203 - \frac{(137.3900)^2}{12}$$

$$= 43.619$$

and

$$S_{xy} = 890.9842 - \frac{(73.0800 \times 137.3900)}{12}$$

$$= 54.279$$

Step 3.   The values of $\hat{\beta}_1$, $\hat{\beta}_0$, and s are now calculated as follows:

$$\hat{\beta}_1 = \frac{S_{xy}}{S_{xx}} = \frac{54.279}{73.703} = .7365$$

$$\hat{\beta}_0 = \bar{y} - \hat{\beta}_1 \bar{x} = 11.449 - .7365 \times 6.090 = 6.964$$

The estimate of σ is

$$s = \sqrt{\frac{S_{yy} - \hat{\beta}_1 S_{xy}}{n-2}}$$

$$= \sqrt{\frac{43.619 - (.7365 \times 54.279)}{10}} = .6035$$

The estimated regression function is then

$$\hat{\mu}(x) = \hat{\beta}_0 + \hat{\beta}_1 x = 6.964 + .7365x$$

Graphing the Regression Line

Graphically, this equation corresponds to the straight line $y = 6.964 + .7365x$. Thus, for example, if $x = 3.0$, the y coordinate is

$$y = 6.964 + .7365 \times 3.0 = 9.1735$$

If $x = 10.0$,

$$y = 6.964 + .7365 \times 10.0 = 14.329$$

The easiest way to graph this line is to pick two values of x at the extremes of the range of observed x's and calculate the corresponding y values, as we have just done. Plot the two points with these coordinates on the graph and draw a line between them, using a straightedge. The regression line we are discussing could be drawn through the computed points

$$(3.0, 9.17) \quad \text{and} \quad (10.0, 14.33)$$

for example. This line is shown superimposed on a scatter diagram of the data in display 14.4. (See chapter 4 for a discussion of scatter diagrams).

Display 14.4

Least squares line and scatter diagram of data of display 14.3; estimation for predicted Y value at x = 7 also shown

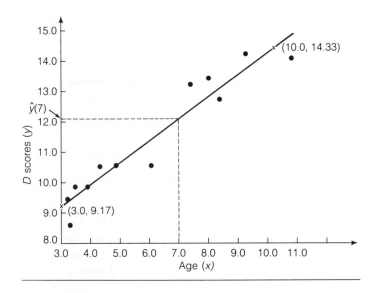

Interpreting the Regression Line and Some Important Inference Problems

Note that the line in display 14.4 appears to fit the data rather well in the sense of passing through the center of the scatter of points. The estimated slope $\hat{\beta}_1$ represents the average rate of increase of D scores with age (or, more properly, the rate of increase of average D scores). We see that for each year of age, the D score increases about .737 units, on the average. However, $\hat{\beta}_1$ is simply an estimate of the true rate β_1, and it would be desirable to have a measure of the error of estimation for this sample. This task is what we have used confidence intervals for in past chapters, and what is required here is a confidence interval for β_1. This interval will be presented in section 15.2.

If we assume the correctness of the linear model $\mu(x) = \beta_0 + \beta_1 x$ for the regression function, the value of $\mu(x)$ for a particular x could be estimated by the least-squares estimator $\hat{\mu}(x)$ given above. Thus, for example, we would estimate the average D score for children of age $x = 7.0$ years to be

$$\hat{\mu}(7) = 6.964 + .7365 \times 7 = 12.120$$

(See display 14.4.) A confidence interval for assessing how different $\hat{\mu}(7)$ and the true mean $\mu(7)$ will be provided in section 15.3.

Often we will want confidence intervals for more than a single x value. It may be desirable, for example, to obtain confidence intervals for average D scores for children of several ages. Moreover, we will want to guarantee a prescribed overall confidence for all these intervals simultaneously. This situation presents us with a multiplicity problem that will be neatly solved by providing a method for constructing confidence intervals for $\mu(x)$ that will have guaranteed confidence for *all* possible values of x simultaneously.

Another important regression problem is that of predicting the Y value for a future observation with a given X value. For example, we might want to predict the D score for a child of age $x = 7$ years. Intuitively, since it is assumed that the Y values vary randomly and unpredictably around the mean $\mu(x)$, the best prediction would be the mean itself. However, we must estimate the mean from the data, and this estimated mean will be used as the predicted value. That is, if $\hat{y}(x)$ denotes the predicted value of Y at $X = x$, then

$$\hat{y}(x) = \hat{\beta}_0 + \hat{\beta}_1 x$$

Thus for a child of age 7 years, we would predict the D score of

$$\hat{y}(7) = 12.120$$

Now it would be helpful to be able to put bounds on the possible error of this prediction. That is, we would like to know how far from the actual (future) observed value y the predicted value $\hat{y}(x)$ can be. It might be thought that since the prediction is the same as the estimated mean, the errors of estimation should be the same for both. However, this is not the case. To the uncertainty of estimating $\mu(x)$ by $\hat{\mu}(x)$ must be added the uncertainty due to the random variation of Y around $\mu(x)$. This variation leads to a longer "confidence interval" for y than for $\mu(x)$. This interval will be given in section 15.4.

EXERCISES

14.1   In this set of exercises we will see how to predict college performance from high school grade-point averages. It is reasonable to assume that overall grade-point averages (GPAs) of college students should be (roughly) proportional to their high school GPAs. This idea suggests that linear regression could be used to predict students' college performances from their high school GPA scores. Toward this end, the records of six students were selected at random from a colleges' files and the following scores were found. (In both high school and college the GPA score is based on the letter grade equivalent: A = 4.0, B = 3.0, etc.)

STUDENT	HIGH SCHOOL GPA (X)	COLLEGE GPA (Y)
1	2.0	1.7
2	2.6	2.8
3	2.8	2.1
4	3.2	2.3
5	3.3	2.6
6	3.6	3.0

Make a scatter diagram of the data. Does the assumption of linearity appear justified on the basis of this graph?

14.2   For the data of exercise 14.1, obtain the least-squares parameter estimates $\hat{\beta}_0$ and $\hat{\beta}_1$ and the estimate s of σ for model 2. What is the average increase in college GPA per unit increase in high school GPA?

14.3   Superimpose the graph of the least-squares line for the data of exercise 14.1 on the scatter diagram you drew in exercise 14.1. Check visually that your line passes through the center of the scatter of points. (This procedure is a good rough check on the correctness of your calculations.)

14.4   Write down the expression $\hat{y}(x) = \hat{\beta}_0 + \hat{\beta}_1 x$ for predicting college GPAs, using the estimates $\hat{\beta}_0$ and $\hat{\beta}_1$ calculated in exercise 14.2 from the data. What college GPA would be predicted for a student who had a high school GPA of 3.0?

◢ SECTION 14.5

A Descriptive Measure of Linearity:
The Coefficient of Determination

The sums of squares SS_1 and SS_2, defined on pages 457 and 458, are the sums of the squared deviations of the observed values of Y from the least-squares-fitted regression functions for model 1 and model 2, respectively. As such, they are useful measures of the goodness of fit of these models to the data. Thus, for example, if model 1 provides a good fit to the data, SS_1 would be small. On the other hand, a small value of SS_2 would represent a good fit for model 2.

Model 1 completely ignores any possible relationship between the variables X and Y by leaving the values of X out of the model entirely. Consequently, SS_1 measures the basic, total variability of Y, taking nothing else into account. On the other hand, SS_2 represents the variation of Y assuming a linear regression of Y on X. That is, it measures the variability of Y around a straight line of the form hypothesized by model 2.

Model 2 will be an improvement over model 1 in explaining the variation of Y only if SS_2 is substantially smaller than SS_1. In fact, the additional amount of variation explained by the linear regression model can be measured by the difference

$$SS_1 - SS_2$$

For the present application it is desirable to normalize this measure by dividing it by the total variation SS_1. The resulting ratio,

$$\frac{SS_1 - SS_2}{SS_1}$$

coefficient of determination   called the **coefficient of determination**, then represents that proportion by which the total variation of Y can be reduced if model 2 is used instead of model 1. In the usual terminology we say that it is *the proportion of the total variation of Y explained by the linear regression of Y on X.*

The coefficient of determination is often represented by r^2. The reason this notation is used is that it can be shown algebraically that

$$\frac{SS_1 - SS_2}{SS_1} = r^2$$

(Pearson product-moment) correlation coefficient

where r is the classical (**Pearson product-moment**) **correlation coefficient**. The correlation coefficient is defined by the expression

$$r = \frac{S_{xy}}{\sqrt{S_{xx}S_{yy}}}$$

This quantity is a commonly used sample measure of the association between two measurement variables X and Y. Because of its dependence on familiar summary statistics, the coefficient of determination r^2 (often called the model r-square) is easily calculated after step 2 of the regression line computations.

The computing formula is

$$r^2 = \frac{S_{xy}^2}{S_{xx}S_{yy}}$$

This calculation will be included hereafter as a regular addendum to step 3.

Interpretation of r^2

An important property of the correlation coefficient is that its values are restricted to the range from -1 to 1. It follows that the coefficient of determination satisfies the inequalities

$$0 \le r^2 \le 1$$

The value $r^2 = 1$ is achieved when $SS_2 = 0$, which can happen only if all of the points (x_i, y_i) lie exactly on a line. The computation of the last section would then show this line to be the least-squares regression line. Thus $r^2 = 1$ corresponds to perfect or complete linearity of the data. For values of r^2 near 1, the points of a scatter diagram would necessarily lie near a straight line.

On the other hand, a value of r^2 near 0 indicates that SS_1 and SS_2 are nearly equal. Consequently, the scatter of points about the regression line is nearly the same as the scatter of points about $\bar{y}$, indicating a weak linear relationship between X and Y. In this case the least-squares regression line is nearly horizontal—that is, the slope $\hat{\beta}_1$ of the line is nearly 0. The extreme value $r^2 = 0$ corresponds to a complete absence of linear regression.

The most useful interpretation of the numerical values of r^2 when it lies strictly between 0 and 1 is the one given above—namely, it is the proportion of the variation of Y attributable to its linear regression on X. This statement concerns an observed relationship, and consequently care must be exercised not to interpret it in cause-and-effect terms.

We next look at a numerical example.

EXAMPLE 14.1

Let us calculate the coefficient of determination for the development score data of display 14.3 (p. 459). In section 14.4 the step 2 computations for this data set yielded

$$S_{xx} = 73.703 \qquad S_{yy} = 43.619 \qquad S_{xy} = 54.279$$

Consequently,

$$r^2 = \frac{54.279^2}{73.703 \times 43.619} = .9164$$

Thus 91.64% of the total D score variation is explained by the linear regression of D scores on age. This result indicates a rather high degree of linearity for the data, which is reflected in the tight clustering of the observed data points near the regression line in display 14.4. One would expect to be able to predict D scores rather well from children's ages by using the regression line. But to quantify how well, we will need statistical inference.

EXERCISES

14.5   Calculate and interpret the coefficient of determination for the following data.

x	y
1.5	0.8
2.1	0.7
0.8	1.0
2.3	1.3

14.6   Calculate and interpret the coefficient of determination for the data of exercise 14.1.

Interpretation of the Correlation Coefficient as a Measure of Association

Because of its familiarity and ease of calculation, the Pearson correlation coefficient r is commonly used as the sole measure of association between two measurement variables. Since this association can be very complex, it is unreasonable to believe that it can be characterized by a single number. Consequently, it is important to understand just what the correlation coefficient does measure.

Except for the fact that r can be negative, r and r^2 measure the same thing, namely, the extent of the *linear* relationship between the variables. That is, r characterizes the degree to which the data can be fitted by a straight line. The sign of r simply indicates the direction of the linear association. If r is positive, the slope of the regression line is positive, and the values of the two variables increase or decrease together. If r is negative, one variable increases as the other decreases.

It is not uncommon for an investigator to drop a project that has been undertaken to determine the association between two variables upon finding a value of r not significantly different from zero. He concludes that if the variables are uncorrelated, they are not associated. However, this result only indicates that they are not *linearly* associated. Among other things, an interesting nonlinear regression may exist. If the linear trend term (see display 14.2, p. 455) is absent, this regression will be missed by a test based on r. Consequently, it is neither safe nor reasonable to base an evaluation of association on r (or r^2) alone. A scatter diagram of the data should always be obtained. When this diagram is excessively noisy (i.e., has a great deal of variability), an exploratory regression analysis of the type described in chapter 4 should be carried out. An example of this situation is given next.

EXAMPLE 14.2

The amount of biologically active (free) testosterone in a woman's blood is the difference between the total amount of testosterone (T) and the amount made inactive by being bound to certain proteins through the action of an agent called testosterone-binding globulin (TBG). In order for the amount of free testosterone to remain proportionately constant (relative to the total amount of testosterone), one would expect a roughly linear relationship between T and TBG. The following data represent the T and TBG values for a sample of 23 women under 45 years of age taken from a study by Purifoy and co-workers [28].

I.D.	TBG (X)	T (Y)	I.D.	TBG (X)	T (Y)
1	22	42	13	18	23
2	22	29	14	20	14
3	56	14	15	35	45
4	16	31	16	18	17
5	17	35	17	31	28
6	34	23	18	22	48
7	11	16	19	20	30
8	25	38	20	28	45
9	25	42	21	32	37
10	27	32	22	32	45
11	14	19	23	25	49
12	44	22			

With T as the dependent variable Y and TBG as the independent variable X, the estimated parameters for the least-squares line are

$$\hat{\beta}_0 = 31.445$$
$$\hat{\beta}_1 = .00128$$
$$s = 11.763$$

The r^2 value is

$$r^2 = .0000029$$

This value casts serious doubt on the linearity of the relationship between T and TBG. In fact, as will be shown in the next chapter, the statistical test for a linear relationship is highly insignificant. However, some other kind of association should not be discounted without further investigation.

Display 14.5 shows the scatter diagram of the data and the least-squares line. The line is almost horizontal, as the analysis suggests. The scatter diagram does show some additional structure, however.

Display 14.5

Scatter diagram, fitted least-squares regression line, and exploratory regression curve for testosterone and TBG data

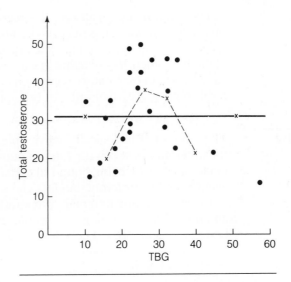

A curvilinear relationship is suggested, with a rise in testosterone for small TBG values, then a leveling off, and a decrease for larger values of TBG. An exploratory regression curve, which supports this indication, is shown in display 14.5. Since this evidence is based on relatively few points, one would want to see the results of a more comprehensive study before drawing firm conclusions.

◼ SECTION 14.6

Use of Residuals for the Study of Regression Curvilinearity

residuals

A method for studying structure in the data beyond that explained by a fitted regression curve is to form the **residuals**

$$r_i = y_i - \hat{y}_i$$

of the observed values y_i from the values $\hat{y}_i$ predicted from the fitted regression expression. This technique has the effect of isolating and magnifying the variations of the observations from the fitted curve. A scatter diagram of the residuals r_i against the corresponding values x_i of the independent variable will display additional structure in the regression of Y on X more clearly than the original scatter diagram. In particular, when the fitted regression is linear, an examination of the scatter diagram of residuals, aided by an exploratory regression analysis, will often make it possible to see indications of curvilinearity not obvious in the original scatter diagram.

Variations in dispersion can also be more easily seen in a residual plot. However, the full potential of the residual plot will not be realized until it is used as a diagnostic tool for multiple regression in chapter 16.

EXAMPLE 14.3

In this example we will investigate curvilinearity for the regression of weight versus height. In chapter 4 it was seen by studying the data of display 1.1 (p. 9) that there is an apparent curvilinearity in the regression of weight on height when the data of both sexes are combined. It was not clear whether this result was due to the mixing of two reasonably linear regressions with different slopes or to a curvilinear relationship within each sex category. In chapter 16 we will test the equality of the slopes for least-squares regression lines fitted separately to the data for the two sexes. Here we will look for a possible curvilinearity of regression within sex categories. Again, the data for the women will be used in this example in order to leave the data for the less numerous male group for an exercise.

The least squares prediction equation for weight (Y) versus height (X) for the women of display 1.1 is

$$\hat{y}(x) = -73.398 + 3.061x$$

(The data have not been corrected for the height recording error for student number 27, discovered in chapter 2).

The calculation of residuals can be conveniently summarized in a table such as the following one. (Only the first five entries of the table are given.)

HEIGHT (x)	WEIGHT (y)	PREDICTED WEIGHT $(\hat{y})$	RESIDUAL (r)
66	114	128.6	−14.6
65	115	125.5	−10.5
62	122	116.4	5.6
69	125	137.8	−12.8
63	105	119.4	−14.4

The computation for the first entry involves the calculation of the predicted weight for height $x = 66$:

$$\hat{y}(66) = -73.398 + 3.061 \times 66 = 128.6$$

(to one decimal place, which is enough for plotting purposes). Then the residual is

$$r = y - \hat{y} = 114 - 128.6 = -14.6$$

This calculation is carried out for each of the $n = 22$ women.

Display 14.6 consists of (a) the original height-weight scatter diagram with the least-squares regression line and (b) the scatter diagram of weight residuals versus heights. Although there are obvious similarities between the two diagrams, note that the full scale of the residual plot is devoted to emphasizing the deviations of the observations from the linear fit. An exploratory (median) plot is superimposed on part (b). Since the median regression plot is resistant to outliers, the indication of curvilinearity seen in the dashed curve is not just an artifact of the few large residuals. The result suggests that, at least for women, the regression of weight on height is curvilinear.

Display 14.6

Diagrams of weight versus height. (a) Scatter diagram of original data and corresponding least-squares regression line. (b) Scatter diagram of weight residuals versus height with superimposed curve of medians.

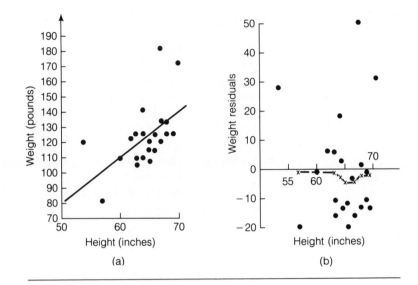

(a)

(b)

EXERCISES

14.7   Do a residual analysis for the data of exercise 14.1 (p. 463).

14.8   Complete the table in example 14.3 of predicted values and residuals for the women's height-weight data. Verify the correctness of the residual scatter diagram and exploratory regression curve of display 14.6(b).

14.9   Repeat the residual analysis given above for the men's data of display 1.1 (p. 9). (Recall that observation 20 is assumed to be male. Save the least-squares line analysis for chapter 15.) Discuss the curvilinearity of the weight versus height regression. How does the shape of the regression curve for males (including linear trend) appear to differ from that for females?

Chapter 14 Quiz

1.   What is a linear regression function in a linear model with one independent variable?

2.   What is the graphical interpretation of the regression coefficients for this model?

3.   What are some justifications for the use of linear regression functions in practice?

4.   What is a linear model? Give an example of a linear model for which the regression function is not a linear function.

5.   What are the two linear models that form the basis for inference in simple linear regressions?

6.   What is the least-squares criterion for obtaining the estimators of the regression coefficients in a linear model? Give an example.

7. What is the strategy for obtaining an estimator of the standard deviation σ of the errors e_i in a linear model? Give an example.

8. What is the hand-computing strategy for calculating the estimates of the regression coefficients and standard deviation for model 2? Give an example.

9. What is a simple method for plotting a fitted regression line on a graph? Give an example in which the fitted line is plotted on a scatter diagram of the data. (Note that this plot provides a rough check on the accuracy of the calculations.)

10. What is the coefficient of determination? What does it measure?

11. What are residuals? How are they used to look for curvilinearity in regression?

SUPPLEMENTARY EXERCISES

In some of the problems that follow, you will be asked to *analyze the data*. The analysis will be understood to consist of the following steps:

1. Make a scatter diagram of the data and assess, visually, the reasonableness of a linear fit.

2. From the data, or from the summarized data, obtain the least-squares estimates of β_0, β_1, and σ.

3. Calculate the coefficient of determination and compare the indication of data linearity given by this statistic with your assessment based on the scatter diagram.

4. Visually assess the accuracy of your computations by plotting the least-squares line on the graph containing the scatter diagram.

14.10 The accompanying two data sets represent levels of two hormones (called catecholamines) important in the regulation of metabolism, epinephrine (E) and norepinephrine (N), obtained from a runner before and after exercise on ten different days. During a resting state these chemicals are produced at different sites and by different mechanisms in the body. However, under physical stimulation (here, a 10-mile run) large additional amounts are produced by the adrenal gland.

(a) What might be expected to happen to the correlation between the levels of these chemicals during exercise?

(b) Verify your expectations by analyzing both data sets and comparing the coefficients of determination.

(c) In previous chapters we have discussed the fact that, in general, an association between two variables cannot be interpreted in causal terms. Thus, for example, we cannot attribute variations in the levels of N (taken to be the dependent variable) to the observed changes in the levels of E. In many instances strong correlations can be generated by a common dependence of both variables on a third variable that is not even measured. Discuss the differences in correlations seen in these two data sets in this light. What is the unmeasured common factor?

RESTING		STIMULATED	
E	N	E	N
70	521	212	3287
64	482	163	2418
60	478	193	3321
54	486	165	2562
61	582	212	3506
62	511	179	2635
71	506	255	3446
69	497	205	3035
63	452	171	2763
67	460	188	2823

Data courtesy of David Schade, M.D., School of Medicine, University of New Mexico, Albuquerque.

14.11   Various chemical components of air pollution are measured from the volumes of deposits caught in special collectors. This is the case for sulfur (S) in the atmospheric chemistry data given in exercise 2.30 and repeated below. Rain is the principal agent by which sulfur is removed from the atmosphere and deposited in the collectors. Consequently, it is to be expected that the amount of sulfur deposited in a given month will depend not only on the concentrations over that month but also on the amount of rainfall measured by the precipitation (PREC.) variable.

MONTH	PREC.	S
1	35	55
2	25	30
3	12	25
4	36	43
5	81	135
6	19	38
7	55	63
8	63	93
9	69	64
10	23	17
11	52	34
12	35	34

(a)   Analyze the data for S (the dependent variable) versus PREC. Is the linear relationship between these variables sufficiently impressive to support the contention that the amount of the sulfur deposited is associated with the amount of rainfall? Does this analysis prove that rainfall causes sulfur to be deposited?

(b)   Even though no causal implication is intended, it is common to speak of variation in one variable *due to* the other. We adopt this terminology here. Variations in sulfur concentrations not due to precipitation are largely obscured by the magnitude of the variations due to precipitation. An important use of regression is to correct one variable for its (linear) relationship with another by subtracting the predicted values, based on the regression, from the observed values—that is, by forming the residuals as outlined in section 14.6. For example, the sulfur residuals formed by subtracting the predicted values due to precipitation, based on the line calculated in part (a), will greatly magnify the variations in sulfur concentrations due to causes other than rainfall. Form these residuals and plot them against month of the year. Assuming the atmospheric sulfur comes primarily from coal-burning industries upwind from the recording station, what could account for the (gross) pattern shown in this plot of residuals?

14.12   Refer to the data of exercise 1.23 (p. 33).

(a) Use the following summary statistics to analyze the algebra scores (Y) versus arithmetic scores (X).

$$
\begin{aligned}
n &= 44 & \Sigma xy &= 2{,}974 \\
\Sigma x &= 639 & \Sigma y &= 174 \\
\Sigma x^2 &= 10{,}101 & \Sigma y^2 &= 1{,}224
\end{aligned}
$$

(b) What algebra score would be predicted for an individual with an arithmetic score of 8? Of 15?

(c) What is the estimated average algebra score of students having arithmetic scores of 12? Of 19?

(d) Compute the sample mean of the algebra scores for those students who obtained an arithmetic score of 19. Does this compare reasonably well with the estimate from the regression line obtained in part (c)?

14.13   It was seen in chapter 10 that, because of a positive correlation between the numbers of cigarettes smoked by men (Y) and women (X) in the same families, the paired-difference test for a difference in means was more sensitive than the two-sample test for the following data.

SMOKER				FAMILY										
	1	2	3	4	5	6	7	8	9	10	11	12	13	14
Father (Y)	11	20	20	20	10	20	20	30	9	20	5	20	20	20
Mother (X)	20	10	6	4	6	30	4	20	6	20	2	20	18	10

(a) Analyze the data from the regression viewpoint.
(b) From the coefficient of determination obtained in part (a), how strong does the correlation between the two variables appear to be?

14.14   A proposed measure for comparing the volume of household electricity used from one month to another is the consumption rate, in kilowatts per household, at the time of peak demand for the entire city (system) during the month. Since peak household usage will not always occur at the system peak, the effectiveness of this quantity as a measure of electricity consumption is in question. One way to check effectiveness is to look at the correlation of the measures for two consecutive months for which the lighting and heating conditions are nearly the same. A high correlation, along with a least-squares line that predicts about the same value for the second month as for the first ($\hat{\beta}_0 \approx 0$ and $\hat{\beta}_1 \approx 1$), would indicate that the measure is probably suitable for comparing electrical consumption for other months. The accompanying data represent values of this measure for 18 households for September and October of a given year.

(a) Analyze the data with the aid of the given summary statistics.

(b) Check the parameter estimates obtained in part (a) against the criteria above to indicate the suitability of the measure.

SEPTEMBER (X)	OCTOBER (Y)	SEPTEMBER (X)	OCTOBER (Y)
2.56	2.16	.68	.68
2.08	.64	.36	.36
.60	.40	.24	.52
.28	.76	3.96	4.12
1.32	1.00	.12	.24
.28	.16	1.92	1.16
1.92	2.56	.32	.36
.00	.16	.76	.80
.88	.52	.80	.56

Data courtesy of B. Waldman, Public Service Co. of New Mexico.

Summary statistics:

$$n = 18 \qquad \Sigma xy = 34.510$$
$$\Sigma x = 19.080 \qquad \Sigma y = 17.160$$
$$\Sigma x^2 = 38.952 \qquad \Sigma y^2 = 34.011$$

14.15 In medical experiments involving the effects of a drug on a particular internal organ of an animal, it is usually necessary to regulate the amount of the drug administered to the size of the organ. Clearly, for experiments on living animals, the size of the organ must be estimated from such observable measurements as total body weight. The accompanying experimental data present the possibility of developing a predicted heart weight for (domestic) cats from body weight. The data are extracted from a study of lethal doses of digitalis on cats in which the heart weights were obtained from autopsies. Body weights (X) are in kilograms and heart weights (Y) are in grams.

(a) Analyze the data for both sexes and plot the lines and scatter diagrams on the same graph. [Note: Code the points for females by crosses (+) and those for males by circles (O).]

(b) Evaluate the linearity of the two data sets from the coefficients of determination. Does the linear prediction of heart weights from body weights appear promising? For which sex are the data more linear?

(c) Predict the heart weights of 2.3-kilogram male and female cats.

FEMALES		MALES		SUMMARY STATISTICS		
BODY WEIGHT (X)	HEART WEIGHT (Y)	BODY WEIGHT (X)	HEART WEIGHT (Y)	VARIABLE	FEMALES	MALES
2.3	9.6	2.9	9.4	n	15	16
3.0	10.6	2.4	9.3	Σx	34.4	44.6
2.9	9.9	2.2	7.2	Σx^2	80.08	125.84
2.4	8.7	2.9	11.3	Σxy	304.55	489.60
2.3	10.1	2.5	8.8	Σy	131.40	173.30
2.0	7.0	3.1	9.9	Σy^2	1176.56	1940.43
2.2	11.0	3.0	13.3			
2.1	8.2	2.5	12.7			

(table continued on next page)

FEMALES		MALES		SUMMARY STATISTICS		
BODY WEIGHT (X)	HEART WEIGHT (Y)	BODY WEIGHT (X)	HEART WEIGHT (Y)	VARIABLE	FEMALES	MALES
2.3	9.0	3.4	14.4			
2.1	7.3	3.0	10.0			
2.1	8.5	2.6	10.5			
2.2	9.7	2.5	8.6			
2.0	7.4	2.8	10.0			
2.3	7.3	3.1	12.1			
2.2	7.1	3.0	13.8			
		2.7	12.0			

Source: Halek, Kimura, and Bartels [15]. Copyright 1946, American Pharmaceutical Association. All rights reserved. Adapted with permission.

14.16   The summary data for the verbal scores (Y) versus age (X) (example 4.3), tabulated in display 4.9 (p. 105), are given below for the two age groups 3–5 and 6–10.

	AGES	
	3–5	6–10
n	32	55
Σx	129	443
Σx^2	541	3,675
Σxy	39,201	163,048
Σy	9,134	19,841
Σy^2	3,063,002	7,344,971

(a)   Combine the data by adding together corresponding sums to obtain summary data for the entire data set. (For example, the Σx sum for the entire set is 129 + 443 = 572.) Analyze the combined data (omitting the scatter diagram) and plot the least-squares line on the same graph as the exploratory regression curve plotted from the medians given in example 4.2. Note how the linear regression averages the curvilinear features of this curve.

(b)   Analyze separately the data for the two age ranges (again omitting scatter diagrams) and plot the least-squares lines on the graph constructed in part (a). Compare the separate coefficients of determination with the one for the overall line. Note the improvement in linearity for the individual lines. Can you reconcile the differences in r^2's for the two age groups with the linearity of the two parts of the exploratory regression curve?

14.17   From the exploratory analysis of androstenedione (Y) versus age (X) carried out in exercise 4.24 (p. 117), it was seen that beyond the age of 25 the regression curve has a strong decreasing linear component. As a first approximation, this linear component could be used to estimate average androstenedione levels for various ages or to predict the androstenedione level for a woman of given age whose level is not known. The inference procedures of the next chapter will make it possible to determine error bounds for these estimates and predictions. The accompanying summary data are for the data set labeled "From Normal Women."

Summary statistics for androstenedione data from normal women:

$$n = 53 \qquad\qquad \Sigma xy = 269{,}806$$
$$\Sigma x = 2{,}439 \qquad\qquad \Sigma y = 6{,}549$$
$$\Sigma x^2 = 125{,}735 \qquad\qquad \Sigma y^2 = 959{,}505$$

(a) Analyze these data. [A scatter diagram will be available if you did part (a) of exercise 4.24.]

(b) Plot the regression line on a copy of the graph showing the exploratory regression curve for the entire data set. Over which age ranges is the line likely to provide good estimates of the average androstenedione level?

(c) In statistics the term *bias* is applied to the estimation of a curvilinear regression function by a linear regression function (or the wrong curvilinear function). Over various ranges of the independent variable, the regression estimates based on the line will necessarily differ consistently from the values on the regression curve simply because a straight line cannot follow the oscillations of a curve. Again, what we are estimating (a linear function) is not what we want to estimate—or think we are estimating (the regression curve). For which regions of age is this bias likely to be the greatest?

14.18 It is well known that people tend to slow down physically with age. How does this well-known fact apply to bicyclists—at least, that is, to bicyclists with sufficient dedication to undertake a reasonably strenuous bicycle tour? If by *slowing down* we mean having a slower average speed on the tour, we can check this observational hypothesis by looking at the regression of average speed versus age. If the hypothesis is true, the regression curve should have an overall decreasing trend (at least for ages greater than 15 or so). This trend would be represented by a least-squares line with negative slope. The accompanying data were subsampled from display 1.17 (p. 25), the sixth annual Tour of the Rio Grande Valley data. (Both sexes and all distances are present in the subsample.) Analyze the data for the hypothesized decreasing trend. How strong does the trend appear to be? (We will address this question inferentially in chapter 15.)

AGE (X)	SPEED (Y)	SUMMARY DATA
22	12.0	$n = 18$
17	12.6	$\Sigma x = 544$
54	11.4	$\Sigma x^2 = 19{,}028$
39	9.0	$\Sigma xy = 6{,}883.8$
45	11.9	$\Sigma y = 232$
30	17.9	$\Sigma y^2 = 3{,}122.48$
17	12.7	
42	13.0	
42	16.7	
36	9.2	
20	9.5	
22	18.0	
49	11.4	
21	14.6	
15	15.9	
23	10.1	
30	12.8	
20	13.3	

14.19 The accompanying data represent the energy (in coded units) used for heating and cooling a sample of 10 small office buildings during a given year in Louisville, Kentucky. Analyze the data. Does there appear to be any relationship between the heating and cooling requirements for small office buildings in this city? Discuss any unusual features of the data.

HEATING (X)	COOLING (Y)
3.15	9.33
2.87	11.08
3.56	4.39
1.75	6.74
1.78	7.73
6.66	7.99
3.12	11.82
2.84	11.18
2.91	5.86
1.66	8.40

Data courtesy of D. Hadfield, R. & D. Associates, Albuquerque, New Mexico.

14.20 Two measures of automobile pollution, obtained in the study described in exercise 4.18 (p. 115), are carbon monoxide (CO), in percentage of total emissions, and hydrocarbons (HC), in parts per million cubic centimeters (ppm), measured with the engine at idling speed. It might be expected that these measures would be positively correlated, since engines in poor condition should generate high levels of both pollutants, while well-tuned engines should produce low levels of both. Check the accuracy of this conjecture by analyzing the accompanying data drawn at random from the data file for the study mentioned above.

CO (X)	HC (Y)
5.8	235
12.6	847
8.7	553
2.6	423
8.1	1035
4.0	294
1.8	1646
5.2	1800
2.6	450
4.6	1140
6.2	2000
4.9	570
3.0	750

Data courtesy of F. Wessling, College of Engineering, University of New Mexico, Albuquerque.

CHAPTER 15

Statistical Inference for Simple Linear Regression

■ SECTION 15.1
Introduction

Statistical inference results will be given in this chapter for simple linear regression problems. These include the topics mentioned in section 14.4: confidence intervals and tests for the slope parameter β_1, confidence intervals for the regression line, and error bounds for predicting values of new observations using the regression line. The interpretation of the hypothesis of zero slope is also discussed, in section 15.3.

The flexibility of the simple linear regression model and fitting method will be demonstrated in section 15.6. By linearizing regression models through the transformation of one or both of the variables, linear regression can be extended to cover a wide range of problems. The power transformations introduced in chapter 12 will be the basis of useful methods, not only to straighten curves but also to correct for inhomogeneities in the variances of the conditional distributions of Y given X.

Finally, the influence of outliers in linear regression will be discussed in section 15.7. A simple strategy for diagnosing the presence of this influence will be given.

■ SECTION 15.2
A Confidence Interval for the Slope Parameter β_1

The most important parameter in a linear regression model (model 2) is the slope parameter β_1. The linear model theory, based on the normality of the errors e_i, specifies that the statistic

$$t = \frac{\hat{\beta}_1 - \beta_1}{\text{SE}_{\hat{\beta}_1}}$$

has Student's t distribution with

$$v = n - 2$$

degrees of freedom, where

$$SE_{\hat{\beta}_1} = \frac{s}{\sqrt{S_{xx}}}$$

The estimate s is the one determined under model 2 assumptions, whose value is calculated in step 3 of the computing strategy given in section 14.4.

A confidence interval for β_1 is now formed in the usual way. The form of the interval is

$$\hat{\beta}_1 - tSE_{\hat{\beta}_1} \leq \beta_1 \leq \hat{\beta}_1 + tSE_{\hat{\beta}_1}$$

where t is obtained from table 3 for the given confidence coefficient and $v = n - 2$ degrees of freedom.

EXAMPLE 15.1

In this example we will determine a 95% confidence interval for the slope β_1 of the D score versus age trend line. The estimated average rate of increase of D scores with age for the data of display 14.3 (p. 459) was shown to be

$$\hat{\beta}_1 = .7365$$

The calculated least-squares statistics given for these data in chapter 14 included

$$s = .6035 \quad \text{and} \quad S_{xx} = 73.703$$

Consequently, the standard error of $\hat{\beta}_1$ is

$$SE_{\hat{\beta}_1} = \frac{s}{\sqrt{S_{xx}}} = \frac{.6035}{\sqrt{73.703}} = .07030$$

The critical value t from table 3 for 95% confidence and $v = 12 - 2 = 10$ degrees of freedom is

$$t = 2.228$$

Thus since $tSE_{\hat{\beta}_1} = 2.228 \times .0703 = .1566$, the confidence limits are

$$L = \hat{\beta}_1 - tSE_{\hat{\beta}_1} = .7365 - .1566 = .5799$$

and

$$U = \hat{\beta}_1 + tSE_{\hat{\beta}_1} = .7365 + .1566 = .8931$$

We conclude with 95% confidence that the true increase in D score for each year of age is between .58 and .89 points, on the average.

EXERCISE

15.1 Find a 95% confidence interval for the slope β_1 of the straight-line relationship between high school and college GPAs of exercise 14.1 (p. 463).

Tests of the Hypothesis H_0: $\beta_1 = 0$

Perhaps the most commonly applied test in regression is of the hypotheses

$$H_0: \beta_1 = 0 \quad \text{versus} \quad H_1: \beta_1 \neq 0$$

As we will see shortly, this is a test for a *linear* relationship (regression) between the two variables X and Y. The hypothesis H_0 is the null hypothesis of no linear regression, while H_1 is the hypothesis that a linear regression of some magnitude exists.

The test can be based on the theory outlined at the beginning of section 15.2. The test statistic for this test is the t statistic given in section 15.2,

$$t = \frac{\hat{\beta}_1}{SE_{\hat{\beta}_1}}$$

with $\beta_1 = 0$, the value specified by H_0. Thus the test of H_0 versus H_1 follows the now familiar pattern: H_0 is rejected if $|t| \geq c$, where c is obtained from the table of Student's t distribution for the given significance level and $v = n - 2$ degrees of freedom. P-values and one-sided tests can also be obtained as they were for tests involving the mean of a normal distribution in chapter 10.

EXAMPLE 15.2

Is the strong linear relationship ($r^2 = .92$) shown between D scores and age in the sample of display 14.3 (p. 459) statistically significant? How can the strength of evidence (significance) for this relationship be measured? The P-value for the test of H_0: $\beta_1 = 0$ versus H_1: $\beta_1 \neq 0$ can be used to answer both questions. To calculate the P-value, we need the observed value of the test statistic t.

In Example 15.1 we showed that

$$SE_{\hat{\beta}_1} = .07030$$

Thus the value of t is

$$t = \frac{\hat{\beta}_1}{SE_{\hat{\beta}_1}} = \frac{.7365}{.07030} = 10.477$$

This is also the value of $|t|$. The degrees of freedom are $v = 10$ as shown earlier.

It is now seen from table 3 that the P-value bound for these degrees of freedom and the computed $|t|$ value is

$$P < .001$$

Thus if $\beta_1 = 0$, a value of $|t|$ this extreme would be seen (much) less frequently than one time in every 1000 samples of size $n = 12$. The evidence for the existence of a linear regression between D scores and age is quite strong. In light of the discussion of chapter 14, this result would mean that a linear trend component is a dominant feature of the regression curve of D

scores with age. It may be the *only* component. However, we cannot be sure of this, simply because the sample size is too small to obtain a reasonable indication of curvilinearity, even with an exploratory analysis. This will often be the case for small samples.

Interpretation of the Test of H_0: $\beta_1 = 0$ as a Test for Linearity

Since the test of H_0: $\beta_1 = 0$ versus H_1: $\beta_1 \neq 0$ is often misinterpreted, it is worthwhile discussing just what can be concluded from the test and what cannot. Formally, the test is used to decide whether model 2 (the linear regression model of chapter 14) provides a better explanation of the data than does model 1. Recall that model 2 reduces to model 1 if $\beta_1 = 0$. Consequently, the hypotheses could equally well be written as

H_0: model 1 holds   versus H_1: model 2 holds

When put in these terms it would seem natural to try to base a test on the statistic $SS_1 - SS_2$, the difference of the goodness-of-fit sums of squares for models 1 and 2 introduced in chapter 14. This statistic is a natural choice because it has the property of being large when H_1 is true and small (actually, near 1) for H_0. It is also the statistic specified by the general linear model theory in this regression context. If we are to obtain a statistic with a known distribution when H_0 is true (the other requirement of a test statistic), the theory suggests we use

$$F = \frac{SS_1 - SS_2}{s^2}$$

where s is the model 2 estimate of σ. This statistic has the F distribution with 1 and $n - 2$ degrees of freedom. To carry out the test, we would reject H_0 if $F \geq c$, where c is found from table 4.

However, there is a simple relationship between the statistics F and t, namely

$$F = t^2$$

In fact, the two-sided t test given above is equivalent to this F test. The advantage of using the t statistic is that it is also possible to construct one-sided tests of H_0 against the one-sided alternative H_1: $\beta_1 > 0$, which specifies a positive linear association—that is, an increasing relationship of Y with X—or against H_1: $\beta_1 < 0$, which corresponds to a negative linear association. In any event, it is a *linear* association that is being tested.

The Test of H_0: $\beta_1 = 0$ versus H_1: $\beta_1 \neq 0$ as a Test of Zero Correlation

There is a tendency to lose sight of the term *linear* when the test of H_0: $\beta_1 = 0$ versus H_1: $\beta_1 \neq 0$ is carried out. This test, or one completely equivalent to it, is sometimes mistakenly used to test the more global hypothesis of independence (no association of *any* kind) between two measurement variables. Unfortunately, there is some statistical theory that, if not properly understood, seems to actually promote this use, as we will show in the following discussion.

In chapter 4 we discussed the idea of joint frequency distributions for categorical variables. The most commonly used statistical model for the joint distribution of two measurement variables is an extension of the normal distribution from one to two (or more) variables. Because of its convenient mathematical properties, this model forms the basis for (most of) the many-variable, or *multivariate*, theory of statistics.

A mathematical consequence of the use of the multivariate normal distribution is that the association between two variables is completely specified by a parameter ρ (Greek letter rho), called the *population correlation coefficient*. When $\rho = 0$, not only are the variables unassociated in the linear sense but they are also independent. Consequently, *for normal variables* a test of H_0: $\rho = 0$ versus H_1: $\rho \neq 0$ is actually a test of independence.

The Pearson product-moment correlation coefficient r, defined in chapter 14, is the natural estimator for ρ. Consequently, a reasonable test statistic for a test of these hypotheses could be based on r. In fact, the usual test is based on

$$t = \frac{r}{\sqrt{\dfrac{1 - r^2}{n - 2}}}$$

This statistic has Student's t distribution with $n - 2$ degrees of freedom when H_0: $\rho = 0$ is true. And well it should, because this t is, algebraically, exactly the same as the statistic based on $\hat{\beta}_1$ given at the beginning of this section. That is, the test for H_0: $\rho = 0$ versus H_1: $\rho \neq 0$ is identical to the test for H_0: $\beta_1 = 0$ versus H_1: $\beta_1 \neq 0$. Thus the test for independence is exactly the test we have claimed to be for linear association only!

Of course, the solution to this paradox lies in the assumption of normality for the test of independence. For normal variables association is always linear. However, as we have seen repeatedly in our data analyses, the normal distribution is not, in general, an accurate model of physical reality. This observation is especially critical in multivariate situations where the association between variables can be substantially more complex than the linear association enforced by the normal model.

Since information about the possible independence of two variables is important, it is interesting to ask what can be concluded about the general dependence of two variables on the basis of the test for linear association. If the hypothesis H_0: $\beta_1 = 0$ is rejected in favor of H_1: $\beta_1 \neq 0$ (or one of the one-sided alternatives), it is reasonable to conclude that the two variables are dependent. This conclusion is valid because a linear association is one form of dependence.

On the other hand, if H_0 is accepted, it is *not* possible to conclude the variables are independent. This implication is the one that causes trouble. Variables can be linearly unassociated yet have some other form of dependence that is not detected by the test. Some forms of dependence are not representable as regressions of the conditional means at all. In other cases the dependence may be representable as a curvilinear regression with no linear trend component. Such is the case in the following example. In cases where the hypothesis of no linear regression is accepted, alternative statistical tests not given in this text would be required in order to pursue further the question of independence of the variables. For our purposes, a follow-up exploratory look at the scatter diagram for hints of other forms of association will suffice.

EXAMPLE 15.3

To what extent does the small r^2 value (.00000129) for the data of example 14.2 (p. 467) indicate the lack of a linear relationship between total testosterone (T) and testosterone-binding globulin (TBG)? We can measure the strength of evidence in the sample for or against a linear relationship by obtaining the P-value of the test for

$$H_0: \beta_1 = 0 \quad \text{versus} \quad H_1: \beta_1 \neq 0$$

The necessary summary values for carrying out the test are

$$\hat{\beta}_1 = .00128 \quad s = 11.763 \quad S_{xx} = 2271.304$$

Then

$$SE_{\hat{\beta}_1} = \frac{11.763}{\sqrt{2271.304}} = .2468$$

and

$$t = \frac{.00128}{.2468} = .00519$$

Entering table 3 with $v = 23 - 2 = 21$ degrees of freedom, we find that the P-value for the test satisfies the inequality

$$P > .50$$

That is, if H_0 is true and there is no linear relationship between the variables, we would see an outcome as extreme as this sample (well) over half the time. The test outcome is highly insignificant, and we would conclude that the sample provides no evidence of a linear relationship between T and TBG. However, an indication of a possible nonlinear relationship was shown in example 14.2, which would be of interest to pursue in another experiment. In particular, as seen from the scatter diagram in display 14.5 (p. 468), it would be especially important to explore further the high-TBG region of the diagram.

EXERCISE

15.2   Show that the expression for t in terms of the correlation coefficient r gives the same value as that obtained in example 15.3 for the data of example 14.2 (p. 467).

■ SECTION 15.4

Confidence Intervals for the Regression Line

It is often of interest to estimate the value $\mu(x) = \beta_0 + \beta_1 x$ of the underlying population regression line at a given value x. It is natural to use the least-squares estimator $\hat{\mu}(x) = \hat{\beta}_0 + \hat{\beta}_1 x$ to do this. For example, in a designed experiment, to determine the parameters of a straight-line relationship $y = \mu(x)$ between two variables, dictated by a physical law, we may use the least-squares estimator $\hat{\mu}(x)$ fitted to the experimental data to estimate the y value of the relationship for any x. Similarly, in an observational experiment, if the linear term is the dominant component of the regression function $\mu(x)$ of Y given X, as indicated by a large value of r^2, for example, a good approximation to the conditional mean of Y for a given value x would be obtained by using the least-squares estimator $\hat{\mu}(x)$ of the linear component.

A reasonable criterion for using the least-squares line for this purpose is a significant test for linearity, that is, the rejection of $H_0: \beta_1 = 0$ in favor of $H_1: \beta_1 \neq 0$ at, say, the 5% significance level. This test can be carried out quickly by using the version (given in the last section) of the t statistic that depends on r, since r is one of the first statistics calculated. Thus the test for $\beta_1 = 0$ can be used to screen the data for sufficient linearity to make it reasonable to proceed with other inference procedures that require the regression to be linear.

The error in estimating $\mu(x)$ can be quantified by a confidence interval. Since such an interval could, in theory, be computed for each x, the lower and upper interval boundaries $L(x)$ and $U(x)$ will sweep out two curves as x varies over the interesting range of the independent variable. See display 15.1 for an example. These curves can, in fact, be computed at a relatively small number of points and drawn on a graph by interpolating through these points. They can then be used to determine the confidence interval for any

Display 15.1

95% confidence intervals for linear relationship of weld break strength (Y) versus velocity (X). Solid curves define individual 95% intervals; dashed curves define simultaneous 95% intervals.

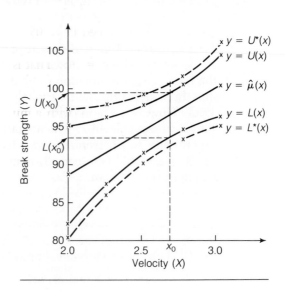

value of x. A vertical line is drawn at the desired x, and horizontal lines are drawn over to the y axis at the points of intersection of this line with the two curves. The endpoints of the interval are then read from the y axis scale. The interval at a value x_0 is shown in display 15.1, using the graphical representation of a confidence interval introduced in chapter 10. This graphical method has been used often in practice and is interesting because it shows how the widths of the confidence intervals vary with x, but the more commonly used procedure today is to have a computer programmed to calculate directly the values of $L(x)$ and $U(x)$ as needed.

As an added bonus, because of the relationship $\beta_0 = \mu(0)$, the confidence interval for the regression line at $x = 0$ is also the appropriate confidence interval for the intercept, β_0. This fact will be used and illustrated later in this section.

Joint Confidence Intervals for $\mu(x)$ at Several Values of x

If the confidence intervals for $\mu(x)$ have a confidence coefficient of 95%, for example, the guarantee is that, for any *single* value of x, $\mu(x)$ will be between $L(x)$ and $U(x)$ with 95% confidence. Suppose that it is desirable to estimate $\mu(x)$ for several values of x. For example, for the data leading to display 15.1 we might wish to estimate the true regression line in the range from $x = 2.0$ to $x = 3.0$ in steps of .1. Thus, estimates would be required for the 11 x values 2.0, 2.1, . . . , 3.0. The curves $L(x)$ and $U(x)$ in display 15.1 provide a 95% confidence interval for $\mu(x)$ at *each* x. That is, the confidence is *individually* 95% at 2.0, 95% at 2.1, and so on. But with what confidence do these intervals *simultaneously* cover all of the parameters $\mu(2.0)$, $\mu(2.1)$, . . . , $\mu(3.0)$?

Unfortunately, we can't say. However, it is possible to get a lower bound on this confidence by using the Bonferroni inequality (see chapter 11). Since for 95% intervals the probability of each interval missing (i.e., not covering) the true regression value is $1 - .95 = .05$, the probability of at least one miss cannot exceed $11 \times .05 = .55$. Thus the probability that all intervals cover, which is one minus the probability of at least one miss, cannot be smaller than $1 - .55 = .45$. That is, the *joint* confidence level is between 45% and 95%. This range of values does not provide a very satisfactory protection guarantee, unfortunately.

We are again faced with a multiplicity problem of some importance. When estimates of $\mu(x)$ are made at several points, it is highly desirable to bound them with prescribed *overall* confidence. One way to do this, which was also used in chapter 13, is to make the Bonferroni inequality work for us. If intervals at k values of x are desired with overall confidence $1 - \alpha$, simply compute the individual intervals at confidence $1 - (\alpha/k)$. Thus, for example, the 11 intervals above will have confidence of at least $.95 = 1 - .05$ if the individual intervals are designed to have confidence $1 - (.05/11) \approx .995$, or 99.5%. However, a special mathematical relationship makes a (usually) better alternative procedure available.

It is possible to compute two curves $L^*(x)$ and $U^*(x)$ such that for each

x, $\mu(x)$ is contained between these two limits with a prescribed confidence that holds for *all* values of x simultaneously. That is, we can have prescribed confidence not only for a fixed number of intervals but for any and all intervals at the same time! Because of this result, the region between these curves, which are shown as dashed curves in display 15.1 (p. 484), is often called a confidence interval, or *confidence region*, for the (entire) regression line. Another way of putting this is to say that any line that can be drawn in this region (without crossing either of the two dashed curves) is a candidate for the true population regression line.

Because the confidence interval for the regression line covers all values of $\mu(x)$ with simultaneous confidence, it might be thought that the individual intervals based on $L^*(x)$ and $U^*(x)$ must be substantially longer than those based on $L(x)$ and $U(x)$. It is seen from display 15.1 that this is not the case. The added width is relatively small. Moreover, the computations for calculating $L^*(x)$ and $U^*(x)$ are no more difficult than those for $L(x)$ and $U(x)$. In fact, with one small alteration, they are the same. The computational details will be given next in the context of an example.

Computing the Bounding Curves for the Confidence Interval for $\mu(x)$

The curves $L(x)$ and $U(x)$ are obtained by a now familiar line of reasoning. Under the assumptions of model 2 (chapter 14), for each x the statistic

$$t = \frac{\hat{\mu}(x) - \mu(x)}{SE_{\hat{\mu}(x)}}$$

has Student's t distribution with $v = n - 2$ degrees of freedom, where the standard error of the least-squares estimator $\hat{\mu}(x) = \hat{\beta}_0 + \hat{\beta}_1 x$ is

$$SE_{\hat{\mu}(x)} = s\sqrt{\frac{1}{n} + \frac{(x - \bar{x})^2}{S_{xx}}}$$

As before, s is the estimate of σ for model 2, calculated in section 14.4. The remaining ingredients of this standard error, n, S_{xx}, and $\bar{x}$, are also available from the calculations shown in section 14.4.

The bounding curves for the individual confidence intervals for $\mu(x)$ are now

$$L(x) = \hat{\mu}(x) - tSE_{\hat{\mu}(x)}$$

and

$$U(x) = \hat{\mu}(x) + tSE_{\hat{\mu}(x)}$$

where t is the critical value from table 3 for $v = n - 2$ degrees of freedom and the given confidence coefficient.

The bounding curves $L^*(x)$ and $U^*(x)$ for the simultaneous confidence intervals are obtained by replacing the critical value t in these expressions by the quantity w, where

$$w = \sqrt{2F}$$

and F is the critical value of the F distribution (table 4) corresponding to the given confidence coefficient, with

$v_1 = 2$ numerator degrees of freedom

and

$v_2 = n - 2$ denominator degrees of freedom

The resulting region is called the *Working-Hotelling interval* after the statisticians who developed the theory for it. (See Neter, Wasserman and Kuttner [24].)

The reason for qualifying the Working-Hotelling interval as being usually better than the one provided by the Bonferroni inequality is that, for a relatively small number of comparisons, the Bonferroni intervals can be shorter. Which intervals are better in any situation can be determined by computing both the w value and the t value for the given number of comparisons. Choose the method corresponding to the smaller value.

EXAMPLE 15.4

The following experiment to test the strength of welds made by an inertial friction method is used as an example by Box, Hunter, and Hunter [3]. Welds are formed between two pieces of metal in the following way. One of the pieces is rotated at a given velocity (X) and then brought to a standstill by forcing it into contact with the other, which is held stationary. The heat generated by friction at the interface of the two pieces produces a hot pressure weld. These welds, for several samples of the metal, are subsequently tested for break strength (Y).

It is known that the relationship between break strength and velocity is (nearly) linear over the range of velocities and for the types of metals of interest. However, the parameters of the relationship must be determined experimentally for each metal. The problem is to estimate the break strengths for certain rotational velocities not actually used in testing. These estimates can be obtained by using the theory above. Both 95% individual and simultaneous confidence intervals will be derived.

The test data for the metal type of interest are given in the following table. (Break strength units are kilograms per square inch, and velocities are in units of 100 feet per minute.)

PIECE NUMBER	VELOCITY (X)	BREAK STRENGTH (Y)
1	2.00	89
2	2.50	97
3	2.50	91
4	2.75	98
5	3.00	100
6	3.00	104
7	3.00	97

Source: Box, Hunter, and Hunter [3].

The quantities required for the confidence interval calculations are as follows:

$$n = 7 \qquad S_{xx} = .83929 \qquad \hat{\beta}_1 = 11.660$$
$$\bar{x} = 2.6786 \qquad \hat{\beta}_0 = 65.340 \qquad s = 2.9536$$

The value of the coefficient of determination, $r^2 = .7234$, would seem to support a linear relationship between the variables. (For the two-sided test of $H_0: \beta_1 = 0$, the calculated t value is $t = 3.616$, yielding $.01 < P < .02$.)

We have two options for calculating the confidence intervals for $\mu(x)$. We can calculate the intervals only for the particular velocities of interest. Or if several intervals but not great accuracy are required, the curves given in display 15.1 (p. 484) can be drawn carefully on graph paper and the desired intervals read from the graph. The curves $L(x)$, $U(x)$, $L^*(x)$, and $U^*(x)$ can be plotted from relatively few points. Those for display 15.1 were plotted from points calculated at the velocity values $x = 2.00, 2.25, 2.50, 2.75$, and 3.00. The calculations are organized in the following table.

x	$\hat{\mu}(x)$	$SE_{\hat{\mu}(x)}$	$L(x)$	$U(x)$	$L^*(x)$	$U^*(x)$
2.00	88.658	2.456	82.344	94.972	80.300	97.016
2.25	91.573	1.776	87.007	96.139	85.529	97.617
2.50	94.488	1.256	91.259	97.717	90.214	98.762
2.75	97.404	1.140	94.471	100.333	93.523	101.281
3.00	100.319	1.523	96.401	104.233	95.134	105.500

The calculation for $x = 2.00$ is as follows: Since $\hat{\mu}(x) = 65.340 + 11.659x$, then

$$\hat{\mu}(2.00) = 65.340 + 11.659 \times 2.00 = 88.658$$

Also,

$$SE_{\hat{\mu}(2.00)} = s \sqrt{\frac{1}{n} + \frac{(2.00 - \bar{x})^2}{S_{xx}}}$$

$$= 2.9536 \sqrt{\frac{1}{7} + \frac{(2.00 - 2.6786)^2}{.83929}} = 2.456$$

The 95% Student's t critical value from table 3 for $v = 7 - 2 = 5$ degrees of freedom is

$$t = 2.571$$

Then

$$L(2.00) = \hat{\mu}(2.00) - tSE_{\hat{\mu}(2.00)} = 88.658 - 2.571 \times 2.456 = 82.344$$

and

$$U(2.00) = \hat{\mu}(2.00) + tSE_{\hat{\mu}(2.00)} = 88.658 + 2.571 \times 2.456 = 94.972$$

Thus with 95% confidence, we have

$$82.344 \leq \mu(2.00) \leq 94.972$$

of the independent variable. In making these estimates we are relying heavily on the validity of the *linear law* relating the two variables. We have been able to check this law in the given velocity range, if somewhat crudely, by looking at the scatter diagram and by calculating the r^2 value for the data. However, the linear law may fail, possibly rather rapidly, outside this range. Consequently, it is risky to extrapolate the estimates and confidence intervals there. Extrapolation is reasonable only when it is known from sources other than the data at hand that the assumed linear relationship is valid in the extended region of the independent variable. Since this knowledge is seldom available, extrapolation is seldom sensible.

A safer course of action is to augment the experiment by testing (obtaining y values) at new values of the independent variable in the extended region. Assuming compatibility of the tests for the two experiments, these results can be combined with those of the first experiment to provide a sample over the extended region of x values. In this way not only can the linearity of the relationship be checked in this region, but also more precise estimates (as indicated by shorter confidence intervals) can often be obtained at the extremes of the region. The importance of this statement can be appreciated by noting in display 15.1 (p. 484) that the least precisely estimated values of $\mu(x)$ occur at the ends of the x value range. Extending the range and increasing the sample size tend both to decrease the standard error of the estimate $\hat{\mu}(x)$ and the critical t value, thus decreasing the confidence interval lengths.

EXERCISES

15.3 Refer to the grade point average (GPA) data of exercise 14.1 (p. 463).
   (a) Obtain 95% confidence intervals for individual estimates of college GPA versus high school GPA.
   (b) Obtain 95% confidence intervals for the regression line of college GPAs versus high school GPAs.
   (c) Plot the curves bounding these intervals on a graph as in display 15.1.

15.4 Refer to exercise 15.3. Estimate, with simultaneous 95% confidence intervals, the average college GPAs of students with high school GPAs of 2.0, 2.5, and 3.0, respectively.

SECTION 15.5

Prediction Based on the Least-Squares Line and a Measure of Its Uncertainty

An important use of a least-squares-estimated regression line is to predict a future value of the dependent variable for a given value of the independent variable. In chapter 14 this idea was applied to the prediction of college GPAs, for students entering college, from their high school GPAs. We now show how to obtain bounds that provide a useful measure of the error or uncertainty for the predictions.

Prediction Interval for $y(x)$

If the linear regression model, model 2, holds, a future value of Y at a given value x will be of the form

$$y(x) = \mu(x) + e$$

where $\mu(x) = \beta_0 + \beta_1 x$ is the true regression line and e is a random disturbance. On the other hand, the estimate $\hat{y}(x)$ of $y(x)$ is the value from the estimated regression line,

$$\hat{y}(x) = \hat{\mu}(x) = \hat{\beta}_0 + \hat{\beta}_1 x$$

The error in estimating $y(x)$ by $\hat{y}(x)$ is then the difference

$$y(x) - \hat{y}(x) = \mu(x) - \hat{\mu}(x) + e$$

That is, the prediction error is the sum of the error $\hat{\mu}(x) - \mu(x)$ of estimating the regression line and the random disturbance e. The standard error of the prediction estimator, which we will denote by $SE_{\hat{y}(x)}$, although it is actually the standard error of $\hat{y}(x) - y(x)$, is

$$SE_{\hat{y}(x)} = s \sqrt{1 + \frac{1}{n} + \frac{(x - \bar{x})^2}{S_{xx}}}$$

Again, s is the estimate of the model 2 standard deviation σ. Compare this expression with the one for the standard error of $\hat{\mu}(x)$. Note the added term under the square root sign, which is the contribution of e.
    Now,

$$t = \frac{\hat{y}(x) - y(x)}{SE_{\hat{y}(x)}}$$

has Student's t distribution with $v = n - 2$ degrees of freedom. For a prescribed confidence coefficient, an interval for $y(x)$ based on the pivotal method will have limits

$$L_y(x) = \hat{y}(x) - t SE_{\hat{y}(x)}$$
$$U_y(x) = \hat{y}(x) + t SE_{\hat{y}(x)}$$

As usual, t is the critical value obtained from table 3 for $v = n - 2$ degrees of freedom and the given confidence coefficient.
    Since $y(x)$ is not a parameter, this interval is not strictly a confidence interval. To avoid misusing the term *confidence interval*, we will call this a *prediction interval*. In the next example we will see how this interval can be used to quantify possible errors of prediction.

EXAMPLE 15.5   Some children suffering from dwarfism can be cured by a prolonged treatment with a human growth factor hGH (human growth hormone). Because for some children this factor is lacking in their systems, and for others it is present but not properly utilized, there is great variability in the way they respond to the drug. For those lacking hGH, its application can quickly return them to normal growth patterns. For those whose systems cannot properly utilize this factor, its application may lead to no improvement whatsoever.

Because a prolonged treatment with hGH is expensive, is restrictive to the patient, and has some unpleasant side effects, it is useful to have a way of predicting how it will affect the growth rate of each child. It was shown in a study by Rudman and co-workers [30] that there is a close linear relationship ($r^2 = .94$) between the growth rate achieved in a long-term treatment schedule and that achieved in an initial trial period in which hGH is administered for ten days. This relationship makes it possible to predict in advance the average long-term-treatment growth rate, Y, for a child who achieved growth rate X during the preliminary trial.

The following data are the average growth rates, in centimeters per week, for the 18 patients studied by Rudman and colleagues.

X	Y	X	Y
.133	.202	.083	.158
.075	.123	.058	.094
.125	.150	.158	.206
.108	.165	.025	.067
.025	.077	.017	.060
.050	.092	.125	.150
.025	.098	.200	.254
.183	.227	.150	.215
.117	.167	.142	.181

Source: Rudman, Kutner, Goldsmith, and Blackston [30]. Copyright 1979, the Endocrine Society.

The values of the statistics needed in calculating the least-squares predictor $\hat{y}(x)$ and its standard error $SE_{\hat{y}(x)}$ are as follows:

$$n = 18 \qquad S_{xx} = .0562 \qquad \hat{\beta}_1 = .9898$$
$$\bar{x} = .0999 \qquad \hat{\beta}_0 = .0503 \qquad s = .01496$$

Given the trial growth rate x, the predicted extended treatment growth rate would be

$$\hat{y}(x) = .0502 + .9898x$$

Note that since the slope is very nearly 1, the predicted growth rate for the extended period is about .05 centimeter per week greater than that observed for the trial period.

Bounds for the error in this prediction will be obtained by calculating the 90% prediction interval. The calculations are very similar to those of the last section. At each value of x, the standard error is

$$SE_{\hat{y}(x)} = .01496 \sqrt{1 + \frac{1}{18} + \frac{(x - .0999)^2}{.0562}}$$

The critical value from table 3 for 90% confidence and $v = 18 - 2 = 16$ degrees of freedom is

$$t = 1.746$$

Consequently, the bounds for the prediction interval are

$$L_y(x) = \hat{y}(x) - 1.746 SE_{\hat{y}(x)}$$

and

$$U_y(x) = \hat{y}(x) + 1.746 SE_{\hat{y}(x)}$$

These values can be calculated as x values are obtained in the experiment. But more useful tools for a physician who doesn't have access to a computer are the curves of $L_y(x)$ and $U_y(x)$ versus x, plotted on graph paper. The prediction limits can then simply be read from the graph for any value of x. The graph for the growth rate data is given in display 15.2. The scatter diagram of the data is also shown on the graph. The calculations from which these curves were plotted are organized in the following table.

x	$\hat{y}(x)$	$SE_{\hat{y}(x)}$	$L_y(x)$	$U_y(x)$
.00	.0502	.0166	.0212	.0792
.02	.0700	.0162	.0418	.0982
.04	.0898	.0158	.0622	.1174
.06	.1096	.0156	.0824	.1368
.08	.1294	.0154	.1025	.1563
.10	.1492	.0154	.1223	.1760
.12	.1690	.0154	.1420	.1959
.14	.1888	.0156	.1616	.2160
.16	.2086	.0158	.1809	.2362
.18	.2284	.0162	.2001	.2566
.20	.2482	.0166	.2191	.2772

For example, if a child achieved a growth rate of .12 centimeter per week during the trial period, with 90% confidence one would predict that his or her growth rate would be between .142 and .196 centimeter per week during the long-term treatment.

Simultaneous Prediction Intervals

Because of the random term e in the expression for $\hat{y}(x) - y(x)$, the Working-Hotelling simultaneous confidence interval for $\mu(x)$ does not extend to the prediction interval for $y(x)$. It is now necessary to use a device like the Bonferroni inequality to achieve prescribed overall confidence for several prediction intervals. This device requires specifying in advance how many intervals are to be covered by the joint confidence coefficient.

As before, if k intervals are to have an overall confidence level of at least $1 - \alpha$, each should have confidence $1 - (\alpha/k)$. The critical t values for a number of values of k can be read from table 3 by rounding to the next larger (or next smaller) confidence coefficient. For example, if $k = 25$ pa-

Display 15.2

Graph for computing predicted average growth rate per week for extended treatment with growth hormone. Prediction line with 90% prediction limits shown; dots indicate data from Rudman study.

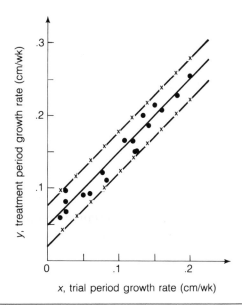

tients are to be tested for treatment with hGH and the physician wants overall confidence 90%, or $.90 = 1 - .10$, for all predictions, the individual intervals should be computed for a confidence coefficient of $1 - (.10/25) = .996$. Since $n = 18$, and thus $v = 16$, for the growth rate data, the critical value $t = 3.252$, corresponding to a confidence coefficient of $.995$, will provide overall confidence of at least $1 - (25 \times .005) = .875$. The critical value $t = 3.686$, corresponding to a confidence coefficient of $.998$, yields overall confidence of at least $1 - (25 \times .002) = .95$. If the stringency of the requirement for 90% confidence is not great, the smaller t value could be used. But for either choice the widths of the intervals do not differ appreciably.

◤
EXERCISES

15.5   Refer to exercise 14.1 (p. 463). Predict the college GPA of an entering college student with a high school GPA of 2.3. Use a 95% prediction interval.

15.6   Refer to exercise 15.5. Predict, with simultaneous confidence of at least 95%, the college GPAs of three students with high school GPAs of 2.0, 2.5, and 3.0, respectively.

◤ SECTION 15.6

Transformations to Straighten Curves and Equalize Dispersions*

Transformations of variables can be used to extend the scope of simple linear regression to certain classes of curvilinear regression models. The

---

* Optional material; not used in subsequent chapters.

curves are "straightened" by transforming one or both of the dependent and independent variables and a linear regression is fitted by the methods of chapter 14. Tests and confidence intervals for parameters and prediction intervals can be applied to the straightened data. Then, a curve fitting the original data can be obtained by reversing the transformation(s) and the inferences can be applied to the resulting regression model.

We will illustrate this methodology for the important power transformations introduced in section 12.3. These transformations can be applied directly, or after preliminary transformation, in a broad class of problems. A simple but effective method of fitting power transformations, attributed to J. W. Tukey, will be given in the next subsection.

A feature sometimes seen in scatter diagrams of regression data is a dispersion that increases (or decreases) with increasing values of the dependent variable. This feature indicates a violation of the model assumptions that tends to desensitize the inference procedures. The ability of the power transformations to equalize dispersions, which was the subject of section 12.3, can be used to correct this problem. Often the same transformation will equalize dispersions and straighten the curve. In other situations, the dispersion is nearly proportional to the magnitude of the dependent variable, and a logarithmic transformation is an effective dispersion stabilizing transformation. An example will be given in this section.

Curve Straightening Using Power Transformations

The power transformations can be applied to problems for which both independent (X) and dependent (Y) variables are strictly positive and for which the regression curve (suggested by the scatter diagram) is either strictly increasing or strictly decreasing with increasing X. For example, the height and weight data for the students of display 1.1 have this feature. The scatter diagram is given in display 15.3. We speculated about the upward curvilinearity of these data in chapter 4 and we will now fit a curvilinear model by using a power transformation to straighten the data.

Recall that the power transformations are of the form

$$
\begin{array}{ll}
y^p & \text{if } p > 0 \\
\log(y) & \text{if } p = 0 \\
-y^p & \text{if } p < 0
\end{array}
$$

where the logarithm is the natural logarithm. Tukey [32, chapter 6] provides both a suggestive terminology and a simple method for fitting these transformations. Our treatment will be rather brief, and further details can be found in the reference.

Tukey views the transformations as forming a *ladder*, with the position on the ladder given by the index p. We will begin at the current position on the ladder, $p = 1$, (the identity transformation that transforms y into $y^1 = y$), and proceed up the ladder (in the direction of increasing p) or down the ladder (decreasing p), as the fitting method directs.

The fitting method depends on the choice of three points. These points are selected from the scatter diagram to be near the center of the

Display 15.3

Scatter diagram of weight versus height for the students of display 1.1 (p. 9)

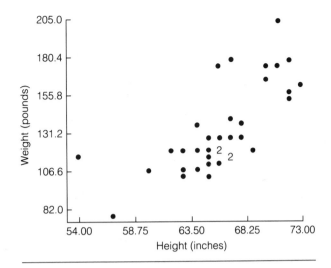

scatter (in the y direction), the first one with x coordinate near the smallest x value in the data, the second with x coordinate about halfway between the smallest and largest x coordinates, and the third near the largest x value in the data. These points can, but need not be, points from the data set itself. The goal is to find points that are representative of the center of the scatter in the middle and at the extremes of the scatter diagram. For example, the points chosen from the height-weight scatter diagram of display 15.3 on the basis of this criterion are

point   1   2   3

(54,90), (64,125), (72,185)

Note that the points (54,90) and (72,185) do not occur in the data set, but rather are constructed to be near the center of the scatter at the ends of the diagram.

The fitting method consists of passing a line through points 1 and 2 and a second line through points 2 and 3. The slopes of the two lines are compared. If they differ, a power transformation is applied to either the x or y coordinates *of just these three points* in an attempt to equalize the slopes. Since the transformation is applied to only three points, the calculations are easily carried out by hand.

The fitting method provides a rule that tells in which direction to change the ladder index to approach equality of slopes. It does not tell by how much to change the index, however, so the method will usually require more than one iteration to equalize the slopes. Even so, it is seldom necessary to make more than three or four iterations, since transformations with ladder indices that are "not too different" will have roughly the same effect on the data. Thus, it is not essential to exactly equalize the slopes.

The following method for determining the direction to change the ladder index at a given iteration is a variant of Tukey's "curve bulging" rule [36, chapter 6]. The two lines passing through the three points can be viewed as forming the head of an arrow with the middle point at the tip. The next iteration of the fitting process moves the ladder index in the direction the arrowhead points. For example, note that the arrow points to the right and downward for the three initial points of the height-weight data plotted in display 15.4. Thus, if we choose to transform the X variable, the next (here, the first) iteration would move up the ladder, that is, use a larger value of p than the current one. If the transformation is performed on the Y variable, the next iteration would move down the ladder to a smaller value of p.

Display 15.4

Lines through the points (54,90), (64,125), (72,185) indicating the ladder index direction for a power transformation of the height-weight data of display 1.1

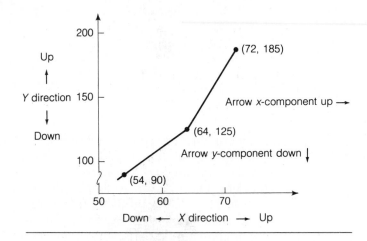

The choice of whether to transform the X variable or the Y variable—or possibly both—is made on the basis of whether, in addition to a curvature, the scatter diagram indicates a trend in dispersion. If it does, then transforming Y will often equalize the dispersion as it straightens the curve. If the dispersion is already reasonably constant, then transforming X will preserve this property while straightening the curve. A mild increase in dispersion is seen in the height-weight data of display 15.3, so we will transform the Y variable.

Denote the three selected points, in the order of increasing x values, by (x_1, y_1), (x_2, y_2), (x_3, y_3). At each stage of the computation we will compute the two slopes

$$S_1 = \frac{y_2 - y_1}{x_2 - x_1}$$

and

$$S_2 = \frac{y_3 - y_2}{x_3 - x_2}$$

It will be possible to make a crude judgment of how large a step to take in the value of p by computing a measure of the relative difference in slopes:

Rel. Diff. $= |S_1 - S_2| / (|S_1| + |S_2|)$

For example, the slopes and relative difference for the initial points of the height-weight data are

$$S_1 = \frac{125 - 90}{64 - 54} = 3.50$$

$$S_2 = \frac{185 - 125}{72 - 64} = 7.50$$

and

$$\text{Rel. Diff.} = \frac{|3.50 - 7.50|}{3.50 + 7.50} = .364$$

The next Y-direction for p is down the ladder, so we will try $p = 0$, which corresponds to the log transform. The three values $y_1 = 90$, $y_2 = 125$, and $y_3 = 185$ are transformed, yielding the new values $y_1 = \log(90) = 4.50$, $y_2 = \log(125) = 4.83$, and $y_3 = \log(185) = 5.22$. Verify that the slopes and relative differences for the new points, (54, 4.50), 64, 4.83), and (72, 5.22), are

$$S_1 = .033 \qquad S_2 = .049 \qquad \text{Rel. Diff.} = .197$$

The slope S_1 is still smaller than S_2, which means that the arrowhead formed from these lines again points downward in Y. Thus, the next step of the process will be to take a still smaller value of p. We take $p = -1$, which corresponds to the reciprocal transformation $-y^{-1} = -(1/y)$. Apply this transformation to the three original y values and verify the following slopes and relative difference:

$$S_1 = .000311 \qquad S_2 = .000324 \qquad \text{Rel. Diff.} = .021$$

We note that the relative difference has again decreased and that S_1 is still smaller than S_2. The downward direction in ladder index is again indicated for the next iteration.

Moving to the ladder index $p = -2$ produces the following.

$$S_1 = 5.95 \times 10^{-6} \qquad S_2 = 4.35 \times 10^{-6} \qquad \text{Rel. Diff.} = .155$$

The slope S_1 is now larger than S_2, which indicates that we have passed up the slope-equalizing index and should next go back up the ladder. Moreover, the large increase in the relative difference indicates that we have come much too far from the index $p = -1$. This suggests that we should go back to a value close to -1.

A calculation at $p = -1.1$ yields a relative difference of .003 and an indication of a downward direction. At $p = -1.2$ the relative difference is .015 and an upward direction is indicated. Thus, $p = -1.1$ would be the best index to the nearest tenth. However, the transformation corresponding to $p = -1$ is nearly as good and has the added advantages of being more familiar and easier to apply.

The next step in the process is to transform all of the original Y values (weights) using the power transformation with $p = -1$. A scatter diagram will then be made of the resulting data to check the effect of the transformation. This scatter diagram is given in display 15.5

Display 15.5

Scatter diagram of the
height-weight data of
display 1.1; the weight
values have been
transformed with the
power transformation
for $p = -1$

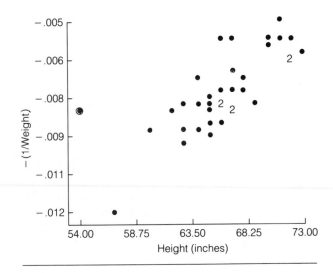

Note that, with the exception of the circled point, the scatter of points now
appears to be both straight and of reasonably uniform dispersion.

The fitting method can be somewhat sensitive to the choice of the
initial three points. However, the quality of the final transformation will be
judged on the basis of the scatter diagram of the transformed data. A transformation that straightens the data and corrects for variation in dispersion
may be missed with a poor choice of points.

On the other hand, even a good choice may not lead to a transformation that does both jobs. In this case, a useful strategy is to correct for
dispersion by transforming Y, then follow with a power transformation of X
to straighten the variation-equalized data. This was not necessary for the
height-weight data.

To obtain a fitted regression curve for the height-weight data, the least-
squares method is used to fit a straight line to the transformed data. (For
reasons to be discussed in the next section, the circled point has been
omitted from the calculation.) The fitted intercept and slope are

$$\hat{\beta}_0 = -.03035 \quad \text{and} \quad \hat{\beta}_1 = .0003422$$

Since weight (y) has been transformed by the expression $-(1/y)$, the fitted
line at height x is

$$-\frac{1}{y} = -.03035 + .0003422x$$

The fitted regression curve is now found by solving this equation for y:

$$y = \frac{1}{.03035 - .0003422x}$$

A plot of this curve, superimposed on the scatter diagram of the original
data, is given in display 15.6. The curve tends to underestimate the points at

Display 15.6

Height-weight scatter diagram with fitted regression curve

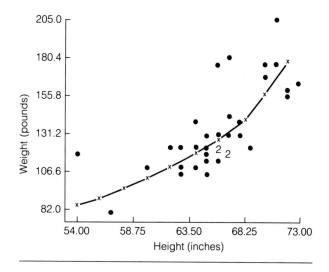

the upper end of the diagram. This is, in part, due to a phenomenon called *transformation bias* (see Miller [43]). However, it also suggests that different curves may be needed to describe the height versus weight "law" for men and women.

An example in which the methods of this chapter are applied to find a confidence interval for a theoretically derived regression curve is given in the next subsection.

Exponential Regression Curves

Many phenomena in the physical and natural sciences can be shown, theoretically, to follow an *exponential regression law* of the form

$$y = \alpha e^{\beta x}$$

From a curve-fitting perspective, this law can be viewed as suggesting that regression curves for data generated by such phenomena can be straightened by the log transformation. The reason for this suggestion is that the log transformation of the exponential law expression yields

$$\log(y) = \log(\alpha) + \beta x$$

This is a linear regression function of the form $\mu(x) = \beta_0 + \beta_1 x$, with intercept $\beta_0 = \log(\alpha)$ and slope $\beta_1 = \beta$.

The various tests and confidence intervals for linear regression can be extended to data following an exponential law by the following strategy. First, linearize the data by applying the log transformation to the y variable. Apply the tests and confidence intervals to the linearized data. Then, use an exponential transformation to return to the original data units. This idea will be illustrated in the next example.

EXAMPLE 15.6

One of the standard engineering tests applied to automobile tires is designed to determine the force with which they grip particular road surfaces. For a given surface the tire under study is mounted on a test trailer and pulled at a specified velocity. By a braking mechanism, a given amount of drag (measured in percentages) is applied to the tire and the force (in tens of pounds) with which it grips the surface is measured. In this example we will estimate the relationship between grip force and drag. Percentage drag will be taken as the independent variable X and grip force as the dependent variable Y.

It is known that the relationship of grip force to drag follows an exponential law over the observed range of the drag variable from 10% to 100% ($X = 0\%$ denotes no braking, and at $X = 100\%$ the brake is locked.) The parameters α and β of the exponential law are then numerical characteristics of the particular tire type and road surface, and they can be used to compare and design tires.

The following data represent the grip forces for 19 tires of a given type tested at six drag values. The trailer velocity for this test was 30 miles per hour.

DRAG (X)	GRIP FORCE (Y)
10	55, 46, 61
20	51, 41, 58
30	47, 36, 48
50	39, 31, 40
70	30, 28, 34
100	25, 20, 20, 20

Data courtesy of R. Schrader, Department of Mathematics and Statistics, University of New Mexico, Albuquerque.

The problem is to find a 95% confidence interval for the regression curve of grip strength versus drag.

Even when data are theoretically determined to follow an exponential law, it is a good idea to check whether this model provides a reasonable fit. Non-ideal experimental conditions can cause data to deviate from their theoretical form. A rough check on the model can be carried out by fitting a power transformation to the data by the method given above. The exponential model would be "confirmed" if the log transformation does a good job of straightening the data.

Verify that, when applied to the three points (10,53), (40,39), and (100,22), the curve-straightening method leads to the ladder index $p = -.2$ (to the nearest tenth). Since this index is close to the value $p = 0$, corresponding to the log transformation, the exponential model should fit the data well.

The limits of the confidence interval for the regression curve will be obtained from the confidence limits for the regression line of $\log(y)$ versus x. If $L(x)$ and $U(x)$ denote the confidence limits for this line, then the confidence limits for the curve will be

$$L'(x) = e^{L(x)} \quad \text{and} \quad U'(x) = e^{U(x)}$$

Display 15.7

Fitted exponential regression curve and 95% confidence interval (region between outside lines) for true regression curve of grip force versus percentage drag for automobile tire data

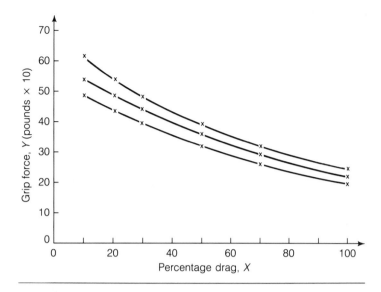

Recall that the confidence limits for the regression line are of the form

$$L(x) = \hat{\mu}(x) - wSE_{\hat{\mu}(x)}$$

and

$$U(x) = \hat{\mu}(x) + wSE_{\hat{\mu}(x)}$$

where $\mu(x)$ is the estimated regression line and $SE_{\hat{\mu}(x)}$ is the standard error for the regression line estimator. The critical value is $w = \sqrt{2F}$, where F is the critical value of the F distribution for the given confidence coefficient and degrees of freedom 2 and $n - 2$.

The computed regression line for the log-transformed data is

$$\hat{\mu}(x) = 4.0967 - .010241x$$

with standard error

$$SE_{\hat{\mu}(x)} = .1234 \sqrt{\frac{1}{19} + \frac{(x - 49.474)^2}{19{,}894.74}}$$

A 95% confidence interval for the line requires the critical value of F from table 4 for 2 and 16 degrees of freedom (since 17 is not listed and we choose to be conservative in this application). We find

$$F = 3.63$$

Thus

$$w = 2.694$$

The calculations are summarized in the following table, and the resulting curves are graphed in display 15.7.

x	$\hat{\mu}$	SE	wSE	L	U	$\hat{\mu}'$	L'	U'
10	3.994	.0447	.1204	3.874	4.114	54.272	48.135	61.191
20	3.892	.0383	.1032	3.789	3.995	49.009	44.212	54.326
30	3.789	.0330	.0889	3.700	3.878	44.212	40.447	48.327
50	3.585	.0283	.0762	3.508	3.661	36.053	33.381	38.900
70	3.380	.0335	.0902	3.290	3.470	29.371	26.843	32.137
100	3.073	.0525	.1414	2.931	3.214	21.607	18.746	24.878

EXERCISE

15.7   Confirm the computations of example 15.6.

Pretransforming Data in Order to Apply the Power Transform Fitting Method

The power transformations impose restrictions on the data to be straightened. The X and Y variables must be strictly positive in order to fit a transformation with $p < 1$. Moreover, the regression curve must be strictly increasing or strictly decreasing. There is also an implication that both variables are unbounded—that is, they are free to take on arbitrarily large values.

These constraints are not as restrictive as they seem, because the data can often be pretransformed to put them in the proper form for a power transformation.

For example, if one or both variables are bounded from below, but the bounds are not 0, then suitable constants can be added to the data to make them positive. Similarly, multiplication by -1 can be used to convert strictly negative variables into positive ones. These operations can be undone after the fitting process to find the regression curve for the original data.

When data are bounded, then an "unbounding" transformation can be used before power transformation. For example, when data are confined to the interval from 0 to 1, the transformation $y/(1 - y)$ converts them to data of the appropriate type; values near 0 remain near 0 with this transformation, but values near 1 can be arbitrarily large, thus removing the upper boundary.

This pretransformation is used in *logistic regression*. The dependent variable is a proportion, y, whose regression on a measurement variable, x, is of interest. The dependent variable is subjected to the *logistic or logit transformation:*

$$\text{logit}(y) = \log\left(\frac{y}{1 - y}\right)$$

A linear regression is now fitted to the logit (y) versus x data. A fitting method called *weighted least squares* is often used for this purpose. We will not pursue this topic.

A data-analytic interpretation of the logit regression strategy leads to a broad and useful class of regression models for data whose dependent variable is a proportion. The log transform is clearly being used to straighten data that have first been made to conform to power transformation requirements by the $y/(1 - y)$ transformation. This suggests that if another power transformation does a better job of straightening the data, it can be used in place of the log transformation in the regression. The proportions can first be transformed by $y/(1 - y)$, and then the curve straightening strategy can be used to determine a more appropriate power transformation. An example will be given shortly.

More generally, if the data are confined to an interval from A to B, say, (but are not equal to either A or B), then the pretransformation

$$\frac{y - A}{B - y}$$

will bring them into conformity with the requirements of the power transformations. For example, the sample correlation coefficient, r, is confined to the range from -1 to 1. In a regression problem for which the correlation coefficient is the dependent variable, the pretransformation $(r + 1)/(1 - r)$ could be applied before determining a power transformation by curve straightening.

R. A. Fisher used this pretransformation, followed by the log transformation, to approximate the distribution of the sample correlation coefficient. The combined transformation yielded a variable that ranged over the entire number line. He then used a normal distribution to approximate the distribution of this variable. This is an example of the use of transformations to create variables with new probability distributions. We saw examples of this use for linear transformations in chapters 5 and 6. Unfortunately, we will not be able to pursue this important topic in this text.

An application of the use of a pretransformation in regression is given in the next example.

EXAMPLE 15.7

A problem that arose in a medical research project is the following. A fluid, initially containing a known amount of insulin, is circulated from a reservoir through a rat's liver and then back into the reservoir (a process called perfusion). During the passage through the liver, some of the insulin is bound to cells and degraded into other chemicals. These degraded products can be identified by radio-assay methods. Consequently, by a periodic sampling of the fluid in the reservoir, the proportion of the initial amount of insulin still remaining (as undegraded insulin) at these times can be calculated. By examining the curves of undegraded insulin versus time, formed under various environmental conditions, we can study the mechanisms affecting binding and degradation of insulin by liver cells in the rat's body—and, by extrapolation, in our bodies as well.

The accompanying data constitute the proportions of undegraded insulin (Y) at nine times (X) (in seconds) after the beginning of the perfusion process. The proportions are averaged from ten different experiments carried out under the same experimental conditions.

TIME (X)	PROPORTION UNDEGRADED INSULIN (Y)
5	.91
10	.88
15	.82
20	.71
30	.63
45	.46
60	.40
75	.36
90	.34

Data courtesy of R. Greenberg, M.D., School of Medicine, University of New Mexico, Albuquerque.

A graph of the data is given in display 15.8.

Display 15.8

Graph of proportion of undegraded insulin (Y) versus time (X)

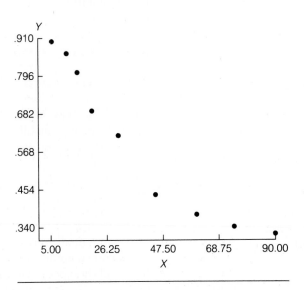

A good regression description will be sought by applying the power transform fitting method to the $y/(1 - y)$ versus x data.

After applying the $y/(1 - y)$ transformation, the three points with x coordinates 5, 30, and 90 can be used in the fitting process. This process leads to a curve-straightening index of $p = -.7$ (to the nearest tenth). The value $p = -1$, corresponding to the reciprocal transformation, $-1/y$, is a more convenient one and will be used for the model.

To compare the result of the fitting process with the use of the logistic regression method in this problem, the coefficient of determination, r^2, can be used as a measure of the straightness of the transformed data. For the log transform, corresponding to the logit fit, this coefficient is $r^2 = .869$, while for the reciprocal transformation it is $r^2 = .982$. Consequently, the reciprocal transformation does a better job of straightening the $y/(1 - y)$ versus x data.

The fitted regression line is

$$-\left(\frac{y}{1 - y}\right)^{-1} = .0590 - .02392x$$

The estimated regression curve for the fitted model can now be found by solving for y:

$$y = \frac{1}{.9410 + .02392x}$$

EXERCISE

15.8 (a) Verify the details of example 15.7.
      (b) Plot the fitted curve on the scatter diagram of the original data.
      (c) Plot the residuals, $y -$ (fitted y), as a function of x and discuss any systematic variation seen in this plot.
      (d) Use the method of this section to straighten the original insulin (y) versus time (x) data. Derive the corresponding fitted model by solving for y and superimpose the resulting values on the plot made in part (b). Justify the use of the $y/(1 - y)$ pretransformation by discussing the unpleasant features this fit has that the fit derived in part (a) does not.

Equalizing Dispersions with Power Transformations

An assumption made for the linear regression model is that the variation of the y values (σ) is the same over the entire range of x values. The construction of the least-squares estimators and their standard errors depend on this assumption, and when it is violated, the statistical performance (usually the sensitivity) of the statistical procedures given earlier is degraded.

A trend in variability can often be diagnosed by means of a scatter diagram of the data. It is usually easy to see variations in the degree of scatter of the points. When the data are exceptionally "noisy" it is often easier to detect a trend in variability by observing the variations in the distance between the quartile lines in an exploratory regression plot of the kind discussed in chapter 4.

The form of dispersion trend seen most often in practice is an increase in dispersion with increasing values of the regression curve. This shows up in a scatter diagram as a triangular or wedge-shaped point scatter. Thus, a diagram that indicates an increasing regression curve from left to right would also display an increase in dispersion from left to right.

Power transformations can often be used to equalize dispersions in regression problems. In particular, the logarithm is often used successfully for this purpose. The transformation is applied to the dependent (y) variable in order to equalize dispersions. It may or may not then be necessary to apply a transformation to the independent variable to achieve (or maintain) a straightened curve.

An illustration of the use of the logarithm to equalize dispersions is given next.

EXAMPLE 15.8

A few years ago the computing center management at the University of New Mexico evaluated a new accounting system, which had several useful features not possessed by the system then in use. At that time each major accounting unit, such as an academic department, was assigned a monthly factor, which was roughly proportional to its total (dollar) computing volume for that month. This factor depended on a complex formula for estimating and combining charges of a variety of services, not all of which were directly observable. One of the tests of the new system was to determine how well it was able to reproduce the factors assigned by the old system.

The following data represent the factors assigned by the old and the new accounting systems for a given month of operation and for those accounting units for which factors were computed.

	FACTOR	
DEPARTMENT	NEW ACCOUNTING SYSTEM (X)	OLD ACCOUNTING SYSTEM (Y)
Anthropology	.0084	.0041
Architecture	.0066	.0060
Art	.0001	.0002
Biology	.0078	.0154
Business and administration science	.0245	.0549
Business research (bureau of)	.0171	.0118
Chemical engineering	.0062	.0157
Chemistry	.0073	.0132
Civil engineering	.0016	.0019
Communicative disorders	.0005	.0002
Computing center	.1759	.1376
Computing and information science	.0574	.0568
Data processing center	.2215	.1757
Economics	.0411	.0540
Education administration	.0062	.0045
Education foundations	.0235	.0200
Electrical engineering	.0517	.0588
Elementary education	.0084	.0025
Geography	.0068	.0053
Geology	.0027	.0023

(table continued on next page)

| | FACTOR | |
DEPARTMENT	NEW ACCOUNTING SYSTEM (X)	OLD ACCOUNTING SYSTEM (Y)
Government research (division of)	.0033	.0022
Guidance and counseling	.0000	.0000
Health, physical education, and recreation	.0001	.0001
History	.0001	.0001
Institute research	.0015	.0015
Law	.0003	.0003
Linguistics	.0040	.0024
Mathematics	.0574	.0818
Mechanical engineering	.0393	.0412
NROTC	.0037	.0044
Physics and astronomy	.0074	.0088
Political science	.0015	.0008
Psychology	.0052	.0044
Secondary education	.0001	.0001
Sociology	.0092	.0069
Speech communication	.0007	.0003

Data courtesy of S. Bell, Department of Computing and Information Science, University of New Mexico, Albuquerque.

A scatter diagram of the data is given in display 15.9.

Display 15.9

Scatter diagram of old
versus new accounting
system factors.
Numbers indicate the
number of points at the
given plot position. The
symbol @ indicates
that more than 9 points
were plotted in the
lower left corner of the
diagram.

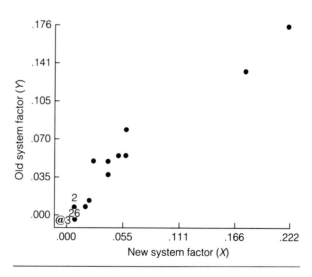

Note the variation in dispersion in the left-hand portion of the diagram. Also note the large gap in the data produced by the two largest computer users. Because the data are already reasonably straight, the log transformation applied to the y variable to equalize dispersion is accompanied by a compensating log transformation of the x variable. The result is shown in display 15.10.

Display 15.10

Scatter diagram of the
logarithms of the old
versus new system
factors (the point for
which $x = y = 0$ has
been deleted)

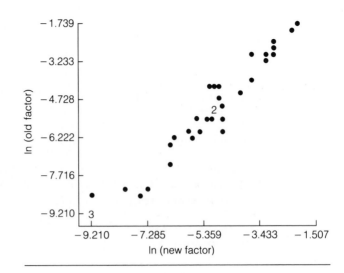

Note that not only have the transformations equalized the dispersion but they have also "tamed" the two outlying points.

The coefficient of variation for the transformed data is $r^2 = .950$, indicating a high degree of linearity. The linear regression parameter estimates are

$$\beta_0 = .000815 \quad \text{and} \quad \beta_1 = 1.0137$$

The regression equation for the transformed data is

$$\log(y) = \beta_0 + \beta_1 \log(x)$$

In terms of the original variables, this equation is

$$y = e^{\beta_0} x^{\beta_1}$$
$$= e^{.000815} x^{1.0137}$$
$$= 1.0008 x^{1.0137}$$

Note that this result is very close to the equation $y = x$, which would hold for the two accounting factors if $\beta_0 = 0$ and $\beta_1 = 1$. The null hypothesis that the parameters have these values, against the alternative that either $\beta_0 \neq 0$ or $\beta_1 \neq 1$, is accepted for these data by a 5% test that will be constructed in exercise 15.20. So we would conclude at the 5% significance level that there is no reason to believe that the new system does not satisfactorily reproduce the accounting factors for the old system.

Our treatment of transformations has been, necessarily, brief. For more complete discussions of this important topic, see the texts by Box, Hunter, and Hunter [3], Neter, Wasserman, and Kutner [24], and Tukey [32].

■ SECTION 15.7

Influence of Outliers on the Regression Line*

The method of least squares is linked to the normal distribution in the sense that least-squares estimates are best by a certain statistical criterion when the dependent variable is normally distributed. From what has been demonstrated in previous chapters concerning the loss of sensitivity of normal theory methods in the presence of outliers, it should not be surprising that outliers can also cause problems for least-squares regression. Not only can they cause serious loss of sensitivity for both hypothesis tests and confidence intervals—by inflating the model 2 standard deviation s—but they can also affect the fit of the line. The reason for this is that outliers implicitly receive more weight in the fitting process than do the main body of points, and thus they tend to pull the line toward themselves.

The effect of a particular point on the fit can be demonstrated by recomputing the least-squares line with that point removed. This technique is used on the four circled points in the residual plot of the weight versus height data for women given in display 15.11(a). This plot was used earlier in

Display 15.11

Residuals of women's weight versus height. (a) From the least-squares line; circled points discussed in text. (b) From robust line fitted by eye.

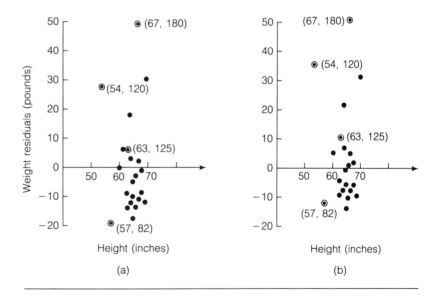

display 14.6(b) as a diagnostic for curvilinearity. Here we will use it to identify special points in the data set.

The first point that looks interesting is (height, weight) = (67, 180), the point with the largest residual. The effect on the least-squares line of removing this point is shown in display 15.12. In fact, the effect on the line is surprisingly modest. The new least-squares line has a somewhat smaller

---

* Optional material; not used in subsequent chapters.

Display 15.12

Original least-squares
line for the women's
height-weight data and
least-squares lines
obtained after deleting
indicated points from
data

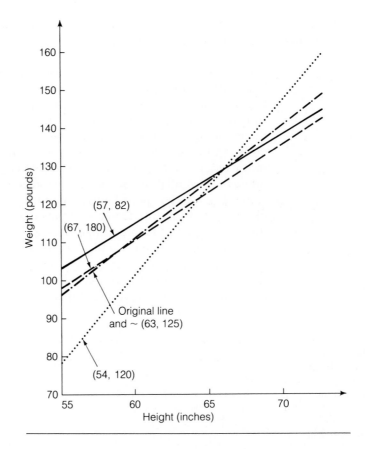

slope, but otherwise it lies reasonably close to the original. The major effect
of removing this point is to decrease the estimated standard deviation s from
17.604 to 13.926. Thus this point is a major contributor to the insensitivity of
confidence intervals and tests for the parameters of the true regression line.

The point (54, 120) does not appear to be seriously out of line relative
to the absolute sizes of the other residuals. However, it has by far the most
disastrous effect on the regression line. With this point removed, the slope
of the line nearly doubles. Moreover, the linearity of the data set, as measured by r^2, improves from .304 to .455. The reason for this phenomenon is
that the influence, or leverage, a point has on the fit depends not only on the
size of the residual but also on how extreme (in the sense of distance from $\bar{x}$)
the x coordinate of the point is. Points near the left or right extremes of the
scatter diagram have relatively larger influences than those in the middle.
This result provides the reason that we looked for outliers in the height (X)
variable in section 4.8. It was discovered that the height of 54 inches was a
recording error and should have, instead, been 64 inches (5 feet 4 inches).
The penalty for not correcting this error now becomes apparent.

In the same exploratory analysis it was shown that the height of 57
inches also constituted an outlier in the height distribution. It is to be

expected that the point (57, 82) will also be a high leverage point in the least-squares analysis. In fact, it is, but in a surprising way. Its presence has actually made the data set look *more* linear rather than less; its removal decreases the r^2 value from .304 to .194. The effect of this point on the least-squares line was to pull the left end of the line down toward itself. This result is seen from the upward shift when the point is deleted. On the other hand, this point has had little effect on the variation around the line as measured by s. ($s = 17.360$ with the point deleted, 17.604 with it included.)

As a benchmark for the discussion above, it is useful to see what kind of influence one of the more usual points has on the least-squares line. For this reason, the computations were performed with the point (63, 125) deleted. It is seen from display 15.11(a) (p. 511) that this point is on the fringe of the main cluster of points; thus it should be at least as influential as any of them singly. In fact, the effect of removing this point is quite small. The resulting regression line is graphically indistinguishable from the original. Moreover, the other estimates are also relatively unaffected by the removal of this point. ($r^2 = .307$ and $s = 18.015$ with the point deleted, compared to $r^2 = .304$ and $s = 17.604$ when it is included.)

Strategies for Locating and Dealing with Outliers

The least-squares-fitting method works best when all the data points are relatively ordinary, like the point (63, 125). This is the reason it works best for the normal distribution, since the normal distribution tends to produce clusters of ordinary points. Moreover, in general, we want the least-squares regression line to describe the flow of the centers of the conditional distributions. Thus it is undesirable to have a few rogue points distracting the line from the center of the main concentration of frequency. Unfortunately, it is not always possible to detect when this situation is happening. Compensating influences of outliers may produce a line that looks fairly reasonable. But it is highly likely that these outliers will seriously desensitize the statistical procedures given in earlier sections.

Thus it is very desirable to be able to identify the points likely to cause problems. It would seem that a plot of the residuals from the least-squares line, such as given in display 15.11(a), along with an outlier diagnostic like the one provided by the box plot, should make this identification possible. However, since the outliers have already affected the least-squares fit in a manner that tends to decrease their residual values, this method does not always work well. What is needed is a robust fitting procedure that is not adversely affected by the outliers. Their presence is then more likely to be seen in an analysis of the residuals from the robust line.

Once the troublesome outliers have been identified, several techniques are possible. First, an intensive study of the origins of the outliers is necessary. Often they will be found to be caused by faulty measurement processes or, more simply, by recording errors, as was the point (54, 120) in the height-weight data. Sometimes different mechanisms or laws are at work in the generation of the data, leading to occasional excessive measurement values. Moreover, one or more members of the sample may be unusual or atypical, leading to out-of-line values of one or both variables. (An example of this phenomenon was given in example 15.8.) In such cases the deviant values would either be corrected or be removed from the data set for separate analysis and description. Then the revised data set could be reanalyzed by the least-squares method and subjected to the inference procedures given earlier in the chapter.

When no unusual cause can be attributed to the outliers, it is likely that the assumed normality of the random errors in the linear regression model (model 2) is sufficiently invalid to cause real problems with the normal theory inference procedures. As seen in the last section, a transformation of variables, such as a log transformation, will sometimes correct the outlier problem by "pulling in," thus downweighting, the outliers.

When it is desirable to retain the original variable units, it will be necessary to downweight the influence of the outliers in another way. Robust inference procedures have been developed to do this. Unfortunately, these procedures require a digital computer to implement and are not yet available in standard statistical computer packages. Presumably this situation will be remedied eventually. In any event, these procedures are beyond the scope of this text.

Whether or not we can solve an outlier problem, it is still important to be able to identify one when it exists. Left unidentified, outliers can hide real effects or lead to seriously erroneous conclusions. The construction of a robust regression line and its use to identify outliers will be discussed next.

Robust Lines and Robust Residuals

The human eye is a remarkably good robust line fitter. Given a scatter diagram of the data on graph paper and a transparent plastic ruler, we can maneuver the ruler around on the graph until the top edge appears to pass well through the center of the cluster of points. The line drawn along this edge will provide a reasonable robust fit. The reason we get this result is that the eye tends to concentrate on the main body of points and ignore, or at least discount, those more distant from it. The author's eye-fit robust line for the height-weight data is shown in display 15.13.

Once this line has been drawn, two points can be selected from it to compute the slope and intercept. For example, the points (55, 87.5) and (70, 138) appear to lie on the eye-fit line of display 15.13. Now the slope β_1^* and intercept β_0^* of this line can be found as follows: If (x_1, y_1) and (x_2, y_2) represent these (or any) two points on the line, then

$$\beta_1^* = \frac{y_2 - y_1}{x_2 - x_1}$$

$$\beta_0^* = \frac{x_2 y_1 - x_1 y_2}{x_2 - x_1}$$

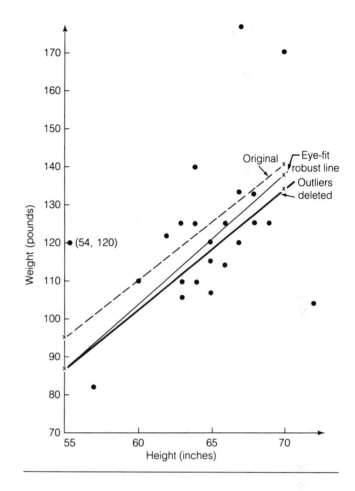

Thus with $x_1 = 55$, $y_1 = 87.5$, $x_2 = 70$, and $y_2 = 138$, we find

$$\beta_1^* = \frac{138 - 87.5}{70 - 55} = 3.37$$

Similarly,

$$\beta_0^* = \frac{(70 \times 87.5) - (55 \times 138)}{70 - 55} = -97.67$$

Thus the predicted values of y for the robust regression line can be computed from the expression

$$y^*(x) = -97.67 + 3.37x$$

Now, the residuals from the robust line,

$$r_i^* = y_i - y^*(x_i)$$

can be obtained. A graph of these residuals for the height-weight data is given in display 15.11(b) (p. 511).

A comparison of the residuals from the robust line with those from the least-squares line shows some interesting differences. The main body of points is now more symmetrically placed around zero. The point (57, 82) has more nearly joined this group, while the points (54, 120) and (67, 180) are more distinctly separated from it.

A diagnosis for outliers can be based on the absolute values of the residuals. A box plot of these quantities provides a somewhat more sensitive indicator of outliers than would a box plot of the residuals themselves. That is, if the residuals were actually normally distributed around zero, as the least-squares theory would like, a value designated as an outlier by the box plot of absolute residuals would be slightly less than a 1-in-20 chance occurrence (its probability is about .047), whereas a residual designated as an extreme outlier by this method would have a probability of about .005 (1 in 200). Because of the large effect even modest outliers in y can have at the extremes of the x range, it is useful to have this somewhat more refined outlier designation.

A box plot of the absolute residuals from the robust line is given in display 15.14, along with the box plot of absolute residuals from the least-squares line for comparison.

Display 15.14

Box plots of absolute weight residuals. (a) From least-squares line. (b) From robust eye-fit regression line. Locations of four points shown in display 15.4 are indicated.

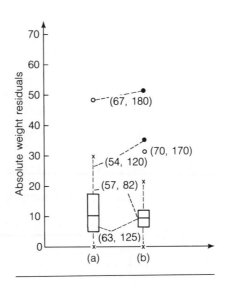

Note that only (67, 180) is classified as an outlier by the distribution of residuals from the least-squares line. On the other hand, the analysis based on residuals from the robust line identifies this and two additional points as outliers. The height error (54, 120) is identified as an extreme outlier. The high leverage point (57, 82) appears as one of the crowd relative to the robust line. The ordinary point (63, 125) becomes even more ordinary.

The least squares line computed with these three outlying points removed from the data set is shown in display 15.13. This line parallels the original least-squares line, shown as dashes, but is roughly 8 pounds in weight below it. Note that it is reasonably close to the robust line and thus better represents the trend of the main cluster of points.

If the outliers could be omitted from the sample on the basis of their not being from the population of interest—or for comparable reasons—the least-squares line could be used to estimate the population regression curve, and the usual tests and confidence intervals could be computed. As would be expected, these procedures will be more sensitive than the usual ones. The value of s is 9.803 for this line, compared to 17.604 for the original, and the linearity of the data set has improved from $r^2 = .304$ to $r^2 = .456$. On the other hand, the sample size is smaller by four than the original, and the value of S_{xx}, which appears in the standard error terms of both estimates, is also smaller. Both of these factors counter the gains in sensitivity somewhat—but not enough to cancel them out.

We can see that the minor outlier (70, 170) detected in the box plot of absolute residuals deserves attention by noting that if only the two extreme outliers are deleted, the data set has an estimate of σ equal to $s = 11.947$. The s value can be improved an additional 18% by removing this point as well.

For those who are put off by the subjectivity of the by-eye, robust-fitting method, another technique based on grouped medians will be given in an exercise.

Chapter 15 Quiz

1. How is a confidence interval for the slope of a regression line computed? How is the interval interpreted? Give an example.

2. What is the interpretation of the null hypothesis of zero slope for a regression line in terms of the two models of chapter 14?

3. In what sense is the (two-sided) test for zero slope a screening test for the fit of the linear regression model? In particular, what would be concluded about the model and what action would be taken if the null hypothesis were accepted?

4. How is the test for zero slope of a regression line for variables X and Y related to the test for zero correlation between these variables? If the hypothesis of zero correlation is accepted in the test, can it be concluded that the variables are independent? Explain.

5. Why is it necessary to make a distinction between finding a confidence interval for a regression function $\mu(x)$ at a single value of (the independent variable) x and finding intervals at more than one value of x?

6. How is a confidence interval for the regression line at a given x computed? How is this interval interpreted? Give an example.

7. What is the Working-Hotelling confidence interval for the regression line? How is it computed? Give an example.

8. How can a confidence interval for the intercept of a regression line be computed? Give an example.

9. How can joint confidence intervals be computed for the intercept and slope of a regression line? Give an example.

10. Why is it dangerous to extrapolate regression line estimates outside the range of the data? What can be done instead?

11. What is a prediction interval for a future observation $y(x)$ at a value x of the independent variable? How is such an interval computed? Give an example.

12. How can prediction intervals be computed at more than one value of x with given joint confidence?

The following questions are for the optional section 15.6.

13. What is the method for fitting a power transformation to straighten a curve? What forms of curves can usually be straightened in this way?

14. How is the curve-straightening method used to fit nonlinear regression curves to data?

15. How can the curve-straightening method be used to obtain confidence intervals for a regression curve?

16. In what circumstances will a pretransformation often improve the performance of the curve-straightening method?

17. Why is it undesirable for the variation in y to change with x? How can this condition be diagnosed?

18. When is a power transformation, such as the logarithm, likely to be useful for equalizing the dispersion of y as a function of x?

These questions are for the optional section 15.7.

19. What are outliers and high influence (leverage) points in the regression context?

20. Why is it desirable to use a robust regression line in order to pinpoint outliers?

21. What is a simple procedure for fitting a robust regression line to a set of data? Give an example.

SUPPLEMENTARY EXERCISES

The scatter diagrams constructed for the exercises of chapter 14 are to be used here to look for outliers and other potentially influential points before proceeding with the inference methods. Points well away from the main scatter should be viewed with suspicion, and separate inferential analyses should be run with these points excluded. (In practice, these points would be studied and described separately.) Complete each exercise by following these steps:

1. Screen each data set for sufficient linearity to proceed with other analyses by performing the test for $H_0: \beta_1 = 0$ versus $H_1: \beta_1 \neq 0$. Continue only for

those data sets for which the test is significant ($P < .05$). Since the coefficient of determination r^2 is readily available from the calculations of the last chapter, it is convenient to use the form

$$|t| = \sqrt{\frac{(n-2)r^2}{1-r^2}}$$

for computing the test statistic.

2. For those samples that pass the screening test, give bounds for the P-value of the test and carry out the additional analyses requested. Interpret the results of the analyses in terms of the original problem statements.

3. For those data sets that fail the screening test, give a lower bound for the P-value of the test and conclude that there is insufficient evidence of a *linear* relationship between the variables to proceed with analyses based on the linear regression model.

4. As before, where significance levels and confidence coefficients are not given, assume 5% and 95%, respectively.

15.9 Refer to exercise 14.10 (p. 471).

(a) Estimate and give a confidence interval for the mean N level of the runner after a 10-mile run for an E level of 200.

(b) Predict the N level for this E level and give the 95% prediction interval.

(c) As a convenience to the runner, construct a graph with the prediction line and upper and lower prediction limits from which he can obtain prediction intervals for N given E levels with *individual* confidence of 95%.

15.10 Refer to exercise 14.12 (p. 473).

(a) Obtain individual 95% algebra score prediction intervals for the prediction requested in exercise 14.10(b).

(b) Find joint 95% confidence intervals for the mean algebra scores at the arithmetic scores given in exercise 14.10(c).

15.11 Refer to exercise 14.13 (p. 473).

(a) Verify that the linear relationship between the number of cigarettes smoked by husbands and wives in the same families is not significant.

(b) Even though the relationship is not significant, obtain a 95% confidence interval for the slope of the least-squares line of number of cigarettes smoked by fathers (Y) versus number smoked by mothers (X).

15.12 Why should an exploratory analysis be made before inference is attempted? As we have seen several times before, acting on the diagnostic information of the exploratory analysis can be critical to the performance of the inferential procedure. The accompanying data set provides another case in point. It represents the energy use, in coded units, of a sample of seven small office buildings in Atlanta, Georgia, for a given year. The goal is to predict cooling energy requirements from energy use for heating by means of a least-squares linear regression.

HEATING (X)	COOLING (Y)
1.67	14.12
.00	11.65
2.05	72.64
.86	11.07
.49	9.39
2.84	19.21
2.11	20.30

Data courtesy of D. Hadfield, R. & D. Associates,
Albuquerque, New Mexico.

(a) Perform the screening test for linearity on this data set. Does there appear to be sufficient linearity to proceed with the analysis?

(b) Make the scatter diagram. A recheck of coding forms indicates that a large office building was mistakenly classified as small. Remove the incorrect observation and redo the screening test. Compare the results with the test of part (a).

(c) Obtain a 90% prediction interval for the cooling energy requirements of a small office building that uses 1.80 units of energy for heating.

15.13   Refer to exercise 14.15 (p. 474).

(a) Obtain a 95% prediction interval for the heart weight of a 2.5-kilogram female cat. A 2.5-kilogram male cat.

(b) Obtain joint 95% confidence intervals for the average heart weights of 2.0-, 2.5-, and 3.0-kilogram male cats.

15.14   Several hormone samples were assayed by using an out-of-date chemical. When the error was discovered, 14 of the samples were redone, using the correct chemical. Since the assays are expensive, it was hoped that a sufficiently strong linear relationship existed between the incorrect and correct results to permit accurate linear prediction of the correct assay values from the incorrect ones. The data are as follows:

RESULT OF INCORRECT ASSAY (X)	RESULT OF CORRECT ASSAY (Y)	SUMMARY DATA
350	227	$n = 14$
130	98	$\Sigma x = 3{,}970$
250	153	$\Sigma x^2 = 1{,}244{,}450$
200	134	$\Sigma xy = 753{,}695$
170	117	$\Sigma y = 2{,}466$
165	114	$\Sigma y^2 = 463{,}060$
390	227	
350	197	
420	238	
360	199	

(table continued on next page)

RESULT OF INCORRECT ASSAY (X)	RESULT OF CORRECT ASSAY (Y)	SUMMARY DATA
285	221	
210	156	
310	191	
380	194	

Data courtesy of F. Purifoy, Ph.D., Department of Anthropology, University of New Mexico, Albuquerque.

(a) Do the exploratory analysis for the data. (See the instructions for the exercises of chapter 14.)

(b) Is the linear association strong enough to support reasonable linear predictions? (Justify this by means of the magnitude of r^2 and the linear association screening test.)

(c) Use individual 95% prediction intervals to predict, and bound the prediction error for, correct hormone levels for specimens with incorrectly assayed levels of 150, 200, and 250.

(d) Use the Bonferroni method to bound simultaneously the predictions for the three values given in part (c), with a joint confidence level of 95%.

(e) Obtain joint 95% confidence intervals for the slope and intercept of the regression line of correct on incorrect hormone levels.

15.15   **A robust-line-fitting method based on medians**   The essence of this method is to divide the data into three groups (of roughly equal sizes), to find the x and y median pairs (x_1, y_1) and (x_2, y_2) of the *first* and *third* groups, and to pass a line through the two points so defined. The slope and intercept of the line can then be obtained from the formulas given in section 15.7.

(a) Compute the robust line for the women's height (X) versus weight (Y) data of display 1.1 (p. 9), using cutpoints 63.5 and 66.5. Verify that $x_1 = 62$, $y_1 = 110$, $x_2 = 67.5$, and $y_2 = 133$, leading to the line with intercept

$$\beta_0^* = \frac{x_2 y_1 - x_1 y_2}{x_2 - x_1} = -149.273$$

and slope

$$\beta_1^* = \frac{y_2 - y_1}{x_2 - x_1} = 4.182$$

(b) Superimpose this line and the by-eye robust line given in section 15.7 (p. 515) on a scatter diagram of the data. Which line appears to fit the data better? What is the largest difference in weights predicted by these two lines over the range of heights from 55 to 70 inches?

15.16   Calculate the robust regression line based on medians for the data given in exercise 15.12, using cutpoints 0.80 and 2.00.

(a) Compare this line with the least-squares line fitted to all the data.

(b) Compare this line with the least-squares line fitted to the data with outlier deleted. How well does the robust line ignore the outlier? Show that a box plot of absolute residuals from the robust line clearly delineates the outlier.

15.17 It is reasonable to expect that the more miles an automobile has been driven, the higher will be the levels of pollutants it emits. As a check on this expectation, the accompanying data were subsampled from the set from which the data of exercise 4.18 were obtained. The distances traveled (X) by the automobiles are given in thousands of miles, and the hydrocarbon emissions (Y) are in parts per million cubic centimeters.

DISTANCE (X)	HYDROCARBON EMISSIONS (Y)	SUMMARY DATA
34	270	$n = 13$
312	2352	$\Sigma x = 954$
84	1058	$\Sigma x^2 = 141{,}598$
50	1035	$\Sigma xy = 1{,}310{,}257$
89	2105	$\Sigma y = 11{,}893$
50	658	$\Sigma y^2 = 16{,}997{,}697$
33	588	
109	611	
38	1700	
1	600	
71	540	
52	247	
31	129	

Data courtesy of F. Wessling, College of Engineering, University of New Mexico, Albuquerque.

(a) Do an exploratory analysis based on a scatter diagram and box plot for the x variable.

(b) Determine the effect of the high leverage point identified by the box plot by carrying out the screening test on the full data set and on the data set with this point removed. (Note: Obtain the sums for the reduced data set by subtracting the data for this point from the sums given above.)

(c) Analyzing the main body of data and the high leverage point separately, does the exploratory analysis show any indication that the expectation mentioned above is true? Is there sufficient information in the sample to confirm this expectation?

15.18 Refer to exercise 14.18 (p. 476). What are the appropriate hypotheses H_0 and H_1 for checking the "slow down with age" conjecture? Test these hypotheses at the 5% significance level.

15.19* Refer to exercise 14.14 (p. 473).

(a) Obtain a 95% confidence interval for the slope β_1 of the regression line of October on September electrical energy consumption.

(b) Obtain a 95% confidence interval for the intercept β_0.

(c) Using the Bonferroni method, obtain joint confidence intervals for β_0 and β_1 with confidence of at least 95%.

---

* Depends on material in section 9.7.

(d) The confidence intervals of part (c) can be used to test the hypothesis $H_0: \beta_1 = 1$ and $\beta_0 = 0$ versus $H_1: H_0$ not true, at level $\alpha = .05$, by adopting the rule to reject H_0 if either the interval for β_1 does not contain 1 or the interval for β_0 does not contain 0. (See section 9.7.) Carry out the test and relate the outcome to the effectiveness of the electrical consumption measure suggested in exercise 14.14.

15.20* Refer to example 15.8 (p. 508). Use least squares to fit a line $\beta_0 + \beta_1 x$ to the $\log(y)$ versus $\log(x)$ data suggested in the example. Use the method described in exercise 15.19 to test the hypothesis $H_0: \beta_0 = 0$ and $\beta_1 = 1$ versus $H: \beta_0 \neq 0$ or $\beta_1 \neq 1$ (or both) at the 5% significance level.

———

* Depends on material in section 15.6.

Multiple Regression

Introduction

In multiple regression the ideas of simple linear regression are extended to problems involving more than one independent variable. It would be difficult to overestimate the importance of this subject in statistics. The multiple regression model, to be introduced in section 16.2, is the most widely used model for describing complex systems of interrelated variables. In an attempt to sort out these relationships, this model will focus attention on a single variable, y, the *dependent* or *response* variable. We will ask how this variable responds to changes in other variables, $x_1, \ldots, x_k$, the *independent* or *predictor* variables.

Among the goals of this chapter will be to find answers to the three basic questions we have considered before:

1. Is there an association (between dependent and independent variables)? If so,
2. What is the nature of the association? Finally
3. How large is the association?

In the multiple regression context, these questions and their answers require careful interpretation. We will first use the method of least squares to fit the model to data from a sample. The question, "Is there an association?" will be answered by a statistical test to see if the fitted model explains a "significant" amount of the observed y variation.

This screening test for model adequacy will determine whether or not we will pursue a more detailed investigation of the relationships implied by the model. If so, then the nature of the relationships will be investigated by looking more carefully at the significance of the individual variables in the model. Confidence intervals, where appropriate, can then be used to answer the question, "How large is the association?" Details of the analysis will be given in section 16.2.

If the global test of model adequacy suggests that the model does not explain enough of the variation of y to be meaningful, further analysis *of that model* will be abandoned. However, as in the case of simple linear regression, the failure of a particular model does not mean that no association exists. Moreover, even if the model explains a significant amount of the y variation, a different model involving the same variables may explain even more. Consequently, it is important to have methods for showing when it is desirable to go beyond the first model to determine if additional structure in the data can be accounted for. Residual plots, introduced in chapter 14, will serve this important diagnostic function.

Once the desirability of taking the modeling process further has been established, strategies for creating new models are required. We will see, in the applications given in this chapter, how the flexibility of the multiple regression model can be exploited to create a wide variety of new models. For example, powers of the original independent variables can be introduced as new variables in order to create models of *polynomial regression*. These models can be used to describe a variety of curvilinear regressions. Model building and the diagnosis of the various ills of multiple regression are complementary activities and will be treated together in section 16.3.

Another feature that adds significantly to the flexibility of the multiple regression model is the fact that, although the dependent variable will be taken to be of measurement type, categorical variables can also be included among the predictors. The *general linear model*, hinted at in chapters 14 and 15, for which a unified inference theory exists, is simply the multiple regression model with this more liberal use of both measurement and categorical predictor variables. The models used in the analysis of variance, seen in chapter 11 and to be seen again in chapter 17, are simply multiple regression models for which the predictor variables are all categorical. This fact will be demonstrated in section 16.4.

While hand-computation was possible (at least for small data sets) for the methods of previous chapters, it is *not* recommended for multiple regression. Here, the number of computations will be excessively large even for small data sets, and the chance for errors is simply too great to make hand-computing feasible. Consequently, for those without access to a computer with a multiple regression capability, this will be largely a "regression appreciation" chapter.

For those with access to a multiple regression package, the primary exercises for this chapter will be to repeat the analyses of the given data sets and to learn where the various items of computer analysis show up on the printed output from your program. Also, if regression diagnostics are provided, try to reconcile their indications with those found by the methods used in this chapter. A few additional exercises will allow you to try out these skills on your own.

More complete treatments of multiple regression are given in the books by Daniel and Wood [41], Draper and Smith [42], and Weisberg [47]. A particularly unique and useful coverage of the topic is provided by Mosteller and Tukey [45].

◢ SECTION 16.2

The Multiple Regression Model and Its Statistical Inference

The multiple regression model will be introduced in this section, along with the standard tests of hypothesis. To illustrate the discussion, we will consider the problem of predicting the tensile strength (y) of specimens of die-cast aluminum from measurements of their hardness (x_1) and density (x_2). To do this, a multiple regression will be fitted to data for ten specimens. The data are given in display 16.1.

Display 16.1

Tensile strength (in thousands of pounds per square inch), hardness (Rockwell E), and density (grams per cubic centimeter) for ten specimens of die-cast aluminum. Source: W. A. Shewhart, *Economic Control of Quality of Manufactured Product*, Van Nostrand, New York, 1931, page 42.

OBS.	TENSILE STRENGTH y	x_0	HARDNESS x_1	DENSITY x_2
1	29.31	1	53.0	2.67
2	34.86	1	70.2	2.71
3	36.82	1	84.3	2.86
4	30.12	1	55.3	2.63
5	34.02	1	78.5	2.58
6	30.82	1	63.5	2.63
7	35.40	1	71.4	2.67
8	31.26	1	53.4	2.65
9	32.18	1	82.5	2.72
10	33.42	1	67.3	2.61

The Multiple Regression Model

The multiple regression model will be written as

$$y = \beta_0 + \beta_1 x_1 + \cdots + \beta_k x_k + e \tag{1}$$

The error term, e, is assumed to be a normally distributed random variable with mean 0 and (unknown) standard deviation σ.

regression function   The **regression function** is

$$\mu(x_1, \ldots, x_k) = \beta_0 + \beta_1 x_1 + \cdots + \beta_k x_k$$

regression coefficients   where $x_1, \ldots, x_k$ are the independent variables. The **regression coefficients**

$$\beta_0, \ldots, \beta_k$$

are parameters with unknown values. They, along with σ, are the parameters that must be estimated from the sample. Note that there are k independent variables but $k + 1$ regression coefficients. The parameter β_0 qualifies as a regression coefficient in that it can be viewed as the coeffcient of an implicitly defined variable x_0 whose values are all 1's.

Equation (1) expresses the dependent variable y as the sum of a (fixed) regression function and a random error. This is in the same form as the simple linear regression model given in chapter 14. It is reasonable that comparable methods of analysis will hold for the two models; in fact, the multiple regression analysis will entail a rather direct extension of the ideas given in chapters 14 and 15.

Expression (1) provides a description of the multiple regression model in terms of the model variables. However, the values of these variables for each observation are also viewed as having been generated according to this model. A more precise specification of the model would add a subscript i, say, to each variable to designate the specific model for each observation. For a random sample of size n ($n = 10$ in display 16.1) the errors $e_1, \ldots, e_n$ are assumed to be independent, with the same normal distribution ($\mu = 0$, unknown σ). Thus, for example, the value $y_1 = 29.31$ for observation number 1 in display 16.1 is assumed to have been generated by the relationship

$$y_1 = \beta_0 \cdot 1 + \beta_1 \cdot 53.0 + \beta_2 \cdot 2.67 + e_1$$

through the random selection of the value of e_1. To avoid cumbersome notation, the observation subscripts, i, will be included only where needed for clarity.

The Least-Squares Fitting Process

The multiple regression least-squares fitting problem is the problem of finding the quantities $\beta_0, \ldots, \beta_k$ that minimize the sum of squares

$$G(\beta_0, \ldots, \beta_k) = \Sigma(y - \beta_0 - \beta_1 x_1 - \cdots - \beta_k x_k)^2$$

where the sum is taken over all observations (that is, over the missing observation subscript i). By an application of differential calculus methods, the solution of this problem is reduced to the simultaneous solution of a set of $k + 1$ linear equations in the unknowns $\beta_0, \ldots, \beta_k$. The resulting solution is the set of least-squares estimates of the regression parameters. Computer packages, such as SAS and MINITAB, use modern methods of numerical linear algebra to carry out this solution efficiently for problems involving large numbers of observations and/or large numbers of variables.

After estimating the regression coefficients, the regression function can be estimated for each observation by the expression

$$\hat{y} = \hat{\beta}_0 + \hat{\beta}_1 x_1 + \cdots + \hat{\beta}_k x_k \tag{2}$$

residuals

Then, the **residuals** can be found by subtracting the regression function estimate for each observation from the observed value of y:

$$r = y - \hat{y}$$

We will use these residuals for diagnostic purpose later. However, as we now show, they also play a key role in the screening test for the model.

The Screening Test for the Regression Model

The goal of the screening test is to determine whether the data provide sufficient support for the assumed regression model to continue with the analysis. This goal can be achieved by means of a hypothesis test.

As in the case of simple linear regression, statistical hypotheses will correspond to the specifications of regression models. The test will then constitute a choice between two such models.

The alternative hypothesis for the screening test specifies the complete model under consideration (called *the full model*). This was designated as model 2 in chapter 14:

$$H_1: y = \beta_0 + \beta_1 x_1 + \cdots + \beta_k x_k + e \qquad \text{(model 2)}$$

Here, and in all tests that we will consider, the null hypothesis (model 1) is formed by setting equal to 0 one or more of the terms in the regression function of model 2. (The result is called *the reduced model*). For the screening test, *all* terms other than the constant term are set equal to 0:

$$H_0: y = \beta_0 + e \qquad \text{(model 1)}$$

Note that, with model 2 specified, the null hypothesis can also be written as

$$H_0: \beta_1 = \beta_2 = \cdots = \beta_k = 0$$

Now, a test of these hypotheses can be based on a measure of the difference in the estimated regression functions under the two models. As was seen in chapter 14, the least-squares estimate of the regression function under model 1 is $\bar{y}$. Thus, again using a measure based on sums of squares, the **regression (or model) sum of squares** is defined to be

regression (or model) sum of squares

$$SSR = \Sigma(\hat{y} - \bar{y})^2$$

where $\hat{y}$ is the regression function estimate (2).

degrees of freedom for regression

The **degrees of freedom for regression,** associated with this sum of squares, are

$$\nu_R = k$$

Note that the number of degrees of freedom equals the number of independent variables. The **regression mean square** is then the regression sum of squares divided by its degrees of freedom

regression mean square

$$MSR = \frac{SSR}{\nu_R}$$

To arrive at a test statistic with known distribution, an estimator of the unknown variance σ^2 is also required. This estimator is derived from the **error sum of squares,** which is simply the sum of squares of the residuals:

error sum of squares

$$SSE = \Sigma r^2$$

degrees of freedom for error

The **degrees of freedom for error** are

$$\nu_E = n - k - 1$$

error mean square

The estimator of σ^2 is then the **error mean square**

$$MSE = \frac{SSE}{\nu_E}$$

Finally, the test statistic is

$$F = \frac{\text{MSR}}{\text{MSE}}$$

When H_0 is true, this statistic has the F distribution (table 4), with k and $n - k - 1$ degrees of freedom. The screening test then uses the rule:

Reject H_0 if $F \geq c$

where c is obtained from table 4 in the usual manner.

Alternatively, if F_{obs} is the value of F computed from the sample, bounds for the P-value, $P(F \geq F_{obs})$, can be found.

Note that to reject H_0 is to decide in favor of model 2, which means that there is evidence that the regression model explains a significant amount of the variation in y. In this case, we would proceed to investigate the model further.

If H_0 is accepted, on the other hand, the implication is that the model is no better at explaining the variation than is model 1, which does not depend on the independent variables at all. In this case, the inference process would stop and we would either look for better models relating the variables under study (by methods to be considered later) or conclude that the sample does not support a regression relationship between y and the independent variables.

As in the case of simple linear regression, both the sums of squares and degrees of freedom satisfy "analyses of variance" relationships. Consequently, the quantities required for the screening test are typically given in an ANOVA table. The following table summarizes the screening test for the tensile strength data of display 16.1.

SOURCE OF VARIATION	SS	df	MS	F
Model (Regression)	34.0088	2	17.0044	
Error	21.0625	7	3.0089	5.651
Total	55.0719	9		

The total sum of squares, SST, which is given in the last row of the table, is, as in earlier chapters, the sum of squares of the y values around their sample mean. This sum of squares has $n - 1$ degrees of freedom.

From table 4, with numerator degrees of freedom = 2, denominator degrees of freedom = 7, and F = 5.651, bounds for the P-value for the screening test are

$.01 < P < .05$

Consequently, we would reject H_0 at the 5% significance level. That is, we would conclude at the 5% level that the data provide enough support for the given regression model to make a more detailed analysis worthwhile.

The Coefficient of Determination

*coefficient of
determination
model R²*

A descriptive measure of how much of the y variation the given model accounts for is provided by the **coefficient of determination** or **model R^2**:

$$R^2 = \frac{SSR}{SST}$$

This statistic is commonly given along with the screening test analysis of variance. As for simple linear regression, this quantity is interpreted as the proportion of the variation of the dependent variable that is accounted for by the linear regression model. For the tensile strength data,

$$R^2 = \frac{34.0088}{55.0719}$$
$$= .6175$$

Thus, the given model of tensile strength versus hardness and density accounts for 61.75% of the total variation in tensile strength.

At this point, we will want to look more closely at the nature of the association between tensile strength and each independent variable. The information required for this study is typically summarized by the computer in another table. We look at this topic next.

Tests of Significance and Confidence Intervals for Regression Coefficients

Tests of hypotheses can be devised to determine which of the individual terms in the regression function contribute "significantly" to explaining the variation in the dependent variable. Presumably, terms that do not make significant contributions can be dropped from the model, thus leading to simpler regression models.

The null hypothesis that expresses the condition that the jth model term makes no contribution to the regression is

$$H_0: \beta_j = 0$$

It is worthwhile noting that this hypothesis is ambiguous without the specification of the model for the alternative hypothesis, model 2. Model 2 specifies what other variables are included in the regression when we ask the question, "Does x_j make a difference?" The above null hypothesis will lead to different reduced models for different model 2's and different tests will be required. It is not surprising, then, that these tests can lead to different answers.

For this reason, it is important to be clear about the model 2 that is used in the standard hypothesis test—the one provided in computer printouts. This model is the original regression model. That is, the question being addressed by the hypothesis test is, "With all other independent variables in the model, does x_j make a difference?"

The standard test is a (two-sided) Student's t test based on the test statistic

$$t = \frac{\hat{\beta}_j}{SE_{\hat{\beta}_j}}$$

The quantities necessary to carry out the test are (generally) tabulated together in a summary table. Among other things, this table will give the regression parameter estimates, the standard errors, and the values of the t statistics for all independent variables.

The variable summary for the tensile strength data of display 16.1 is as follows.

Variable summary

VARIABLE	REG. COEF.	SE	t
Intercept	8.2852	20.9315	0.396
Hardness	.1458	.0567	2.570
Density	5.4740	8.4243	.650

The degrees of freedom for the t statistics are the *error degrees of freedom* in the screening test ANOVA table—namely $n - k - 1$.

The degree of freedom for the tensile strength data = 7. From table 3 for the t distribution, two-sided P-value bounds are seen to be

$.02 < P < .05$ for the hardness variable,

and

$P > .50$ for density and intercept

Consequently, we would conclude that the major source of the association of tensile strength with these variables is with the hardness variable. Put in another way, density accounts for very little of the ability of the model to explain the variation in tensile strength and could be dropped from the model without much damage.

Confidence intervals for the population regression coefficients can also be easily calculated from the information in the summary table. If t is the critical value from table 3 for the given confidence coefficient and $n - k - 1$ degrees of freedom, then limits for the confidence interval for β_j are

$$L = \hat{\beta}_j - t \cdot SE_{\hat{\beta}_j}$$

and

$$U = \hat{\beta}_j + t \cdot SE_{\hat{\beta}_j}$$

For example, to obtain a 95% confidence interval for the tensile strength regression coefficient β_1, the critical value from table 3 for 7 degrees of freedom is $t = 2.365$. The regression coefficient and standard error estimates are read from the variable summary table. Thus, the confidence limits are

$.1458 \pm 2.365(.0567)$

or

$.0117 \le \beta_1 \le .2799$

With 95% confidence, the increase in tensile strength per unit increase in hardness is between .0117 and .2799 units (i.e., between 11.7 and 279.9 pounds per square inch).

Confidence intervals for the regression function and prediction intervals are also available in the multiple regression context. The theory follows closely the ideas given in chapter 15, and we will not cover these topics. See Neter, Wasserman, and Kutner [24] for details. Options for computing these intervals are given in computer programs such as SAS [46].

EXERCISES

16.1   The following data were obtained in a study of the verbal abilities of children (as measured by a test) as a function of their ages and scores on visual and auditory skills tests.

VERBAL	AGE	VISUAL	AUDITORY
10.68	3.25	12.13	13.57
14.02	8.92	14.75	14.93
13.50	7.25	14.30	14.60
14.52	10.75	14.70	14.43
14.55	4.67	13.30	14.83
14.10	5.41	13.26	14.70
13.26	6.33	8.53	13.50
12.90	8.00	12.88	13.77
13.64	8.83	13.58	13.33
13.98	9.91	13.70	14.80
10.93	3.41	10.73	13.00
12.65	4.58	11.43	13.33

(a)   Use the screening test at the 5% significance level to determine whether the regression model with verbal ability as dependent variable and age, visual skill, and auditory skill as independent variables is of interest.

(b)   If the model of part (a) explains a significant amount of the variation in verbal ability, determine which of the variables contribute significantly (5% level) to the model and obtain 95% confidence intervals for their regression coefficients.

16.2   A study of the durability of pottery was made by firing sherds of varying widths at different temperatures, then immersing them in water for a period of time. The dry weight of the sherd was determined before immersion. The wet weight was measured at the end of the test period, and by subtracting the dry weight from it, the amount of water the sherd absorbed during the test can be computed. The break strength of the sherd was then determined. The problem is to determine the relationship of break strength (y) to firing temperature (x_1), amount of water absorbed (x_2), and sherd width (x_3). The data are as follows.

BREAK STRENGTH	FIRING TEMP.	DRY WT.	WET WT.	WIDTH
6.00	750	10.58	12.48	162
10.10	950	15.04	17.29	160
8.4	900	12.67	14.60	165
4.6	600	12.08	14.07	163
35.2	1050	12.62	13.99	165
18.10	750	15.13	17.50	173
38.10	1050	13.99	15.12	200
11.2	950	17.45	20.20	195
7.0	600	14.19	16.51	185
4.3	750	8.87	10.3	187

(a) Find the P-value for the screening test of the regression model of y on x_1, x_2, and x_3.

(b) If the screening test is significant at the 5% level, find the independent variables that are significant at the 5% level and determine 95% confidence intervals for their regression coefficients.

16.3   Data for the study of hand circumferences (y) as a function of height (x_1), weight (x_2), age (x_3), and ponderal index (x_4), from the legal case discussed in exercise 13.4, are as follows. (A ponderal index is a measure of body conformation: Relatively thin individuals would have small ponderal indices, and relatively heavy individuals would have large indices.) Of particular interest in this study was the dependence of hand circumference on age; it is conjectured that hand size increases with age. Consider this conjecture in your conclusions.

y	x_1	x_2	x_3	x_4
22.5	67	140	22	.775
21	69	153	52	.775
20.5	61	103	22	.768
24	62	140	57	.837
23.1	62	125	34	.806
21.4	63	121	48	.785
22.5	67	153	26	.798
22.1	66.5	137	21	.775
22.3	67	158	35	.806
21.7	63.5	115	21	.766
22.7	66	118	21	.743
22.5	67.3	130	24	.753
22.6	67	133	23	.762

(a) Find the P-value for the screening test of the regression model of y on $x_1, \ldots, x_4$.

(b) If the screening test is significant at the 5% level, find the independent variables that are significant at the 5% level and determine 95% confidence intervals for their regression coefficients.

◢ SECTION 16.3

Multiple Regression Problems, Their Diagnosis and (Occasionally) Their Cures

Much of this section will be taken up with graphical methods. One important kind of graph, *the residual plot*, will be useful for detecting when other regression models involving the same variables should be tried. Residual plots* are also useful for detecting points (outliers) that adversely affect a regression analysis and for indicating when special variable transformations might be useful.

Other special graphs, called *partial regression plots*, can be used to indicate those regression coefficients most seriously influenced by outliers. In some instances, outliers can "create" a significant regression coefficient; this effect can be quickly detected with a partial regression plot.

Independent variables will sometimes be associated with one another as well as with the dependent variable, causing a loss in sensitivity to significance tests for certain terms in the model. This problem, called *collinearity*, can also be inadvertently introduced into the model during the model-building process. Simple diagnostics to detect collinearity will be presented in this section, and some suggestions for avoiding the problem in model-building will be made.

More complete treatments of regression diagnostics are given by Belsley, Kuh, and Welsch [38] and by Cook and Weisberg [40]. The Belsley, Kuh, and Welsch diagnostics are implemented in the computer package, SAS [46].

Residual Plots

The residuals, $r = y - \hat{y}$, can be plotted in a number of ways. For example, they can be plotted against observation number (the "variable" Obs. in display 16.1) to see if a systematic trend occurs that is related to the order in which the observations were taken. They can also be plotted against any or all of the independent variables to investigate the form of any residual variation with these variables.

Perhaps the most useful residual plot for detecting any remaining dependence between independent and dependent variables, and for locating points that are likely to have large influence in the least-squares fitting process, is a graph of the residuals against the predicted values, $\hat{y}$. This is the plot we will be primarily concerned with. The plot of residuals against predicted values for the tensile strength data of display 16.1 is given in display 16.2.

This residual plot shows one point that deviates substantially from the group of remaining points. The following table, which represents the data

---

* We will consider only plots of *raw* residuals, as defined in the last section. Some investigators prefer plots of *Studentized residuals*, which correct raw residuals for differences in their standard errors. See reference [38] for details.

Display 16.2

Residual plot showing
the residual $r = y - \hat{y}$
versus the predicted
value $\hat{y}$ for each
observation of tensile
strength (y) from
display 16.1

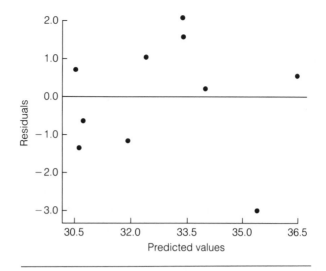

graphed in the residual plot, shows that observation 9 produced the out-of-line value. This would lead us to investigate the generation of this observation. Possibly, experimental conditions were abnormal when it was taken, so that it could legitimately be dropped from the analysis.

Residual Summary

OBS.	Y	PREDICTED Y	RESIDUAL
1	29.31	30.627	−1.317
2	34.86	33.353	1.507
3	36.82	36.229	0.591
4	30.12	30.743	−0.623
5	34.02	33.851	0.169
6	30.82	31.938	−1.118
7	35.40	33.309	2.091
8	31.26	30.576	0.684
9	32.18	35.201	−3.021
10	33.42	32.383	1.037

In any event, to determine the influence of this observation, it can be deleted and the analysis rerun. When this is done, the F statistic for the screening test increases to 13.831 (from 5.651), and the model R^2 increases from .618 to .822. This suggests that one or more outliers could easily lead to a useful regression model flunking the screening test. Consequently, the screening test should *always* be accompanied by a residual plot.

As in chapter 15, outliers can be identified by making a box plot of the absolute residuals.

When a regression model has adequately explained as much of the variation in y as possible with the given independent variables, the residual plot should show an amorphous, unstructured scatter of points with (relatively) even variation from left to right. Because this assessment is somewhat subjective, more objective diagnostic measures have been incorporated in some regression computer packages, such as SAS and MINITAB. However, the human eye remains the best detector of curious features in a regression analysis, and the residual plot is still one of the most important diagnostic tools. Another application is given in the next example.

EXAMPLE 16.1

In exercise 14.24 (p. 117), data were given for the study of the variation of the hormone androstenedione with age in females. In this example, every other point has been chosen, so as to reduce the size of the data set. The data are given in the following table. The first 13 points represent data from the endocrinologist's records and thus correspond to women with possible hormonal imbalances. The remaining points were obtained from normal subjects.

DATA FROM ENDOCRINOLOGIST'S
RECORDS

OBS.	ANDRO (y)	AGE (x)
1	105	25
2	220	24
3	525	24
4	133	16
5	67	9
6	224	19
7	140	19
8	144	23
9	87	15
10	151	19
11	157	25
12	140	17
13	489	21

DATA FOR NORMAL WOMEN

OBS.	ANDRO (y)	AGE (x)
14	45	76
15	175	39
16	102	56
17	165	29
18	116	40
19	137	73
20	172	31
21	227	22
22	74	56
23	158	28
24	56	53
25	108	42
26	84	61

(table continued on next page)

DATA FOR NORMAL WOMEN

OBS.	ANDRO (y)	AGE (x)
27	116	46
28	123	39
29	140	29
30	91	44
31	84	63
32	116	55
33	98	47
34	140	38
35	35	61
36	84	62
37	63	66
38	189	20
39	172	38

A scatter diagram of these data is given in display 16.3.
Initially, a simple linear regression model for andro versus age was used.
This model shows some promise, as the following ANOVA indicates.

ANOVA

SOURCE	SS	df	MS	F
Model	72982	1	72982	
Error	289043	37	7812.0	9.342
Total	362025	38		

Model R-Square = .202

However, the residual plot in display 16.4 shows two points well separated from the remaining cluster of points that are sure to have an adverse influence on the regression analysis. A residual summary (not given) identifies

Display 16.3

Scatter diagram of androstenedione versus age for the data of example 16.1

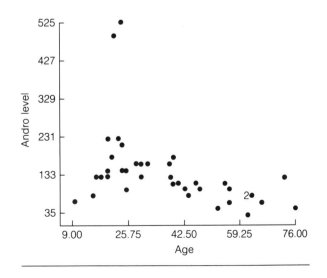

Display 16.4

Residual plot for the simple linear regression model of androstenedione versus age based on the data of example 16.1

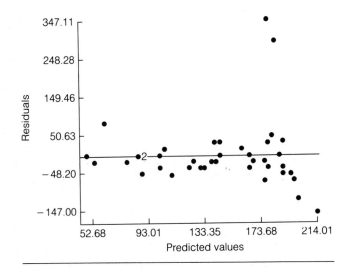

these points as belonging to observations 3 and 13 from the endocrinologist's group of patients. Since these levels are quite abnormal and we are interested in the regression for women with normal androstenedione levels, these points are deleted from the analysis. After the deletion, the curvilinearity in the residual plot is much more clearly evident. This suggests the need for a model that can account for the additional structure seen in the plot.

Polynomial Regression

An important hypothesis for example 16.1 is that andro levels peak for women in their mid-20s and decrease thereafter, with a leveling off in later years. Evidence of this feature is seen in display 16.3. This would suggest the need for a regression function that can represent both the peak and the leveling-off segment of the curve. A polynomial of degree 3 can represent such a curve, with the quadratic (x^2) term defining the peak and the cubic (x^3) term being primarily responsible for defining the leveling-off region.

The flexibility of the multiple regression model makes polynomial regressions easy to formulate. The cubic polynomial model

$$y = \beta_0 + \beta_1 x + \beta_2 x^2 + \beta_3 x^3 + e$$

is seen to be a special case of a multiple regression model with "independent variables" $x_1 = x$, $x_2 = x^2$, and $x_3 = x^3$. Note that at this point some confusion arises over the term *independent variable*. From the viewpoint of the problem, only the variable x is independent. However, from the viewpoint of the regression model, all three powers of x are independent variables. J. W. Tukey has coined the term *carriers* for the independent variables in a regression, in order to avoid this confusion. However, we will proceed with the standard terminology and caution the reader that the term is being used in two senses.

Display 16.5

Residual plot for a cubic polynomial model of androstenedione versus age based on the data of example 16.1

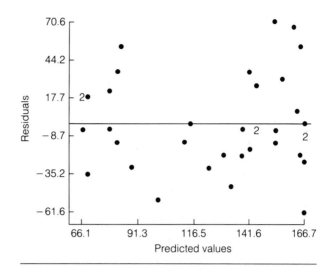

The ANOVA for a cubic polynomial fit for the data of example 16.1 is given in the following table.

ANOVA

SOURCE	SS	df	MS	F
Model	49764.3055	3	16588.1018	15.534
Error	35240.1269	33	1067.8826	
Total	85004.4324	36		
Model R-Square = 0.585				

The simple linear regression model R^2, after the removal of the two outliers, was .364. We see that the inclusion of the new terms in the model has boosted the R^2 to .585. Moreover, as seen in display 16.5, the new model has left the residuals reasonably unstructured.

A basic axiom of model building is to stop with the simplest model that adequately "explains" the data. This and the form of the residuals suggests that we have gone far enough in the model-building process. (Another reason for not going further will be seen in the next subsection.)

The variable summary for the cubic model is as follows.

Variable summary

VARIABLE	REG. COEF.	SE	t
Intercept	−45.9787	56.6960	−0.811
x	18.7218	4.8491	3.861
x^2	−0.5017	0.1224	−4.097
x^3	0.0037	0.0009	3.917

Note that all three powers of the age variable are highly significant ($P <$ .001), thus none could be omitted from the model without adversely affecting the fit. This suggests, among other things, that the leveling-off hypothesis is supported by the data.

Individual Regression Coefficients and the Collinearity Problem

To introduce the collinearity problem, we will look at a simple polynomial regression problem resembling that of example 16.1. The data are given in the following table.

Data

OBS.	Y	$X(1)$
1	10	1
2	14	2
3	19	3
4	15	4
5	13	5
6	16	6
7	20	7

Both a plot of the data and a residual plot based on the simple linear regression model indicate an up-down-up pattern suggestive of a cubic polynomial. A cubic regression model of the type considered in example 16.1 was fitted to the data, with the following results.

ANOVA

SOURCE	SS	df	MS	F
Model	61.0952	3	20.3651	5.912
Error	10.3333	3	3.4444	
Total	71.4286	6		

Model R-Square = .855

Variable summary

VARIABLE	REG. COEF.	SE	t
Intercept	-3.57144		
x	17.51587	5.43631	3.222
x^2	-4.73809	1.52882	-3.099
x^3	.38889	.12628	3.080

It is seen that $.05 < P < .10$ for the screening test, thus the model would be considered for further analysis at the 10% significance level. Moreover, all three powers of x are significant ($P < .10$), indicating that none of them can be removed from the model without seriously affecting the fit.

Adopting the philosophy that if a cubic polynomial fits well, then a quartic polynomial (one containing an x^4 term in addition to those of the cubic) should fit even better, we find the following curious variable summary.

Variable summary

VARIABLE	REG. COEF.	SE	t
Intercept	-2.7142936		
x	16.15653	20.26915	.797
x^2	-4.08332	9.41071	$-.434$
x^3	.26768	1.71438	.156
x^4	.00758	.10671	.071

Not only is the quartic (x^4) term insignificant; now, none of the other terms in the model are significant either. Note that while the regression coefficients have not changed by much, their standard errors have increased substantially, leading to a decrease in the t ratios. The analysis is suffering from a combination of ills that can be best understood through a detailed study of the expression for the standard errors.

The square of the standard error of $\hat{\beta}_j$ can be written in the following form:

$$\text{SE}_{\hat{\beta}_j}^2 = \left(\frac{1}{n - k - 1}\right)\frac{s_y^2(1 - R^2)}{s_{x_j}^2(1 - R_j^2)} \tag{3}$$

where n is the sample size, k is the number of independent variables in the model, s_y^2 is the ordinary sample variance of y (as though it were being considered alone), $s_{x_j}^2$ is the ordinary sample variance of x_j, and R^2 is the model coefficient of determination.

The quantity R_j^2 is the coefficient of determination of x_j relative to the *remaining independent variables*. That is, it measures the amount of the variation of x_j that can be explained by a linear regression of this variable on the other independent variables. If this quantity is near 1, then x_j is very nearly a linear combination of the remaining independent variables. This is *collinearity* what is meant by **collinearity.**

Note that because the term $1 - R_j^2$ appears in the denominator of expression (3), the effect of collinearity is to inflate the standard error of any variable that has a high degree of collinearity with the other independent variables in the regression. The R_j^2 value itself is a useful collinearity diagnostic, and it will be given hereafter in variable summaries.

A second measure of collinearity is the (standard error) *inflation factor*, which shows how much the standard error of the regression coefficient is affected by collinearity:

$$\text{Inflation factor} = \frac{1}{\sqrt{1 - R_j^2}}$$

The R_j^2's and inflation factors for the quartic regression model fitted above are as follows.

VARIABLE	R-SQUARE	INFL. FAC.
x	.999552	47.246
x^2	.999969	179.605
x^3	.999982	235.050
x^4	.999905	102.815

Note that each independent variable is extremely strongly affected by collinearity. The standard errors are inflated by extremely large factors. Thus, the loss in sensitivity of the t tests in going from the cubic to the quartic models can be accounted for by the collinearities among the independent variables alone.

However, expression (3) shows that collinearity is not necessarily the only possible source of diminished sensitivity in the estimation of the regression coefficients. A small model R^2 (which is likely to be detected in the screening test) also increases the standard errors. Moreover, when the number of independent variables in the regression approaches the size of the sample, the factor $1/(n - k - 1)$ becomes large and increases the standard errors. This effect is often caught in the screening test as well. Thus, in the above example, the value of the F statistic for the quartic model was 2.965 $(P > .25)$, whereas it was 5.912 $(.05 < P < .10)$ for the cubic model.

When the sample size is large relative to the number of variables in the model, the effect of collinearity will often be overcome by the dominance of the $1/(n - k - 1)$ factor and the "strength" of the regression. This was the case for the analysis of example 16.1. The R^2's and inflation factors for the variables in the cubic model fit are as follows.

VARIABLE	R-SQUARE	INFL. FAC.
x	0.9963	16.5430
x^2	0.9992	35.3804
x^3	0.9975	19.8413

Yet the strength of the cubic trend as well as the fact that the degrees of freedom for error are $n - k - 1 = 33$ lead to significant regression coefficients.

Collinearity frequently occurs among the independent variables in an observational experiment. One reason for this is that what we may believe are different measures often measure nearly the same thing and thus turn out to be highly correlated—that is, collinear. For example, in a study of the educational achievement of children, one would often measure such background (independent) variables as family income, educational levels of the parents, degree of availability of books in the home, parents' attitudes toward study, and the like. These socioeconomic variables tend to be highly interrelated and, if included in a regression model, could lead to collinearity problems.

A standard procedure when a collinearity problem occurs is to drop one or more of the highly collinear independent variables from the model. Methods for doing this exist in most regression packages. Unfortunately, this deprives the experimenter of potentially useful information about the dropped variables.

Another reason for collinearity in observational experiments is that particular patterns of independent variables tend to be seen more frequently than others. In the educational achievement sample, we are likely to see patterns of high family incomes associated with high parent education, large numbers of books, and so forth. If it were possible to fill in the gaps by sampling the relatively rare combinations of low income, high parent education, high numbers of books, and the like, the collinearity problem would be substantially reduced. An advantage of designed experiments, in which the assignment of independent variable values to experimental units is controlled, is that collinearity can be minimized. However, in observational experiments, collinearity simply becomes one of the limitations on what can be learned from the experiment.

Having said this, it is necessary to back off a bit in the case of polynomial regression. Collinearity is actually introduced into the experiment by the analyst when he or she includes more powers of the independent variable in the model. When the range of the independent variable (the distance between the smallest and largest observed x values) is relatively short compared with the distances of the observed x's from 0, the powers of x are highly collinear.

This collinearity can be substantially reduced by subtracting a constant from the x variable to position its zero point at roughly the middle of the scatter of x values—for example at the median of the x's. For example, the x variable in the last example can be recentered by subtracting 4 from each x value. The variable summary for the quartic polynomial fit to the recentered data is as follows.

Variable summary

VARIABLE	REG. COEF.	SE	t	R-SQUARE	INFL. FAC.
Intercept	15.649351				
x	−1.72222	1.16326	−1.481	.863980	2.711
x^2	−.14394	1.05100	−.137	.944456	4.243
x^3	.38889	.15447	2.518	.863980	2.711
x^4	.00758	.10671	.071	.944456	4.243

Note the substantial decrease in R^2's and inflation factors. This table suggests that the quartic term is not important in the model, but the cubic term is. We would return to the cubic model fitted earlier.

Variable Transformations

The recentering of the x variable is a simple version of a variable transformation; it is a linear transformation applied to the independent variables of the model. It is possible to extend the class of regression models considerably as well as to restructure variables so as to better accommodate the least-squares fitting method by transforming the independent variables (and sometimes the dependent variable as well).

Least-squares methods prefer that both dependent and independent variables appear to be from normal distributions. Box plots of all variables (individually) should always be obtained at the beginning of a regression analysis. When a variable is badly skewed, a power transformation should be applied to symmetrize it. A method for fitting a symmetrizing power transformation, comparable to the curve-straightening method of chapter 15, is given by Tukey [32]. However, with automatic box plot and power transformation programs, it is easier to try a transformation, then check symmetry by box-plotting the transformed variable. It is seldom necessary to go beyond the square root ($p = .5$), log ($p = 0$), and reciprocal ($p = -1$) transformations. The symmetrization of variables will often "tame" outliers as well.

Outliers that remain in the symmetrized variables are potential high-influence points in the multiple regression. Thus, observations with outliers in one or more of the variables should be marked down for further attention. However, observations with no outliers can also correspond to points of high influence. The next subsection treats the topic of partial regression plots that can be used to detect such points.

A trend in dispersion, signalled by a triangular or funnel shape of the point scatter in the residual plot, can usually be cured by applying a power transformation to the dependent variable. The log transformation is often successful in this regard. (See chapter 15 for an example.)

More details of variable transformations are given by Neter, Wasserman, and Kutner [24].

High-Influence Points and Partial Regression Plots

We found scatter diagrams to be an effective diagnostic for high-influence points in simple linear regression. This diagnostic can be extended to multiple regression by reducing the estimation of each regression coefficient to a simple linear regression problem. This is done as follows.

A variable y can be *(linearly) corrected for a variable* x by finding and subtracting from y its predicted value in terms of x. Formally, this is done by forming the regression coefficient

$$b = \frac{\Sigma x_i y_i}{\Sigma x_i^2}$$

The variable y corrected for x is then

$$\bar{y} = y - bx$$

Now the least-squares estimate of the multiple regression coefficient β_j can be obtained by correcting the dependent variable y and independent variable x_j, using the above procedure (one at a time in any order) for all independent variables other than x_j, including the variable x_0. Let $\bar{y}_j$ and $\bar{x}_j$ denote these corrected variables. It can be shown that the least-squares estimate of the simple linear regression intercept for these variables is 0. Moreover the least-squares estimator of the slope is exactly the multiple regression estimator of β_j discussed earlier. Consequently, problems that affect least-squares estimates of this regression coefficient will show up in a scatter diagram of the variables $\bar{y}_j$ versus $\bar{x}_j$. This scatter diagram is the
partial regression (leverage) plot
partial regression (leverage) plot for β_j. Partial regression plots are available in computer packages such as SAS [40]. The following example shows an application of these plots.

EXAMPLE 16.2

The salaries, years of service, and a merit measure (based on teaching performance, research, and so forth) for 12 faculty members in a certain university department are as follows.

OBS.	SALARY ($1000s) (y)	YEARS OF SERVICE (x_1)	MERIT MEASURE (x_2)
1	20.8	21	2.5
2	23.0	2	2.5
3	33.0	5	4.0

(table continued on next page)

OBS.	SALARY ($1000s) (y)	YEARS OF SERVICE (x_1)	MERIT MEASURE (x_2)
4	35.3	8	2.5
5	31.7	12	2.9
6	40.2	7	4.5
7	27.3	10	1.0
8	34.3	2	4.7
9	28.6	5	3.5
10	40.5	9	4.3
11	21.4	24	2.8
12	38.2	10	3.5

A study was made to determine how salaries (y) depend on years of service (x_1) and merit (x_2). The ANOVA and variable summary for a regression analysis based on the model

$$y = \beta_0 + \beta_1 x_1 + \beta_2 x_2 + e$$

is as follows.

ANOVA

SOURCE	SS	df	MS	F
Model	252.3183	2	126.1592	3.949
Error	287.4908	9	31.9434	
Total	539.8092	11		

Model R-Square = .467

Variable summary

VARIABLE	REG. COEF.	SE	t	R-SQUARE	INFL. FAC.
Intercept	22.8379				
x_1	−.29187	.26869	−1.086	.133953	1.075
x_2	3.45762	1.71623	2.015	.133953	1.075

The ANOVA is significant at the 10% significance level, and the effect of the merit variable is also significant at this level. Moreover, the residual plot (not given) is fairly unstructured, suggesting that little more can be said about salaries in terms of these variables.

The investigators were perplexed by the lack of significance of the years of service variable. They expected this variable to be one of the strongest predictors of salaries. Even more puzzling was the negative regression coefficient for this variable, which suggests that salaries go down as years of service increase!

The resolution of this dilemma is seen in the partial regression plot of salaries versus years of service given in display 16.6 Except for two high-influence points in the lower right-hand side of the plot, an increasing trend of salaries versus years of service is actually observed. However, the influence of these two points was sufficiently strong to cause the line to pass near them, thus producing a negative regression coefficient (slope). These points correspond to two individuals with over 20 years of service who were hired

Display 16.6

Partial regression plot of salaries (y) versus years of service (x_1) for the data of example 16.2

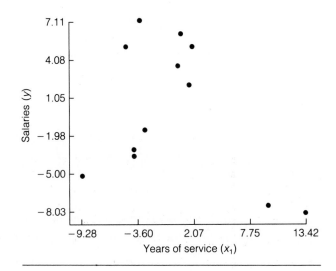

when salaries were low and whose salary increases did not keep up with the "market." If we restrict attention to the more recently hired faculty members by omitting these two individuals from the study, the fitted model is as follows.

ANOVA

SOURCE	SS	df	MS	F
Model	230.0227	2	115.0113	12.350
Error	65.1863	7	9.3123	
Total	295.2090	9		

Model R-Square = .779

Variable summary

VARIABLE	REG. COEF.	SE	t	R-SQUARE	INFL. FAC.
Intercept	12.1524				
x_1	0.9892	0.3176	3.114	0.0853	1.0456
x_2	4.1615	0.9019	4.614	0.0853	1.0456

The model R^2 has increased from .467 to .779, indicating that the deleted observations were causing problems with the model's ability to explain the variation in salaries. The fitted model is now

$$y = 12.152 + 0.989x_1 + 4.161x_2$$

This model indicates an average increase in salary of about $989 for each year of service and an average bonus of $4161 for each point of increase on the merit scale.

Had these ten individuals been selected at random from, say, the population of faculty for this university, then bounds could be placed on the population regression coefficients by means of confidence intervals. As it stands, the model is being used as a descriptive summary of salary information for the given department, and confidence intervals are not meaningful.

EXERCISES

The following steps are to be carried out in exercises 16.4–16.6.

(a) Make box plots of the variables to check for symmetry and outliers. Transform any badly skewed variables with a power transformation and rerun the analysis. Note any remaining outliers.

(b) Make a residual (versus predicted) plot. Note any structure in the plot likely to be indicative of the need for a different regression model. Check for variations in dispersion likely to require a power transformation of the dependent variable.

(c) Make partial regression plots for each regression coefficient and interpret the results of the t tests provided by the variable summary in the light of these plots. In particular, identify any points that have especially strong influence on the outcomes of the tests. Remove these points from the data set and rerun the analysis to determine the extent of their influence.

16.4   Carry out the above steps for the data of exercise 16.1 (p. 532).

16.5   Carry out the above steps for the data of exercise 16.2 (p. 532).

16.6   Carry out the above steps for the data of exercise 16.3 (p. 533).

16.7   The following data are testosterone levels (y) versus age (x) for a group of Caucasian women used as controls in a study of androgen levels.[*]

OBS.	AGE	TESTOSTERONE
1	20	29
2	24	35
3	39	23
4	34	23
5	29	16
6	40	42
7	44	14
8	35	23
9	32	19
10	24	45
11	38	37
12	31	23
13	33	48

(table continued on next page)

[*] From "Serum Androgens by Age in Obese Pima Indian Females," Frances E. Purifoy, Lambert H. Koopmans, Ronald W. Tatum, and Darrel E. Mayes, *Journal of Physical Anthropology* **55**, pp. 491–496, 1981.

OBS.	AGE	TESTOSTERONE
14	42	28
15	38	38
16	40	32
17	20	22
18	37	42
19	39	41
20	31	17
21	29	45
22	34	52
23	40	30
24	43	45
25	31	45
26	22	49
27	46	22
28	23	55
29	35	27
30	42	28
31	24	27

(a) Fit a cubic polynomial model to the data (without recentering the age data) and compute the collinearity diagnostics for the terms in the model.

(b) Transform the age variable by subtracting the median age for the data set from all age values. Repeat the cubic polynomial fit and collinearity computation and compare the results with part (a). What effect has the recentering had on the regression coefficient tests?

(c) Note any high-influence points in the residual plot for the fit of part (b). Investigate the influence of these outliers by rerunning the analysis of part (b) without them.

▰ SECTION 16.4

Indicator Variables and Their Applications

The scope of multiple regression models is greatly increased by the inclusion of categorical variables among the independent variables. This is done by forming variables consisting of 0's and 1's, called *indicator variables*, for each value or level (but one) of the categorical variables. These indicator variables then become independent variables in the model.

Why is this device used instead of simply assigning a numerical code to the values of a categorical variable and using the coded variable in the model? If this were done, the independent variable would receive a single regression coefficient. This regression coefficient would be the increase in the dependent variable per unit increase in the categorical variable code. But the coding of a categorical variable is usually rather arbitrary; the numerical values of the code need have no function other than to label the categories. Thus, each category should have its own separately estimated coefficient. This requires that each level of the categorical variable be represented by a different term in the model.

An indicator variable is formed for a given level of a categorical variable by assigning the value 1 to every observation that has this level of the variable and 0 to all other observations. Now, if an indicator variable were assigned to each level of the categorical variable, then an observation-by-observation sum of all indicator variables would add up to the variable x_0, consisting of all 1's. Since this variable is always (implicitly) in the model when $\beta_0 \neq 0$, we would have created perfect collinearity among the independent variables. Least-squares regression coefficients and their standard errors do not exist in this case, and any attempt to analyze such a model with a computer program unequipped to trap this kind of error would lead to disaster.

To avoid this problem, indicator variables are created for *one fewer* than the number of levels of the categorical variable. One level is singled out as a baseline level, and no indicator variable is constructed for it. The regression coefficient for a second level of the variable will then represent the effect of that level *relative to the baseline level*—that is, the value of y predicted for that level of the variable minus the y value predicted for the baseline level.

The use of indicator variables will be demonstrated through some examples. It was mentioned earlier that the analysis of variance can be viewed as multiple regression in which the independent variables are all categorical and are thus made up entirely of indicator variables. Applications to the one-way ANOVA, which includes the two-sample t test as a special case, will be demonstrated by framing as regression problems some problems seen in earlier chapters. Then, examples in which indicator variables are combined with measurement variables will demonstrate the variety of the models that can be so constructed and how some of them are applied in practice.

The One-Way Analysis of Variance as a Multiple Regression Problem

The categorical variable in a k-sample, measurement variable comparison problem is the variable that labels the populations. For example, in the comparison of head breadths for modern Englishmen and Celts, considered in example 3.1 (p. 71), the categorical variable would have the two levels: *Englishmen* and *Celts*. Choosing *Celts* to be the base level for this variable, there will be one indicator variable, x_1, that takes the value 1 for *Englishmen*, and 0 otherwise. The dependent variable is the head breadth measurement. In regression format, the data are as follows.

Data

OBS.	HEAD BREADTHS (y)	ENGLISHMEN INDICATOR (x_1)
1	141	1
2	148	1
3	132	1
4	138	1

(table continued on next page)

OBS.	HEAD BREADTHS (y)	ENGLISHMEN INDICATOR (x_1)
5	154	1
6	142	1
7	150	1
8	146	1
9	155	1
10	158	1
11	150	1
12	140	1
13	147	1
14	148	1
15	144	1
16	150	1
17	149	1
18	145	1
19	133	0
20	138	0
21	130	0
22	138	0
23	134	0
24	127	0
25	128	0
26	138	0
27	136	0
28	131	0
29	126	0
30	120	0
31	124	0
32	132	0
33	132	0
34	125	0

Now, note that the model

$$y = \beta_0 + \beta_1 x_1 + e$$

is

$$y = \beta_0 + \beta_1 + e$$

for group 1, the modern Englishmen, because the indicator variable has the value 1 for this group. If μ_1 denotes the mean for this group, it follows that $\mu_1 = \beta_0 + \beta_1$.

On the other hand, for the Celts, the model is

$$y = \beta_0 + e$$

since the indicator variable is 0 for this group. If the group mean for Celts is μ_2, then $\mu_2 = \beta_0$. It follows that $\beta_1 = \mu_1 - \mu_2$, and the two-sample problem hypotheses

$$H_0: \mu_1 = \mu_2$$

and

$$H_1: \mu_1 \neq \mu_2$$

are equivalent to the regression hypotheses

$$H_0: \quad \beta_1 = 0 \quad \text{(model 1)}$$

and

$$H_1: \quad \beta_1 \neq 0 \quad \text{(model 2)}$$

The screening test ANOVA and the t test for the significance of β_1 are equivalent in this case. The information required to carry out the test is given in the following regression summary.

ANOVA

SOURCE	SS	df	MS	F
Model	2101.24	1	2101.24	
Error	1135.50	32	35.484	59.216
Total	3236.74	33		
Model R-Square = .649				

Variable summary

VARIABLE	REG. COEF.	SE	t	R-SQUARE	INFL. FAC.
Intercept	130.75				
x_1	15.75000	2.0466	7.696	.000000	1.000

Note that the estimate of β_1, its standard error, and the t statistic are exactly the same as the estimate of $\mu_1 - \mu_2$, its standard error, and the value of the two-sample t statistic found in example 10.4 (p. 313). That is, the regression test is completely equivalent to the two-sample t test.

A one-way analysis of variance example is considered next.

EXAMPLE 16.3

In example 11.3 (p. 354), the insulin production for pancreatic cells was studied over a three-week period. In the regression context, the categorical variable that labels the weeks will have levels *week 1*, *week 2*, and *week 3*. We will take week 1 as baseline and measure the effects on insulin production (y) relative to week 1. The data, in regression format, are as follows.

Data

OBS.	y	x_1	x_2
1	2.02	0	0
2	3.83	0	0
3	6.67	0	0
4	5.38	0	0
5	5.49	0	0

(table continued on next page)

OBS.	y	x_1	x_2
6	3.50	0	0
7	5.90	0	0
8	4.89	0	0
9	1.49	1	0
10	2.67	1	0
11	4.62	1	0
12	4.18	1	0
13	2.78	1	0
14	2.56	1	0
15	4.46	1	0
16	3.79	1	0
17	.33	0	1
18	1.67	0	1
19	4.67	0	1
20	2.45	0	1
21	2.29	0	1
22	1.95	0	1
23	.49	0	1
24	1.81	0	1

Note that indicator variable x_1 is the indicator of week 2, and x_2 is the indicator of week 3. The (full) regression model for this problem is

$$y = \beta_0 + \beta_1 x_1 + \beta_2 x_2 + e \tag{4}$$

Since $x_1 = x_2 = 0$ for observations in week 1, the model reduces to

$$y = \beta_0 + e$$

Thus, if μ_1 is the mean for week 1, $\mu_1 = \beta_0$.

For observations in week 2 the model is

$$y = \beta_0 + \beta_1 + e$$

Consequently, the mean for week 2 is $\mu_2 = \beta_0 + \beta_1$. It follows that $\beta_1 = \mu_2 - \mu_1$.

Similarly, from the model for week 3 we see that $\mu_3 = \beta_0 + \beta_2$. Consequently, the k-sample hypotheses

$$H_0: \mu_1 = \mu_2 = \mu_3$$

and

$$H_1: \text{not } H_0$$

are equivalent to the regression null hypothesis

$$H_0: \beta_1 = \beta_2 = 0 \quad \text{(model 1)}$$

and H_1 specifies the model given by expression (4). But model 1 is the model with no independent variables (except x_0), so the test of this hypothesis is the basic screening test given by the regression ANOVA. It follows that the usual k-sample ANOVA and the regression screening ANOVA are the same test. The correspondence between sources of variation are Between (Weeks) = Model and Within (Weeks) = Error.

The regression analysis output is as follows.

ANOVA

SOURCE	SS	df	MS	F
Model	30.306	2	15.153	
Error	37.030	21	1.7633	8.593
Total	67.336	23		
Model R-Square = .450				

Variable summary

VARIABLE	REG. COEF.	SE	t	R-SQUARE	INFL. FAC.
Intercept	4.71000				
x_1	−1.39125	.66395	−2.095	.250000	1.155
x_2	−2.75250	.66395	−4.146	.250000	1.155

Except for small differences due to round-off errors, this ANOVA is identical to the one given in example 11.3. The intercept in the variable summary is the sample mean for week 1, while the regression coefficient estimates for x_1 and x_2 are the week 2 minus week 1 and week 3 minus week 1 differences in sample means, respectively.

The residual plot for this regression is given in display 16.7. The predicted values are the mean insulin production values for the three weeks. Because the data exhibit a downward trend, the smallest mean occurs for week 3 and the largest for week 1. The residual plot gives somewhat the same information as the schematic diagram, display 11.1 (p. 355), except that the spreads are not quantified by box plots. The minor outlier detected in week 3 is seen as the high data point in the first group of points in display 16.7. The residual plot also shows the differences in dispersion among the three samples seen in display 11.1.

Mixing Categorical and Measurement Variables:
An Analysis of Covariance Example

We saw in the last section that it is desirable to keep the number of independent variables as small as possible, because of the decreased sensitivity and increased potential for collinearity that are natural results of having many variables. Is it also possible to have too few variables? The answer is yes, if variables likely to have sizable influence on the dependent variable are excluded from the model. The penalties for excluding such a variable from the model are twofold.

First, the y variability that would be accounted for by this variable appears in the sum of squares for error rather than in the model sum of

Display 16.7

Residual plot for example 16.3 (predicted values are the mean insulin production values for the three weeks)

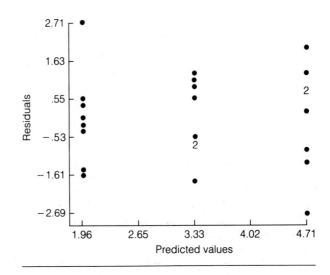

squares. Thus an opportunity to sensitize both the screening test and the tests for the individual regression coefficient is lost.

Second, the exclusion of this term from the model may cause the y variation due to the variable to be attributed to a related variable in the model, thus leading to incorrect conclusions. This phenomenon is seen in the next example.

EXAMPLE 16.4

Measures of skills development or D scores for six boys and six girls are given in regression format in the following table. Girls have been taken as the control category for the sex variable, so the indicator variable is 1 for boys.

Data

OBS.	D SCORES (y)	AGE (x_1)	SEX (x_2)
1	8.61	3.33	0
2	9.40	3.25	0
3	9.86	3.92	0
4	9.91	3.50	0
5	10.53	4.33	1
6	10.61	4.92	0
7	10.59	6.08	1
8	13.28	7.42	1
9	12.76	8.33	1
10	13.44	8.00	0
11	14.27	9.25	1
12	14.13	10.75	1

The ANOVA and variable summary for a regression analysis of D scores versus sex (as though the age variable were not present) is as follows.

ANOVA

SOURCE	SS	df	MS	F
Model	15.709	1	15.709	
Error	27.910	10	2.791	5.629
Total	43.619	11		
Model R-Square = .360				

Variable Summary

VARIABLE	REG. COEF.	SE	t	R-SQUARE	INFL. FAC.
Intercept	10.305				
Sex	2.28833	.96454	2.372	.000000	1.000

The test that the regression coefficient $\beta_2 = 0$, which is equivalent to the two-sample t test of equality of D score means for the two sexes, is significant at the 5% level ($P < .05$). Consequently, we would be led to believe that boys have larger D scores than do girls, on the average.

An important missing variable in this analysis is age. These are the same D scores whose simple regression against age was calculated in chapter 14 (p. 459ff). A scatter diagram of D scores versus age is given in display 16.8, with scores for boys coded with B's and those for girls with G's. Note that if

Display 16.8

Scatter diagram of D scores versus age for the data of example 16.4; girls are coded G and boys B

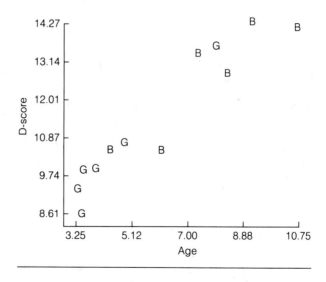

the data points were projected horizontally onto the y axis, the points for boys and girls separate into nearly two distinct groups, with the D scores for boys larger on the average than those for girls. This is the difference detected by the above test.

However, this test ignores what the graph clearly shows, namely that the boys are, in general, older than the girls. It is possible that the difference in D scores can be attributed to age differences rather than a difference in sex. This can be tested by an ANOVA on a model that includes both the sex (x_2) and age (x_1) variables. The regression model is

$$y = \beta_0 + \beta_1 x_1 + \beta_2 x_2 + e$$

Intuitively, this model assumes a linear "law" of increase in D scores for both boys and girls. The two lines are assumed to have the same slope β_1. It follows that the "boy minus girl" D score effect β_2 measures the (constant) difference in mean D scores for the two sexes. The regression ANOVA and variable summary for these data are as follows.

ANOVA

SOURCE	SS	df	MS	F
Model	40.002	2	20.001	
Error	3.617	9	.4019	49.765
Total	43.619	11		

Model R-Square = .917

Variable summary

VARIABLE	REG. COEF.	SE	t	R-SQUARE	INFL. FAC.
Intercept	6.9270				
Age	.75290	.09684	7.775	.418549	1.311
Sex	−.12598	.48000	−.262	.418549	1.311

It is seen that the addition of the age variable has greatly improved the fit of the model, as the scatter diagram suggested it would. But even more important is the fact that almost all of the variation in D scores is now accounted for by the age variable. That is, after correcting D scores for the ages of the children, the effect of the sex variable is insignificant. Without the age variable in the model, we were led to a wrong conclusion.

The regression coefficient summary indicates that, on the average, D scores increase for both sexes about 0.753 points for each year of age. This increase could be further quantified by a confidence interval if desired. The interpretation of the sex regression coefficient as the age corrected difference in average D scores for boys relative to girls suggests that, in fact, D scores are higher for girls than boys. This is a consequence of the negative sign of the regression coefficient. However, the inference procedure leads to the conclusion, at any reasonable significance level, that a difference in average D scores by sex is not supported by the data.

It was obvious that age would be an important variable in the developmental scores for children in the last example. Knowing this fact, the age variable could be taken into account, or *controlled*, in designing the experiment. Children could first be grouped both by age (e.g., 3-year-olds in one group, 4-year-olds in a second, and so forth) and by sex. An equal number of children could then be randomly selected from each age-sex category. The D score comparisons would then be made on these children by a two-way analysis of variance, a method to be considered in the next chapter. In effect, age is held constant by this design while the effect of sex on D scores is assessed.

When a variable is not controlled in an analysis but is measured or observed along with the dependent variable, as was age, it is called a *covariate*. Analyses like the one in the above example, in which the dependent variable is corrected for a covariate before comparing the effects of other, categorical factors, fall under the heading of the analysis of covariance (ANCOVA). Thus, ANCOVA utilizes regression models that mix both categorical and measurement variables among the independent variables.

Interaction Terms and a Test for the Equality of Slopes for Two Regression Lines

Another model that contains both categorical and measurement independent variables will be given in the next example. This model will take advantage of *interaction terms*, which are simply new independent variables formed by taking the products, observation by observation, of other independent variables in the model. The inclusion of interaction terms further increases the flexibility of a regression model and makes possible the investigation of the joint influence of two or more variables on the dependent variable. Interaction terms will be more fully treated in the ANOVA context in the next chapter. Here, they make possible the testing of the equality of slopes for two regression lines.

EXAMPLE 16.5

An apparent curvilinearity was seen in the relationship between weight (y) and height (x) in display 4.13 (p. 110) for the data of display 1.1 (p. 9). One possible explanation of this feature is that the weight versus height regressions are linear but are different for the two sexes. Because the line for women represents the lower part of the curve and that for men the upper part, the apparent curvilinearity is simply due to a difference in slopes for the two lines.

To test this hypothesis, we will construct a multiple regression model that reduces to simple linear regression models with different slopes and intercepts for men and women. Then we will test the hypothesis that the slopes are the same. This is done as follows. Let y denote weight and x_1 denote height. Taking men as the baseline level of the sex variable, we construct an indicator sex variable x_2 with value 1 for each woman.

In addition, we form an interaction variable $x_3 = x_1 \cdot x_2$ for each observation by multiplying together the values of x_1 and x_2 for that observation. The resulting data set for display 1.1 is as follows. (The height for observation 27 has been corrected from 54 to 64 inches.)

Data

OBS.	WEIGHT (y)	HEIGHT (x_1)	SEX (x_2)	HEIGHT · SEX (x_3)
1	180	72	0	0
2	114	66	1	66
3	115	65	1	65
4	122	62	1	62
5	205	71	0	0
6	125	69	1	69
7	133	66	0	0
8	105	63	1	63
9	110	63	1	63
10	175	71	0	0
11	160	72	0	0
12	133	67	1	67
13	145	67	0	0
14	163	73	0	0
15	180	67	1	67
16	125	65	0	0
17	120	67	0	0
18	125	66	1	66
19	110	60	1	60
20	140	68	0	0
21	133	68	1	68
22	107	65	1	65
23	140	64	1	64
24	125	64	1	64
25	125	63	1	63
26	170	70	1	70
27	120	64	1	64
28	110	64	1	64
29	125	66	1	66
30	120	65	1	65
31	130	65	0	0
32	175	70	0	0
33	120	67	1	67
34	175	66	0	0
35	155	72	0	0
36	82	57	1	57

The regression model can be written in the form

$$y = \beta_0 + \beta_1 x_1 + \beta_2 x_2 + \beta_3 x_3 + e \tag{5}$$
$$= (\beta_0 + \beta_2 x_2) + (\beta_1 + \beta_3 x_2) \cdot x_1 + e$$

Now, for men ($x_2 = 0$) the regression model becomes

$$y = \beta_0 + \beta_1 x_1 + e$$

while for women ($x_2 = 1$) it is of the form

$$y = (\beta_0 + \beta_2) + (\beta_1 + \beta_3) x_1 + e$$

Thus, the parameter β_3 represents the difference in slopes between the regression lines for men and women. It follows that the null hypothesis of equal slopes is equivalent to the hypothesis

$$H_0: \quad \beta_3 = 0$$

The alternative hypothesis is the full model given by expression (5). The test of these hypotheses is then simply the t test from the variable summary for variable x_3. The ANOVA and variable summary for this problem are as follows.

ANOVA

SOURCE	SS	df	MS	F
Model	16394.5	3	5464.83	
Error	9416.5	32	294.26	18.571
Total	25811.0	35		
Model R-Square = .635				

Variable summary

VARIABLE	REG. COEF.	SE	t	R-SQUARE	INFL. FAC.
Intercept	−225.6072				
x_1	5.53316	1.62872	3.397	.745900	1.984
x_2	39.78152	139.92569	.284	.998243	23.859
x_3	−.75831	2.07538	−.365	.998106	22.978

The model is seen to be highly significant ($P < .001$). However, the model variable x_3, representing the difference in slopes, is quite insignificant ($P > .5$), so we would conclude that there is no evidence of a difference in slopes for the regression lines for men and women.

Testing More General Hypotheses

Note that the coefficient β_2 in the last example represents the difference in intercepts for the two regression lines. To test the hypothesis that there is no difference in the linear "laws" of weight versus height for the two sexes, we would want to adopt the null hypothesis that the difference in both slopes and intercepts are 0:

$$H_0: \quad \beta_2 = \beta_3 = 0$$

The alternative hypothesis would again specify the model of expression (5).

Note that this pair of hypotheses is different from those tested by the screening test and by the test for individual model coefficients. The null hypothesis (model 1) sets equal to zero more than one but not all of the regression coefficients of the model of expression (5) (model 2).

Most regression packages have options for testing more general models such as this. However, a test can be easily obtained by using the screening test alone. The procedure requires two applications of the screening test.

The first application is the usual one, in which model 2 is taken as the alternative hypothesis. This test yields an error sum of squares SSE_2 and error degrees of freedom ν_2.

Next, the screening test is applied with alternative hypothesis specified by model 1. The data for this test can be obtained simply by deleting the variables corresponding to the regression coefficients set equal to zero to arrive at model 1. The screening test will yield an error sum of squares SSE_1 and error degrees of freedom ν_1.

Now, the test statistic for the test of H_0: model 1 versus H_1: model 2 is

$$F = \frac{(SSE_1 - SSE_2)/(\nu_1 - \nu_2)}{SSE_2/\nu_2}$$

When H_0 is true, this statistic has the F distribution with $\nu_1 - \nu_2$ numerator and ν_2 denominator degrees of freedom. This statistic can then be used as the basis for the test of these hypotheses. The test and P-value computation are carried out just as for the screening test, but with the above degrees of freedom.

For example, to test the hypothesis of equal straight-line regressions for men and women in the above height-weight example, we would first read the error sum of squares and degrees of freedom from the initial screening test: $SSE_2 = 9416.5$ and $\nu_2 = 32$.

The corresponding quantities for model 1 are found by dropping the variables x_2 and x_3 from the data set and repeating the screening test. The resulting ANOVA table is the following.

ANOVA

SOURCE	SS	df	MS	F
Model	15635.4	1	15635.4	
Error	10175.5	34	299.3	52.243
Total	25810.9	35		
Model R-Square = .60576705				

Note that $SSE_1 = 10175.5$ and $\nu_1 = 34$. The F statistic for the test is then

$$F = \frac{(SSE_1 - SSE_2)/(\nu_1 - \nu_2)}{SSE_2/\nu_2}$$

$$= \frac{(10175.5 - 9416.5)/(34 - 32)}{9416.48/32}$$

$$= 1.290$$

The degrees of freedom for this statistic are $34 - 32 = 2$ and 32. From table 4 we find that the P-value for this test exceeds .25. We conclude that there is no evidence of a difference in the lines describing the increase of weight with height for men and women.

EXERCISES

16.8   Formulate the two-sample problem of exercise 10.13 (p. 321) as a regression problem with indicator variables and carry out the regression analysis. Show that the screening test is equivalent to the two-sample t test.

16.9   Refer to exercise 16.7 (p. 547). The data in that exercise are to be used as a control for the study of the rate of change of testosterone with age for Pima Indian women. The following data for Pimas was used in the study referenced in exercise 16.7.

OBS.	AGE	TESTOSTERONE
1	43	20
2	38	21
3	36	19
4	35	18
5	29	51
6	27	37
7	27	68
8	26	28
9	25	52
10	58	18
11	25	19
12	22	50
13	19	43
14	44	13
15	34	19
16	30	23
17	29	27
18	26	31
19	25	37
20	22	31

(a)   Use the method shown in example 16.5 to test the hypothesis that the slopes of straight lines describing the variation of testosterone with age for the controls of exercise 16.7 and the Pima women are the same.

(b)   Use the method of the last subsection to test for the equality of the lines of part (a).

Chapter 16 Quiz

1.   What is the multiple regression model? Give the form of the regression function and the assumptions about the error terms.

2.   What is the process by which estimators are obtained for the parameters of the regression function?

3.   How are the residuals defined and computed?

4.   What is the screening (ANOVA) test for the regression model?

5. What does the rejection of the null hypothesis in the screening test signify? What action would be taken in that event? What action would be taken if the null hypothesis is accepted?

6. What is the coefficient of determination? How is it interpreted?

7. What alternative hypothesis (model 2) are the t tests for the individual regression coefficients based on? How would you interpret the rejection of the null hypothesis using such a test?

8. What is the estimator of the standard deviation, σ, of the error term in a regression model? Where can it be found in the standard regression printout?

9. How can the tabular summary for the regression coefficients of the individual variables be used to construct confidence intervals for the regression coefficients?

10. What is a residual plot? What are its uses?

11. What are the independent variables in a polynomial regression?

12. What is the collinearity problem?

13. How are the independent variable R-squares and inflation factors interpreted?

14. How can collinearity be reduced in polynomial regressions?

15. Why should box plots be obtained for all variables before carrying out a multiple regression?

16. What is a partial regression plot? What are its uses?

17. What is an indicator variable? How is a categorical variable represented by indicator variables in a regression model?

18. What is the regression model corresponding to the two-sample comparison problem model for measurement variables? How are the regression coefficients related to the means of the two populations?

19. What are the disadvantages of leaving important explanatory variables out of a regression analysis?

20. What is a covariate? How are covariates used in the analysis of covariance?

21. What is an interaction term in a regression model?

22. How can one test for the equality of slopes in a linear regression by the inclusion of interaction terms in the model?

23. How can the basic screening test be used to test a null hypothesis that sets equal to zero more than 1 but not all regression coefficients in a multiple regression model?

◢

SUPPLEMENTARY EXERCISES

16.10 Data were obtained from a random sample of junior high schools to study the relationship of student performance in mathematics (y) to performance in reading

(x_1), language (x_2), and social studies (x_3). The following are average scores for the ten sampled schools.

MATHEMATICS	READING	LANGUAGE	SOCIAL STUDIES
42	47	49	51
81	71	71	78
68	64	69	69
47	38	39	41
59	59	53	63
75	60	59	63
38	38	41	41
57	54	53	63
40	40	39	43
59	54	53	66

(a) Find the P-value for the screening test of the regression model of y on x_1, x_2, and x_3.

(b) If the screening test is significant at the 5% level, find the independent variables that are significant at the 5% level and determine 95% confidence intervals for their regression coefficients.

16.11 The following data are to be used in a study of the relationship between serum iron levels (y) in women as a function of age (x_1), vitamin intake (x_2), and cholesterol levels (x_3).

SERUM IRON	AGE	VITAMIN	CHOLESTEROL
112	29	76	160
65	32	41	160
83	66	33	155
74	72	74	173
107	45	80	174
101	36	50	160
103	23	38	220
94	36	68	245
74	27	35	165
123	57	58	192
121	52	59	185
89	48	82	210
45	21	48	180
47	23	52	133

(a) Find the P-value for the screening test of the regression model of y on x_1, x_2, and x_3.

(b) If the screening test is significant at the 5% level, find the variables that are significant at the 5% level and determine 95% confidence intervals for their regression coefficients.

16.12 Carry out the steps requested for exercises 16.4–16.6 (p. 547) for the data of exercise 16.10.

16.13 Carry out the steps requested for exercises 16.4–16.6 (p. 547) for the data of exercise 16.11.

16.14 Formulate the k-sample comparison problem of exercise 11.9 (p. 363) as a regression problem with indicator variables and carry out the regression analysis. Show that the screening test is equivalent to the one-way ANOVA F test.

Experimental Design and the Two-Way Analysis of Variance

Introduction

In this final chapter, a brief glimpse will be given of the important topic called *experimental design*. This includes the design aspect of sample size selection that we have considered in what have been called *observational experiments*. In addition, a *designed experiment* requires that the assignment of the treatments whose effects are to be assessed be under the control of the experimenter. How is this done in various experimental situations? Why is it done the way it is? How are the results of the experiment analyzed and interpreted? These questions form the content of this chapter.

Many experiments are carried out to observe the relationships between changes in independent variables and the corresponding changes in (one or more) dependent or response variables. As before, we will restrict attention to a single measurement-type response variable. One or more of the independent variables will correspond to treatments whose influences on the response variable are to be compared. Other independent variables will account for aspects of the experimental situation that are being held constant (that is, are being *controlled*) while comparisons of treatments are made. For example, environmental conditions such as temperature and humidity are often controlled in chemical or physical experiments.

A primary motivation for performing designed experiments is that, because the treatments can be varied while holding other experimental factors constant, changes in the response variable can be attributed to changes in treatments. That is, a *causal relationship* between treatment and response can be deduced. This kind of conclusion *requires* the intervention of the experimenter; it cannot be made by simply noting concurrent changes in independent and dependent variables in an observational experiment. A quote attributed to the eminent statistician G. E. P. Box is, "The only way to find out what will happen when a complex system is disturbed is to disturb the system, not to merely observe it passively." A designed experiment is necessary (if not always sufficient) to establish causality.

As we have seen in previous chapters, a given computational method often serves for more than one type of problem. The general linear model and the associated *t* tests and analyses of variance have been used for comparison problems in observational experiments from the beginning of our study of statistical methodology. These same methods also form the computational foundation for designed experiments. However, in the experimental design context, these methods take on new richness and variety. Sensible principles of experimental design will be seen to have important consequences for the analyses and their interpretations. For example, it will be possible to break down the analysis for several independent variables into pieces, each of which can be analyzed and interpreted separately. For the problems to be dealt with in this chapter, the pieces will be small enough for hand-computation; a simple scheme will be given for carrying out the analyses.

We will consider two types of designs that can be analyzed by a *two-way analysis of variance*. The first design, the simplest *randomized block design*, will be contrasted with the so-called *completely randomized design* in the next section in order to introduce a number of design concepts. *Replicated two-factor designs* will then be used to study the important concept of *interaction* in section 17.3.

SECTION 17.2

Completely Randomized versus Randomized Block Designs

We will consider two ways to design and analyze the following simple experiment. A gasoline additive is to be tested to see whether it will increase the mileage of current models of automobiles. Money is available for testing the additive on ten cars. The first experimental design we will look at is the completely randomized design.

The Completely Randomized Design

If the question were to determine whether the additive ever made any difference in mileage, it would be sensible to test it for a particular type of automobile. If mileage figures were available for this type of automobile for gasoline without additive, one strategy would be to use gasoline with the additive in all ten cars, then make the appropriate comparison. This is not a desirable strategy, however, since it is seldom possible to guarantee the comparability of the experimental conditions under which the "additive" and "no additive" data are obtained. For example, if a new blend had been formulated for the gasoline in the time between experiments, or an unpublicized modification in the automobiles had been made, it would not be possible to distinguish an effect due to the additive from the influences of the other changes. Controls for this type of influence can be best exercised if data for the comparison are obtained in the same experiment.

completely randomized design
experimental units

In a **completely randomized design,** the ten automobiles (**experimental units**) allocated to the experiment would be as alike as possible—for instance, from the same manufacturing run from the factory. (Experimental units with this property are called *homogeneous.*) Five cars would then be randomly selected to receive gasoline containing the additive (treatment 1), and the remaining five would get gasoline without the additive (treatment 2). For example, one might put tags with car I.D. numbers in a basket and draw tags one at a time, without replacement. The first five cars could be assigned to treatment 1 and the remainder to treatment 2.

Note that treatment 2 is no treatment at all. The cars receiving this treatment constitute the *control group* for the experiment, while those receiving treatment 1 are the *treatment group*.

The randomization serves two functions. First, every effort is made to ensure that treatment and control groups are as alike as possible, so that any differences seen can be attributed to the treatment (additive). All other *known* influences would be similarly controlled (held constant) or eliminated. However, in complex experiments there are almost always potential influences that are either unknown or mistakenly thought to be unimportant. By randomizing treatments, these influences are likely to be eliminated or at least greatly diffused; the randomly dispersed influences appear as part of the "noise" or uncontrolled variation in the experiment.

The second function served by the randomization process is to make possible a statistical analysis of the experiment. In many experiments, this randomization process will be the lone source of randomness. However, we will adopt a linear model for such experiments and will use the appropriate least-squares fitting method and inference procedures. This least-squares analysis can be shown to be a good approximation to the form of analysis that is actually correct when the randomization process is the only source of randomness in the experiment. In this sense, the randomization process "validates" the method of statistical analysis to be used.

The form of the model traditionally used in experimental design involves a change in both the notation and terminology of chapter 16. The independent variables are categorical. They are either primary variables, called **factors,** or are formed from combinations of these primary variables. The values or *levels* of factors are designated by integers. Each factor is allotted an *index*, which designates its levels in the model. Thus, in the gasoline additive example, the two treatments "additive" and "no additive" are viewed as two levels of a treatment factor. This factor will be assigned the index i in the model. The index value $i = 1$ will represent the gasoline with additive, and $i = 2$ will correspond to the treatment without additive. More generally, the treatment factor will have levels (i values) $1, 2, \ldots, I$, where I is the number of treatments.

factors

The randomization process in the above experiment can be viewed in a different way. The ten experimental units are first randomly grouped into five pairs. Then, the two treatments are randomly assigned within each pair.

replications

In this way, the complete experiment constitutes five independent repetitions or **replications** of the simple experimental design in which the treatments are assigned to pairs of experimental units. Replication is viewed as a factor in the experiment and is assigned an index, j. In the gasoline additive example, j ranges from 1 to $J = 5$.

Note that by specifying a given treatment i and replicate j, a unique experimental unit is singled out. One method for specifying the different experimental units is thus to specify the values of their factor level combinations.

The *response* (value of the dependent variable) of the experimental unit receiving levels i and j of the treatment and replication factors will be denoted by x_{ij}. A table containing the (fictitious) results of the above gasoline additive experiment is given in display 17.1

Display 17.1

Fictitious mileage for ten automobiles of the same type using gasoline with and without an additive that is purported to increase mileage

| | TREATMENT (i) | |
REPLICATE (j)	WITH (1)	WITHOUT (2)
1	15.8	15.4
2	15.9	15.1
3	15.3	15.1
4	15.6	14.9
5	15.4	15.3

For example, the response of the car in replicate 1 that received the gasoline with additive ($i = 1$) is $x_{11} = 15.8$ miles per gallon, while for the car in the same replicate receiving gasoline without additive it is $x_{21} = 15.4$.

The linear model for the completely randomized design is as follows.

$$x_{ij} = \mu + \alpha_i + e_{ij}, \qquad i = 1, \ldots, I, \qquad j = 1, \ldots, J \qquad (1)$$

overall mean

treatment effect

The term μ represents the **overall** (or *grand*) **mean** of the x's. The parameter α_i represents the ith **treatment effect**—the contribution (relative to the grand mean) of the ith treatment to the response.

Finally, the e_{ij}'s are random variables which represent the *uncontrolled variation* in the experiment—the variation not explained by other terms in the model. These random variables are assumed to be independent and normally distributed, with 0 means and (common) standard deviation σ.

In the last chapter, it was seen that if indicator variables are assigned to every level of a categorical variable, then an exact collinearity is created with the intercept variable, which is a variable of 1's. This collinearity was avoided by leaving out the indicator variable for one factor level. In the present context, we will avoid the problem by imposing the following constraint on the α_i's:

$$\sum_i \alpha_i = 0$$

where the sum is over the range from $i = 1$ to I.

The model (1) is a thinly disguised version of the model used in the two-sample problem of chapter 10 when the number of treatments is $I = 2$. It is also equivalent to the model for the k-sample problem of chapter 11 when $I > 2$. Since the theory is the same, it follows that the test of the hypothesis of no treatment effect,

$$H_0: \quad \alpha_1 = \cdots = \alpha_I = 0$$

in the completely randomized design is simply (equivalent to) the two-sample t test when $I = 2$ and the one-way analysis of variance (ANOVA) when $I > 2$. (When $I = 2$, the ANOVA is equivalent to a two-sided t test.) Consequently, no new computing methods are needed for the completely randomized design.

The comparison of the procedures of this and the next subsection are best made in terms of the analysis of variance. Consequently, the one-way ANOVA table for the data of display 17.1 is given in display 17.2.

Display 17.2

One-way analysis of variance table for the data of display 17.1

ANOVA table

SOURCE	SS	df	MS	F
Treatment	.4840	1	.4840	9.398
Error	.4120	8	.0515	
Total	.8960	9		

The "between population" sum of squares in the k-sample problem is the *treatment* sum of squares in the completely randomized design, and the "within population" sum of squares measures the uncontrolled *error* in the experiment. The usual F test is then a test of significant treatment effects.

From table 4, we find that the F value of 9.398 for 1 and 8 degrees of freedom leads to P-value bounds $.01 < P < .05$. Thus, we would conclude at the 5% significance level that there is a difference in treatment effects. An inspection of treatment means, 15.6 for the gasoline with additive and 15.16 for the gasoline without, would lead to the conclusion that the additive improves gas mileage.

A Limitation of the Completely Randomized Design

If the intent of the gasoline additive study is to demonstrate that the additive is good for general use, then a weakness of the completely randomized design is that we have shown an improvement in mileage for only one type of automobile. To increase the scope of the study, we would have to include several makes and models of automobiles in the experiment; it would be unfortunate indeed if the additive were particularly suited to the type of automobile used in the experiment but did nothing to improve mileage for other types of cars.

We cannot test the additive for all possibilities, since a ten-car limit has been established for the experiment. However, we can select a variety of cars of different sizes and makes that will be more representative of the general mix of cars currently in use. If the mileage improvement stands up for this cross-section of cars, we will be more confident of the additive's worth; the "generalizability" of the experiment will have been enhanced.

An experimental design with the appropriate features is the randomized block design, discussed in the next subsection.

Randomized Block Design

blocks

Given a ten-car limit, a reasonable strategy for testing the gasoline additive would be to use five different types of cars, gathered in two car groups or **blocks.** The cars within each block would be of the same type and as closely matched as possible in all respects. (That is, they would be homogeneous.)

Now, the treatments are randomly assigned within each block. That is, the gasoline with additive would be tested on one car within a block and the gasoline without additive would be tested on the other. The reasons for randomizing the treatments within blocks are as before. This practice accounts for the name *randomized block design.*

Fictitious data for the gasoline additive problem, using different types of cars, are shown in display 17.3.

Display 17.3

Fictitious mileage for five types of automobiles arranged in blocks, with two cars of the same type within each block

| | TREATMENT (*i*) | |
BLOCK (*j*)	WITH (1)	WITHOUT (2)
1	15.8	15.4
2	17.9	17.1
3	20.3	20.1
4	15.6	14.9
5	12.4	12.3

In the randomized block model, blocks replace replicates as a factor in experiments. The index *j* will now label the levels of the blocking factor, distinguishing the different car types.

The data structure in display 17.3 is similar to that of display 17.1. There is a temptation to use the one-way ANOVA (or two-sample *t* test) to test for a treatment effect. This analysis leads to immediate disaster, however, as is seen in display 17.4.

Display 17.4

One-way analysis of variance for the data of display 17.3

ANOVA table

SOURCE	SS	df	MS	F
Treatment	.4840	1	.4840	.057
Error	67.7320	8	8.4665	
Total	68.2160	9		

Compare the results of this analysis with display 17.2. Note that while the treatment sums of squares are the same for the two analyses, the error sum of squares in display 17.4 has been greatly increased, leading to an insignificant test outcome. Since the only difference in the experiments is that the second has been made on different types of cars, the increase in sum of squares must be due to the variation in mileage from one type of car to another.

This difficulty with the one-way ANOVA suggests an alternative procedure. Because the cars are grouped into pairs by the blocking process, we can make the treatment comparison within blocks, then "average" the comparisons over blocks. In chapter 10, the paired-difference t was used (in the case of two treatments) to replace the two-sample t test for this purpose.

When applied to the data of display 17.3, the paired-difference t statistic has a value of 3.226, yielding P-value bounds $.02 < P < .05$ for a two-sided test. In contrast, the two-sample t statistic (which is the square root of the F statistic in display 17.4) has a value of .2391 ($P > .50$).

The extreme difference in the outcomes of these two tests can be explained as follows. In the two-sample t test, the block-to-block variation is included along with the within-sample variation in the error sum of squares that figures in the denominator of the t statistic. In the paired-difference method, on the other hand, the block contribution to the variation is eliminated by forming the pairwise differences. Thus, only the intrinsic variation among the individual automobiles contributes to the error sum of squares.

Put in another way, the paired difference method isolates the block-to-block variation and removes it from the error sum of squares. This process is basic to the two-way analysis of variance, which is the extension of the paired-difference method to more than two treatments. The two-way ANOVA is equivalent to the two-sided paired-difference t test for two treatments. We will apply it to the gasoline additive problem for purposes of illustration.

The Two-Way Analysis of Variance for Randomized Blocks

The two-way analysis of variance is a consequence of applying the general linear model theory to the following model for randomized blocks. The treatment factor is again indexed by i, and the blocking factor is indexed by j. Thus, x_{ij} represents the value of the response variable for the experimental unit in block j that received treatment i. The model is

$$x_{ij} = \mu + \alpha_i + \beta_j + e_{ij}, \qquad i = 1, \ldots, I, \qquad j = 1, \ldots, J \qquad (2)$$

where μ is, again, the overall mean, α_i is the effect due to the ith treatment, and e_{ij} represents random error. These errors are again assumed to be independent and normally distributed, with 0 mean and common standard deviation σ.

block effect

The blocking factor contributes the new term in the model: β_j is the jth **block effect,** which is the contribution (relative to the overall mean) of the jth block to the response. This term represents the average response for experimental units in the jth block. It is thus capable of accounting for the differences in mileage for different types of automobiles.

Constraints on the treatment and block effects are

$$\sum_i \alpha_i = 0 \quad \text{and} \quad \sum_j \beta_j = 0$$

The Parameter Estimates and Two-Way ANOVA

Denote the regression term of the model, $\mu + \alpha_i + \beta_j$, by μ_{ij}. Then the above constraints on the treatment and block effects imply the following relationships:

$$\mu = \bar{\mu}_{..}$$
$$\alpha_i = \bar{\mu}_{i.} - \bar{\mu}_{..}$$
$$\beta_j = \bar{\mu}_{.j} - \bar{\mu}_{..}$$

The notation $\bar{\mu}_{i.}$ denotes the average (arithmetic mean) of μ_{ij} over all values of the index j, and $\bar{\mu}_{.j}$ is the average over the index i. The parameter $\bar{\mu}_{..}$ is the average of μ_{ij} over both i and j.

The least-squares estimation procedure leads to especially simple parameter estimators for this model. The estimators can be found by replacing μ_{ij} by x_{ij} in these relationships among parameters:

$$\hat{\mu} = \bar{x}_{..}$$
$$\hat{\alpha}_i = \bar{x}_{i.} - \bar{x}_{..}$$
$$\hat{\beta}_j = \bar{x}_{.j} - \bar{x}_{..}$$

Sums of squares for treatments and blocks are the sums of squares of these estimated effects:

$$SS_T = \sum_i \sum_j \hat{\alpha}_i^2 = J \sum_i \hat{\alpha}_i^2$$

and $\qquad\qquad\qquad\qquad\qquad\qquad\qquad\qquad\qquad\qquad\qquad$ (3)

$$SS_B = \sum_i \sum_j \hat{\beta}_j^2 = I \sum_j \hat{\beta}_j^2$$

The *degrees of freedom* associated with these sums of squares are

$$\nu_T = I - 1$$

and

$$\nu_B = J - 1$$

The values of x_{ij} predicted by the model are

$$\hat{x}_{ij} = \hat{\mu} + \hat{\alpha}_i + \hat{\beta}_j$$

and the residuals are

$$r_{ij} = x_{ij} - \hat{x}_{ij}$$

The *error sum of squares* is then the sum of squares of residuals

$$SS_E = \sum_i \sum_j r_{ij}^2 \qquad\qquad\qquad\qquad\qquad\qquad (4)$$

and the *degrees of freedom for error* are

$$\nu_E = (I - 1)(J - 1)$$

The *total sum of squares* is

$$SS_{Tot} = \sum_i \sum_j (x_{ij} - \bar{x}_{..})^2$$

$$= \sum \sum x_{ij}^2 - IJ\bar{x}_{..}^2$$

and the *total degrees of freedom* are

$$\nu_{Tot} = IJ - 1$$

Note that since the number of observations (experimental units) is equal to the number of blocks, J, times the number of experimental units in each block, I, the total degrees of freedom are the number of observations minus 1, as before.

The results from multiple regression, given in the last chapter, guarantee a decomposition (ANOVA) of the total sum of squares into the sum of a model or regression sum of squares and the error sum of squares:

$$SS_{Tot} = SS_{Model} + SS_E$$

Special features of the randomized block model provide a further decomposition of the model sum of squares into treatment and block sums of squares:

$$SS_{Model} = SS_T + SS_B$$

The same decomposition holds for the degrees of freedom.

The test for the presence of treatment effects is an F test that uses an F statistic formed as the ratio of treatment and error mean squares. In the randomized block context, the attitude usually adopted is that treatments are randomly selected but the blocks are not, and a test for the presence of block effects is therefore inappropriate. The block sum of squares plays its role in the adjustment of the error mean square. The computations can be organized in an ANOVA table, which has the following form.

ANOVA table

SOURCE OF VARIATION	SUMS OF SQUARES	DEGREES OF FREEDOM	MEAN SQUARES	F
Treatments	SS_T	$I - 1$	$MS_T = SS_T/(I - 1)$	MS_T/MS_E
Blocks	SS_B	$J - 1$	$MS_B = SS_B/(J - 1)$	
Error	SS_E	$(I - 1)(J - 1)$	$MS_E = SS_E/(I - 1)(J - 1)$	
Total	SS_{Tot}	$IJ - 1$		

The two-way ANOVA table for the gasoline additive problem, using the data of display 17.3, is given in display 17.5.

The statement that the two-way ANOVA isolates the block effect from the error term of the one-way ANOVA can now be seen by comparing the error sum of squares from display 17.4 with the blocks and error sums of squares of display 17.5. Note that the one-way ANOVA error sum of squares

Display 17.5

ANOVA table for the two-way analysis of variance for the data of display 17.3

SOURCE	SS	df	MS	F
Treatments	.4840	1	.4840	10.409
Blocks	67.5460	4	16.8865	
Error	.1860	4	.0465	
Total	68.2160	9		

is the sum of the block and error sums of squares in the two-way ANOVA, with the bulk of the variation attributed to error in the one-way analysis now attributed to blocks. The effect of including the blocking factor in the model has been to absorb most of the previously unexplained error into the new block sum of squares, thus greatly reducing the magnitude of the unexplained error. Since this error is the basis of the F test for treatment effect, the sensitivity of the test has been greatly improved.

The F statistic for the gasoline additive test has the value 10.409, which, for degrees of freedom 1 and 4, is significant at the 5% level ($0.1 < P < .05$). Thus, we would conclude at the 5% level that the gasoline additive has had some effect on mileage for the five types of cars.

The magnitude of the effect can be determined by a confidence interval, as will be seen presently. Before covering this topic, we will present a simple tabular method for computing the estimates of effects to use in the two-way ANOVA table.

A Method for Finding the Effects in a Two-Way ANOVA Model

The estimated effects and residuals for a two-way analysis can be easily calculated by a two-stage procedure. The method will be demonstrated for the gasoline additive data of display 17.3. The original data table is as follows.

	TREATMENTS	
BLOCKS	1	2
1	15.8	15.4
2	17.9	17.1
3	20.3	20.1
4	15.6	14.9
5	12.4	12.3

An intermediate table is constructed with the same format as the original table in the first stage. The arithmetic means of the rows of the original table are written down in a new, last column. Each entry in the body of the intermediate table is then the corresponding entry from the original table *minus* its row mean (from the last column). The gasoline additive intermediate table is as follows.

Intermediate table

.20	−.20	15.60
.40	−.40	17.50
.10	−.10	20.20
.35	−.35	15.25
.05	−.05	12.35

Note that the mean for the first row of the original table is 15.6, and the values from this row minus the row mean are .200 and −.200. Similarly, the values from the second row minus the row mean of 17.5 are .400 and −.400, and so forth.

effects table   The final table, called the **effects table,** is formed in the second stage. This table has the same format as the intermediate table, but with a new, last row. This row contains the means of the *columns* of the intermediate table. Note that the new column in the intermediate table is treated exactly like the other columns in this computation. Each entry in the body of the effects table is now the corresponding entry of the intermediate table minus its column mean.

Effects table

	RESIDUALS (r)		BLOCK EFFECTS ($\hat{\beta}$)
	−.02	.02	−.58
	.18	−.18	1.32
	−.12	.12	4.02
	.13	−.13	−.93
	−.17	.17	−3.83
Treatment effects ($\hat{\alpha}$)	.22	−.22	Overall mean ($\hat{\mu}$) 16.18

Note, for example, that the mean of the first column of the intermediate table is .22; subtracting this number from the value .20 in the first row yields −.02. Similarly, .40 − .22 = .18, and so forth.

The effects table contains all of the relevant parameter estimates as well as the residuals. The estimate of the overall mean, $\hat{\mu} = 16.18$, is in the lower right-hand corner. The treatment effects $\hat{\alpha}_1 = .22$ and $\hat{\alpha}_2 = −.22$ are in the bottom row, and the block effects $\hat{\beta}_1 = −.58, \ldots , \hat{\beta}_5 = −3.83$ are in the last column. The residuals, r_{ij}, are in the body of the table. For example, the residual for treatment 1 and block 3 is the number in the first column and third row: $r_{13} = −.12$.

Calculating the Two-Way ANOVA Table from the Effects Table

The effects table contains all of the information needed to construct the sums of squares for the two-way ANOVA table. Thus, from expression (3), the treatment and block sums of squares are

$$SS_T = J\sum_i \hat{\alpha}_i^2 = 5(.22^2 + (−.22)^2) = .4840$$

and

$$SS_B = I\sum_i \hat{\beta}_i^2 = 2((-.58)^2 + 1.32^2 + \cdots + (-3.83)^2) = 67.546$$

From expression (4), the error sum of squares is the sum of squares of the residuals

$$SS_E = \sum_i \sum_j r_{ij}^2 = (-.02)^2 + .02^2 + \cdots + (-.17)^2 + .17^2 = .1860$$

The ANOVA table in display 17.5 can now be constructed from these values.

Interpreting the Entries of the Effects Table

In addition to providing the information for the ANOVA table, the effects table has important descriptive uses. For example, the difference of effect estimates, say $\hat{\alpha}_i - \hat{\alpha}_{i'}$, is equal to the estimate of the difference in means for that factor (averaged over all levels of the other factor). Since the effect for treatment 1, which is the gasoline with additive, is .22, and that for treatment 2 is $-.22$, we conclude that the additive has increased mileage by an average of $.22 - (-.22) = .44$ miles per gallon.

Similarly, by inspecting the block effects, we see that cars of type 3 have the best mileage, while cars of type 5 have the worst. The mean difference in mileage for these two types of cars is $4.02 - (-3.83) = 7.85$ miles per gallon.

The treatment and block means are equal to the overall mean plus the corresponding effect. These means can then be estimated from the effects table. For example, since the overall mileage of the cars in the experiment is $\hat{\mu} = 16.18$ miles per gallon, and the cars of type 3 get an additional 4.02 miles per gallon, the mean mileage for this type of car is 20.20 miles per gallon. Since cars of type 5 got 3.83 miles per gallon less than the overall mean, their average mileage is 12.35 miles per gallon.

Similarly, cars using the gasoline with additive got an average mileage (over all car types) of $16.18 + .22 = 16.40$ miles per gallon, while the average mileage of cars using gasoline without the additive was $16.18 - .22 = 15.96$ miles per gallon.

EXERCISES

17.1   Calculate the effects table and two-way ANOVA table for the following data.

| | TREATMENT | |
BLOCK	1	2
1	5	13
2	22	10
3	10	13
4	12	16
5	8	3
6	17	14

Find the *P*-value for the treatment factor. If a significant difference is found (at the 5% level), use the effects table to describe it.

17.2 The famous naturalist Charles Darwin performed an experiment to determine whether self- or cross-pollinated plants produce the more vigorous seeds. He matched plants in pairs according to genetic backgrounds, randomly selecting one member of the pair for cross-pollination and the other for self-pollination. Plants from seeds of each pair were then grown in the same pot and their heights measured after maturing. It is assumed that the more vigorous seeds produce taller plants.

The following are heights in inches for ten pairs of plants grown in Darwin's experiment.

CROSS-POLLINATED	SELF-POLLINATED
23.500	17.375
21.000	20.000
22.000	20.000
19.125	18.375
21.500	18.625
22.125	18.625
20.375	15.250
18.250	16.500
21.625	18.000
23.250	16.250

(a) Construct the effects table for these data.

(b) Construct the ANOVA table and test for a treatment effect at the 5% significance level. Give your conclusions.

Confidence Intervals for Contrasts

contrast

A **contrast** for the effects of a factor is a linear combination of these effects with coefficients that sum to 0. For example, a contrast for the treatment effects $\alpha_1, \ldots, \alpha_I$ is of the form

$$C = k_1\alpha_1 + \cdots + k_I\alpha_I$$

where the coefficients $k_1, \ldots, k_I$ satisfy the condition $k_1 + \cdots + k_I = 0$.

The natural estimator for a contrast is the same linear function $\hat{C}$ of the estimated effects. However, because $\bar{x}_{..}(k_1 + \cdots + k_I) = 0$, this estimator can be written in terms of the means $\bar{x}_{i.}$:

$$\hat{C} = k_1\hat{\alpha}_1 + \cdots + k_I\hat{\alpha}_I$$
$$= k_1(\bar{x}_{1.} - \bar{x}_{..}) + \cdots + k_I(\bar{x}_{I.} - \bar{x}_{..})$$
$$= k_1\bar{x}_{1.} + \cdots + k_I\bar{x}_{I.}$$

Because the means $\bar{x}_{i.}$ are independent random variables, the rules for variance derived in chapter 5 make it possible to calculate the variance of $\hat{C}$:

$$\text{var}(\hat{C}) = k_1^2\,\text{var}(\bar{x}_{1.}) + \cdots + k_I^2\,\text{var}(\bar{x}_{I.})$$

But, since $\bar{x}_{i.}$ is the arithmetic mean of J identically distributed terms with variance σ^2, it follows that

$$\text{var}(\bar{x}_{i.}) = \frac{\sigma^2}{J}$$

Thus, the standard error of $\hat{C}$ is

$$\sigma_{\hat{C}} = \sigma \sqrt{\frac{k_1^2 + \cdots + k_I^2}{J}}$$

This standard error can be estimated by replacing σ by the pooled estimate of standard deviation based on the error mean square, MS_E, from the two-way ANOVA table. That is,

$$SE_{\hat{C}} = \hat{\sigma} \sqrt{\frac{k_1^2 + \cdots + k_I^2}{J}}$$

where

$$\hat{\sigma} = \sqrt{MS_E}$$

Now, the statistic

$$t = \frac{\hat{C} - C}{SE_{\hat{C}}}$$

has Student's t distribution with degrees of freedom equal to the error degrees of freedom $\nu_E = (I - 1)(J - 1)$. Thus, the pivotal method of chapter 8 can be applied to find a $100(1 - \alpha)\%$ confidence interval for the contrast. The interval limits are

$$L = \hat{C} - tSE_{\hat{C}}$$

and

$$U = \hat{C} + tSE_{\hat{C}}$$

where t is the critical value obtained from table 3 for (two-sided test) probability α and ν_E degrees of freedom.

Most of the interesting parameters in the analysis of variance can be written as contrasts. For example, if $k_1 = 1$, $k_2 = -1$, and all other k_i's $= 0$, then the contrast is the difference in effects $\alpha_1 - \alpha_2$. The limits of a confidence interval for this difference are then

$$L = \hat{\alpha}_1 - \hat{\alpha}_2 - t\hat{\sigma}\sqrt{\frac{2}{J}}$$

and

$$U = \hat{\alpha}_1 - \hat{\alpha}_2 + t\hat{\sigma}\sqrt{\frac{2}{J}}$$

where $\hat{\sigma} = \sqrt{MS_E}$.

Thus, to obtain a 95% confidence interval for the difference in treatment effects in the gasoline additive problem, we require the critical value,

t, corresponding to a two-sided test probability of .05 for $\nu_E = 4$ degrees of freedom. (See the ANOVA table given in display 17.5.) From table 3, we find $t = 2.776$. The estimate of σ is $\sqrt{MS_E} = \sqrt{.0465} = .216$, and the difference in estimated effects is $\hat{\alpha}_1 - \hat{\alpha}_2 = .22 - (-.22) = .44$ from the effects table. Thus, the confidence limits are

$$L = .44 - 2.776 \cdot .216 \cdot \sqrt{\tfrac{2}{5}} = .061$$

and

$$U = .44 + 2.776 \cdot .216 \cdot \sqrt{\tfrac{2}{5}} = .819$$

We conclude with 95% confidence that, on the average, the additive will increase mileage between .061 and .819 miles per gallon. This average is over the types of cars used in the experiment, which we have chosen to be representative of the cars in current use. If the additive is reasonably inexpensive so that no appreciable increase in the price of gasoline is required, this magnitude of mileage increase might well be attractive to the gasoline manufacturer as a selling point for the product.

Fisher's Least Significant Difference Method

Confidence intervals for differences of effects can be used to implement Fisher's least significant difference method for the corresponding factor effects. If the procedure is to be carried out at significance level α, the confidence intervals are computed with confidence coefficient $100(1 - \alpha)\%$. Two factor levels are put in the same group if the confidence interval for the difference in their effects contains 0. Otherwise, they are put in different groups.

EXAMPLE 17.1

In an experiment to determine when in the growing season to apply nitrogen to a certain variety of wheat, four fields (blocks) were subdivided into four plots each. Wheat was planted in all fields and the plots within each field were randomly selected to receive one of the four treatments: *None* (no nitrogen applied), *Early* (nitrogen applied early in the growing season), *Middle* (mid-season application), and *Late* (late-season application). A measure of the response of the wheat to treatment based on the wet and dry weights of the resulting grain is given in the following table. Large values correspond to good responses.

	NITROGEN APPLICATION (TREATMENT)			
BLOCK	NONE	EARLY	MIDDLE	LATE
1	.718	.732	.734	.792
2	.725	.781	.725	.716
3	.704	.799	.763	.758
4	.726	.765	.738	.781

The test of the no-treatment (season) effect hypothesis is significant at the 10% level. The treatments, in increasing order of their effects are None, Middle, Late, and Early. The following are 90% confidence limits for the differences of pairs of treatment effects.

EFFECTS	L	U
None minus Early	−.084	−.018
None minus Middle	−.055	.012
None minus Late	−.077	−.010
Early minus Middle	−.004	.063
Early minus Late	−.026	.041
Middle minus Late	−.055	.012

Note that the confidence interval for the None minus Middle effects contains 0. Thus, these effects are in the same group. However, the None and Late treatments are in different groups, since the interval for the difference of their effects does not contain 0. Continuing this process leads to the 10% FSD groupings

{None, Middle} < {Middle, Late, Early}

Thus, we would conclude at the 10% significance level that late and early applications of nitrogen are superior to no application, but application in mid-season cannot be distinguished from the other treatments.

EXERCISES

17.3 Calculate a 95% confidence interval for the difference in treatment effects for the data of exercise 17.1 (p. 576).

17.4 Find a 95% confidence interval for the differences in treatment effects for the Darwin data of exercise 17.2. Interpret the results.

The Additivity of Effects Assumption

additivity of effects

The randomized block model has an important assumption built in: that of **additivity of effects.** If μ_{ij} denotes the regression function for this model, then the influence of treatments and blocks on the response is taken to be the sum of the effects of the individual factor effects:

$$\mu_{ij} = \mu + \alpha_i + \beta_j$$

This is what is meant by the additivity of these effects.

An important implication of this assumption is that the "action" of a factor does not depend on the level of the other factor. For example, the difference in mean responses, $\mu_{ij} - \mu_{i'j}$, attributable to any two treatments i and i' within the same block j, is the same for all blocks:

$$\mu_{ij} - \mu_{i'j} = (\mu + \alpha_i + \beta_j) - (\mu + \alpha_{i'} + \beta_j)$$

$$= \alpha_i - \alpha_{i'}$$

Similarly, differences in mean block responses for a given treatment do not depend on the particular treatment.

While additivity may be assumed in the model, it may not actually hold for the data. When this is the case, the effects are said to be *nonadditive*. Nonadditivity will be studied in the next section through the introduction of a new model element, the interaction term.

SECTION 17.3
A Two-Way ANOVA Model with Interaction

The main topic of this section will be *interaction*, the tendency for the response variable in a problem to react differently to a change of levels of a given independent variable at different levels of other independent variables. In a randomized block situation, for example, the presence of interaction is undesirable since it complicates the interpretation of treatment effects. On the other hand, interaction is expected for many types of problems and its study is a principal reason for carrying out an experiment.

In a two-factor study, the factors might be two components of a fertilizer, say nitrogen and iron, being blended for a particular variety of wheat. The levels of the factors are different amounts of each constituent, and the goal of the study is to determine how the size of the wheat kernels varies with the different combinations of the two fertilizer components. In particular, the combination that yields the largest kernels would be of interest.

Because certain combinations may actually poison the plants, one would not necessarily expect the best combination to be the one corresponding to the best level of nitrogen (averaged over all levels of iron) and the best level of iron (averaged over all nitrogen levels), the combination predicted by an additive model. It is necessary to study the interaction of the two factors to sort out the interesting influences.

In this section we will study the simplest model that permits the estimation of a general interaction term—the *replicated two-factor model*. The model and estimates of the terms will be given in the next subsection. An ANOVA table containing the necessary statistics for testing interactions and the additive components of the model, now called *main effects*, will then be given. The computing method of the last section, using an effects table, will be extended to provide the estimates of main effects and interactions to be used in the ANOVA table.

A useful graphical device called an *interaction plot* will be introduced for displaying and interpreting interactions and main effects.

The Replicated Two-Factor Model with Interaction

The levels of the two factors in the model will be indexed by indices $i = 1,$ . . . , I and $j = 1,$ . . . , J, as in the randomized block model. The two-factor design is replicated n times. Replication is taken to be a model factor with levels $k = 1,$ . . . , n. The value of the response variable at levels i, j, and k of the three factors will be denoted by x_{ijk}. It is assumed that the response is generated according to the following model

$$x_{ijk} = \mu + \alpha_i + \beta_j + (\alpha\beta)_{ij} + e_{ijk}$$

The sum composed of the first four terms on the right-hand side represents the regression component μ_{ij} of the model. This component is the mean of x_{ijk} for all replicates of the ijth factor combination, so it does not depend on the replication index k. The error term e_{ijk} accounts for the random, unexplained variations in response. As before, the e_{ijk}'s are assumed to be independent, normally distributed random variables, with 0 mean and (common) standard deviation σ.

main effects
interaction

The components of the regression term are the overall mean μ, the factor 1 and factor 2 **main effects** α_i and β_j, respectively, and the factor 1–factor 2 **interaction** $(\alpha\beta)_{ij}$. These parameters are assumed to satisfy the following constraints:

$$\sum_i \alpha_i = 0, \qquad\qquad \sum_j \beta_j = 0$$

$$\sum_j (\alpha\beta)_{ij} = 0 \quad \text{for each } i \quad \text{and} \quad \sum_i (\alpha\beta)_{ij} = 0 \quad \text{for each } j$$

These constraints impose the following relationships between the regression term μ_{ij} and its constituent parameters:

$$\mu = \bar{\mu}_{..}$$
$$\alpha_i = \bar{\mu}_{i.} - \bar{\mu}_{..}$$
$$\beta_j = \bar{\mu}_{.j} - \bar{\mu}_{..}$$
$$(\alpha\beta)_{ij} = \mu_{ij} - \bar{\mu}_{i.} - \bar{\mu}_{.j} + \bar{\mu}_{..}$$

The factor combinations determine *cells*, and the parameter μ_{ij} represents the population mean for the ijth cell. The least-squares estimators of the regression parameters are obtained by replacing the population cell means by the corresponding sample cell means $\bar{x}_{ij}$ in the above relationships. This strategy produces the following results.

$$\hat{\mu} = \bar{x}_{...}$$
$$\hat{\alpha}_i = \bar{x}_{i..} - \bar{x}_{...}$$
$$\hat{\beta}_j = \bar{x}_{.j.} - \bar{x}_{...} \qquad\qquad (5)$$
$$\widehat{(\alpha\beta)}_{ij} = \bar{x}_{ij.} - \bar{x}_{i..} - \bar{x}_{.j.} + \bar{x}_{...}$$

regression sums of
squares

The **regression sums of squares** and their degrees of freedom are then

a.   for the factor 1 main effect   (6)

$$SS_A = \sum_i \sum_j \sum_k \hat{\alpha}_i^2 = nJ \sum_i \hat{\alpha}_i^2$$

$$\nu_A = I - 1$$

b. for the factor 2 main effect

$$SS_B = \sum_i \sum_j \sum_k \hat{\beta}_j^2 = nI \sum_j \hat{\beta}_j^2$$

$$\nu_B = J - 1$$

c. for interaction

$$SS_{AB} = \sum_i \sum_j \sum_k \widehat{(\alpha\beta)}_{ij}^2 = n \sum_i \sum_j \widehat{(\alpha\beta)}_{ij}^2$$

$$\nu_{AB} = (I - 1)(J - 1)$$

The value of x_{ijk} predicted by the model,

$$x_{ijk} = \hat{\mu} + \hat{\alpha}_i + \hat{\beta}_j + \widehat{(\alpha\beta)}_{ij}$$

is seen from expressions (5) to be simply $\bar{x}_{ij.}$, the ijth cell mean. Thus, the residuals are

$$r_{ijk} = x_{ijk} - \bar{x}_{ij.}$$

error sum of squares   It follows that the **error sum of squares** is

$$SS_E = \sum_i \sum_j \sum_k r_{ijk}^2 \tag{7}$$

$$= \sum_i \sum_j (n_{ij} - 1)s_{ij}^2$$

where n_{ij} is the sample size and s_{ij} is the sample standard deviation for the ijth cell.

In the model given above, the cell sample sizes n_{ij} are all equal to the number of replicates, n. The design is said to be *balanced* in this case. A simpler expression for computing the error sum of squares exists for balanced designs. However, we will look at unbalanced designs later and the more general expression (7) will make possible the use of the same computing strategy in both cases.

degrees of freedom for   For the same reason, the **degrees of freedom for error** are defined by
error   the expression

$$\nu_E = N - IJ$$

where

$$N = \sum_i \sum_j n_{ij}$$

which is the total sample size for all cells.

It follows that when the design is balanced, $N = nIJ$, so

$$\nu_E = IJ(n - 1)$$

total sum of squares   The **total sum of squares**

$$SS_{Tot} = \sum_i \sum_j \sum_k (x_{ijk} - \bar{x}_{...})^2$$

$$= \sum_i \sum_j \sum_k x_{ijk}^2 - N\bar{x}_{...}^2$$

can again be computed as an independent check on the computations of the other sums of squares, if desired.

The ANOVA table for this model is as follows.

ANOVA table

SOURCE OF VARIATION	SUMS OF SQUARES	DEGREES OF FREEDOM	MEAN SQUARES	F
Factor 1 Main Effect	SS_A	$I - 1$	$MS_A = SS_A/(I - 1)$	MS_A/MS_E
Factor 2 Main Effect	SS_B	$J - 1$	$MS_B = SS_B/(J - 1)$	MS_B/MS_E
Interaction	SS_{AB}	$(I - 1)(J - 1)$	$MS_{AB} = SS_{AB}/(I - 1)(J - 1)$	MS_{AB}/MS_E
Error	SS_E	$N - IJ$	MS_E	
Total	SS_{Tot}	$N - 1$		

A method for calculating the effects that are used in this table will be given in the next subsection.

The Estimation of Effects

The method for computing the estimated effects for the randomized block model can be used to calculate the effects estimates for the present model as well. However, instead of applying the method to the responses directly, it is applied to the cell means. The computation will be illustrated for the data of display 17.6.

Display 17.6

Data for the illustration of two-way ANOVA computations. There are two factors with $I = 2$ and $J = 3$ levels and $n = 2$ replicates per cell.

		FACTOR 2 LEVELS		
FACTOR 1 LEVELS		1	2	3
1		1	7	6
		2	9	8
2		3	3	8
		5	4	9

First, the summary data n_{ij}, $\bar{x}_{ij.}$, and s_{ij} are computed for each cell. For display 17.6, all sample sizes are equal to the number of replicates, $n_{ij} = 2$.

The sample standard deviations, which are needed for the error sum of squares, are given in the following table.

Cell standard deviations for display 17.6

	FACTOR 2		
FACTOR 1	1	2	3
1	.707	1.414	1.414
2	1.414	.707	.707

The effects table will be generated from the following table of cell means. The intermediate table and final table in this computation are given below.

Table of means for display 17.6

1.500	8.000	7.000
4.000	3.500	8.500

Intermediate table

−4.000	2.500	1.500	5.500
−1.333	−1.833	3.167	5.333

Effects table

	INTERACTIONS			FACTOR 1 MAIN EFFECTS
	−1.333	2.167	−.833	.083
	1.333	−2.167	.833	−.083
Factor 2 main effects	−2.667	.333	2.333	Overall mean 5.417

The only difference between this effects table and the one given for randomized blocks is that the *interaction effects* rather than the residuals appear in the body of the table.

The sums of squares for main effects and interactions can now be computed from this table using expressions (6 a–c). Thus, for example,

$$SS_B = nI\Sigma\hat{\beta}_j^2 = 2 \cdot 2((-2.667)^2 + .333^2 + 2.333^2)$$
$$= 50.667$$

The error sum of squares is calculated from the cell sample sizes and the above table of standard deviations using expression (7):

$$SS_E = \sum_i \sum_j (n_{ij} - 1)s_{ij}^2$$
$$= (2 - 1)(.707^2 + 1.414^2 + \cdots + .707^2)$$
$$= 7.500$$

The completed ANOVA table for these data, obtained by finishing the computations indicated in the above ANOVA table form, is given in display 17.7.

Interpreting Interaction by Means of an Interaction Plot

The strategy for interpreting a two-way ANOVA analysis is to first look at the test for interaction. The value of the F statistic for interaction is com-

Display 17.7

ANOVA table for the data of display 17.6

SOURCE	SS	df	MS	F
Factor 1 Main Effect	.0833	1	.0833	.067
Factor 2 Main Effect	50.6667	2	25.3333	20.267
Interaction	28.6667	2	14.3333	11.467
Error	7.5000	6	1.2500	
Total	86.9167	11		

pared with the values from table 4 for numerator degrees of freedom equal to the interaction degrees of freedom, and denominator degrees of freedom equal to the degrees of freedom for error. If this test is significant at the desired level, the interaction is described. Main effects, which we will see to be averages of interaction effects, are of secondary importance.

On the other hand, if the interaction test is insignificant, then the main effects provide the desired description of the data. An interpretation based on an additive model of the type considered in the last section would then be adequate.

In display 17.7, the interaction test produces an F value of 11.467, which, for 2 and 6 degrees of freedom, has P-value bounds $.001 < P < .01$. Thus, we would reject the hypothesis of no interaction effects at the 5% significance level. Because of this, we are obliged to describe the nature of the interaction—that is, how the influences of factor 1 on the response variable vary over the levels of factor 2.

An extremely useful graphical display of interaction information is the *interaction plot*

interaction plot. It is simply a plot of the *cell means* for each level of one factor, say factor 1, against the levels of the other factor. The plots for all levels of factor 1 are drawn on the same graph for comparison purposes.

For example, an interaction plot for the data of display 17.6 can be obtained by plotting the cell means in each row (factor 1 level) of the table of means against the levels of factor 2. The table of means is repeated here.

Cell means for the data of display 17.6

FACTOR 1 LEVELS	FACTOR 2 LEVELS		
	1	2	3
1	1.5	8.0	7.0
2	4.0	3.5	8.5

The "curve" for level 1 of factor 1 would be obtained by graphing the three points (1, 1.5), (2, 8.0), and (3, 7.0). Straight lines would then be drawn to connect these points. This process would then be repeated for the second row, joining the points (1, 4.0), (2, 3.5), and (3, 8.5) with straight lines. Interaction plots are easily constructed by hand. However, they are also standard options in statistical computer packages. A computer-generated interaction plot for the data of display 17.6 is given in display 17.8.

Display 17.8

Interaction plot for the
data of display 17.6.
Curve 1 corresponds to
level 1 of factor 1 and
curve 2 to level 2 of
this factor. The
horizontal coordinates
are the levels of factor
2, while the vertical
coordinates are the cell
mean values. The
dashed line is the curve
of averages over levels
of factor 1 that
represents the factor 2
main effect.

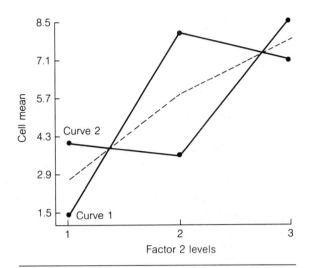

The interaction plot actually contains information about both interactions and main effects. When there is no interaction, the interaction plot consists of parallel (broken) lines. The difference in main effects for two levels of factor 1 (the row factor) would then be equal to the (constant) difference between the graphs for these factor levels. The difference in main effects for two levels of factor 2 (the column factor) would be the difference in heights of any curve at the two points on the horizontal axis corresponding to these levels. Thus, the separations between curves measure the factor 1 main effects, and the amounts of rise and fall in the curves measure the factor 2 main effects.

When interactions are present, the plots are no longer parallel. Interaction is measured by the difference in trends of the curves from point to point on the horizontal axis. Thus, the strong interaction in the data of display 17.6 can be described as follows. For level 1 of factor 1 (curve 1), we see a rise in cell means from level 1 to level 2 of factor 2, then a decline to level 3. However, the trend for level 2 (curve 2) of this factor is quite different, consisting of an initial decline, then a rise in means.

When interactions are present, the main effects are *averages* of the effects of one factor over the levels of the other. Thus, whereas the factor 1 main effect is the (constant) difference between curves when interaction is absent, it becomes the average difference between curves (from level to level of factor 2) when interaction is present.

Note that since the curves in display 17.8 cross each other, the differences between curves 1 and 2 (in that order) are negative at levels 1 and 3 of factor 2 but positive at level 2. When these three differences are averaged, the result is nearly 0. This result produces the insignificant main effect for factor 1 seen in display 17.7.

On the other hand, the factor 2 main effects are the averages (over levels of factor 1) of the mean values at the various factor 2 levels. These averages can be viewed as forming a curve of averages. Thus, if this average curve tends to rise and/or fall from point to point, a factor 2 main effect is indicated. In display 17.8 it is seen that the average values (dashed line) at the three levels of factor 2 show a distinct upward trend. This accounts for the strong factor 2 main effect found in the ANOVA of display 17.7.

This example demonstrates why it is necessary to test for and interpret interactions before moving on to main effects. Had we simply ignored the interactions, we would have been led, by the insignificant factor 1 main effect, to believe that factor 1 had no influence on the response variable. In fact, the influence of factor 1 is quite strong, but it depends in both its magnitude and direction on the level of factor 2.

An application of this type of analysis to a medical study will be given next.

EXAMPLE 17.2

The potassium levels in the portal and jugular veins of rats fed different diets was studied to obtain baseline data for a more complex study. Eight rats were fed normal diets, and eight were given potassium-depleted diets. Four rats of each type were then randomly selected for measurement of potassium levels in the portal vein and four for measurement of levels in the jugular. We want to study the dependence of potassium levels on the Diet and Vein factors. The data are as follows.

| | VEIN | |
DIET	JUGULAR	PORTAL
Normal	4.3	5.9
	4.4	4.4
	5.3	5.8
	4.4	7.3
Potassium-depleted	3.4	3.6
	4.9	3.9
	5.0	4.0
	3.9	3.6

The tables used in the calculation of the effects table are as follows.

Table of means

| | VEIN | |
DIET	JUGULAR	PORTAL
Normal	4.600	5.850
Potassium-depleted	4.300	3.775

Intermediate table

Normal	−.625	.625	5.225
Potassium-depleted	.263	−.263	4.038

Effects table

	VEIN		
DIET	JUGULAR	PORTAL	
Normal	−.444	.444	.594
	.444	−.444	−.594
Potassium-depleted	−.181	.181	4.631

The standard deviations for the four cells are as follows.

Standard deviations

	VEIN	
DIET	JUGULAR	PORTAL
Normal	.4690	1.1846
Potassium-depleted	.7789	.2062

Verify the following ANOVA table for these data.

ANOVA table

SOURCE	SS	df	MS	F
Vein	.5256	1	.5256	.925
Diet	5.6406	1	5.6406	9.928
Interaction	3.1506	1	3.1506	5.546
Error	6.8175	12	.5681	
Total	16.1344	15		

Both interaction and the Diet main effect are significant at the 5% level ($P < .05$ for interaction, $P < .01$ for Diet). The following interaction plot shows that the interaction is due to a difference in trends in potassium levels from the jugular to the portal veins for rats fed the different diets.

Interaction plot

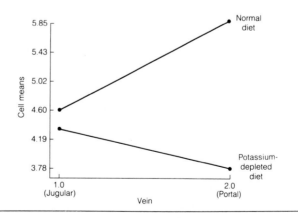

For rats fed normal diets, the potassium level is higher in the portal vein than in the jugular, but the reverse is true for rats fed potassium-depleted diets. No vein main effect shows up because the average potassium levels for rats fed the two types of diets are nearly the same at the two veins. On the other hand, the average potassium level over the two veins is higher for rats fed a normal diet than those fed a potassium-depleted diet, which shows up as a strong diet main effect.

EXERCISES

17.5 The following data are from a two-factor experiment.

FACTOR 1	FACTOR 2		
	1	2	3
1	13.7	15.9	3.7
	14.1	16.2	5.2
	14.0	15.4	4.6
2	12.4	8.7	12.5
	11.4	9.3	10.9
	13.2	7.9	11.6
3	16.2	14.5	12.8
	13.5	15.3	10.8
	14.8	19.0	10.2

(a) Form the effects table for the data.

(b) Carry out a two-way ANOVA with tests for both main effects and interaction at the 5% significance level.

(c) Draw the interaction plot for these data and interpret the results of part (b) in terms of this plot. What are the trends observed in the factor levels?

**Two-Way Analysis of Variance for Tables
with Unequal Sample Sizes**

In two-factor observational experiments as well as replicated two-factor designed experiments, it is not uncommon to end up with an unequal number of observations for different factor combinations. For example, in experiments on animals, even if the numbers of experimental units are initially taken to be equal, some of the animals may die before the responses can be measured. In observational experiments, the numbers of experimental units falling in different categories of the independent factors will often be a matter of chance. Thus, unequal sample sizes occur frequently in practice.

balanced   Recall that a two-factor experiment with equal numbers of observations per cell is called **balanced.** Balanced designs are intuitively appealing because comparisons of levels of the same factor are based on averages over the same numbers of elements. However, even more compelling reasons for balance arise from the mathematics of the least-squares estimation of factor effects and from the corresponding hypothesis tests of their significance.

We saw in chapter 16 that estimates of regression coefficients in a multiple regression model and, thus, the tests of their significance, depended on what other terms were taken to be in the model. This has not been a concern in this chapter until now, because the basic effect of balance in an ANOVA model is to make the estimates of the effects for a given factor independent of what other factors are in the model. Thus, for example, the estimates of main effects in the two-factor model will be the same whether or not we assume the interaction terms to be present in the model. Moreover, the order in which the tests of main effects and interactions are carried out makes no difference. These features vanish for unbalanced designs.

There is no problem with obtaining least-squares estimates and hypothesis tests for factors in unbalanced ANOVA designs. The multiple regression methods of chapter 16 can (and usually must) be applied. However, estimates of effects can vary depending on the order in which hypothesis tests are carried out; this order must be carefully specified. The topic of unbalanced designs is still being developed in the statistical literature, so we leave the general topic to more advanced treatises. However, a good approximation to the more complex situation in the two-factor model closely resembles what we have done for the balanced design.

method of unweighted means

The method, called the **method of unweighted means,** essentially ignores the problems we have mentioned and continues to follow, with minor modifications, the procedures for estimating and testing effects given above. The summary data n_{ij}, $\bar{x}_{ij.}$, and s_{ij} are obtained for each cell, and the effects table is constructed from the cell means exactly as before. However, in expressions (6) for the interaction and main effects sums of squares, the number of replicates, n, is replaced by the following quantity, called the *harmonic mean* of the n_{ij}'s:

$$n_H = \frac{1}{\dfrac{1}{IJ} \displaystyle\sum_i \sum_j \dfrac{1}{n_{ij}}}$$

That is, the harmonic mean is formed by taking the arithmetic mean of the reciprocals $1/n_{ij}$ over all cells, and then taking the reciprocal of this mean. The sums of squares are then computed from the expressions

$$SS_A = n_H J \sum_i \hat{\alpha}_i^2$$

$$SS_B = n_H I \sum_j \hat{\beta}_j^2$$

$$SS_{AB} = n_H \sum_i \sum_j \widehat{(\alpha\beta)}_{ij}^2$$

The degrees of freedom for these sums of squares are the same as before.

The error sum of squares and its degrees of freedom now require the somewhat more complicated form (7). These expressions are repeated here:

$$SS_E = \sum_i \sum_j (n_{ij} - 1)s_{ij}^2$$

and

$$\nu_E = N - IJ$$

where

$$N = \sum_i \sum_j n_{ij}$$

The rest of the analysis, including the ANOVA table, is the same as before.

EXAMPLE 17.3

In a continuation of the study considered in example 17.2, the response of insulin levels in the jugular and portal veins of rats was to be studied as a function of time after injection with a dose of insulin. Observations were gathered opportunistically over a period of time, leading to the following data. A baseline reading was made at the time of the injection, then readings were made at 30 and 60 minutes after the injection, leading to levels 0, 30, and 60 for the time factor. The vein factor is as before.

VEIN	0		30		60	
Jugular	18		61		18	
	36		116		133	
	12		63		33	
	24		132			
	43		68			
			37			
Portal	96	120	146	115	132	196
	72	49	193	199	110	195
	34	92	78	253	141	84
	41	111	127	338	204	105
	98	99	136		69	71
	77	94	144		152	83

The computation of the effects table from the table of means is as follows.

Table of means

VEIN	0	30	60
Jugular	26.600	79.500	61.333
Portal	81.917	172.900	128.500

Intermediate table

Jugular	−29.211	23.689	5.522		55.811
Portal	−45.856	45.128	.728		127.772

Effects table

				VEIN MAIN EFFECTS
Jugular	8.322	−10.719	2.397	−35.981
Portal	−8.322	10.719	−2.397	35.981
Time Main Effects	−37.533	34.408	3.125	91.792

The cell standard deviations are given in the following table.

	TIME		
VEIN	0	30	60
Jugular	12.759	36.446	62.517
Portal	27.747	76.118	49.718

The ANOVA table can be calculated from these values noting that the harmonic mean of sample sizes is

$$n_H = \frac{1}{(\frac{1}{6})(\frac{1}{5} + \frac{1}{12} + \frac{1}{6} + \frac{1}{10} + \frac{1}{3} + \frac{1}{12})}$$
$$= 6.207$$

This number replaces the number of replicates n in the expressions for the main effects and interaction sums of squares in the table. The sum of squares for error is

$$SS_E = \sum_i \sum_j (n_{ij} - 1)s_{ij}^2$$
$$= (5 - 1)12.759^2 + \cdots + (12 - 1)49.718^2$$
$$= 102914.18$$

The results of the calculations are given in the following ANOVA table.

ANOVA table

SOURCE	SS	df	MS	F
Time Main Effect	32306.2760	2	16153.1380	6.592
Vein Main Effect	48212.6960	1	48212.6960	19.676
Interaction	2357.5347	2	1178.7674	.481
Error	102914.1800	42	2450.3376	

Both the Time and Vein main effects are seen to be quite strong ($P < .01$ for Time and $P < .001$ for Vein). The interaction, on the other hand, is quite insignificant. The following interaction plot shows what is going on.

Interaction plot

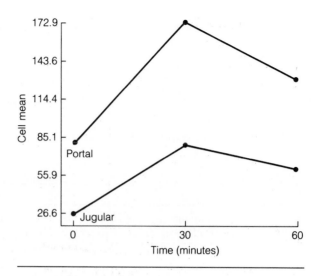

Note that the insulin levels rise in concert from the time of the injection to about 30 minutes, then fall in concert to intermediate values at 60 minutes. The near parallelism of the curves accounts for the insignificance of the test for interaction. On the other hand, the rise and fall of the average curve, which simply parallels the two given curves, accounts for the significant time effect. Moreover, the substantial difference between the curves accounts for the significant vein effect. It is reasonable to attribute a constant difference in insulin levels to the two veins, independent of time. This is the difference in vein main effects.

The differences in time main effects accurately represent the simultaneous rise and fall of the two curves. Thus, the parameters of an additive model describe the behavior of the means for these data quite well.

It makes sense to quantify the differences in main effects in the last example by confidence intervals. As was seen earlier, the factor 1 main effect at level 1, for example, can be written as

$$\alpha_1 = \bar{\mu}_{1.} - \bar{\mu}_{..}$$

Thus, the difference in these main effects for levels 1 and 2 is

$$\alpha_1 - \alpha_2 = \bar{\mu}_{1.} - \bar{\mu}_{2.}$$

It can be shown that this difference is simply a contrast for the cell means μ_{ij}. As mentioned earlier, virtually all parameters of interest in the analysis of variance can be written as contrasts of these means. Consequently, a unified approach to finding confidence intervals for contrasts will be useful. This approach will be given in the next subsection.

◪
EXERCISE

17.6   Data for a two-way ANOVA are as follows.

FACTOR 1	FACTOR 2			
	1	2	3	4
1	10.5	3.7	5.8	8.4
	8.3		6.2	7.7
	9.1			7.3
2	2.6	5.2	12.9	4.0
	1.5	6.2	10.9	5.2
		6.1		6.2
		5.8		
3	7.2	3.1	4.2	1.9
	9.4	4.1	0.4	0.2
	8.8		1.3	2.1
				2.0

(a)   Make an effects table for these data and compute the cell standard deviations.

(b)   Make the two-way ANOVA table for the data and carry out the tests for main effects and interaction at the 5% significance level by the method of un-weighted means.

(c)   Make an interaction plot and describe the trends in the data. Use this plot to explain the results of the tests in part (b).

Confidence Intervals for Contrasts of Cell Means

The natural estimator for the contrast

$$C = \sum_i \sum_j k_{ij}\mu_{ij}, \qquad \sum_i \sum_j k_{ij} = 0$$

is the same linear function of the sample cell means:

$$\hat{C} = \sum_i \sum_j k_{ij}\bar{x}_{ij.}$$

Because the sample cell means are independent random variables, the rule for computing variances given in chapter 5 yields

$$\text{var}(\hat{C}) = \sum_i \sum_j k_{ij}^2 \left(\frac{\sigma^2}{n_{ij}}\right)$$

where σ is the common standard deviation of the x's. It follows that the standard error of the estimator is

$$SE_{\hat{C}} = \hat{\sigma}\sqrt{\sum_i \sum_j \frac{k_{ij}^2}{n_{ij}}}$$

where, as before, we use the error mean square from the ANOVA table for the estimate of σ:

$$\hat{\sigma} = \sqrt{MS_E}$$

The statistic

$$t = \frac{\hat{C} - C}{SE_{\hat{C}}}$$

has Student's t distribution with degrees of freedom equal to the degrees of freedom for error, $\nu_E = N - IJ$. Thus, a $100(1 - \alpha)\%$ confidence interval for C is

$$\hat{C} - tSE_{\hat{C}} \leq C \leq \hat{C} + tSE_{\hat{C}}$$

where the critical value t is the two-sided α value for $N - IJ$ degrees of freedom from table 3. An example of the computations is as follows.

EXAMPLE 17.4

The data for a two-factor experiment are given in the following table.

FACTOR 1	FACTOR 2	
	1	2
1	12	2
		4
2	23	45
	21	47
	25	

The table of means is as follows.

Table of means

FACTOR 1	FACTOR 2	
	1	2
1	12.000	3.000
2	23.000	46.000

The ANOVA table is

ANOVA table

SOURCE	SS	df	MS	F
Factor 2	84.0000	1	84.0000	28.000
Factor 1	1249.7143	1	1249.7143	416.571
Interaction	438.8572	1	438.8572	146.286
Error	12.0000	4	3.0000	
Total	1784.5715	7		

A strong interaction is indicated. From the table of means, it is evident that a larger difference in means exists between the responses to factor 1 at level 2 of factor 2 than at level 1.

A 95% confidence interval for the difference in means for factor 1 at level 2 of factor 2 would be a confidence interval for the contrast

$$C = -1\mu_{12} + 1\mu_{22}$$

The estimate of this contrast can be computed from the table of means:

$$\hat{C} = -1\bar{x}_{12.} + 1\bar{x}_{22.} = -1(3.000) + 1(46.000)$$
$$= 43.000$$

The standard error for this contrast is

$$SE_{\hat{C}} = \hat{\sigma} \sqrt{\frac{(-1)^2}{n_{12}} + \frac{(1)^2}{n_{22}}}$$

$$= \sqrt{3.000} \sqrt{\frac{(-1^2)}{2} + \frac{(1)^2}{2}}$$

$$= 1.732$$

where we have used the square root of the error mean square to estimate σ. The critical value from table 3 for two-sided probability .05 and 4 degrees of freedom (the degrees of freedom for error) is $t = 2.776$. Thus, the confidence limits for C are obtained by subtracting and adding $tSE_{\hat{C}} = 2.776(1.732) = 4.808$ from $\hat{C}$:

$$L = \hat{C} - tSE_{\hat{C}} = 43 - 4.808 = 38.192$$
$$U = \hat{C} + tSE_{\hat{C}} = 43 + 4.808 = 47.808$$

Thus, with 95% confidence the mean responses for the two levels of factor 1 (in the order 2 minus 1) at level 2 of factor 2 differ by an amount between 38.192 and 47.808.

Unreplicated Two-Way ANOVA and Additivity Diagnostics for Randomized Blocks

The two-way analysis of variance that was applied to the randomized block model in section 17.2 can be equally well applied to a two-factor problem in which only one observation is made for each cell. The blocking factor now becomes a second treatment or environmental factor. The test of hypothesis for this factor, which was suppressed in the randomized block model, can now be carried out, simply by using the F statistic equal to the mean square for blocks divided by the error mean square.

A more delicate question concerns a test for interaction. No test for a general interaction term exists in the unreplicated case. However, J. W. Tukey devised a test for the existence of an interaction of the form

$$(\alpha\beta)_{i,j} = D\alpha_i\beta_j$$

where D is an unknown constant and α_i and β_j are the main effects for the ith and jth levels of factors one and two, respectively. This test, known as *Tukey's one degree of freedom test for additivity*, can be found in the book by Neter, Wasserman, and Kutner [24].

An exploratory look at the interaction can be obtained by means of an interaction plot, just as in the case of the replicated design. This plot can also be used to check the additivity assumption for randomized blocks. When the presence of interaction between treatments and blocks signals a violation of additivity, an alternative nonparametric procedure, such as the Friedman test (see [24]), can be used.

EXERCISE

17.7   Refer to example 17.4 (p. 596).

(a)   Find a 95% confidence interval for the contrast

$$C = \mu_{11} - \mu_{12}$$

(b)   The difference in main effects for the two levels of factor 1 can be written as the contrast

$$C = (\tfrac{1}{2})\mu_{11} + (\tfrac{1}{2})\mu_{12} - (\tfrac{1}{2})\mu_{21} - (\tfrac{1}{2})\mu_{22}$$

Find a 95% confidence interval for this contrast.

Chapter 17 Quiz

1.   What is assumed about the assignment of treatments to experimental units in a designed experiment?

2.   Why can causality often be deduced from a designed experiment?

3.   What is a completely randomized design? How would such a design be implemented in an experiment?

4.   What is the purpose of randomizing treatment assignments in a completely randomized design?

5.   What is the replication of an experimental design?

6.   What is the model for the completely randomized design? Give the names for the various terms in the model.

7.   What form of analysis is applied to the completely randomized design?

8.   What is a limitation of the completely randomized design? How can the randomized block experiment be used to overcome this weakness?

9.   What is a randomized block design? How would such a design be implemented in an experiment?

10.   What feature of a randomized block design makes the one-way analysis of variance generally a poor method for testing for treatment effects? How does the two-way analysis of variance overcome this difficulty?

11. What is the model for the randomized block experiment? How does it differ from the model for the completely randomized design?

12. How is the ANOVA table for the two-way analysis of variance for randomized blocks related to the ANOVA table for the one-way analysis of variance for the completely randomized design? What features are the same? What is the relationship between the sums of squares for blocks and the error sums of squares for the two designs?

13. How is the effects table constructed from a two-way table of data in a randomized block experiment? Give an example and identify the estimates. Interpret these estimates.

14. How are the sums of squares for the two-way analysis of variance ANOVA table calculated from the effects table? Give an example.

15. What is a contrast (of effects)?

16. How can a confidence interval be constructed for a contrast?

17. How is the answer to question 16 used to obtain a confidence interval for the difference of two effects?

18. How are confidence intervals for the differences of pairs of effects used in the Fisher least significant difference method to construct groupings of effects?

19. What is meant by the additivity of effects assumption? What does this assumption imply about the difference in treatment effects for the various blocks in a randomized block model?

20. What is meant by interaction?

21. What is the replicated two-factor model with interaction term?

22. What is the method for computing the effects table? How are the components of this table used to construct the ANOVA table for the two-way analysis of variance with interaction?

23. What is an interaction plot? How is it constructed?

24. What would an interaction plot look like if there were no interaction between factors? Draw an example of such a plot.

25. What are the main effects in terms of features of an interaction plot? Draw an interaction plot for which there is an interaction but no factor 1 main effect. Draw a plot for which there is interaction but no factor 2 main effect.

26. When is a two-way analysis of variance balanced? What is the main benefit of balance in an ANOVA model?

27. How is the method of unweighted means applied to unbalanced two-way ANOVA models? Give an example.

28. How are confidence intervals calculated for contrasts of means in the two-way ANOVA model? Give an example.

SUPPLEMENTARY EXERCISES

17.8   The following data represent measurements of damage to various grain varieties by a plant virus at a number of growing sites (blocks). Form the effects table and carry out a two-way ANOVA on the data.

		VARIETY			
GROWING SITE	1	2	3	4	5
1	0.05	0.05	1.30	1.50	1.50
2	0.30	3.00	7.50	1.00	12.70
3	2.50	0.00	20.00	37.50	26.25
4	0.01	25.00	55.00	5.00	40.00

Test at the 5% level whether any variety is significantly more resistant to the virus than others. From the effects table, which appears to be the most resistant? The least resistant? (Note: Large values indicate more damage.)

17.9   A goal in the field of communicative disorders is to design tests which, when applied to children with speech problems, yield audio-taped results that can be evaluated consistently by judges who are experts in the field. One criterion of reliability is that the judges be able to reproduce their own evaluations with high reliability after a length of time has elapsed. A test for which all judges are able to do this would be considered a good test.

The following data* are the proportions of responses that six judges reproduced correctly after the assigned time lapse for four different tests. Think of the tests as treatments and the judges as blocks and do a randomized block analysis of the data.

		TEST		
JUDGE	1	2	3	4
1	.81	.72	.48	.77
2	.88	.76	.51	.85
3	.90	.85	.70	.86
4	.89	.80	.85	.78
5	.89	.80	.97	.87
6	.90	.74	.74	.83

(a)   At the 5% significance level, is there evidence of a difference between tests?

(b)   Make an interaction plot consisting of a broken-line graph of the proportions over the four test levels for each of the six judges. Using this plot, discuss the quality of the tests both from the viewpoint of high judge reliability and of the variation in reliability from one judge to another. Which test appears best? Which appears worst?

17.10   Apply the FSD procedure at the 5% level to group the tests of exercise 17.9 according to overall average judge reliability.

---

* Data courtesy of L. P. Evans, Albuquerque Aphasia and Speech Consultants.

17.11   Five varieties of wheat were grown at each of four agricultural stations. A sample of each variety was taken from each station and milled into flour, and 2 loaves of bread were baked from each sample. The coded loaf volumes (coded by subtracting 500 from the volume in cubic centimeters, then dividing the result by 10) are given in the following table.

	STATION			
WHEAT VARIETY	1	2	3	4
1	7.5	15.5	16.6	19.0
	4.5	14.0	14.5	18.6
2	12.5	20.0	15.0	23.8
	13.2	18.5	14.0	24.4
3	7.0	10.0	15.5	17.8
	1.0	8.0	14.0	18.5
4	1.5	13.0	8.5	14.8
	2.0	15.0	9.0	16.6
5	28.0	19.5	10.5	22.0
	29.0	16.0	12.0	24.8

(a)   Compute the table of means and the table of standard deviations for these data.

(b)   Find the effects table from the table of means.

(c)   Use the expressions for sums of squares to construct the ANOVA table. Find the P-value bounds for the tests of main effects and interactions.

(d)   If the test for interactions is significant at the 5% level, use the table of means to construct an interaction plot. Describe the results of the analysis as shown by this plot. Is any wheat variety better than all others at all stations? Which variety appears to have the most consistent performance at all stations?

(e)   Use the effects table to answer the following questions. Which wheat variety has the best average performance over all stations? Which has the worst? At which station did all varieties perform best, on the average? Where did they perform worst? What is the average loaf volume for all loaves of bread in the study? (Remember to decode the average by multiplying by 10, then adding 500.)

17.12   Refer to example 17.3 (p. 592).

(a)   Write the difference of portal minus jugular vein main effects as a contrast of cell means.

(b)   Find a 95% confidence interval for the contrast of part (a).

(c)   The rate of increase in insulin levels (per minute) in the first 30 minutes after injection can be expressed as the slope of the curve of averages for the two veins for this time period—that is, the slope of the line joining the first two levels of the time main effects β_1 and β_2:

$$\text{slope} = (\beta_2 - \beta_1)/30$$

Write this slope as a contrast of the means μ_{ij} and find a 95% confidence interval for this contrast.

17.13   In a study of drug prescription prices, samples of the prices for hydrocortisone prescriptions were obtained from pharmacies carrying 1, 2, and 3 or more brands of the drug. The prescription was filled with either a brand name or a generic drug brand. The prescription prices in cents are given in the following table.

(a)   Construct the tables of means and standard deviations for these data.

(b)   Compute the effects table.

(c)   Determine the harmonic mean n_H of the sample sizes and construct the sums of squares for the ANOVA table, based on the method of unweighted means.

(d)   Complete the ANOVA table and discuss the results of the tests for interaction and main effects. Use the effects table to describe the nature of any differences indicated by the tests.

(e)   Make an interaction plot for these data and confirm the description of part (d) in terms of the plot.

	NUMBER OF AVAILABLE BRANDS						
BRAND TYPE	1		2			3 or more	
Brand name	975	575	515	835	505	595	
	560		559	510	274	415	
	485		525	510	556	540	
	439		449	960	489	530	
	650		545	439	598		
	342		360			249	330
	395		395			515	277
Generic						420	330
						420	
						310	

Data courtesy of L. Chilton, M.D., Department of Pediatrics, University of New Mexico School of Medicine.

17.14   Refer to exercise 17.13. Find a 95% confidence interval for the difference in mean drug prices for brand name and generic prescriptions of hydrocortisone for the population of drug stores from which the sample was drawn.

Appendix Tables

Table 1A
Standard normal
probabilities

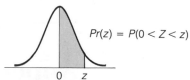

$Pr(z) = P(0 < Z < z)$

z	.00	.01	.02	.03	.04	.05	.06	.07	.08	.09
0.0	.0000	.0040	.0080	.0120	.0160	.0199	.0239	.0279	.0319	.0359
0.1	.0398	.0438	.0478	.0517	.0557	.0596	.0636	.0675	.0714	.0753
0.2	.0793	.0832	.0871	.0910	.0948	.0987	.1026	.1064	.1103	.1141
0.3	.1179	.1217	.1255	.1293	.1331	.1368	.1406	.1443	.1480	.1517
0.4	.1554	.1591	.1628	.1664	.1700	.1736	.1772	.1808	.1844	.1879
0.5	.1915	.1950	.1985	.2019	.2054	.2088	.2123	.2157	.2190	.2224
0.6	.2257	.2291	.2324	.2357	.2389	.2422	.2454	.2486	.2517	.2549
0.7	.2580	.2611	.2642	.2673	.2704	.2734	.2764	.2794	.2823	.2852
0.8	.2881	.2910	.2939	.2967	.2995	.3023	.3051	.3078	.3106	.3133
0.9	.3159	.3186	.3212	.3238	.3264	.3289	.3315	.3340	.3365	.3389
1.0	.3413	.3438	.3461	.3485	.3508	.3531	.3554	.3577	.3599	.3621
1.1	.3643	.3665	.3686	.3708	.3729	.3749	.3770	.3790	.3810	.3830
1.2	.3849	.3869	.3888	.3907	.3925	.3944	.3962	.3980	.3997	.4015
1.3	.4032	.4049	.4066	.4082	.4099	.4115	.4131	.4147	.4162	.4177
1.4	.4192	.4207	.4222	.4236	.4251	.4265	.4279	.4292	.4306	.4319
1.5	.4332	.4345	.4357	.4370	.4382	.4394	.4406	.4418	.4429	.4441
1.6	.4452	.4463	.4474	.4484	.4495	.4505	.4515	.4525	.4535	.4545
1.7	.4554	.4564	.4573	.4582	.4591	.4599	.4608	.4616	.4625	.4633
1.8	.4641	.4649	.4656	.4664	.4671	.4678	.4686	.4693	.4699	.4706
1.9	.4713	.4719	.4726	.4732	.4738	.4744	.4750	.4756	.4761	.4767
2.0	.4772	.4778	.4783	.4788	.4793	.4798	.4803	.4808	.4812	.4817
2.1	.4821	.4826	.4830	.4834	.4838	.4842	.4846	.4850	.4854	.4857
2.2	.4861	.4864	.4868	.4871	.4875	.4878	.4881	.4884	.4887	.4890
2.3	.4893	.4896	.4898	.4901	.4904	.4906	.4909	.4911	.4913	.4916
2.4	.4918	.4920	.4922	.4925	.4927	.4929	.4931	.4932	.4934	.4936
2.5	.4938	.4940	.4941	.4943	.4945	.4946	.4948	.4949	.4951	.4952
2.6	.4953	.4955	.4956	.4957	.4959	.4960	.4961	.4962	.4963	.4964
2.7	.4965	.4966	.4967	.4968	.4969	.4970	.4971	.4972	.4973	.4974
2.8	.4974	.4975	.4976	.4977	.4977	.4978	.4979	.4979	.4980	.4981
2.9	.4981	.4982	.4982	.4983	.4984	.4984	.4985	.4985	.4986	.4986
3.0	.4987	.4987	.4987	.4988	.4988	.4989	.4989	.4989	.4990	.4990
3.1	.4990	.4991	.4991	.4991	.4992	.4992	.4992	.4992	.4993	.4993
3.2	.4993	.4993	.4994	.4994	.4994	.4994	.4994	.4995	.4995	.4995
3.3	.4995	.4995	.4995	.4996	.4996	.4996	.4996	.4996	.4996	.4997
3.4	.4997	.4997	.4997	.4997	.4997	.4997	.4997	.4997	.4997	.4998

z	Pr(z)	z	Pr(z)	z	Pr(z)
3.5	.49977	4.0	.499968	4.5	.4999960
3.6	.49984	4.1	.499979	4.6	.4999979
3.7	.49989	4.2	.499987	4.7	.4999987
3.8	.499928	4.3	.4999915	4.8	.4999992
3.9	.499952	4.4	.4999946	4.9	.4999995

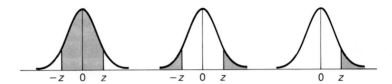

Table 1B

Critical values for the
standard normal
distribution

CONFIDENCE INTERVALS $P(\lvert Z\rvert \leq z)$	TWO-SIDED TESTS $P(\lvert Z\rvert \geq z)$	ONE-SIDED TESTS $P(Z \geq z)$	CRITICAL VALUE z
.10	.90	.45	.126
.20	.80	.40	.253
.30	.70	.35	.385
.40	.60	.30	.524
.50	.50	.25	.674
.60	.40	.20	.842
.70	.30	.15	1.036
.80	.20	.10	1.282
.90	.10	.05	1.645
.95	.05	.025	1.960
.98	.02	.01	2.326
.99	.01	.005	2.576
.995	.005	.0025	2.807
.999	.001	.0005	3.290
.9995	.0005	.00025	3.480
.9999	.0001	.00005	3.890
.99999	.00001	.000005	4.420
.999999	.0000001	.0000005	4.900

Source: D. B. Owen and D. T. Monk, *Tables of the Normal Probability Integral,* Sandia Corporation Technical Memo 64-57-51 (March 1957).

Table 2

Critical values for the chi-square distribution

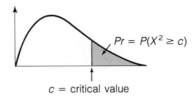

$Pr = P(X^2 \geq c)$

c = critical value

				Pr				
ν	.500	.250	.100	.050	.025	.010	.005	.001
1	.455	1.323	2.706	3.841	5.024	6.635	7.879	10.83
2	1.386	2.773	4.605	5.991	7.378	9.210	10.60	13.82
3	2.366	4.108	6.251	7.815	9.348	11.34	12.84	16.27
4	3.357	5.385	7.779	9.488	11.14	13.28	14.86	18.47
5	4.351	6.626	9.236	11.07	12.83	15.09	16.75	20.52
6	5.348	7.841	10.64	12.59	14.45	16.81	18.55	22.46
7	6.346	9.037	12.02	14.07	16.01	18.48	20.28	24.32
8	7.344	10.22	13.36	15.51	17.53	20.09	21.96	26.12
9	8.343	11.39	14.68	16.92	19.02	21.67	23.59	27.88
10	9.342	12.55	15.99	18.31	20.48	23.21	25.19	29.59
11	10.34	13.70	17.28	19.68	21.92	24.72	26.76	31.26
12	11.34	14.85	18.55	21.03	23.34	26.22	28.30	32.91
13	12.34	15.98	19.81	22.36	24.74	27.79	29.82	34.53
14	13.34	17.12	21.06	23.68	26.12	29.14	31.32	36.12
15	14.34	18.25	22.31	25.00	27.49	30.58	32.80	37.70
16	15.34	19.37	23.54	26.30	28.85	32.00	34.27	39.25
17	16.34	20.49	24.77	27.59	30.19	33.41	35.72	40.79
18	17.34	21.60	25.99	28.87	31.53	34.81	37.16	42.31
19	18.34	22.72	27.20	30.14	32.85	36.19	38.58	43.82
20	19.34	23.83	28.41	31.41	34.17	37.57	40.00	45.32
21	20.34	24.93	29.62	33.67	35.48	38.93	41.40	46.80
22	21.34	26.04	30.81	33.92	36.78	40.29	42.80	48.27
23	22.34	27.14	32.01	35.17	38.08	41.64	44.18	49.73
24	23.34	28.24	33.20	36.42	39.36	42.98	45.56	51.18
25	24.34	29.34	34.38	37.65	40.65	44.31	46.93	52.62
26	25.34	30.43	35.56	38.89	41.92	45.64	48.29	54.05
27	26.34	31.53	36.74	40.11	43.19	46.96	49.64	55.48
28	27.34	32.62	37.92	41.34	44.46	48.28	50.99	56.89
29	28.34	33.71	39.09	42.56	45.72	49.59	52.34	58.30
30	29.34	34.80	40.26	43.77	46.98	50.89	53.67	59.70
40	39.34	45.62	51.81	55.76	59.34	63.69	66.77	73.40
50	49.33	56.33	63.17	67.50	71.42	76.15	79.49	86.66
60	59.33	66.98	74.40	79.08	83.30	88.38	91.95	99.61
70	69.33	77.58	85.53	90.53	95.02	100.4	104.2	112.3
80	79.33	88.13	96.58	101.9	106.6	112.3	116.3	124.8
90	89.33	98.65	107.6	113.1	118.1	124.1	128.3	137.2
100	99.33	109.1	118.5	124.3	129.6	135.8	140.2	149.4

Source: Abridged from Table 8 of *Biometrika Tables for Statisticians*, Vol. 1, edited by E. S. Pearson and H. O. Hartley (London: Cambridge University Press, 1962).

Table 3

**Critical values for
Student's *t* distribution**

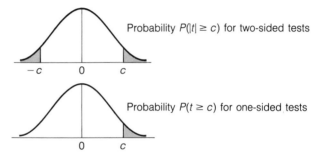

Probability $P(|t| \ge c)$ for two-sided tests

$-c$ 0 c

Probability $P(t \ge c)$ for one-sided tests

0   c

PROBABILITY

ν	.50	.20	.10	.05	.02	.01	.005	.002	.001	TWO-SIDED TESTS
	.25	.10	.05	.025	.01	.005	.0025	.001	.0005	ONE-SIDED TESTS
1	1.000	3.078	6.314	12.706	31.821	63.637	127.32	318.31	636.62	
2	.816	1.886	2.920	4.303	6.965	9.925	14.089	22.326	31.598	
3	.765	1.638	2.353	3.182	4.541	5.841	7.453	10.213	12.924	
4	.741	1.533	2.132	2.776	3.747	4.604	5.598	7.173	8.610	
5	.727	1.476	2.015	2.571	3.365	4.032	4.773	5.893	6.869	
6	.718	1.440	1.943	2.447	3.143	3.707	4.317	5.208	5.959	
7	.711	1.415	1.895	2.365	2.998	3.499	4.020	4.785	5.408	
8	.706	1.397	1.860	2.306	2.896	3.355	3.833	4.501	5.041	
9	.703	1.383	1.833	2.262	2.821	3.250	3.690	4.297	4.781	
10	.700	1.372	1.812	2.228	2.764	3.169	3.581	4.144	4.537	
11	.697	1.363	1.796	2.201	2.718	3.106	3.497	4.025	4.437	
12	.695	1.356	1.782	2.179	2.681	3.055	3.428	3.930	4.318	
13	.694	1.350	1.771	2.160	2.650	3.012	3.372	3.852	4.221	
14	.692	1.345	1.761	2.145	2.624	2.977	3.326	3.787	4.140	
15	.691	1.341	1.753	2.131	2.602	2.947	3.286	3.733	4.073	
16	.690	1.337	1.746	2.120	2.583	2.921	3.252	3.686	4.015	
17	.689	1.333	1.740	2.110	2.567	2.898	3.222	3.646	3.965	
18	.688	1.330	1.734	2.101	2.552	2.878	3.197	3.610	3.922	
19	.688	1.328	1.729	2.093	2.539	2.861	3.174	3.579	3.883	
20	.687	1.325	1.725	2.086	2.528	2.845	3.153	3.552	3.850	
21	.686	1.323	1.721	2.080	2.518	2.831	3.135	3.257	3.189	
22	.686	1.321	1.717	2.074	2.508	2.819	3.119	3.505	3.792	
23	.685	1.319	1.714	2.069	2.500	2.807	3.104	3.485	3.767	
24	.685	1.318	1.711	2.064	2.492	2.797	3.091	3.467	3.745	
25	.684	1.316	1.708	2.060	2.485	2.787	3.078	3.450	3.725	
26	.684	1.315	1.706	2.056	2.479	2.779	3.067	3.435	3.707	
27	.684	1.314	1.703	2.052	2.473	2.771	3.057	3.421	3.690	
28	.683	1.313	1.701	2.048	2.467	2.763	3.047	3.408	3.674	
29	.683	1.311	1.699	2.045	2.462	2.756	3.038	3.396	3.659	
30	.683	1.310	1.697	2.042	2.457	2.750	3.030	3.385	3.646	
40	.681	1.303	1.684	2.021	2.423	2.704	2.971	3.307	3.551	
60	.679	1.296	1.671	2.000	2.390	2.660	2.915	3.232	3.460	
120	.677	1.289	1.658	1.980	2.358	2.617	2.860	3.160	3.373	
∞	.674	1.282	1.645	1.960	2.326	2.576	2.807	3.090	3.291	

Source: Abridged from Table 12 of *Biometrika Tables for Statisticians*, Vol. 1, edited by E. S. Pearson and H. O. Hartley (London: Cambridge University Press, 1962).

Table 4
Critical values for the F distribution

$Pr = P(F \geq c)$

c = critical value

NUMERATOR DEGREES OF FREEDOM

Denom. df	Pr	1	2	3	4	5	6	8	10	20	40	∞
1	.25	5.83	7.50	8.21	8.58	8.82	8.98	9.19	9.32	9.58	9.71	9.85
	.10	39.86	49.50	53.59	55.83	57.24	58.20	59.44	60.19	61.74	62.53	63.33
	.05	161.4	199.5	215.7	224.6	230.2	234.0	238.9	241.9	248.0	251.1	254.3
2	.25	2.57	3.00	3.15	3.23	3.28	3.31	3.35	3.38	3.43	3.45	3.48
	.10	8.53	9.00	9.16	9.24	9.29	9.33	9.37	9.39	9.44	9.47	9.49
	.05	18.51	19.00	19.16	19.25	19.30	19.33	19.37	19.40	19.45	19.47	19.50
	.01	98.50	99.00	99.17	99.25	99.30	99.33	99.37	99.40	99.45	99.47	99.50
	.001	998.5	999.0	999.2	999.2	999.3	999.3	999.4	999.4	999.4	999.5	999.5
3	.25	2.02	2.28	2.36	2.39	2.41	2.42	2.44	2.44	2.46	2.47	2.47
	.10	5.54	5.46	5.39	5.34	5.31	5.28	5.25	5.23	5.18	5.16	5.13
	.05	10.13	9.55	9.28	9.12	9.01	8.94	8.85	8.79	8.66	8.59	8.53
	.01	34.12	30.82	29.46	28.71	28.24	27.91	27.49	27.23	26.69	26.41	26.13
	.001	167.0	148.5	141.1	137.1	134.6	132.8	130.6	129.2	126.4	125.0	123.5
4	.25	1.81	2.00	2.05	2.06	2.07	2.08	2.08	2.08	2.08	2.08	2.08
	.10	4.54	4.32	4.19	4.11	4.05	4.01	3.95	3.92	3.84	3.80	3.76
	.05	7.71	6.94	6.59	6.39	6.26	6.16	6.04	5.96	5.80	5.72	5.63
	.01	21.20	18.00	16.69	15.98	15.52	15.21	14.80	14.55	14.02	13.75	13.46
	.001	74.14	61.25	56.18	53.44	51.71	50.53	49.00	48.05	46.10	45.09	44.05
5	.25	1.69	1.85	1.88	1.89	1.89	1.89	1.89	1.89	1.88	1.88	1.87
	.10	4.06	3.78	3.62	3.52	3.45	3.40	3.34	3.30	3.21	3.16	3.10
	.05	6.61	5.79	5.41	5.19	5.05	4.95	4.82	4.74	4.56	4.46	4.36
	.01	16.26	13.27	12.06	11.39	10.97	10.67	10.29	10.05	9.55	9.29	9.02
	.001	47.18	37.12	33.20	31.09	29.75	28.84	27.64	26.92	25.39	24.60	23.79
6	.25	1.62	1.76	1.78	1.79	1.79	1.78	1.78	1.77	1.76	1.75	1.74
	.10	3.78	3.46	3.29	3.18	3.11	3.05	2.98	2.94	2.84	2.78	2.72
	.05	5.99	5.14	4.76	4.53	4.39	4.28	4.15	4.06	3.87	3.77	3.67
	.01	13.75	10.92	9.78	9.15	8.75	8.47	8.10	7.87	7.40	7.14	6.88
	.001	35.51	27.00	23.70	21.92	20.81	20.03	19.03	18.41	17.12	16.44	15.75

DENOMINATOR DEGREES OF FREEDOM

df	p											
7	.25	1.65	1.66	1.67	1.69	1.70	1.71	1.71	1.72	1.72	1.70	1.57
	.10	2.47	2.54	2.59	2.70	2.75	2.83	2.88	2.96	3.07	3.26	3.59
	.05	3.23	3.34	3.44	3.64	3.73	3.87	3.97	4.12	4.35	4.74	5.59
	.01	5.65	5.91	6.16	6.62	6.84	7.19	7.46	7.85	8.45	9.55	12.25
	.001	11.70	12.33	12.93	14.08	14.63	15.52	16.21	17.19	18.77	21.69	29.25
8	.25	1.58	1.59	1.61	1.63	1.64	1.65	1.66	1.66	1.67	1.66	1.54
	.10	2.29	2.36	2.42	2.54	2.59	2.67	2.73	2.81	2.92	3.11	3.46
	.05	2.93	3.04	3.15	3.35	3.44	3.58	3.69	3.84	4.07	4.46	5.32
	.01	4.86	5.12	5.36	5.81	6.03	6.37	6.63	7.01	7.59	8.65	11.26
	.001	9.33	9.92	10.48	11.54	12.04	12.86	13.49	14.39	15.83	18.49	25.42
9	.25	1.53	1.54	1.56	1.59	1.60	1.61	1.62	1.63	1.63	1.62	1.51
	.10	2.16	2.23	2.30	2.42	2.47	2.55	2.61	2.69	2.81	3.01	3.36
	.05	2.71	2.83	2.94	3.14	3.23	3.37	3.48	3.63	3.86	4.26	5.12
	.01	4.31	4.57	4.81	5.26	5.47	5.80	6.06	6.42	6.99	8.02	10.56
	.001	7.81	8.37	8.90	9.89	10.37	11.13	11.71	12.56	13.90	16.39	22.86
10	.25	1.48	1.51	1.52	1.55	1.56	1.58	1.59	1.59	1.60	1.60	1.49
	.10	2.06	2.13	2.20	2.32	2.38	2.46	2.52	2.61	2.73	2.92	3.28
	.05	2.54	2.66	2.77	2.98	3.07	3.22	3.33	3.48	3.71	4.10	4.96
	.01	3.91	4.17	4.41	4.85	5.06	5.39	5.64	5.99	6.55	7.56	10.04
	.001	6.76	7.30	7.80	8.75	9.20	9.92	10.48	11.28	12.55	14.91	21.04
12	.25	1.42	1.45	1.47	1.50	1.51	1.53	1.54	1.55	1.56	1.56	1.46
	.10	1.90	1.99	2.06	2.19	2.24	2.33	2.39	2.48	2.61	2.81	3.18
	.05	2.30	2.43	2.54	2.75	2.85	3.00	3.11	3.26	3.49	3.89	4.75
	.01	3.36	3.62	3.86	4.30	4.50	4.82	5.06	5.41	5.95	6.93	9.33
	.001	5.42	5.93	6.40	7.29	7.71	8.38	8.89	9.63	10.80	12.97	18.64
14	.25	1.38	1.41	1.43	1.46	1.48	1.50	1.51	1.52	1.53	1.53	1.44
	.10	1.80	1.89	1.96	2.10	2.15	2.24	2.31	2.39	2.52	2.73	3.10
	.05	2.13	2.27	2.39	2.60	2.70	2.85	2.96	3.11	3.34	3.74	4.60
	.01	3.00	3.27	3.51	3.94	4.14	4.46	4.69	5.04	5.56	6.23	8.86
	.001	4.60	5.10	5.56	6.40	6.80	7.43	7.92	8.62	9.73	11.78	17.14
16	.25	1.34	1.37	1.40	1.45	1.46	1.48	1.48	1.50	1.51	1.51	1.42
	.10	1.72	1.81	1.89	2.03	2.09	2.18	2.24	2.33	2.46	2.67	3.05
	.05	2.01	2.15	2.28	2.49	2.59	2.74	2.85	3.01	3.24	3.63	4.49
	.01	2.75	3.02	3.26	3.69	3.89	4.20	4.44	4.77	5.29	6.23	8.53
	.001	4.06	4.54	4.99	5.81	6.19	6.81	7.27	7.94	9.00	10.97	16.12
18	.25	1.32	1.35	1.38	1.42	1.43	1.45	1.46	1.48	1.49	1.50	1.41
	.10	1.66	1.75	1.84	1.98	2.04	2.13	2.20	2.29	2.42	2.62	3.01
	.05	1.92	2.06	2.19	2.41	2.51	2.66	2.77	2.93	3.16	3.55	4.41
	.01	2.57	2.84	3.08	3.51	3.71	4.01	4.25	4.58	5.09	6.01	8.29
	.001	3.67	4.15	4.59	5.39	5.76	6.35	6.81	7.46	8.49	10.39	15.38

DENOMINATOR DEGREES OF FREEDOM

Table 4
(continued)

DENOMINATOR DEGREES OF FREEDOM

NUMERATOR DEGREES OF FREEDOM

	Pr	1	2	3	4	5	6	8	10	20	40	∞
20	.25	1.40	1.49	1.48	1.46	1.45	1.44	1.42	1.40	1.36	1.33	1.29
	.10	2.97	2.59	2.38	2.25	2.16	2.09	2.00	1.94	1.79	1.71	1.61
	.05	4.35	3.49	3.10	2.87	2.71	2.60	2.45	2.35	2.12	1.99	1.84
	.01	8.10	5.85	4.94	4.43	4.10	3.87	3.56	3.37	2.94	2.69	2.42
	.001	14.82	9.95	8.10	7.10	6.46	6.02	5.44	5.08	4.29	3.86	3.38
30	.25	1.38	1.45	1.44	1.42	1.41	1.39	1.37	1.35	1.30	1.27	1.23
	.10	2.88	2.49	2.28	2.14	2.05	1.98	1.88	1.82	1.67	1.57	1.46
	.05	4.17	3.32	2.92	2.69	2.53	2.42	2.27	2.16	1.93	1.79	1.62
	.01	7.56	5.39	4.51	4.02	3.70	3.47	3.17	2.98	2.55	2.30	2.01
	.001	13.29	8.77	7.05	6.12	5.53	5.12	4.58	4.24	3.49	3.07	2.59
40	.25	1.36	1.44	1.42	1.40	1.39	1.37	1.35	1.33	1.28	1.24	1.19
	.10	2.84	2.44	2.23	2.09	2.00	1.93	1.83	1.76	1.61	1.51	1.38
	.05	4.08	3.23	2.84	2.61	2.45	2.34	2.18	2.08	1.84	1.69	1.51
	.01	7.31	5.18	4.31	3.83	3.51	3.29	2.99	2.80	2.37	2.11	1.80
	.001	12.61	8.25	6.60	5.70	5.13	4.73	4.21	3.87	3.15	2.73	2.23
60	.25	1.35	1.42	1.41	1.38	1.37	1.35	1.32	1.30	1.25	1.21	1.15
	.10	2.79	2.39	2.18	2.04	1.95	1.87	1.77	1.71	1.54	1.44	1.29
	.05	4.00	3.15	2.76	2.53	2.37	2.25	2.10	1.99	1.75	1.59	1.39
	.01	7.08	4.98	4.13	3.65	3.34	3.12	2.82	2.63	2.20	1.94	1.60
	.001	11.97	7.76	6.17	5.31	4.76	4.37	3.87	3.54	2.83	2.41	1.89
120	.25	1.34	1.40	1.39	1.37	1.35	1.33	1.30	1.28	1.22	1.18	1.10
	.10	2.75	2.35	2.13	1.99	1.90	1.82	1.72	1.65	1.48	1.37	1.19
	.05	3.92	3.07	2.68	2.45	2.29	2.17	2.02	1.91	1.66	1.50	1.25
	.01	6.85	4.79	3.95	3.48	3.17	2.96	2.66	2.47	2.03	1.76	1.38
	.001	11.38	7.32	5.79	4.95	4.42	4.04	3.55	3.24	2.53	2.11	1.54
∞	.25	1.32	1.39	1.37	1.35	1.33	1.31	1.28	1.25	1.19	1.14	1.00
	.10	2.71	2.30	2.08	1.94	1.85	1.77	1.67	1.60	1.42	1.30	1.00
	.05	3.84	3.00	2.60	2.37	2.21	2.10	1.94	1.83	1.57	1.39	1.00
	.01	6.64	4.61	3.78	3.32	3.02	2.80	2.51	2.32	1.88	1.59	1.00
	.001	10.83	6.91	5.42	4.62	4.10	3.74	3.27	2.96	2.27	1.84	1.00

Source: Abridged from Table 18 of Biometrika Tables for Statisticians, Vol. 1, edited by E. S. Pearson and H. O. Hartley (London: Cambridge University Press, 1962.)

References

1. Andrews, D. F., P. J. Bickel, F. R. Hampel, P. J. Huber, W. H. Rogers, and J. W. Tukey (1972). *Robust Estimates of Location: Survey and Advances*. Princeton University Press, New Jersey.

2. Arrhenius Laboratory, Department of Meteorology, University of Stockholm, Sweden (1975). "Data from the European Atmospheric Chemistry Network."

3. Box, G. E. P., W. G. Hunter, and J. S. Hunter (1978). *Statistics for Experimenters*. Wiley, New York.

4. Box, G. E. P., and G. C. Tiao (1973). *Bayesian Inference in Statistical Analysis*. Addison-Wesley, Reading, Massachusetts.

5. Brownlee, K. A. (1965). *Statistical Theory and Methodology in Science and Engineering*. Wiley, New York.

6. Carmer, S. G., and M. R. Swanson (1973). "An Evaluation of Ten Pairwise Multiple Comparison Procedures by Monte Carlo Methods. *Journal of the American Statistical Association*, **68**, pp. 66–74.

7. Cochran, W. G. (1952). "The χ^2 Test of Goodness of Fit." *Annals of Mathematical Statistics*, **23**, pp. 315–345.

8. Comess, L. J., P. H. Bennett, T. A. Burch, and M. Miller (1969). "Congenital Anomalies and Diabetes in the Pima Indians of Arizona." *Diabetes*, **18**, pp. 471–476.

9. Dillman, D. A. (1978). *Mail and Telephone Surveys: The Total Design Method*. Wiley, New York.

10. Eaton, R. P., J. Abrams, H. Ellis, L. H. Koopmans, and R. Allen (1981). "Prevalence of Plasma Lypoprotein Abnormalities in a Free-Living Population in New Mexico: Relationship to HDL Cholesterol." *Human Nutrition, Clinical and Biological Aspects*, Proceedings of the 4th A. O. Beckman Conference in Clinical Chemistry. American Association for Clinical Chemistry, pp. 109–120.

11. Enrick, N. L. (1972). *Cases in Management Statistics*. Holt, Rinehart and Winston, New York.

12. Federer, W. T. (1973). *Statistics and Society*. Dekker, New York.

13. Freiman, J. A., T. C. Chalmers, H. Smith, Jr., and R. R. Kuebler (1978). "The Importance of Beta, the Type II Error, and Sample Size in the Design and Interpretation of the Randomized Control Trial." *New England Journal of Medicine*, **299,** pp. 690–694.

14. Hald, A. (1952). *Statistical Theory with Engineering Applications.* Wiley, New York.

15. Halek, H. G. O., K. K. Kimura, and B. Bartels (1946). "Effects of the Anesthetic and the Rate of Injection of Digitalis Upon It: Lethal Dose in Cats." *Journal of the American Pharmaceutical Association*, **35,** pp. 366–370.

16. Hollander, M., and D. A. Wolfe (1973). *Nonparametric Statistical Methods.* Wiley, New York.

17. Iman, R. L. (1976). "An Approximation to the Exact Distrbution of the Wilcoxon-Mann-Whitney Random Test Statistic." *Communications in Statistics*, Series A, **5,** pp. 587–598.

18. Iman, R. L., and J. M. Davenport (1976). "New Approximations to the Exact Distributions of the Kruskal-Wallis Test Statistic." *Communications in Statistics*, Series A, **5,** pp. 1335–1348.

19. Jenkins, F. A., and H. E. White (1950). *Fundamentals of Optics.* McGraw-Hill, New York.

20. Kempthorne, O. (1965). "The Classical Problem of Inference—Goodness of Fit." *Proceedings of the 5th Berkeley Symposium on Mathematical Statistics and Probability.* University of California Press, Berkeley.

21. Koopmans, L. H., and J. Fullilove (1979). "Analysis of Pulmonary Arteriole Medial Thickness and the Sudden Infant Death Syndrome." University of New Mexico Department of Mathematics and Statistics Technical Report No. 363.

22. Mayer, L. S. (1978). "Estimating the Effects of the Onset of the Energy Crisis on Residential Energy Demand." *Resources and Energy*, **1,** pp. 57–92.

23. Mielke, P. W. (1973). "Another Family of Distribution for Describing and Analyzing Precipitation Data." *Journal of Applied Meteorology*, **12,** pp. 275–280.

24. Neter, J., W. Wasserman, and M. Kutner (1983). *Applied Linear Regression Models.* Irwin, Homewood, Illinois.

25. Noether, G. E. (1976). *Introduction to Statistics: A Nonparametric Approach.* 2nd edition. Houghton Mifflin, Boston.

26. Odeh, R. E., and M. Fox (1975). *Sample Size Choice: Charts for Experiments with Linear Models.* Dekker, New York.

27. Purifoy, F. E., and L. H. Koopmans (1979). "Androstenedione, Testosterone, and Free Testosterone Concentration in Women of Various Occupations." *Social Biology*, **26,** pp. 179–188.

28. Purifoy, F. E., L. H. Koopmans, and R. W. Tatum (1980). "Steroid Hormones and Aging: Free Testosterone Concentration, Testosterone

and Androstenedione in Normal Females Aged 20–87 Years." *Human Biology*, **52**, pp. 181–191.

29. Rand Corporation (1955). *A Million Random Digits with 100,000 Normal Deviates*. Free Press, Glencove, Illinois.

30. Rudman, D., M. H. Kutner, M. A. Goldsmith, and R. D. Blackston (1979). "Predicting the Response of Growth Hormone-deficient Children to Long-term Treatment with Growth Hormone." *Journal of Clinical Endocrinology and Metabolism*, **48**(3), pp. 472–477.

31. Stigler, S. M. (1977). "Do Robust Estimators Work with *Real* Data?" *Annals of Statistics*, **5**, pp. 1055–1264.

32. Tukey, J. W. (1977). *Exploratory Data Analysis*. Addison-Wesley, Reading, Massachusetts.

33. Tukey, J. W. (1960). "A Survey of Sampling from Contaminated Distributions." In *Contributions to Probability and Statistics: Essays in Honor of Harold Hotelling*. Edited by I. Olkin et al. Stanford University Press, California. Chapter 39.

34. Tukey, J. W., and D. H. McLaughlin (1963). "Less Vulnerable Confidence and Significance Procedures for Location Based on a Single Sample: Trimming/Winsorization." *1. Sankyā Series* A, **25**, pp. 334–352.

35. Wainer, H. (1976). "Robust Statistics: A Survey and Some Prescriptions." *Journal of Educational Statistics*, **1**(4), pp. 285–312.

36. Yerushalmy, J. (1975). Data from Child Health and Development Study, Kaiser Foundation Hospital, Oakland, California. In *Statlab: An Empirical Introduction to Statistics*. J. L. Hodges, D. Krech, and R. Crutchfield. McGraw-Hill, New York.

37. Yuen, K., and W. J. Dixon (1973). "The Approximate Behavior and Performance of the Two-sample Trimmed *t*." *Biometrika*, **60**(2), p. 369.

Supplementary References

38. Belsley, D. A., E. Kuh, and R. E. Welsch (1980). *Regression Diagnostics*. Wiley, New York.

39. Blyth, C. R., and H. A. Still (1983). "Binomial Confidence Intervals." *Journal of the American Statistical Association*, **78**, pp. 108–116.

40. Cook, R. D., and S. Weisberg (1982). *Residuals and Influence in Regression*. Chapman and Hall, New York.

41. Daniel, C., and F. S. Wood (1971). *Fitting Equations to Data*. Wiley-Interscience, New York.

42. Draper, N. R., and H. Smith (1981). *Applied Regression Analysis*. Wiley, New York.

43. Miller, D. M. (1984). "Reducing Transformation Bias in Curve Fitting." *The American Statistician*, **38**(2), pp. 124–126.

44. Minitab Project (1984). *Minitab.* 215 Pond Laboratory, Pennsylvania State University, University Park, Pennsylvania, 16802.

45. Mosteller, F., and J. W. Tukey (1977). *Data Analysis and Regression.* Addison-Wesley, Reading, Massachusetts.

46. SAS Institute, Inc. (1983). *SAS (Statistical Analysis System).* P. O. Box 800, Cary, North Carolina, 27511.

47. Weisberg, S. (1980). *Applied Linear Regression.* Wiley, New York.

Solutions to Selected Exercises

Chapter 1

1.1 (a) categorical: *Protestant* (with several sub-category possibilities), *Catholic, Jewish, Other*

   (c) categorical: Capricorn, Aries, etc. (12 categories)

   (e) categorical: 1 A.M., 2 A.M. . . . (24 categories)

   (g) measurement: degrees Celsius or Fahrenheit

1.2

I.D. NUMBER	CLASS	GROUPED CLASS
1	Jun.	Upper+
2	Fr.	Fr.
3	Soph.	Soph.
4	Fr.	Fr.
5	Grad.	Upper+
6	Fr.	Fr.
7	Soph.	Soph.
8	Fr.	Fr.
9	Soph.	Soph.
10	Fr.	Fr.
11	Soph.	Soph.
12	Soph.	Soph.
13	Fr.	Fr.
14	Soph.	Soph.
15	Grad.	Upper+

GROUPED CLASS	ABSOLUTE f	RELATIVE f (%)
Fr.	6	40
Soph.	6	40
Upper+	3	20
	$n = 15$	

1.4

```
0 | 9
1 | 757577
2 | 3262957248206089
3 | 6379561090
4 | 306092
5 | 4
```

1.5

CLASS LIMITS	CLASS BOUNDARIES	f	RELATIVE f (%)
0–9	−.5–9.5	1	2.5
10–19	9.5–19.5	6	15.0
20–29	19.5–29.5	16	40.0
30–39	29.5–39.5	10	25.0
40–49	39.5–49.5	6	15.0
50–59	49.5–59.5	1	2.5
	$n = 40$		

1.6

1.7 A hint of a bimodality appears. The separation of the value 9 from the rest of the plot becomes more apparent.

```
0* |
 . | 9
1* |
 . | 757577
2* | 32224200
 . | 69578689
3* | 3100
 . | 679569
4* | 3002
 . | 96
5* | 4
```

CLASS LIMITS	CLASS BOUNDARIES
5–9	4.5–9.5
10–14	9.5–14.5
15–19	14.5–19.5
20–24	19.5–24.5
25–29	24.5–29.5
30–34	29.5–34.5
35–39	34.5–39.5
40–44	39.5–44.5
45–49	44.5–49.5
50–54	49.5–54.5

1.11 The two versions on the next page would both be reasonable solutions.

615

Standard Version

To regain original units, move decimal two places to the left.

```
-0 | 43,37,44,34,85
 0 | 94,90,52,87,24,62,98,50,71
 1 | 35,23
 2 | 11
 3 | 45,15
 4 | 11
```

Two Lines per Stem Alternative Version

To regain original units, move decimal two places to the left.

```
  · | 85
-0* | 44,43,37,34
 0* | 24
  · | 50,52,62,71,87,90,94,98
 1* | 23,35
  ·
 2* | 11
  ·
 3* | 15,45
  ·
 4* | 11
  ·
```

1.13 (a)
```
0* | 0
 · | 8
1* | 02334
 ·
2* | 4
 ·
3*
 ·
4*
 · | 6
```

(b) skewed toward large values (to the right)

(c) The value 4.6 appears to be well separated from the rest of the distribution. The value 2.4 also appears somewhat separated from the central values.

(d)
```
0* | 00
 · | 89
1* | 00,10,14,14,18
 · | 55
2* | 14
```

The distribution now appears to be reasonably symmetric but possibly somewhat long-tailed (peaked). No separated values are apparent now. The transformation has made the distribution symmetric.

1.15 (a) The bar chart is shown in the figure for part (b).

(b)

	SEATING = 1		
SEX	CODE	f	RELATIVE f (%)
Male	0	15	37.5
Female	1	25	62.5
		n = 40	

	SEATING = 2		
SEX	CODE	f	RELATIVE f (%)
Male	0	16	40.0
Female	1	24	60.0
		n = 40	

	SEATING = 3		
SEX	CODE	f	RELATIVE f (%)
Male	0	23	63.9
Female	1	13	36.1
		n = 36	

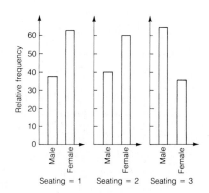

(c) The front two-thirds of the class are predominantly female, while males dominate the back of the class.

(d) The gradient of decreasing femaleness and increasing maleness as we progress from the front to the back of the class is clear from these graphs. Note that for a two-valued variable one of these two graphs would suffice, since the relative

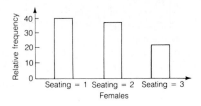

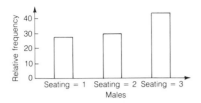

frequences for the other are simply 100% minus the given frequencies. This kind of graph is often used to compare distributions for two-valued variables.

1.17 There is not a clear trend toward younger students as we progress from the front to the rear of the class, as we might expect. It is true that the older students appear to favor the front of the class. However, the section dominated by the youngest students is the middle section of the classroom. The predominant age group at the back of the classroom is the 20–24-year-old group.

Observable trends from the front to the back of the class are (1) a decrease in proportions of students 25 years of age and older, (2) a minor decrease, then an increase in the 20–24-year-old age group, and (3) a peak in the proportion of 17–19-year-old students in the middle of the class.

```
SEATING = 1            SEATING = 2
1·| 999998998889989   1·| 98998899899999999899798
2*| 000001320100400    2*| 0021022111024
 ·| 768                 ·| 9677
3*| 0000
 ·| 9
4*|
 ·| 6
5*|
 ·| 5
```

```
            SEATING = 3
        1·| 99998999999999
        2*| 0112011012042202240
         ·| 776
```

1.19 The diversity of abilities (as measured by speed) of the cyclists selecting the three distance options is apparent in the differences in dispersion. Those selecting 50 and 75 miles are much more alike than those selecting 100 miles. The bimodality of the 100-mile speed distribution indicates two somewhat distinct ability groups, racers and tourists, with the tourists being the more numerous. The

speeds of the racers far exceed those of the other groups, whereas the tourists who elected 100 miles are more comparable to the 50- and 75-mile riders. Even so, the 100-mile tourist group is faster, on the average, than either of the other two, suggesting that they are more experienced cyclists.

To regain original units, move decimal one place to the left.

50 MILES		75 MILES		100 MILES	
7		7	2	7	
8	17	8	80	8	162
9	03000	9	5	9	308823050
10	28	10		10	824241563
11	2	11	3	11	1404497078
12		12		12	0463573172
13		13		13	2083
14		14		14	769
15		15		15	82
16		16		16	5470
17		17		17	6988
18		18		18	0
19		19		19	7
20		20		20	2
21		21		21	1

1.21 The out-of-towners are a more serious (faster) group of cyclists in terms of average speed. Note that this group appears to break down into two separate groups (tourists and racers), as demonstrated by the bimodality of the distribution. The local residents, on the other hand, are basically tourists, although a core of speedy cyclists is also present (the fastest time is from in town). However, the higher proportion of racers in the out-of-town group will cause the average speed to be higher for this group than for the local residents.

ALBUQUERQUE RESIDENTS		OUT-OF-TOWNERS	
7	2	7	
8	117820	8	6
9	0303250090050	9	883
10	82422241563	10	8
11	10490	11	4243778
12	463532	12	0717
13	083	13	2
14	69	14	7
15	8	15	2
16	57	16	940
17	8	17	68
18		18	0
19	7	19	
20	2	20	
21	1	21	

1.23

ARITHMETIC SCORES		ALGEBRA SCORES	
0*		0*	000100101000110
t	2	t	322223
f	5	f	5454455555
s	6	s	77777767
·	8899	·	999
1*	1110	1*	
t	33	t	2
f	45444445	f	4
s	7666666776		
·	99989998899		
2*	00		

(a) The arithmetic score distribution is skewed to the left.

(b) The algebra score distribution is skewed to the right. Both distributions are pushed up against one of the limits imposed by the grading scales. The arithmetic score is stacked up at the upper limit of 20 and the algebra score at the lower limit of zero. There are hints of bimodalities in both distributions, suggesting somewhat distinct ability groups.

(c) Because of the stacking at extremes, it would appear that the arithmetic exam was too easy and the algebra exam too hard. An ideal sought in testing (but seldom achieved) is to design exams so that the scores have normal distributions.

1.25

CATEGORY CODE	RELATIVE f (%)
0	17
1	11
2	7
3	13
4	10
5	10
6	9
7	7
8	9
9	7
	$n = 100$

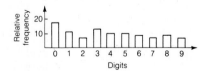

There appears to be reasonable support for equal theoretical frequencies.

Chapter 2

2.1 $\bar{X} = 12.86$

2.3 To regain original units, move decimal one place to the left.

0	08
1	02334
2	4
3	
4	6

$\bar{X} = 1.556$

2.5

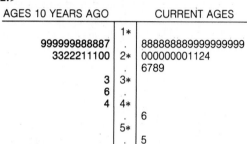

AGES 10 YEARS AGO		CURRENT AGES
	1*	
999999888887	·	888888889999999999
3322211100	2*	000000001124
	·	6789
3	3*	
6	·	
4	4*	
	·	6
	5*	
	·	5

$\bar{X}_{Current} = 22.03$ years
$\bar{X}_{Past} = 21.96$
Difference in mean ages $= .07$ years
There appears to be little evidence that students are, on the average, older now than ten years ago.

2.6 $M = 12.0$

2.8 $M = 1.3$ years. This value is smaller than the sample mean and seems more representative of the center of the frequency.

2.10 $Q_1 = 10.0$, $Q_3 = 15.5$, IQR $= 5.5$

2.12 $Q_1 = 1.0$, $Q_3 = 1.4$, IQR $= 0.4$

2.14 $s = 6.01$ for both (a) and (b).

2.16 $s = 1.302$

2.20

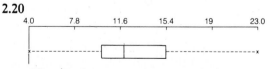

The distribution is nearly symmetric.

2.22

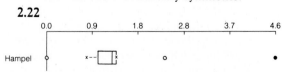

This distribution has long tails and is skewed to the right.

2.25 **(a)** $n_T = 3$, $\overline{X}_T = 3.0$, $s_T = 1.41$

2.27 $n_T = 6$, $\overline{X}_T = -.267$, $s_T = 5.427$

2.29 **(c)**

Trimming fraction

	0%	10%	20%	50%
n_T	9	7	5	
$\overline{X}_T$	1.556	1.343	1.240	1.300
s_T	1.302	.700	.253	

Both means and standard deviations decrease as the trimming fraction increases and the outliers are "accounted for" by the procedure. Because the central part of the distribution is skewed to the right, the 20% trimmed mean is somewhat smaller than the median.

2.31 **(a)**

VAR. NAME	SAMPLE SIZE	MEDIAN	QUARTILES	ADJACENTS	IQR
S	12	40.5	32.0	17.0	31.5
			63.5	93.0	
Minor outliers	135				

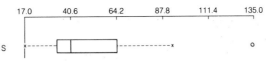

(b) $\overline{X} = 52.6$, $s = 33.3$

(c) The distribution is skewed to the right.

(d) The distribution is slightly long-tailed (one minor outlier).

(e) $M = 40.5$. The skewness and the outlier have made the mean larger than the median.

2.33 **(a)**

VAR. NAME	SAMPLE SIZE	MEDIAN	QUARTILES	ADJACENTS	IQR
Mg	12	2.5	2.0	1.0	3.5
			5.5	9.0	
Minor outliers	12				

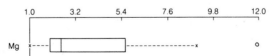

(b) $\overline{X} = 4.08$, $s = 3.37$

(c) Skewed to the right

(d) Somewhat long-tailed (minor outlier)

(e) The skewness and the outlier cause the mean (4.08) to be larger than the median (2.5).

2.35 **(a)**

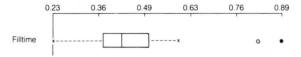

(b) $n = 27$, $\overline{X} = .453$, $s = .1411$
$M = .43$, IQR $= .125$
The outliers have pulled the mean to a value above the median.

(c) A 5% trimmed mean yields $n_T = 23$, $\overline{X}_T = .436$, $s_T = .088$.

(d) The trimmed mean is much closer to the median, while the standard deviation has been decreased by nearly a factor of 2 by trimming.

2.37 Arithmetic

Algebra: To regain original units, move decimal one place to the left.

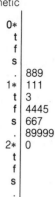

```
0*
 t
 f
 s
 .  889
1*  111
 t  3
 f  4445
 s  667
 .  89999
2*  0
 t
 f
 s
 .
```

```
0 | 00000
1 | 00
2 | 000
3 | 0
4 | 000
5 | 00
6 |
7 | 00
8 |
9 | 00
```

ARITHMETIC

VAR. NAME	SAMPLE SIZE	MEDIAN	QUARTILES	ADJACENTS	IQR
Arithmetic	20	14.5	11.0	8.0	7.5
			18.5	20.0	

```
      8.0      10.4     12.8     15.2     17.6     20.0
```

Arithmetic

ALGEBRA

VAR. NAME	SAMPLE SIZE	MEDIAN	QUARTILES	ADJACENTS	IQR
Algebra	20	2.5	0.5	0.0	4.5
			5.0	9.0	

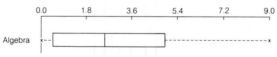

```
      0.0      1.8      3.6      5.4      7.2      9.0
```

Algebra

The stem and leaf plots best show the shape of the central part of the distribution, whereas the box plots best display tail length information.

2.39 (a)

VAR. NAME	SAMPLE SIZE	MEDIAN	QUARTILES	ADJACENTS	IQR
Relative heights	20	0.790	-0.050	-0.850	1.340
			1.290	3.150	
Minor outliers	3.45	4.11			

```
     -0.85     0.14     1.13     2.13     3.12     4.11
```

Relative heights

(b) A 10% trimmed mean eliminates the outliers and yields
$$\bar{X}_T = .811, \quad s_T = 1.323$$
(c) $\bar{X} = .962$, $s = 1.342$

Because the outliers were not very extreme, trimming has had rather little effect here. Because of the "thin" upper tail, a more extreme trimming has a much larger effect. For example, a 20% trim yields $\bar{X}_T = .710$ and $s_T = .863$.

2.41 (a) $\Sigma X_i^2 =$
$$\begin{array}{r} 1002000 \\ +1004000 \\ +1006000 \\ \hline 3012000 \end{array}$$

$$\Sigma X_i = 3006$$

Thus, $(\Sigma X_i)^2 = 9036000/3 = 3012000$, and it follows that expression (2) yields $s = 0$.

(b) $\bar{X} = 1002$, thus
$$\begin{array}{l} X_1 - \bar{X} = -1 \\ X_2 - \bar{X} = 0 \\ X_3 - \bar{X} = 1 \end{array}$$
and $\Sigma(X_i - \bar{X})^2 = 2$. From this we obtain the correct result $s = 1$.

(c) Correcting the observations for their mean before squaring keeps them within the accuracy of the calculator and no significant digits are lost in squaring.

Chapter 3

3.1

SUMMARY

VAR. NAME	SAMPLE SIZE	MEDIAN	QUARTILES	ADJACENTS	IQR
Statistics	8	140.0	94.5	70.0	84.5
			179.0	195.0	
Applied math	11	110.0	103.0	85.0	23.5
			126.5	147.0	
Minor outliers	48				
Pure math and math education	15	104.0	91.0	44.0	86.5
			177.5	203.0	

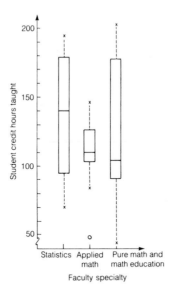

There is an indication of a higher average teaching load for statistics instructors than for the other two faculty categories. However, there is considerable overlap in the distributions. One member of the applied math faculty appears to teach "significantly" fewer hours than the others. (This faculty member teaches primarily small, specialized graduate courses.)

3.3

VAR. NAME	SAMPLE SIZE	MEDIAN	QUARTILES	ADJACENTS	IQR
Anthro	8	57.0	56.0 58.0	55.0 59.0	2.0
Math	6	56.0	55.0 57.0	54.0 58.0	2.0
Psych	6	60.0	59.0 61.0	57.0 63.0	2.0

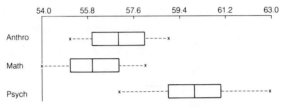

The data indicate that psychologists are substantially more politically conservative, on the average, than are either mathematicians or anthropologists, with mathematicians showing somewhat smaller values of the variable than the anthropologists. There is a suggestion of just two distinct groups, however.

3.5 (a)

VAR. NAME	SAMPLE SIZE	MEDIAN	QUARTILES	ADJACENTS	IQR
Lee 1	8	555.0	532.0 591.5	517.0 612.0	59.5
Lee 2	8	485.5	470.5 524.0	461.0 580.0	53.5

The diagram suggests a difference in the two wells, since the boxes don't overlap. That is, the difference is large relative to the dispersions for both wells.

(b)

VAR. NAME	SAMPLE SIZE	MEDIAN	QUARTILES	ADJACENTS	IQR
Lee 1 minus Lee 2	8	71.0	9.5 102.0	−26.0 146.0	92.5

Note that the zero point for the differences falls outside of the box (middle half of the differences), which suggests that zero is not one of the "typical" values for the distribution of differences. This is another indication that the distributions are different.

3.7

DISTANCE

SHOWUP		100	50	OTHER	ROW SUMS AND RELATIVE FREQUENCY
No show	Obs.	11	6	5	22
	Rel. f	15.1	37.5	45.5	22.0
Show	Obs.	62	10	6	78
	Rel. f	84.9	62.5	54.5	78.0
Column sums		73	16	11	100

COMPARISON BAR GRAPHS FOR VARIABLE SHOWUP

```
                                                              Rel. f (%)
SHOWUP 0    10    20    30    40    50    60    70    80    90    100
No show |AAAAAAAAAAA
        |BBBBBBB|
        |CCCCCCCCCCCCCCCCCCC
        |DDDDDDDDDDDDDDDDDDDDDDDDD|

  Show  |AAAAAAAAAAAAAAAAAAAAAAAAAAAAAAAAAAAAAAAAAAAA
        |BBBBBBBBBBBBBBBBBBBBBBBBBBBBBBBBBBBBBBBBBBBB|
        |CCCCCCCCCCCCCCCCCCCCCCCCCCCCCCCC|
        |DDDDDDDDDDDDDDDDDDDDDDDDDDDDD|
```

LEGEND: DISTANCE

A = Average
B = 100
C = 50
D = Other

The graph indicates that those individuals indicating the shorter distances (50 and other) were no-shows in larger proportions than the 100-milers. Apparently, the more serious cyclists choose the longer distance and are more likely to show up for the tour.

3.9

ETHNIC GROUP

AGE		ANGLO	HISPANIC	NATIVE AMERICAN	ROW SUMS AND RELATIVE FREQUENCY
20–29	Obs.	34	25	11	70
	Rel. f	27.0	50.0	78.6	36.8
30–39	Obs.	26	9	2	37
	Rel. f	20.6	18.0	14.3	19.5
40–49	Obs.	23	6	1	30
	Rel. f	18.3	12.0	7.1	15.8
50–59	Obs.	22	6	0	28
	Rel. f	17.5	12.0	0.0	14.7
>59	Obs.	21	4	0	25
	Rel. f	16.7	8.0	0.0	13.2
Column sums		126	50	14	190

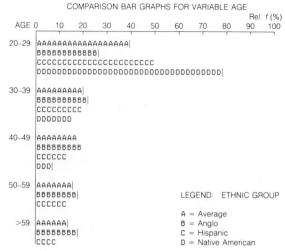

COMPARISON BAR GRAPHS FOR VARIABLE AGE

```
                                                        Rel. f (%)
   AGE  0  10  20  30  40  50  60  70  80  90  100
 20–29 |AAAAAAAAAAAAAAAAAA|
        BBBBBBBBBBB|
        CCCCCCCCCCCCCCCCCCCCCCCC
        DDDDDDDDDDDDDDDDDDDDDDDDDDDDDDDDDDDDDD|

 30–39 |AAAAAAAAA|
        BBBBBBBBBB|
        CCCCCCCCC
        DDDDDDD

 40–49 |AAAAAAAA
        BBBBBBBBB
        CCCCC
        DDD|

 50–59 |AAAAAAA|
        BBBBBBBB|                   LEGEND:  ETHNIC GROUP
        CCCCCC
                                    A = Average
   >59 |AAAAAA|                     B = Anglo
        BBBBBBBB|                   C = Hispanic
        CCCC                        D = Native American
```

The proportions of Native Americans and Hispanics committing suicide between ages 20 and 29 far exceeds the average (for the pooled sample), with nearly 80% of all Native American suicides occurring in this age range. On the other hand, suicides at ages 30 years and over occur much more frequently among Anglos than the other ethnic groups.

3.11

NUMBER OF BRANDS AVAILABLE

BRAND TYPE		1	2	3 OR MORE	ROW SUMS AND RELATIVE FREQUENCY
Brand	Obs.	6	17	4	27
	Rel. f	75.0	89.5	33.3	69.2
Generic	Obs.	2	2	8	12
	Rel. f	25.0	10.5	66.7	30.8
Column sums		8	19	12	39

COMPARISON BAR GRAPHS FOR VARIABLE BRAND TYPE

```
                                                      Rel. f (%)
BRAND TYPE  0  10  20  30  40  50  60  70  80  90  100
   Brand |AAAAAAAAAAAAAAAAAAAAAAAAAAAAAAAAAAAA|
          BBBBBBBBBBBBBBBBBBBBBBBBBBBBBBBBBBBB|
          CCCCCCCCCCCCCCCCCCCCCCCCCCCCCCCCCCCCCCC|
          DDDDDDDDDDDDDDD|
                                         LEGEND: NUMBER OF
 Generic |AAAAAAAAAAAAAA|                        BRANDS AVAILABLE
          BBBBBBBBBBB|
          CCCCC|                         A = Average
          DDDDDDDDDDDDDDDDDDDDDDDDDDDDDDDDDD|     B = 1
                                         C = 2
                                         D = 3 or more
```

The data indicate that brand name drugs are prescribed much more frequently than are generics by pharmacists who have only one or two types of a drug in stock than by pharmacists with access to several types.

3.13 (a)

VAR. NAME	SAMPLE SIZE	MEDIAN	QUARTILES	ADJACENTS	IQR
West side	18	4.960	4.500 6.040	2.800 6.700	1.540
Ease side	23	1.450	1.140 2.000	0.700 2.700	0.860
Minor outliers	3.90				

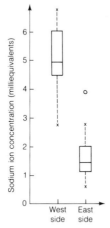

The data suggest a far higher concentration of sodium ions on the west bank of the river than on the east bank, on the average. (Except for the outlier, the two sample distributions do not overlap.)

3.15 (a)

SUMMARY

VAR. NAME	SAMPLE SIZE	MEDIAN	QUARTILES	ADJACENTS	IQR
Mill A	22	0.195	−0.070 0.320	−0.320 0.350	0.390
Minor outliers	1.01	1.30			
Major outliers	1.94	4.27			
Mill B	22	0.615	0.380 0.900	−0.130 1.640	0.520
Minor outliers	−1.16	−0.60 1.70	1.97		
Major outliers	7.02				
Mill C	19	0.470	0.245 0.850	−0.390 1.210	0.605
Mill D	19	0.710	0.515 1.020	0.070 1.450	0.505
Mill E	13	1.130	0.800 1.680	−0.370 2.400	0.880
Minor outliers	−0.93				

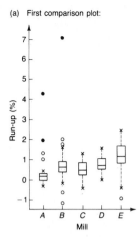

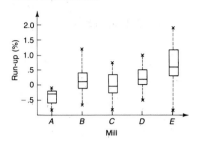

(b) Based on medians alone, mill A would be judged to have the best average run-up performance.

(c) Although there seems to be little difference in IQRs, mill A would appear to have a slight edge over the other mills in this category as well.

(d) Because mills A and B appear to have occasional large lapses (outliers) in quality, mill C would appear to offer the best compromise between consistency and average quality, with mill D a close second. Barring outliers, the mills appear to fall roughly into three "average quality" groups: {A}, {B,C,D}, and {E}.

3.17 (a)

		ETHNIC GROUP				ROW SUMS AND RELATIVE
EXAM RESULT		ANGLO	NATIVE AMERICAN	HISPANIC	BLACK	FREQUENCY
Passed	Obs.	118	1	7	1	127
	Rel. f	85.5	20.0	30.4	50.0	75.6
Failed	Obs.	20	4	16	1	41
	Rel. f	14.5	80.0	69.6	50.0	24.4
Column sums		138	5	23	2	168

COMPARISON BAR GRAPHS FOR VARIABLE EXAM RESULT

```
   EXAM                                                        Rel. f (%)
 RESULT 0    10    20    30    40    50    60    70    80    90    100
Passed  AAAAAAAAAAAAAAAAAAAAAAAAAAAAAAAAAAAAAAAAAA
        BBBBBBBBBBBBBBBBBBBBBBBBBBBBBBBBBBBBBBBBBBBB
        CCCCCCCCCC
        DDDDDDDDDDDDDDD
        EEEEEEEEEEEEEEEEEEEEEEEEE

Failed  AAAAAAAAAAAA
        BBBBBBB
        CCCCCCCCCCCCCCCCCCCCCCCCCCCCCCCCCCCCCCCCCC
        DDDDDDDDDDDDDDDDDDDDDDDDDDDDDDDDDDDDD
        EEEEEEEEEEEEEEEEEEEEEEEEEE
```

LEGEND: ETHNIC GROUP

A = Average
B = Anglo
C = Native American
D = Hispanic
E = Black

(b) The pass rates for minority candidates appear to be well below average, and that for Anglos is somewhat above average.

(c) While the data suggest a cause for concern, they do not *prove* that the exam itself is discriminatory. Other possible contributing factors would have to be ruled out before this conclusion could be reached.

3.19 **(a)**

			SUMMARY		
VAR. NAME	SAMPLE SIZE	MEDIAN	QUARTILES	ADJACENTS	IQR
50 miles	9	9.00	9.00 9.30	8.70 9.30	0.30
Minor outliers Major outliers	8.1 11.2	10.2			
75 miles	5	8.80	8.00 9.50	7.20 11.30	1.50
100 miles	15	12.40	11.25 15.60	8.10 21.10	4.35

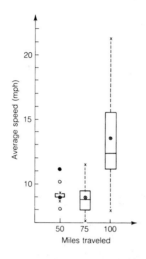

All distributions are somewhat skewed to the right, and the 50-mile speed distribution is long-tailed, as indicated by the outliers.

The means and medians agree well except for the 100-mile distribution, where the distribution skewness displaces the mean upward by about one mile per hour.

(b)

	SPEED		
	50	75	100
n	9	5	15
$\bar{X}$	9.28	8.96	13.62
s	.91	1.57	3.68

3.21

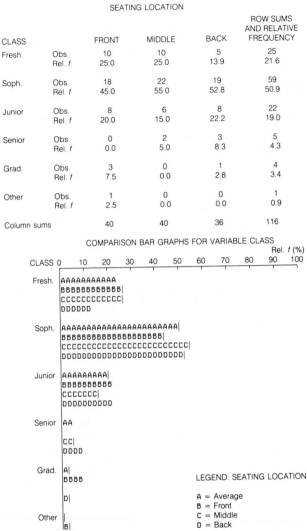

SEATING LOCATION

CLASS		FRONT	MIDDLE	BACK	ROW SUMS AND RELATIVE FREQUENCY
Fresh.	Obs. Rel. f	10 25.0	10 25.0	5 13.9	25 21.6
Soph.	Obs. Rel. f	18 45.0	22 55.0	19 52.8	59 50.9
Junior	Obs. Rel. f	8 20.0	6 15.0	8 22.2	22 19.0
Senior	Obs. Rel. f	0 0.0	2 5.0	3 8.3	5 4.3
Grad.	Obs. Rel. f	3 7.5	0 0.0	1 2.8	4 3.4
Other	Obs. Rel. f	1 2.5	0 0.0	0 0.0	1 0.9
Column sums		40	40	36	116

COMPARISON BAR GRAPHS FOR VARIABLE CLASS

```
                                                    Rel. f (%)
CLASS 0   10  20   30  40   50   60  70   80  90  100
Fresh. AAAAAAAAAA
       BBBBBBBBBB|
       CCCCCCCCCCC|
       DDDDDD

Soph.  AAAAAAAAAAAAAAAAAAAAAAAA|
       BBBBBBBBBBBBBBBBBBBBBB|
       CCCCCCCCCCCCCCCCCCCCCCCCCC|
       DDDDDDDDDDDDDDDDDDDDDDDDDDD|

Junior AAAAAAAAA|
       BBBBBBBBBB
       CCCCCCC|
       DDDDDDDDDD

Senior AA

       CC|
       DDDD

Grad.  A|
       BBBB

       D|

Other  ||
       B|
```

LEGEND: SEATING LOCATION

A = Average
B = Front
C = Middle
D = Back

Freshmen appear to favor the front and middle of the classroom, while sophomores favor the middle and back. Upperclassmen tend to favor (slightly) the back of the classroom. Graduates and others sit primarily in the front.

3.23 (a)

LAYOFF STATUS

MINORITY STATUS		LOW	MEDIUM	HIGH	ROW SUMS AND RELATIVE FREQUENCY
Minority	Obs.	95	13	19	127
	Rel. f	44.2	61.9	100.0	49.8
Nonminority	Obs.	120	8	0	128
	Rel. f	55.8	38.1	0.0	50.2
Column sums		215	21	19	255

COMPARISON BAR GRAPHS FOR VARIABLE MINORITY STATUS

```
MINORITY                                              Rel. f (%)
 STATUS 0   10   20   30   40   50   60   70   80   90   100

Minority |AAAAAAAAAAAAAAAAAAAAAAAA
         |BBBBBBBBBBBBBBBBBBBBB
         |CCCCCCCCCCCCCCCCCCCCCCCCCCCCC
         |DDDDDDDDDDDDDDDDDDDDDDDDDDDDDDDDDDDDDDDDDDDDDDDDD

Nonminority |AAAAAAAAAAAAAAAAAAAAAAAA          LEGEND: LAYOFF STATUS
            |BBBBBBBBBBBBBBBBBBBBBBBBBBB
            |CCCCCCCCCCCCCCCCCC                  A = Average
                                                 B = Low
                                                 C = Medium
                                                 D = High
```

There is a rather large difference in minority status distributions for the three layoff groups. Minority workers predominate in the medium and high distributions, suggesting a layoff policy that discriminates against minority workers.

(b) While the data suggest discrimination against minority workers, they do not prove it. Other possible reasons would have to be ruled out in order to make this case.

3.25

CLASS

HOURS STUDIED		FRESH.	SOPH.	ROW SUMS AND RELATIVE FREQUENCY
1–10	Obs.	48	3	51
	Rel. f	32.0	12.5	29.3
11–26	Obs.	73	15	88
	Rel. f	48.7	62.5	50.6
>26	Obs.	29	6	35
	Rel. f	19.3	25.0	20.1
Column sums		150	24	174

COMPARISON BAR GRAPHS FOR VARIABLE HOURS STUDIED

```
HOURS                                                 Rel. f (%)
STUDIED 0   10   20   30   40   50   60   70   80   90   100

 1–10 |AAAAAAAAAAAAA|
      |BBBBBBBBBBBBBB
      |CCCCCC|

11–26 |AAAAAAAAAAAAAAAAAAAAAAAAA|
      |BBBBBBBBBBBBBBBBBBBBBBBBB|
      |CCCCCCCCCCCCCCCCCCCCCCCCCCCCCC|        LEGEND: CLASS

  >26 |AAAAAAAAAA
      |BBBBBBBBB|                              A = Average
      |CCCCCCCCCC|                             B = Fresh.
                                               C = Soph.
```

(a) Sophomores appear to put in more study time than do freshmen; they have a lower frequency of between 1 and 10 study hours per week and higher frequencies in the two study categories over 11 hours.

(b) About 25% of the sophomores and 19% of the freshman do so.

(c) Since students usually find themselves in the dean's office most frequently when they are having academic difficulties, this is not likely to be the best place to obtain a representative sample of all students.

3.27 (a)

SUMMARY

VAR. NAME	SAMPLE SIZE	MEDIAN	QUARTILES	ADJACENTS	IQR
Old wrench	96	6.0	5.0 7.0	2.0 10.0	2.0
New wrench	96	6.0	6.0 7.0	5.0 8.0	1.0
Minor outliers	3	3	4 4	4 4	

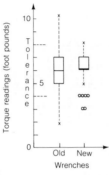

The medians of the two distributions are the same and, since the distributions are reasonably symmetric, the means will be nearly the same as well. Thus, a comparison based on location measures will not show any differences in distributions.

(b) Except for six minor outliers, the new wrench has the tighter distribution (IQR of 1.0 against the old wrench's value of 2.0).

WRENCH

TOLERANCE		OLD	NEW	ROW SUMS AND RELATIVE FREQUENCY
In	Obs.	81	94	175
	Rel. f	84.4	97.9	91.1
Out	Obs.	15	2	17
	Rel. f	15.6	2.1	8.9
Column sums		96	96	192

COMPARISON BAR GRAPHS FOR VARIABLE TOLERANCE

```
                                                              Rel. f (%)
TOLERANCE 0   10   20   30   40   50   60   70   80   90   100
In  |AAAAAAAAAAAAAAAAAAAAAAAAAAAAAAAAAAAAAAAAAAAAAAAAAA|
    |BBBBBBBBBBBBBBBBBBBBBBBBBBBBBBBBBBBBBBBBBBBB
    |CCCCCCCCCCCCCCCCCCCCCCCCCCCCCCCCCCCCCCCCCCCCCCCCCC

Out |AAAA|                              LEGEND: WRENCH
    |BBBBBBBB
    |C                                  A = Average
                                        B = Old
                                        C = New
```

(c) The distribution of torques for the new wrench has a higher proportion of in-tolerance readings than does that for the old wrench. On this basis, the new wrench would be preferred.

3.29

VAR. NAME	SAMPLE SIZE	MEDIAN	QUARTILES	ADJACENTS	IQR
Condition 1	11	26.70	19.20	7.80	11.55
			30.75	34.20	
Condition 2	8	21.80	15.70	6.60	11.55
			27.25	40.40	

The distributions don't have any unusual features, with the possible exception of a mildly short upper tail for condition 1. This distribution also has a somewhat higher median value than the other, although there is substantial overlap in the two distributions.

Chapter 4

4.1, 4.2

Frequency distribution table

HEIGHT		WEIGHT ≤120	121–140	≥141	ROW SUMS & ROW MARGINAL FREQUENCIES
≤64	Obs.	6	4	0	10
	Jnt.	16.7	11.1	0.0	
	C|R.	60.0	40.0	0.0	
	R|C.	50.0	30.8	0.0	27.8
65–67	Obs.	6	6	3	15
	Jnt.	16.7	16.7	8.3	
	C|R.	40.0	40.0	20.0	
	R|C.	50.0	46.2	27.3	41.7
≥68	Obs.	0	3	8	11
	Jnt.	0.0	8.3	22.2	
	C|R.	0.0	27.3	72.7	
	R|C.	0.0	23.1	72.7	30.6
Column sums		12	13	11	36
Column marginal frequencies		33.3	36.1	30.6	

4.3 From the joint distribution, we see that 16.7% of the students in the sample have weight ≤120 pounds and are ≤64 inches (5 feet 4 inches) tall. 0% have weights ≤120 pounds and are ≥68 inches (5 feet 8 inches) tall. 33.3% have weights ≤120 pounds (column marginal) and 41.7% are between 65 and 67 inches tall (row marginal).

4.6 Refer to the C|R (column variable given row variable) row of the table given with the solutions to 4.1–4.3. The following bar graph gives the conditional distributions of weight given height.

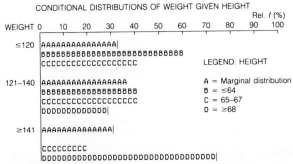

CONDITIONAL DISTRIBUTIONS OF WEIGHT GIVEN HEIGHT

It is seen that among people of height ≤ 64 inches, frequencies in the weight categories decrease with increasing weight, while the reverse is true for individuals of height ≥ 68 inches. This suggests that height increases with weight in this sample of individuals—a trend we would expect to persist in the population.

4.7 Refer to the R|C (row variable given column variable) row in the table given in the solutions to 4.1–4.3. The following bar graph gives the conditional distributions of height given weight.

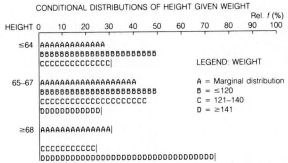

CONDITIONAL DISTRIBUTIONS OF HEIGHT GIVEN WEIGHT

The same trend of decreasing frequencies with increasing height in the low weight (≤ 120 pound) category and the reverse trend in the high weight category supports the conclusion of exercise 4.6.

4.10

VAR. NAME	SAMPLE SIZE	MEDIAN	QUARTILES	ADJACENTS	IQR
Fall	27	15.0	0.0	0.0	18.0
			18.0	39.0	
Minor outliers	46				
Winter	34	15.0	0.0	0.0	15.0
			15.0	34.0	
Minor outliers	47				
Major outliers	87				
Spring	17	18.0	15.0	0.0	19.0
			34.0	47.0	
Major outliers	109				
Summer	27	21.0	15.0	0.0	18.0
			33.0	60.0	
Minor outliers	71				
Major outliers	106	150			

The trend in activity (medians) is constant from fall to winter, then increases into the spring and summer. The variation, as exhibited by the quartile lines, is reasonably constant throughout the year. However, a few mice display considerable activity (outliers) which also appears to depend on season.

4.13 (a)

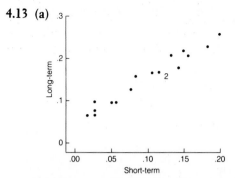

There appears to be a slight upward trend in the exploratory plot, although the scatter diagram suggests that a straight line would fit the data well.

(c) From the scatter diagram, the trend in dispersion appears to be reasonably constant. A slightly larger dispersion at the left end of the diagram is indicated by the quartile lines. These are rather ideal data from the viewpoint of the linear regression, which will be taken up in chapter 14.

Cutpoints at .07 and .13 yield 6 points in each of three groups.

(b)   EXPLORATORY PLOT SUMMARY

GROUP	SAMPLE SIZE	X MEDIAN	Y Q1	Y MEDIAN	Y Q3
1	6	.025	.067	.0845	.094
2	6	.1125	.15	.154	.165
3	6	.154	.202	.2105	.227

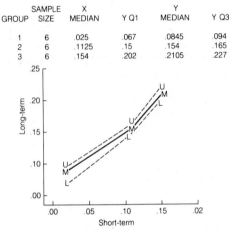

4.15 (a)

Frequency distributions

		BIRTH DEFECT STATUS		
MOTHER'S DIABETIC STATUS		≥1	NONE	ROW SUMS & MARGINAL FREQUENCIES
Nondiabetic	Obs.	31	754	785
	Jnt.	2.6	62.5	
	C|R.	3.9	96.1	
	R|C.	58.5	65.3	65.0
Prediabetic	Obs.	13	362	375
	Jnt.	1.1	30.0	
	C|R.	3.5	96.5	
	R|C.	24.5	31.4	31.1
Diabetic	Obs.	9	38	47
	Jnt.	0.7	3.1	
	C|R.	19.1	80.9	
	R|C.	17.0	3.3	3.9
Column sums		53	1154	1207
Column marginal frequencies		4.4	95.6	

4.4% of the children are born with at least one birth defect and 3.9% of the mothers are diabetic. 0.7% of the children are born to diabetic mothers and have at least one birth defect.

(b)

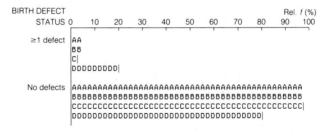

CONDITIONAL DISTRIBUTIONS
OF BIRTH DEFECT STATUS GIVEN MOTHER'S DIABETIC STATUS

LEGEND: MOTHER'S DIABETIC STATUS

A = Marginal distribution
B = Nondiabetic
C = Prediabetic
D = Diabetic

The highest frequency of children with birth defects occurs for diabetic mothers. This frequency is substantially larger than the average, as given by the marginal frequency (19.1% compared with 4.4%). This suggests a link between maternal diabetes and an increased risk of birth defects among children.

(c)

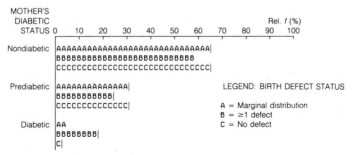

CONDITIONAL DISTRIBUTIONS
OF MOTHER'S DIABETIC STATUS GIVEN BIRTH DEFECT STATUS

LEGEND: BIRTH DEFECT STATUS

A = Marginal distribution
B = ≥1 defect
C = No defect

Among children with at least one birth defect, the proportions of nondiabetic and prediabetic mothers are somewhat smaller than average, while the proportion of diabetic mothers is much larger than average. This again suggests the diabetes–birth-defect link.

(d) Yes. Statistical inference will be needed to determine the degree to which these indications support this association in the population, but the sample is large and the evidence appears strong.

4.17

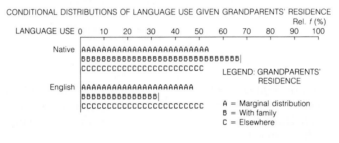

GRANDPARENTS' RESIDENCE

LANGUAGE USE		WITH FAMILY	ELSEWHERE	ROW SUMS AND RELATIVE FREQUENCY
Native	Obs.	10	23	33
	Jnt.	16.4	37.7	
	C|R.	30.3	69.7	
	R|C.	66.7	50.0	54.1
English	Obs.	5	23	28
	Jnt.	8.2	37.7	
	C|R.	17.9	82.1	
	R|C.	33.3	50.0	45.9
Column sums		15	46	61
Column marginal frequency		24.6	75.4	

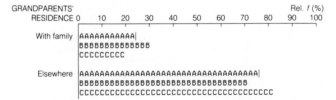

CONDITIONAL DISTRIBUTIONS OF LANGUAGE USE GIVEN GRANDPARENTS' RESIDENCE

Rel. *f* (%)
LANGUAGE USE 0   10   20   30   40   50   60   70   80   90   100

Native   AAAAAAAAAAAAAAAAAAAAAAAAAAAAA
         BBBBBBBBBBBBBBBBBBBBBBBBBBBBBBBBBBBBBB|
         CCCCCCCCCCCCCCCCCCCCCCCCCCC

English AAAAAAAAAAAAAAAAAAAAAAAAA
         BBBBBBBBBBBBBBBB|
         CCCCCCCCCCCCCCCCCCCCCCCCCCC

LEGEND: GRANDPARENTS'
RESIDENCE

A = Marginal distribution
B = With family
C = Elsewhere

CONDITIONAL DISTRIBUTIONS OF GRANDPARENTS' RESIDENCE GIVEN LANGUAGE USE

GRANDPARENTS'   Rel. *f* (%)
RESIDENCE 0   10   20   30   40   50   60   70   80   90   100

With family AAAAAAAAAA|
             BBBBBBBBBBBBBB
             CCCCCCCCC

Elsewhere   AAAAAAAAAAAAAAAAAAAAAAAAAAAAAAAAAAAAAAAA|
             BBBBBBBBBBBBBBBBBBBBBBBBBBBBBBBBBB
             CCCCCCCCCCCCCCCCCCCCCCCCCCCCCCCCCCCCCCCCCCCCCC

LEGEND: LANGUAGE USE

A = Marginal distribution
B = Native
C = English

From the conditional distributions of language use given grandparents' residence, it is seen that the use of the native language is greater among families with the grandparents living in the same residence than among families with the grandparents living elsewhere. From the conditional distribution of grandparents' residence given language use it is seen that the frequency of grandparents living with the family is greatest among families in which the native language predominates.

Although the data cannot conclusively establish it, the reason that is most plausible for the suggested association for these variables is the degree of traditionalism in the families. More traditional families will maintain both the traditions of language and the sharing of homes with their elders, while English and separate residences are more likely to be more prevalent in less traditional homes.

4.19

NPR AWARENESS

SEX		AWARE	NOT AWARE	ROW SUMS AND RELATIVE FREQUENCY
Male	Obs.	81	8	89
	Jnt.	59.6	5.9	
	C|R.	91.0	9.0	
	R|C.	65.3	66.7	65.4
Female	Obs.	43	4	47
	Jnt.	31.6	2.9	
	C|R.	91.5	8.5	
	R|C.	34.7	33.3	34.6
Column sums		124	12	136
Column marginal frequencies		91.2	8.8	

CONDITIONAL DISTRIBUTIONS OF SEX GIVEN NPR AWARENESS

```
                                                    Rel. f (%)
 SEX 0    10    20    30    40    50    60    70    80    90   100
     |----+----+----+----+----+----+----+----+----+----+
Male |AAAAAAAAAAAAAAAAAAAAAAAAAAAAAAAAA|
     |BBBBBBBBBBBBBBBBBBBBBBBBBBBBBBBBB|
     |CCCCCCCCCCCCCCCCCCCCCCCCCCCCCCCCC|

Female |AAAAAAAAAAAAAAAAA|
       |BBBBBBBBBBBBBBBBB|                    LEGEND: NPR AWARENESS
       |CCCCCCCCCCCCCCCCC|
                                               A = Marginal distribution
                                               B = Aware
                                               C = Not aware
```

CONDITIONAL DISTRIBUTIONS OF NPR AWARENESS GIVEN SEX

```
 NPR                                                Rel. f (%)
AWARENESS 0  10    20    30    40    50    60    70    80    90   100
          |----+----+----+----+----+----+----+----+----+----+
   Aware  |AAAAAAAAAAAAAAAAAAAAAAAAAAAAAAAAAAAAAAAAAAAAAAA|
          |BBBBBBBBBBBBBBBBBBBBBBBBBBBBBBBBBBBBBBBBBBBBBBB|
          |CCCCCCCCCCCCCCCCCCCCCCCCCCCCCCCCCCCCCCCCCCCCCCC|

Not aware |AAAA|
          |BBBB|
          |CCCC|                             LEGEND: SEX

                                               A = Marginal distribution
                                               B = Male
                                               C = Female
```

It would be difficult to find an example in which the two variables are less associated than sex and NPR awareness appear to be in this exercise. The degree of awareness appears to be the same for both sexes.

4.21

Frequency distributions

HOUSEHOLD INCOME		NPR AWARENESS		ROW SUMS & MARGINAL FREQUENCIES
		AWARE	NOT AWARE	
<10k	Obs.	25	4	29
	Jnt.	18.4	2.9	
	C|R.	86.2	13.8	
	R|C.	20.2	33.3	21.3
10–14k	Obs.	23	1	24
	Jnt.	16.9	0.7	
	C|R.	95.8	4.2	
	R|C.	18.5	8.3	17.6
15–19k	Obs.	16	1	17
	Jnt.	11.8	0.7	
	C|R.	94.1	5.9	
	R|C.	12.9	8.3	12.5
20–29k	Obs.	30	2	32
	Jnt.	22.1	1.5	
	C|R.	93.8	6.3	
	R|C.	24.2	16.7	23.5
≥30k	Obs.	20	2	22
	Jnt.	14.7	1.5	
	C|R.	90.9	9.1	
	R|C.	16.1	16.7	16.2
No response	Obs.	10	2	12
	Jnt.	7.4	1.5	
	C|R.	83.3	16.7	
	R|C.	8.1	16.7	8.8
Column sums		124	12	136
Column marginal frequencies		91.2	8.8	

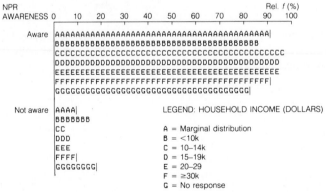

CONDITIONAL DISTRIBUTIONS
OF NPR AWARENESS GIVEN HOUSEHOLD INCOME

```
NPR                                                    Rel. f (%)
AWARENESS 0   10   20   30   40   50   60   70   80   90   100
    Aware |AAAAAAAAAAAAAAAAAAAAAAAAAAAAAAAAAAAAAAAAAAAAAA|
          |BBBBBBBBBBBBBBBBBBBBBBBBBBBBBBBBBBBBBBBBBBBBB
          |CCCCCCCCCCCCCCCCCCCCCCCCCCCCCCCCCCCCCCCCCCCCCCC
          |DDDDDDDDDDDDDDDDDDDDDDDDDDDDDDDDDDDDDDDDDDDDDD
          |EEEEEEEEEEEEEEEEEEEEEEEEEEEEEEEEEEEEEEEEEEEEE
          |FFFFFFFFFFFFFFFFFFFFFFFFFFFFFFFFFFFFFFFFFFFFFF|
          |GGGGGGGGGGGGGGGGGGGGGGGGGGGGGGGGGGGGGGGGGGG|

Not aware |AAAA|              LEGEND: HOUSEHOLD INCOME (DOLLARS)
          |BBBBBBB
          |CC                 A = Marginal distribution
          |DDD                B = <10k
          |EEE                C = 10–14k
          |FFFF|              D = 15–19k
          |GGGGGGGG|          E = 20–29
                             F = ≥30k
                             G = No response
```

Awareness appears to be lowest for low-income families and those who would not divulge their incomes, and highest for middle- and high-income families.

4.25   EXPLORATORY PLOT SUMMARY (FEMALES)

GROUP	SAMPLE SIZE	X MEDIAN	Y Q1	Y MEDIAN	Y Q3
1	15	23	140	162	181.5
2	26	44	165	200	228
3	21	65	214	251	286

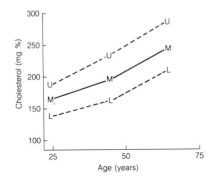

EXPLORATORY PLOT SUMMARY (MALES)

GROUP	SAMPLE SIZE	X MEDIAN	Y Q1	Y MEDIAN	Y Q3
1	12	25.5	155.5	190.5	212
2	24	41	180	220	251
3	14	63.5	212	219	250

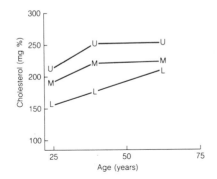

The cholesterol levels for women appear to be increasing with age at a fairly constant rate and with roughly the same variation over the observed age range. On the other hand, the trend for men increases to middle age, then seems to become constant. The variability appears to be smaller among older men. It is likely that this phenomenon is due to an age-dependent selection process. Men with potentially high cholesterol levels in old age have already died or become too incapacitated to go to state fairs. It was primarily the healthier older men with relatively low cholesterol levels who participated in the study.

Chapter 5

5.1 (a) A = {[KC, KH], [KC, QH], [QC, KH], [QC, QH]}
B = {[KC, QC], [KC, KH], [KC, QH], [QC, KH], [QH, KH]}
(b) (1) (not A) = {[KC, QC], [QH, KH]}
(2) (A or B) = S
(3) (A and B) = {[KC, KH], [KC, QH], [QC, KH]}
(4) (not B) = {[QC, QH]}
(5) (A and (not B)) = {[QC, QH]}
(c) (1) and (3), (1) and (4), (1) and (5), (3) and (4), (3) and (5)
(d) ((A and (not B)) or (A and B)) = {[QC, QH], [KC, KH], [KC, QH], [QC, KH]} = A.

5.3 (a) $P(A$ or $B)$ = .5 + .6 − .2 = .9
(c) $P((\text{not }A)$ and $B)$ = $P(B) − P(A$ and $B)$ = .6 − .2 = .4
(e) $P((\text{not }A)$ or $(\text{not }B))$ = $P(\text{not}(A$ and $B))$ = 1 − $P(A$ and $B)$ = 1 − .2 = .8

5.7 Each outcome has probability $\frac{1}{6}$.
(a) $N(A)$ = 4, so $P(A)$ = $\frac{4}{6}$ = $\frac{2}{3}$
(d) $N(A$ and $B)$ = 3, so $P(A$ and $B)$ = $\frac{1}{2}$
(e) By the general addition law, $P(A$ or $B)$ = $P(A) + P(B) − P(A$ and $B)$ = $\frac{4}{6} + \frac{5}{6} − \frac{3}{6}$ = 1. This checks, since A or B = S.

5.9 (a) Let S = {0, 1, . . . , 9} and equate *heads* with {0, 1, . . . , 6} and *tails* with {7, 8, 9}.
(c) S = {00, 01, . . . , 99}, *heads* = {00, . . . , 24}, *tails* = {25, . . . , 99}
(e) S = {000, 001, . . . , 999}, *heads* = {000, . . . , 124}, *tails* = {125, . . . , 999}

5.10 (a) $P(2$ red) = $P(R_1$ and $R_2)$ = $P(R_1)P(R_2|R_1)$ = $(\frac{3}{7})(\frac{2}{8})$ = $\frac{1}{4}$
(b) $P(2$nd ball different color from 1st) = $P(W_1$ and $R_2)$ or $(R_1$ and $W_2))$ = $P(W_1)P(R_2|W_1) + P(R_1)P(W_2|R_1)$ = $(\frac{4}{7})(\frac{3}{8})$ + $(\frac{3}{7})(\frac{4}{8})$ = $\frac{4}{7}$

5.12 (a) $P(A$ and $B)$ = $P(A)P(B)$ = .3 × .5 = .15
(c) $P(A$ and (not $B))$ = $P(A)P(\text{not }B)$ = $P(A)(1 − P(B))$ = .3 × .5 = .15
(f) $P((\text{not }A)$ or (not $B))$ = $P(\text{not}(A$ and $B))$ = 1 − $P(A$ and $B)$ = 1 − .15 = .85

5.13 (a) $P(W_2)$ = $P((W_1$ and $W_2)$ or $(R_1$ and $W_2))$ = $P(R_1)P(W_2|R_1)$ + $P(W_1)P(W_2|W_1)$ = $(\frac{2}{5})(\frac{5}{8})$ + $(\frac{3}{5})(\frac{6}{8})$ = $\frac{7}{10}$
(b) By Bayes' rule, $P(W_1|R_2)$ =

$P(W_1)P(R_2|W_1)/(P(W_1)P(R_2|W_1)$ +
$P(R_1)P(R_2|R_1)) = (\frac{3}{5})(\frac{2}{8})/((\frac{3}{5})(\frac{2}{8}) + (\frac{2}{5})(\frac{3}{8})) = \frac{1}{2}$

5.15 (c) The events A_1 = "fresh on first two selections," A_2 = "fresh on first and sour on second," and A_3 = "sour on first and fresh on second" form a partition of the sample space for this problem. If B = "sour on third selection," then $P(B|A_1) = 1$, while $P(B|A_1)$ and $P(B|A_2)$ are both 0, since the sour carton has already been selected. Thus, $P(B) = P(A_1)P(B|A_1) + 0 + 0 = (\frac{2}{3})(\frac{1}{2})(1) = \frac{1}{3}$.

5.18

VALUE x OF X	EVENT	$P(X = x)$
2	{(1, 1)}	$\frac{1}{36}$
3	{(1, 2), (2, 1)}	$\frac{2}{36}$
4	{(1, 3), (2, 2), (3, 1)}	$\frac{3}{36}$
5	{(1, 4), (2, 3), (3, 2), (4, 1)}	$\frac{4}{36}$
6	{(1, 5), (2, 4), (3, 3), (4, 2), (5, 1)}	$\frac{5}{36}$
7	{(1, 6), (2, 5), (3, 4), (4, 3), (5, 2), (6, 1)}	$\frac{6}{36}$
8	{(2, 6), (3, 5), (4, 4), (5, 3), (6, 2)}	$\frac{5}{36}$
9	{(3, 6), (4, 5), (5, 4), (6, 3)}	$\frac{4}{36}$
10	{(4, 6), (5, 5), (6, 4)}	$\frac{3}{36}$
11	{(5, 6), (6, 5)}	$\frac{2}{36}$
12	{(6, 6)}	$\frac{1}{36}$
		1

5.20 $E(X) = 2(\frac{1}{36}) + 3(\frac{2}{36}) + \cdots + 12(\frac{1}{36}) = 7$

5.22 (a) $E(Y) = -1(\frac{1}{4}) + 0(\frac{1}{4}) + 1(\frac{1}{2}) = \frac{1}{4}$
   (b) $E(Y^2) = (-1)^2(\frac{1}{4}) + 0^2(\frac{1}{4}) + 1^2(\frac{1}{2}) = \frac{3}{4}$
   (c) $E(-3Y^2 + 5Y + 2) = -3E(Y^2) + 5E(Y) + 2 = -3(\frac{3}{4}) + 5(\frac{1}{4}) + 2 = 1$

5.23 (a) $E(X) = 2$
   (b) $E(X^2) = 1^2(\frac{1}{3}) + 2^2(\frac{1}{3}) + 3^2(\frac{1}{3}) = \frac{14}{3}$
       $\text{var}(X) = E(X^2) - (E(X))^2 = \frac{14}{3} - 2^2 = \frac{2}{3}$
   (c) $\sigma_X = \sqrt{\frac{2}{3}} = .8165$
   (d) $E(Y) = 2E(X) + 8 = 12$
       $\text{var}(Y) = 2^2\text{var}(X) = \frac{8}{3} = 2.667$
       $\sigma_Y = \sqrt{\frac{8}{3}} = 1.633$

5.26 (a)

x	y	$(X = x)$ and $(Y = y)$	$P(X = x$ and $Y = y)$
0	0	O	0
0	1	{[QC, QH]}	$\frac{1}{6}$
0	2	O	0
1	0	{[KC, QC]}	$\frac{1}{6}$
1	1	{[KC, QH], [QC, KH]}	$\frac{2}{6}$
1	2	{[QH, KH]}	$\frac{1}{6}$
2	0	O	0
2	1	{[KC, KH]}	$\frac{1}{6}$
2	2	O	0

(b), (c)

	VALUE OF X			
VALUE OF Y	0	1	2	Y MARGINAL
0	0	$\frac{1}{6}$	0	$\frac{1}{6}$
1	$\frac{1}{6}$	$\frac{2}{6}$	$\frac{1}{6}$	$\frac{4}{6}$
2	0	$\frac{1}{6}$	0	$\frac{1}{6}$
X marginal	$\frac{1}{6}$	$\frac{4}{6}$	$\frac{1}{6}$	1

(d)

	VALUE OF X		
VALUE OF Y	0	1	2
0	0	$\frac{1}{4}$	0
1	1	$\frac{1}{4}$	1
2	0	$\frac{1}{4}$	0
	1	1	1

$P(Y = 0|X = 1) = \frac{1}{4}$

5.27 W = number of *heads* in two tosses, $p = \frac{1}{2}$

$$P(W = 1) = \binom{2}{1} (\frac{1}{2})^1(1 - \frac{1}{2})^1 = 2(\frac{1}{4}) = \frac{1}{2}$$

5.29 W = number of women in two draws, $p = \frac{7}{20}$

$$P(W = 1) = \binom{2}{1} (\frac{7}{20})(1 - \frac{7}{20}) = .455$$

5.31 $p = \frac{1}{2}$, W = number of heads in three tosses.

$$P(W = 2) = \binom{3}{2} (\frac{1}{2})^2(1 - \frac{1}{2})$$
$$= 3(\frac{1}{2})^3 = \frac{3}{8}$$

5.33 (non-zero terms only)
$$E(X^2Y) = 1^2(1)(\frac{2}{6}) + 2^2(1)(\frac{1}{6}) + 1^2(2)(\frac{1}{6})$$
$$= \frac{4}{3}$$

5.35 (a) $p = \frac{2}{20} = \frac{1}{10}$
   (b) $\hat{p} = (X_1 + X_2 + \cdots + X_5)/5$
       $E(\hat{p}) = p = \frac{1}{10}$
       $\text{var}(\hat{p}) = p(1 - p)/5 = (\frac{1}{5})(\frac{1}{10})(\frac{9}{10}) = .018$,
       so $\sigma_{\hat{p}} = .134$

5.38 (a) X has a binomial distribution with $n = 3$

and $p = \frac{1}{10}$, and Y has a binomial distribution with $n = 3$ and $p = \frac{1}{50}$. Thus, since $W = X + Y$,

$$E(W) = E(X) + E(Y) = \frac{3}{10} + \frac{3}{50} = \frac{18}{50} = .36$$

(b) $C = 200X + 800Y$

(c) $E(C) = 200E(X) + 800E(Y) = 200(\frac{3}{10}) + 800(\frac{3}{50}) = \$108$

5.40 (a) The marginal distributions of X and Y are both $P(X = -1) = \frac{1}{4}$, $P(X = 0) = \frac{1}{2}$, $P(X = 1) = \frac{1}{4}$.

5.41 (a)

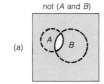

5.43 If C_x denotes the number of colds the individual has during time period x, since

$$C_{year} = C_{summer} + C_{winter} + C_{spring} + C_{fall}$$

then

$$E(C_{year}) = E(C_{summer}) + E(C_{winter}) + E(C_{spring}) + E(C_{fall})$$

But, for each season, the probability of more than one cold is zero, so $E(C_x) = P(C_x = 1)$. It follows that $E(C_{year}) = .3 + 2 \times .1 + .05 = .55$.

5.45 (a) $S = \{(1, 2, 3), (1, 3, 2), (2, 1, 3), (2, 3, 1), (3, 1, 2), (3, 2, 1)\}$

(b) Equally likely model, which assigns probability $\frac{1}{6}$ to each outcome. $P((1, 2, 3)) = \frac{1}{6}$

(c) $P((1, 2, 3), (1, 3, 2)) = \frac{1}{3}$

(d) $P((1, 3, 2), (3, 2, 1), (2, 1, 3)) = \frac{1}{2}$

(e) $P(1, 2, 3), (1, 3, 2), (3, 2, 1), (2, 1, 3)) = \frac{2}{3}$

5.47 (a) $P(\text{1st digit} = 1) = \frac{6}{24} = \frac{1}{4}$

(b) $P((1, 2, 3, 4)) = \frac{1}{24}$

(c) $P(\text{no digits in proper order}) = \frac{9}{24} = \frac{3}{8}$

(d) $P(\text{exactly three digits in proper order}) = 0$

5.49 (a) $P(A)P(\text{not } B) = .5 \times (1 - .6) = .2 = P(A \text{ and (not } B))$. Thus, A and (not B) are independent, which implies that A and B are independent.

(b) $P(A \text{ or } B) = P(A) + P(B) - P(A)P(B) = .5 + .6 - .3 = .8$

5.51 $P(Y = 0) = \frac{9}{10}$, $P(Y = 1) = \frac{1}{10}$

(b) For example, $P(X = 0)P(Y = 0) = (\frac{1}{2})(\frac{1}{2})$, but the joint probability is $P(X = 0$ and $Y = 0) = 0$. Consequently, not all joint probabilities equal the products of the marginals, as required for independence.

(c) $E(X) = E(Y) = 0$

(d) $E(W) = E(XY) = 1(\frac{1}{8}) + (-1)(\frac{1}{8}) + (-1)(\frac{1}{8}) + 1(\frac{1}{8}) = 0$

5.53 (a)

VALUES OF X	PROBABILITY
0	$\frac{1}{20}$
1	$\frac{8}{20}$
2	$\frac{3}{20}$
3	$\frac{1}{20}$
4	$\frac{1}{20}$
5	$\frac{2}{20}$
6	$\frac{0}{20}$
7	$\frac{2}{20}$
8	$\frac{0}{20}$
9	$\frac{0}{20}$
10	$\frac{2}{20}$

$E(X) = 3.25$

(b) $E(X^2) = 19.650$, $var(X) = 9.088$, $\sigma_X = 3.015$

(c) They are the same. That is, the mean of the population frequency distribution equals the expectation and, similarly, the variances are equal.

(d) $Y = 1.609X$, $E(Y) = 5.229$, $\sigma_Y = 4.851$

Chapter 6

6.1 (a) $P(X \le .7) = .7$

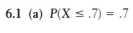
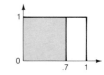

(c) $P(X < .3 \text{ or } X > .7) = P(X < .3) + P(X > .7) = .3 + .3 = .6$

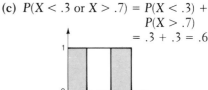

6.2 (a) $P(X < 8) = P((X - 3)/5 < (8 - 3)/5) = P(Z < 1) = .5 + .3413 = .8413$
(c) $P((X < 4)$ or $(X > 7)) = P(X < 4) + P(X > 7) = P(Z < .20) + P(Z > .80) = .7912$

6.4 (a) W is normal with $\mu_W = 3\mu_X + \mu_Y + 10 = 7$ and $\sigma_W = \sqrt{9\sigma_X^2 + \sigma_Y^2} = 9.055$. Thus, $P(8 < W < 14) = P(.11 < Z < .77) = .2356$

6.6 (a) $P((X - 1)/5 \geq 1.645) = P(X \geq 1 + 5(1.645)) = P((X - 3)/5 \geq (1 + 5(1.645) - 3)/5) = P(Z \geq 1.24) = .1075$.

6.7 (a) $P(|Z| \leq 2.45) = 2P(0 < Z \leq 2.45) = .9858$

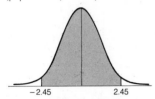

(e) $P(1.04 < |Z| < 2.56) = 2P(1.04 < Z < 2.56) = .2880$

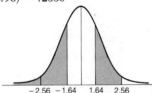

6.8 (a) $P(|X + 6| > 5) = P(|X - (-6)|/3 > \frac{5}{3}) = P(|Z| > 1.67) = .095$

6.9 (a) $P(|(X - 1)/5| \geq 1.96) = P(X - 1 \geq 5(1.96)) + P(X - 1 \leq -5(1.96)) = P((X - 3)/5 \geq (5(1.96) - 2)/5) + P((X - 3)/5 \leq (-5(1.96) - 2)/5) = P(Z \geq 1.56) + P(Z \leq -2.36) = .0685$

6.10 (a)

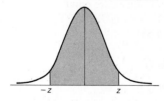

(i) $z = 1.960$
(c)

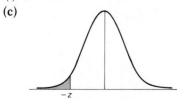

(i) $z = 2.326$

6.11 (a) $.80 < P(|Z| \leq 1.59) < .90$

6.12 (a) $z = 1.645$

6.13 (a) $P(X > 18000) = P(Z > (18000 - 15000)/1250) = .0082$

6.15 (a) If X = exchange rate and Y = dollars received, then $Y = 100X - 15$, which is normal, with $\mu_Y = 100(2.5) - 15 = 235$ and $\sigma_Y = 100\sigma_X = 8$. Thus, $P(230 < Y < 245) = P(-.62 < Z < 1.25) = .6268$.

6.17 (a) $P(|Z| > $ inner fence$) = P(|Z| > 3(.674)) = .0434$

Chapter 7

7.1 $\sigma_{\bar{X}} = 4/\sqrt{50} = .566$
$P(19.3 \leq \bar{X} \leq 20.7) = P(-1.24 \leq Z \leq 1.24) = .785$

7.3 (a) $\sigma_{\bar{X}} = .913$
$P(|\bar{X} - \mu| < .1) \approx P(|Z| < .1/.913) = P(|Z| < .11) = .0876$

7.4 (a) $P(-1.5 < t < 1.7) \approx P(-1.5 < Z < 1.7) = .8886$

7.5 $\sigma_{\hat{p}} = .05332$
$P(.24 < \hat{p} < .42) \approx P(-2.06 < Z < 1.31) = .8852$

7.7 (a) $p = .95$. The smaller of $np/(1 - p)$ and $n(1 - p)/p$ is $n(.05)/.95$. Thus $n \geq 9(.95)/.05 = 171$.

7.9 $\sigma_{\hat{p}} = \sqrt{(150 - 80)/(150 - 1)} \times \sqrt{.35(1 - .35)/80} = .03655$ $P(.24 < \hat{p} < .42) \approx P(-3.01 < Z < 1.92) = .9713$

7.11 (a) $P(|\bar{X} - \mu| < .5) \approx P(|Z| < .0625\sqrt{n})$. For this probability to be at least .95, we must have $.0625\sqrt{n} \geq 1.960$ or $n \geq (1.960/.0625)^2 = 983.45$. Rounding up, $n = 984$.

7.13 $p = .52$, so $P(.505 < \hat{p} < .535) \approx P(-.26 < Z < .26) = .2052$

Chapter 8

8.2 (a) $24.276 \leq \mu \leq 25.044$

8.5 $.6333 \leq \mu \leq .6421$. With 99% confidence, the true mean capacitance for the lot is between .6333 and .6421 μf.

8.8 $.3535 \leq p \leq .4465$. With 90% confidence, the proportion possessing property A is between 35.35% and 44.65%.

8.10 $.0168 \leq p \leq .1832$

8.12 $n = 79 + 117 = 196.$ $\hat{p} = 117/196 = .597.$ $.5283 \leq p \leq .6656.$ With 95% confidence, his correct-choice probability is between 52.8% and 66.6%.

8.14 (a) $.0045 \leq p \leq .0125.$ With 95% confidence, the incidence of AIDS for this city is between 4.5 and 12.5 per 1000 people.
  (b) $.0051 \leq p \leq .0139.$ The more exact 95% confidence limits are 5.1 and 13.9 per 1000.
  (c) The approximation (a) has shifted the interval slightly towards smaller proportions. Even so, the approximation is quite good.

8.15 $\mu \geq .6337.$ With 99% confidence, the true mean capacitance of the lot is at least .6337 μf. Since the confidence interval contains only values greater than .6, the data support a mean capacitance greater than .6 μf.

8.17 $p \geq .5393.$ With 95% confidence, the reporter's correct-choice probability exceeds 53.9%. Since this interval contains only values greater than 50%, it is reasonable to believe that his performance is better than using a fair coin to make the choices.

8.18 $n = (1.96 \times 12.34/.6)^2 = 1624.95,$ which rounds up to $n = 1625$

8.22 (a) $n = 1.960^2 \times .25(1 - .25)/.01^2 = 7203.$ Since this is already a whole number, $n = 7203.$

8.24 $n = 1.960^2 \times .55(1 - .55)/.005^2 = 38031.84,$ which rounds to $n = 38032.$ It is likely that the election staff will have to settle for a less ambitious length in order to gain a more realistic sample size.

8.26 (a) $.6342 \leq \mu \leq .6411$
  (b) $n = 97$ (I.e., 97% of the population must be sampled to achieve the prescribed interval length.)

8.29 With 99% confidence, between 47.38% and 52.82% of people in the city are in the age range from 18 to 35 years.

8.31 With 95% confidence, the mean consumption rate for nonresidential buildings is between 20.62 and 34.32.

8.33 With 95% confidence, the proportion aware of NPR is between 86.4% and 95.9%.

8.35 With 95% confidence, the mean travel distance for mice during winter months is between 9.27 and 20.79 meters.

8.37 (a) $\bar{Y} = 26.2121$ and $s_Y = 10.7453,$ so $\bar{X} = 24.82621,$ $s_X = .010745$
  (b) Uncoded 95% confidence limits: $L = 24.8236,$ $U = 24.8288$
  (c) Coded 95% confidence limits: $L = 23.6197,$ $U = 28.8045.$ The decoded limits are the same as those in part (b).

Chapter 9

9.1 $V_{obs} = (.43 - .5)/\sqrt{.5(1 - .5)/100} = -1.4.$ $P\text{-value} \approx P(|Z| \geq 1.4) = .162.$ The hypothesis is plausible.

9.5 $t_{obs} = (4.8 - 3.8)/(2.1/\sqrt{120}) = 5.216.$ P-value $< .000001.$ The hypothesis is extremely implausible.

9.8 $H_0: p = .5,$ $H_1: p \neq .5.$ Rule for 5% test: Reject H_0 if $|V| \geq 1.960,$ where $V = (\hat{p} - .5)/\sqrt{.5(1 - .5)/100}.$ Since $V_{obs} = -1.4,$ accept $H_0.$ Conclude at the 5% significance level that there is insufficient evidence to say that the coin is unfair.

9.12 $H_0: \mu = 12,$ $H_1: \mu \neq 12.$ Rule for 5% test: Reject H_0 if $|t| \geq 1.960,$ where $t = (\bar{X} - 12)/(s/\sqrt{150}).$ Since $t_{obs} = -7.265,$ reject $H_0.$ Conclude at the 5% significance level that the machine is not filling cans with an average of 12 fluid ounces. Because $\bar{X} = 11.79,$ the evidence suggests that the machine is underfilling cans, on the average.

9.13 (a) p = proportion of infested trees in the county.
  (b) From the agency's perspective, the most important error to avoid is to use the spray (conclude $p > .10$) when this action is unwarranted ($p \leq .10$). Thus, $H_0: p \leq .10,$ $H_1: p > .10.$
  (c) Rule for 1% test: Reject H_0 if $V \geq 2.326,$ where $V = (\hat{p} - .10)/\sqrt{.10(1 - .10)/500}.$
  (d) $\hat{p} = 74/500 = .148,$ so $V_{obs} = 3.578.$ Decision: Reject $H_0.$
  (e) $p \geq \hat{p} - 2.326\sqrt{\hat{p}(1 - \hat{p})/n} = .148 - 2.326\sqrt{.148(1 - .148)/500} = .111.$
  (f) Conclude at the 1% significance level that proportion of infested trees exceeds 10%, so the trees should be sprayed. With 99% confidence, the proportion of infested trees actually exceeds 11.1%.

9.17 (a) μ = mean weight loss in the first month for individuals who use the new diet pill.

(b) The consumer organization will take action if the advertiser's claim is false ($\mu < 15$). Thus,
$$H_0: \mu \leq 15$$
$$\text{vs } H_1: \mu < 15$$

(c) Reject H_0 if $t \geq -1.282$

(d) Since $-11.15 < -1.282$, reject H_0

(e) 90% confidence limits for μ are $L = \bar{X} - 1.645s/\sqrt{n} = 7.199$ and $U = \bar{X} + 1.645s/\sqrt{n} = 9.201$.

(f) Conclude at the 10% significance level that the evidence does not support the advertiser's claim. In fact, with 90% confidence, the actual average weight loss in the first month for users of this product is between 7.2 and 9.2 pounds.

9.19 $\hat{p} = 110/300 = .367$. Thus V_{obs} for the hypothesis $H_0: p = .28$ (versus H_1 $p \neq .28$) is 3.34. P-value $= P(|Z| \geq 3.34) = .0008$. Since $P < .01$, reject H_0 at the 1% significance level. Conclude that the proportion of students taking at least one business course has changed ($P = .0008$ for the hypothesis of no change). The evidence supports an increase in the proportion. This agrees with the Neyman-Pearson test outcome, for which the 1% test rule is to reject H_0 if $|V| \geq 2.576$.

9.23 The hypotheses are $H_0: \mu \leq 15$ and $H_1: \mu > 15$. Thus, the P-value is $P = P(Z \geq t_{obs}) = P(Z \geq -11.17) \approx 1.0$. The hypothesis that the mean weight gain does not exceed 15 pounds is (highly) plausible. The null hypothesis H_0 would be accepted at any significance level. This agrees with the Neyman-Pearson solution (see the solution to exercise 9.17 above).

9.25 Significance level $= 1 - .90 = .10$. Decision: Reject H_0, since $\mu = 69$ does not fall in the interval.

9.27 Since $p = .05$ falls in the interval, the hypothesis $H_0: p = .05$ (versus $H_1: p \neq .05$) would be accepted.

9.29 $\delta = 2.90$ for a one-sided test and minimum power .90. Thus, $n = (6.89 \times 2.90/(56 - 55))^2 = 399.2$, which rounds up to $n = 400$.

9.33 For $\pi = .95$, the two-sided test $\delta \approx 3.6$. Thus $B = (.354 \times 3.6/.05)^2 = 649.638$

For $N = 5000$,
$$n = 5000 \times B/(5000 + B - 1) = 575.040$$
Thus, the required finite population sample size is 576, whereas for an infinite population the sample size would be 650.

9.35 (a) p = actual probability that the interval will cover the true parameter value. The hypotheses to test are $H_0: p = .95$ versus $H_1: p \neq .95$. $V_{obs} = (.96 - .95)/\sqrt{.95(1 - .95)/100} = .46$ (rounded). $P = P(|Z| \geq .46) = .6456 \approx .65$.

9.37 $t_{obs} = (14.491 - 15.34)/(1.079/\sqrt{37}) = -4.786 \approx -4.8$. $P \approx .0000016$.

9.39 p = true correct prediction probability. $H_0: p = .5$, $H_1: p \neq .5$. Rule: Reject H_0 if $|V| \geq 1.960$. $n = 117 + 79 = 196$, so $\hat{p} = 117/196 = .597$ and $V_{obs} = 2.77$. Conclude at the 5% significance level that the true correct-prediction probability is different from .5. The evidence supports the contention that the reporter's predictions are better than could be made by using a fair coin.

9.41 (a) $n = 53$, $\bar{X} = 8.616$, $s = .749$.

(b) Rule: Reject H_0 if $|t| \geq 1.960$. $t_{obs} = (8.616 - 8.798)/(.749/\sqrt{53}) = -1.768$. Decision: Accept H_0. Conclude at the 5% significance level that there is insufficient evidence to claim that the currently accepted parallax value is inconsistent with Short's data.

9.43 μ = mean assembly time using the new tool. $H_0: \mu \geq 13.9$ versus $H_1: \mu < 13.9$. Rule: Reject H_0 if $t \leq -1.645$. Since $t_{obs} = -4.714$, H_0 is rejected. The one-sided 95% confidence interval is $\mu \leq \bar{X} + 1.645s/\sqrt{n} = 13.1 + 1.645(1.2)/\sqrt{50} = 13.379$. Conclude at the 5% significance level that the mean assembly time using the new tool is shorter than that using the old one. With 95% confidence, the time is at least $13.9 - 13.379 = .521$ minutes shorter per assembly, on the average.

9.45 $H_0: \mu = 6.04$ versus $H_1: \mu \neq 6.04$. $t_{obs} = (9.10 - 6.04)/(2.150/\sqrt{128}) = 16.102$. P-value $< .000001$. The hypothesis that the mean arteriole thickness for this SIDS victim is the same as that for a normal child is highly implausible ($P < .000001$). The evidence supports a larger than normal average thickness for the SIDS victim.

9.47 $H_0: \mu \le 15$ versus $H_1: \mu > 15$. $t_{obs} = (15.3 - 15)/(2.5/\sqrt{150}) = 1.47$. P-value $= P(Z \ge 1.47) = .5 - .4292 = .0708$. Since $P > .05$, the hypothesis H_0 is accepted. Conclude that there is insufficient evidence $(P = .07)$ to support the contention that motorists are exceeding the speed limit, on the average.

Chapter 10

10.1 (a) $H_0: \mu = .5$ versus $H_1: \mu \ne .5$. The value of the one-sample t statistic is $t = -1.732$, and the P-value is between .05 and .10. Consequently, H_0 is accepted, and we conclude at the 5% significance level that the evidence does not support a mean fill time different from .5 seconds.
(b) With 95% confidence, the mean fill time is between .397 and .509 seconds.

10.3 (a) $n = 11$, $\bar{X} = 51.273$, $s = 8.259$. Computed $t = 0.511$ and $P > .5$.
(b) With 95% confidence, the mean age at first transplant is between 45.72 and 56.82 years.

10.5 (a) μ = true mean completion time in hours for 100-milers. $H_0: \mu \ge 8.74$ versus $H_1: \mu < 8.74$.
(b) The one-sided P-value, $P = P(t \le t_{obs}) = P(t \le -2.404) < .025$ (for 16 degrees of freedom). Thus, the null hypothesis is rejected and we conclude at the 10% significance level that the 100-milers are speedier, on the average.
(c) With 90% confidence $\mu \le 8.408$ hours. Thus, the 100-milers have completion times at least $8.74 - 8.408 = .332$ hours shorter than the 50-milers, on the average.

10.7 (a)

VAR. NAME	SAMPLE SIZE	MEDIAN	QUARTILES	ADJACENTS	IQR
Glucagon	10	122.5	60.0 183.0	40.0 270.0	123.0

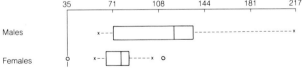

Aside from a slight skewness and short-tailedness, the data appear well suited for the analysis of this chapter.
(b) μ = mean glucagon level for children given this particular course of treatments. Since the most serious error is to discontinue treatments (conclude $\mu < 200$) when they should not be discontinued $(\mu \ge 200)$, the appropriate hypotheses are $H_0: \mu \ge 200$ and $H_1: \mu < 200$.
(c) $n = 10$, $\bar{X} = 135.1$, and $s = 76.604$,

yielding $t = -2.679$. The P-value is $P(t \le -2.679) < .025$. Consequently, we conclude at the 5% level that the mean glucagon level for treated children is less than 200.
(d) The 95% upper confidence limit is 179.50. Thus, with 95% confidence, the mean glucagon level for the treated children is at least $200 - 179.50 = 20.50$ units below the normal level of 200 units.

10.8 (a)

VAR. NAME	SAMPLE SIZE	MEDIAN	QUARTILES	ADJACENTS	IQR
Males	14	118.5	70.0 135.0	59.0 217.0	65.0
Females	18	77.0	66.0 84.0	56.0 101.0	18.0
Minor outliers	35	112			

The difference in dispersions and minor outliers for females suggest that a backup procedure (chapter 12) should be run to support the classical analysis for this problem.

(b) The summary data are as follows.

	MALES	FEMALES
n	14	18
$\bar{X}$	112.5	75.83
s	42.755	17.236

The two-sample t statistic is $t = 3.320$, with two-sided test P-value $< .005$.

(c) The 95% two-sided confidence limits are 14.12 and 59.22. Thus, we would conclude with 95% confidence that the mean androstenedione level for diabetic men exceeds that for diabetic women by an amount between 14.12 and 59.22 units.

10.10 (a) $H_0: \mu_S \leq \mu_N$ versus $H_1: \mu_S > \mu_N$. Since the value of the two-sample t statistic is 6.714, the one-sided P-value is less than .0005. The null hypothesis is rejected at the 5% level. Conclude at the 5% significance level that the mean viewspread for sites exceeds that for nonsites.

(b) The 95% lower (one-sided) confidence bound is 93.75. With 95% confidence conclude that the mean viewspread for sites exceeds that for nonsites by at least 93.75 degrees.

10.12 (a) If μ_2 is taken to be the mean algebra score for students making 14 or less on the arithmetic test (to avoid lower one-

sided tests and confidence intervals) and μ_1 is the mean score for those with arithmetic scores of 15 or above, then the hypotheses are $H_0: \mu_1 \leq \mu_2$ versus $H_1: \mu_1 > \mu_2$. The value of the two-sample t statistic is 5.275, which yields a one-sided P-value less than 0.005. The null hypothesis would be rejected at the 5% level. However, a stronger statement is also appropriate: We conclude that the sample evidence strongly supports $(P < .0005)$, the hypothesis that the mean algebra score for the poorer arithmetic students is less than that for the better ones.

(b) The lower 95% confidence limit for $\mu_1 - \mu_2$ is 2.98. Thus, with 95% confidence, the mean algebra score for the better arithmetic students exceeds that for the poorer ones by at least 2.98 points.

10.14 The hypotheses implied by this exercise are $H_0: \mu_S - \mu_N \leq 90$ versus $H_1: \mu_S - \mu_N > 90$. The null hypothesis would be rejected at the 5% significance level if none of the differences specified by H_0 lie in the 95% one-sided confidence interval obtained in the solution to exercise 10.10, above. The values in that interval are $\mu_S - \mu_N \geq 93.7$. Since no difference smaller than 90 falls in this interval, H_0 is rejected. More simply, if the evidence supports (at the 95% confidence level) a difference of 93.7 degrees or more, it also supports a difference larger than 90 degrees.

10.15 (a)

VAR. NAME	SAMPLE SIZE	MEDIAN	QUARTILES	ADJACENTS	IQR
Difference	10	-1.25	-1.80 -1.00	-2.40 0.20	0.80
Major outliers	-4.6				

The only interesting feature is the extreme outlier. However, this feature will make the support of a backup procedure (chapter 12) highly desirable for this exercise.

(b) The hypotheses are $H_0: \mu_A = \mu_B$ versus $H_1: \mu_A \neq \mu_B$. The value of the paired-difference (A minus B) t statistic is t

$= -3.78$, leading to a P-value of less than .005. A difference in mean sleep gain is supported by the data at the 5% level, with a larger mean gain achieved with remedy B.

(c) With 95% confidence, the mean gain in sleep for remedy B exceeds that for remedy A by between .61 and 2.43 hours.

10.17 (a)

VAR. NAME	SAMPLE SIZE	MEDIAN	QUARTILES	ADJACENTS	IQR
Difference	12	−0.060	−0.225	−0.490	0.305
			0.080	0.420	
Minor outliers	0.72				

−0.490	−0.248	−0.006	0.236	0.478	0.720

Difference

The sample contains one minor outlier. A robust backup is desirable, but will not change the test outcome for this problem.

(b) The value of the paired difference t statistic is −.243. The P-value for the two-sided t test exceeds .5. There is no evidence of a difference in rates at any interesting significance level.

10.20 (a) The 95% confidence interval for the difference in mean heights is $-.011 \leq \mu_C - \mu_S \leq 5.211$.

10.25 (a)

VAR. NAME	SAMPLE SIZE	MEDIAN	QUARTILES	ADJACENTS	IQR
Summer	12	31.60	19.45	6.70	28.05
			47.50	80.00	
Nonsummer	14	14.65	10.50	3.40	13.30
			23.80	40.00	
Major outliers	66.7				

3.4	18.7	34.0	49.4	64.7	80.0

Summer

Nonsummer

The outlier in the nonsummer data, plus the difference in dispersion.

(b) The value of the two-sample t statistic is $t = 2.022$, leading to a P-value between .05 and .10 for the two-sided test. The hypothesis of no difference in mean storm intensities would be accepted at the 5% level.

(c) With 95% confidence, the difference in 5-minute rainfall percentages for summer and nonsummer storms is between −.32 and 30.84.

10.27 Since the data are paired by subject, a paired difference test is appropriate. The

(b) Since the interval of part (a) contains 0, the hypothesis of no difference in mean heights would be accepted at the 5% significance level.

10.23 μ = "before" minus "after" mean tension difference. The value of the one-sample t statistic for testing the hypotheses $H_0: \mu \leq 0$ versus $H_1: \mu > 0$ is $t = -.670$. The P-value, $P(t \geq t_{obs})$, exceeds .5, consequently we conclude at the 5% significance level that there is no evidence of a reduction in tension due to the swimming therapy.

value of the paired t statistic is 2.118 for test 1 minus test 2 differences. The P-value is less than .10, so the hypothesis of no difference in tests is rejected at the 10% level. The evidence favors a higher mean percentage agreement rate for test 1.

10.29 (a) The value of the paired-difference t statistic is −.920, leading to a (two-sided test) P-value greater than .10. We conclude at the 10% significance level that there is insufficient evidence to conclude a difference in mean consumption rates exists.

10.31 (a)

VAR. NAME	SAMPLE SIZE	MEDIAN	QUARTILES	ADJACENTS	IQR
Lengthwise	10	0.895	0.870 0.930	0.840 0.950	0.060
Major outliers	1.15				
Crosswise	10	0.775	0.690 0.800	0.670 0.890	0.110
Minor outliers	0.46				

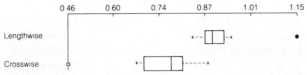

Outliers in both samples suggest the need for a backup procedure.

(b) Note that there is no natural pairing of data in this problem. The two-sample t test is the appropriate procedure. The value of the two-sample t statistic is 3.746, and the two-sided test P-value is smaller than .002.

(c) With 90% confidence, the mean strength for lengthwise-cut specimens exceeds that for cross-cut specimens by an amount between .0945 and .2575 foot pounds.

Chapter 11

11.1

Summary statistics

VARIABLE	SAMPLE SIZE	MEAN	STANDARD DEVIATION
Statistics	8	136.5000	46.8463
Applied math	11	110.0909	27.9587
Pure math	15	124.4000	51.9145

ANOVA table

SOURCE	SS	df	MS	F
Between	3315.520	2	1657.7602	0.844
Within	60910.509	31	1964.8551	
Total	64226.029	33		

5% test rule: Reject H_0 if $F \geq 3.32$ (using degrees of freedom 2 and 30). Decision: Accept H_0. Conclude at the 5% significance level that there is insufficient evidence in the sample to support differences in average credit hours for the three faculty groups.

11.2 $P > .25$

11.4 (a)

VAR. NAME	SAMPLE SIZE	MEDIAN	QUARTILES	ADJACENTS	IQR
Week 0	9	31.20	27.20 33.40	26.40 40.20	6.20
Minor outliers	17.6				
Major outliers	72.0				
Week 1	9	22.20	20.00 24.00	18.40 24.00	4.00
Minor outliers	32.2				
Major outliers	7.8	37.2			
Week 2	9	55.20	40.00 65.60	15.80 80.40	25.60
Week 3	9	52.00	40.60 62.00	22.40 69.20	21.40

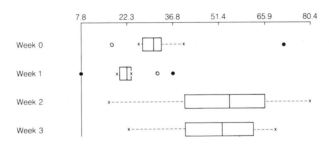

There appears to be an initial drop from week 0 to week 1, then a sharp rise from week 1 to week 2, followed by a leveling off of the median insulin value.

(b) The trend differs considerably from that seen for lean rats, in which the insulin levels decrease over the three-week period. Outliers in weeks 0 and 1, which indicate a violation of the normality assumption, plus rather large dispersion differences from week to week, which suggest that the equal variation condition is violated, are likely to complicate the analysis.

11.5 The FSD groupings at the 5% significance level are

{week 0, week 1} < {week 2, week 3}

Thus, the analysis supports the strong rise in insulin levels from week 1 to week 2 seen in the schematic diagram.

11.6 (a)

VAR. NAME	SAMPLE SIZE	MEDIAN	QUARTILES	ADJACENTS	IQR
Caucasian	12	5.750	5.375 6.000	4.500 6.750	0.625
Minor outliers	7.75				
Black	13	6.500	6.000 7.000	5.000 8.000	1.000
Native American and Oriental	14	6.000	5.000 6.250	4.000 8.000	1.250

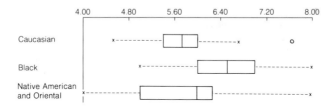

Except for a single outlier, the data appear to be in reasonable conformity with the assumptions of the analysis. The Glabella thicknesses for blacks appears to be somewhat larger than for the other ethnic groups, on the average.

(b)

Summary statistics

VARIABLE	SAMPLE SIZE	MEAN	STANDARD DEVIATION
Caucasian	12	5.8125	0.8334
Black	13	6.4615	0.8947
Native American and Oriental	14	5.8571	1.1168

ANOVA table

SOURCE	SS	df	MS	F
Between	3.398	2	1.6991	1.828
Within	33.461	36	0.9295	
Total	36.859	38		

$.10 < P < .25$. Conclude at the 5% significance level that there is insufficient evidence to support a difference in Glabella thicknesses for the three ethnic groups.

11.7 The 95% confidence limits for the difference in means for group 2 and (minus) group 1 are $L = 10.89$ and $U = 32.57$ units.

With 95% confidence, the mean rise in insulin levels between the groups representing the first two weeks and the second two weeks is between 10.89 and 32.57 units.

11.9

Summary statistics

VARIABLE	SAMPLE SIZE	MEAN	STANDARD DEVIATION
Anthro	8	57.0000	1.3093
Math	6	56.0000	1.4142
Psych	6	60.0000	2.0000

ANOVA table

SOURCE	SS	df	MS	F
Between	52.800	2	26.4000	10.686
Within	42.000	17	2.4706	
Total	94.800	19	P-value < 0.001	

Fisher's least significant difference test

Critical value = 2.110
LSD Groupings:
{Anthro, Math} < {Psych}

The data support a rather strong difference in conservatism among the three groups ($P < .001$). At the 5% significance level, psychology students appear to be more strongly conservative, on the average, than mathematics and anthropology students. No significant difference is seen between the latter two majors.

11.11 (a)

VAR. NAME	SAMPLE SIZE	MEDIAN	QUARTILES	ADJACENTS	IQR
Fall	27	15.0	0.0 18.0	0.0 39.0	18.0
Minor outliers	46				
Winter	34	15.0	0.0 15.0	0.0 34.0	15.0
Minor outliers	47				
Major outliers	87				
Spring	17	18.0	15.0 34.0	0.0 47.0	19.0
Major outliers	109				
Summer	27	21.0	15.0 33.0	0.0 60.0	18.0
Minor outliers	71				
Major outliers	106	150			

Summary statistics

VARIABLE	SAMPLE SIZE	MEAN	STANDARD DEVIATION
Fall	27	13.0370	13.3747
Winter	34	15.0294	17.1331
Spring	17	26.9412	26.0636
Summer	27	30.8148	33.4504

ANOVA table

SOURCE	SS	df	MS	F
Between	6099.299	3	2033.0996	3.782
Within	54298.949	101	537.6134	
Total	60398.248	104	.01 < P-value < .05	

Fisher's least significant difference test

Critical value = 1.980
LSD Groupings:
{fall, winter, spring} < {spring, summer}

A difference in seasonal activity is supported by the data at the 5% significance level. The trend appears to be an increase in activity, on the average, as the weather warms up from winter to summer. At the 5% level, the FSD method supports the difference seen in the schematic diagram between summer and {fall and winter}, but it is unable to distinguish either group from the transitional spring season.

(b) There are many outliers, which strongly suggests that a robust backup method (chapter 12) should be used to determine whether any further information can be gleaned from the data.

11.13

VAR. NAME	SAMPLE SIZE	MEDIAN	QUARTILES	ADJACENTS	IQR
Method 1	5	8.30	7.80 8.30	7.60 8.40	0.50
Method 2	3	7.10	6.25 7.25	5.40 7.40	1.00
Method 3	2	7.25	6.40 8.10	6.40 8.10	1.70
Method 4	3	9.50	8.70 9.75	7.90 10.00	1.05
Method 5	1	7.10	7.10 7.10	7.10 7.10	0.00

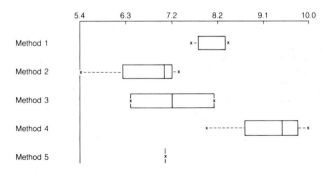

Summary statistics

VARIABLE	SAMPLE SIZE	MEAN	STANDARD DEVIATION
Method 1	5	8.0800	0.3564
Method 2	3	6.6333	1.0786
Method 3	2	7.2500	1.2021
Method 4	3	9.1333	1.0970
Method 5	1	7.1000	0.0000

ANOVA table

SOURCE	SS	df	MS	F
Between	10.903	4	2.7257	3.669
Within	6.686	9	0.7429	
Total	17.589	13		P-value $< .05$

Fisher's least significant difference test

Critical value = 2.262
LSD Groupings:
{method 2, method 3, method 5} < {method 1, method 3, method 4, method 5}

The schematic diagram suggests that methods 1 and 4 are distinctly better than the other preservation methods. The data support a difference in methods at the 5% significance level, as seen from the ANOVA.

However, the LSD grouping shows that there is only enough information in the data to distinguish method 2 from methods 1 and 4 at the 5% level.

11.15 (a) The schematic diagram is repeated here:

VAR. NAME	SAMPLE SIZE	MEDIAN	QUARTILES	ADJACENTS	IQR
Housewife	11	2.00	1.30	0.80	1.00
			2.30	2.30	
Minor outliers	3.9	4.2			
Office	11	2.20	2.05	1.80	0.30
			2.35	2.50	
Major outliers	1.1	5.9			
Professional	24	3.40	2.80	1.20	2.25
			5.05	7.80	

Summary statistics

VARIABLE	SAMPLE SIZE	MEAN	STANDARD DEVIATION
Housewife	11	2.1182	1.0889
Office	11	2.4273	1.2125
Professional	24	3.7875	1.6337

ANOVA table

SOURCE	SS	df	MS	F
Between	26.863	2	13.4314	6.567
Within	87.944	43	2.0452	
Total	114.807	45	P-value $< .01$	

Fisher's least significant difference test

Confidence coefficient = 95.0%
Critical value = 2.000
LSD Groupings:
{housewife, office} < {professional}

The rise in median testosterone level with occupational categories (ranked in order of required "drive") from office worker to professional is supported by the FSD procedure at the 5% level.

(b) The outliers and differences in dispersion.

11.17 (a)

VAR. NAME	SAMPLE SIZE	MEDIAN	QUARTILES	ADJACENTS	IQR
50 miles	9	9.00	9.00	8.70	0.30
			9.30	9.30	
Minor outliers	8.1	10.2			
Major outliers	11.2				
75 miles	5	8.80	8.00	7.20	1.50
			9.50	11.30	
100 miles	15	12.40	11.25	8.10	4.35
			15.60	21.10	

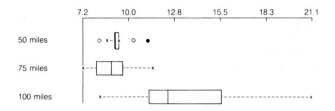

Summary statistics

VARIABLE	SAMPLE SIZE	MEAN	STANDARD DEVIATION
50 miles	9	9.2778	0.9066
75 miles	5	8.9600	1.5662
100 miles	15	13.6200	3.6768

ANOVA table

SOURCE	SS	df	MS	F
Between	144.091	2	72.0453	9.108
Within	205.652	26	7.9097	
Total	349.742	28	P-value < .001	

Fisher's least significant difference test

Critical value = 1.706
LSD Groupings:
{50 miles, 75 miles} < {100 miles}

We see that $P < .001$. At the 10% significance level, the 100-milers are faster than the 50- and 75-milers, on the average.

(b) The normality and equal variation assumptions. The extreme differences in dispersion plus the outliers in the 50-mile group suggest that it would be appropriate to back up this analysis with a robust method.

11.19

VAR. NAME	SAMPLE SIZE	MEDIAN	QUARTILES	ADJACENTS	IQR
Wheat	7	4.20	3.65 4.50	3.00 5.00	0.85
Rice	11	7.80	7.20 8.20	6.70 8.70	1.00
Minor outliers	5.6				
Canna	11	10.40	8.90 11.90	6.30 12.50	3.00
Corn	8	6.65	6.25 7.25	5.80 8.00	1.00
Potato	10	11.90	10.70 13.00	9.70 13.80	2.30
Dasheen	8	6.40	5.90 6.80	5.30 7.10	0.90
Sweet potato	4	9.20	8.30 10.00	7.60 10.60	1.70

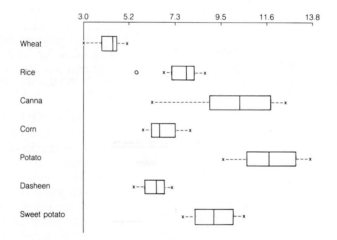

Summary statistics

VARIABLE	SAMPLE SIZE	MEAN	STANDARD DEVIATION
Wheat	7	4.0714	0.6921
Rice	11	7.5909	0.8916
Canna	11	10.1364	2.0791
Corn	8	6.7625	0.7230
Potato	10	11.7900	1.4224
Dasheen	8	6.3250	0.6159
Sweet potato	4	9.1500	1.2369

ANOVA table

SOURCE	SS	df	MS	F
Between	341.768	6	56.9613	35.617
Within	83.162	52	1.5993	
Total	424.929	58	P-value $< .001$	

Fisher's least significant difference test

Critical value = 2.000
LSD Groupings:
{Wheat} < {Corn, Dasheen} < {Rice, Corn} < {Canna, Sweet Potato} < {Potato}

The data strongly support differences in film thicknesses for the different starches ($P <$ .001). At the 5% significance level, the FSD groupings suggest that potato starch film is thickest, on the average, and wheat starch film the thinnest. The groupings support those seen in the schematic diagram.

11.21 (a)

Summary statistics

VARIABLE	SAMPLE SIZE	MEAN	STANDARD DEVIATION
With	5	16.4000	2.9351
Without	5	15.9600	2.8841

ANOVA table

SOURCE	SS	df	MS	F
Between	0.484	1	0.4840	0.057
Within	67.732	8	8.4665	
Total	68.216	9		

(b) Two-sample $t = .239$ and $.239^2 = .057$.

Chapter 12

12.1 (a)

VAR. NAME	SAMPLE SIZE	MEDIAN	QUARTILES	ADJACENTS	IQR
Tooth measurement	20	37.5	36.5 39.0	34.0 40.0	2.5
Major outliers	50				

A 5% trimming fraction will remove the one outlier.

(b) The 95% two-sided confidence limits are $L = 36.47$ and $U = 39.53$, yielding an interval length of 3.06.

(c) The 95% confidence limits for the 5% trimmed mean are $L = 39.69$ and $U = 38.42$. Interval length $= 1.73$.

(d) By trimming, the interval length has been cut almost in half.

12.3 (a)

VAR. NAME	SAMPLE SIZE	MEDIAN	QUARTILES	ADJACENTS	IQR
Glabella	20	5.750	5.000 6.000	4.000 7.000	1.000
Minor outliers	2.00	8.00 8.25			
Major outliers	9.50				

The data contain four outliers. A 15% trimming fraction will account for them.

(b) The 15% trimmed t statistic $= 3.740$, $P < .005$, thus it would be concluded at the 5% significance level that the (trimmed) mean Glabella measurement for Caucasians differs from 4.75, with evidence favoring a larger value.

(c) The 95% trimmed mean confidence limits are $L = 5.127$ and $U = 6.159$. Thus, with 95% confidence, evidence favors a (trimmed) mean Glabella value that exceeds 4.75 by an amount between .38 and 1.41 mm.

12.6 (a) The value of the two-sample (Special minus Control) trimmed t statistic $= -2.920$, and the P value is less than .005.

(b) The 95% confidence limits for Special minus Control trimmed mean difference are $L = -5.622$ and $U = -1.066$. Thus, with 95% confidence, the (trimmed) mean correctness score for Control students exceeds that for Special students by from 1.07 to 5.62 points.

12.8 (a) The P-value is less than .01, based on the classical (uncorrected) test.

(b) The approximate degrees of freedom using the approximation are 47, and the value of the t statistic is 2.348, yielding a P-value larger than .02. The classical test gives an overly sensitive (invalid) test result for these data.

(c) Because the sample size occurring with the smaller standard deviation is much larger than the other sample size.

12.12 The 20% trimmed mean procedure summary is as follows.

Trimmed mean summary statistics

VARIABLE	SAMPLE SIZE	MEAN	STANDARD DEVIATION
Year 1	15	44.0667	12.4366
Year 2	15	45.0667	6.2276
Year 3	15	45.9333	12.4366

12.14 (a)

VAR. NAME	SAMPLE SIZE	MEDIAN	QUARTILES	ADJACENTS	IQR
Week 0 minus week 1	9	8.20	7.20 11.20	4.20 12.80	4.00
Major outliers	34.8				

A major outlier occurs among the differences.

(b) The classical 95% paired-difference confidence limits are $L = 4.272$ and $U = 18.394$. Length $= 14.122$.

(c) The 5% trimmed mean 95% confidence limits are $L = 6.082$ and $U = 11.918$,

ANOVA table

SOURCE	SS	df	MS	F
Between	26.178	2	13.0889	0.113
Within	4873.680	42	116.0400	
Total	4899.858	44	P-value > 0.25	

Note that the P-value exceeds .25, whereas the P-value for the outlier-deleted data was smaller than .01. The use of the large trimming fraction has greatly desensitized the test. This suggests that when large trimming fractions are required to eliminate outliers, another robust procedure may be preferred.

length $= 5.836$. At the 5% significance level, it would be concluded that there is a difference in average insulin levels, with the levels for week 1 exceeding those for week 0 by between 6.08 and 11.92 units on the average.

12.16 (a) A line through the points (ln(median), ln(iqr)) has slope close to .75. The slope of the least-squares line, discussed in chapter 14, is .766, and $1 - .75 = .25$.

The schematic diagram for the data power transformed with index .25 is as follows.

VAR. NAME	SAMPLE SIZE	MEDIAN	QUARTILES	ADJACENTS	IQR
Walker	10	0.56234	0.41618	0.31623	0.27634
			0.63252	0.91141	
Lombard	10	0.82997	0.78514	0.78514	0.09862
			0.88376	1.00000	
Minor outliers	0.5623	0.6223			
Uwet	9	0.70711	0.67695	0.63246	0.15104
			0.82799	1.04664	
Quillagua	10	0.82711	0.75212	0.62233	0.12799
			0.88011	0.88011	
Minor outliers	1.0878	1.2179			
Coahuila	11	1.34781	1.24789	1.08776	0.23610
			1.48399	1.69221	
Tocopilla	12	1.38682	1.29341	1.10668	0.19575
			1.48917	1.60886	
Hex River	7	1.64933	1.42720	1.31607	0.23836
			1.66556	1.77828	
Minor outliers	0.8944				

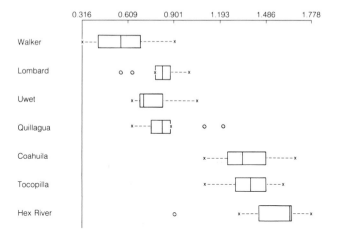

(b)

Classical ANOVA
Summary statistics

VARIABLE	SAMPLE SIZE	MEAN	STANDARD DEVIATION
Walker	10	0.5834	0.2030
Lombard	10	0.8135	0.1344
Uwet	9	0.7749	0.1589
Quillagua	10	0.8538	0.1795
Coahuila	11	1.3609	0.1856
Tocopilla	12	1.3839	0.1458
Hex River	7	1.5011	0.3047

ANOVA table

SOURCE	SS	df	MS	F
Between	7.617	6	1.2694	36.422
Within	2.161	62	0.0349	
Total	9.778	68	P-value < .001	

Fisher's least significant difference test
Significance level = 5%
Critical value = 2.000
LSD Groupings:
{Walker} < {Lombard, Uwet, Quillagua} < {Coahuila, Tocopilla, Hex River}

Both the schematic diagram and FSD procedure suggest three distinct groupings. (It should be noted that the three smallest Walker County measurements were marked questionable in the original data, and it is likely that the two lower groupings should actually be combined.)

(c)

VAR. NAME	SAMPLE SIZE	MEDIAN	QUARTILES	ADJACENTS	IQR
Walker	10	−2.30259	−3.50656 −1.46968	−4.60517 −0.37106	2.03688
Lombard	10	−0.74583	−0.96758 −0.49430	−0.96758 0.00000	0.47329
Minor outliers	−2.3026	−1.8971			
Uwet	9	−1.38629	−1.56065 −0.75502	−1.83258 0.18232	0.80563
Quillagua	10	−0.76071	−1.13943 −0.51083	−1.89712 0.33647	0.62861
Minor outliers	0.7885				
Coahuila	11	1.19392	0.87547 1.57893	0.33647 2.10413	0.70346
Tocopilla	12	1.30797	1.02898 1.59069	0.40547 1.90211	0.56171
Hex River	7	2.00148	1.41069 2.04046	1.09861 2.30259	0.62977
Minor outliers	−0.4463				

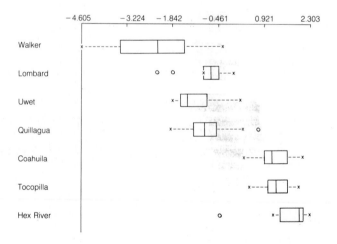

Summary statistics

VARIABLE	SAMPLE SIZE	MEAN	STANDARD DEVIATION
Walker	10	−2.3782	1.4196
Lombard	10	−0.8799	0.7132
Uwet	9	−1.0886	0.7676
Quillagua	10	−0.7067	0.8003
Coahuila	11	1.1982	0.5514
Tocopilla	12	1.2790	0.4281
Hex River	7	1.5372	0.9550

ANOVA table

SOURCE	SS	df	MS	F
Between	130.429	6	21.7382	30.825
Within	43.724	62	0.7052	
Total	174.153	68	P-value < 0.001	

Fisher's least significant difference test
Significance level = 5%
Critical value = 2.000
LSD Groupings:
{Walker} < {Lombard, Uwet, Quillagua} < {Coahuila, Tocopilla, Hex River}

The log transformation does as well as the $q = .25$ power transform in distinguishing groupings in the data. The log transformation has gone somewhat too far, however, as seen from the out-of-line dispersion of the Walker County data. This has had only a modest effect on the value of the F statistic.

12.18 The value of the mill A minus mill C two-sample t statistic for the ranked data is $t = -1.815$, which yields a two-sided test P-value between .05 and .10. With a 20% trimming fraction, the two-sample trimmed t statistic is $t = -2.363$, with P-value $< .05$. The trimmed mean method appears to provide more sensitivity here.

12.21

Summary statistics for ranked data

VARIABLE	SAMPLE SIZE	MEAN	STANDARD DEVIATION
Walker	10	11.3000	11.6242
Lombard	10	23.8500	10.0859
Uwet	9	20.3333	10.7209
Quillagua	10	25.7000	10.9473
Coahuila	11	52.4545	8.7391
Tocopilla	12	53.0833	7.5523
Hex River	7	58.5000	13.1466

ANOVA table

SOURCE	SS	df	MS	F
Between	20802.131	6	3467.0218	32.778
Within	6557.869	62	105.7721	
Total	27360.000	68	P-value < 0.001	

Fisher's least significant difference test

Significance level = 5%
Critical value = 2.000
LSD Groupings:
{Walker, Uwet} < {Lombard,   Uwet,   Quillagua}   < {Coahuila, Tocopilla, Hex River}

The rank analysis provides much the same result as the power transformation method, except that it doesn't separate Walker County from Uwet in the FSD groupings at the 5% level.

12.23 The value of the paired difference t statistic for the ranked data is $t = 4.711$, which yields a P-value $< .002$. This gives the same conclusion as the classical and trimmed t tests at the 5% significance level.

12.25 The value of the signed-rank t statistic is $t = 5.477$, with P-value $< .001$. This test produces the same conclusion at the 5% level as does the method based on differences of ranks.

12.27 The two-sample trimmed t statistic for a 5% trimming fraction is $t = 2.744$, with $.01 < P$-value $< .02$. The two-sample t statistic based on ranks is $t = 2.246$, with $.02 < P$-value $< .05$.

12.29 (a)

VAR. NAME	SAMPLE SIZE	MEDIAN	QUARTILES	ADJACENTS	IQR
Anglo	44	42.0	27.0 52.5	19.0 90.0	25.5
Hispanic	34	29.0	22.0 48.0	14.0 76.0	26.0
Native American	15	23.0	20.5 25.5	16.0 26.0	5.0
Minor outliers	35	36			
Major outliers	48				

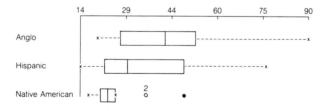

The schematic diagram suggests different averages for the three groups. However, differences in dispersion and outliers will complicate the analysis.

(b)

Summary statistics

VARIABLE	SAMPLE SIZE	MEAN	STANDARD DEVIATION
Anglo	44	41.6591	16.8177
Hispanic	34	35.0588	16.3798
Native American	15	25.0667	8.5060

ANOVA table

SOURCE	SS	df	MS	F
Between	3202.094	2	1601.0468	6.541
Within	22028.702	90	244.7634	
Total	25230.796	92	P-value < 0.01	

Fisher's least significant difference test
Significance level = 5%
Critical value = 2.000
LSD Groupings:
{Native American} < {Anglo, Hispanic}

The standard analysis separates the Native Americans from Anglos and Hispanics at the 5% level but is unable to separate Anglos and Hispanics.

(c)

Summary statistics for ranked data

VARIABLE	SAMPLE SIZE	MEAN	STANDARD DEVIATION
Anglo	44	56.1023	23.9062
Hispanic	34	44.3235	28.4730
Native American	15	26.3667	19.5196

ANOVA table

SOURCE	SS	df	MS	F
Between	10275.036	2	5137.5179	8.160
Within	56662.464	90	629.5829	
Total	66937.500	92	P-value $< .001$	

Fisher's least significant difference test
Significance level = 5%
Critical value = 2.000
LSD Groupings:
{Native American} < {Hispanic} < {Anglo}

The rank analysis has no difficulty separating the three ethnic groups into separate categories at the 5% significance level.

(d) The power transformation fitting method yields a power of about $q = -1.5$. The fitting method does not work as well as it should on these data; the power transform with this index goes "too far." That is, the tendency for power transformations to magnify differences in values near 0 causes the data for Hispanics and Native Americans to "blur together." The less extreme reciprocal ($q = -1$) and log ($q = 0$) transformations both give results identical to the rank method. The summary for the reciprocal transformation is as follows.

Summary statistics

VARIABLE	SAMPLE SIZE	MEAN	STANDARD DEVIATION
Anglo	44	−0.0280	0.0110
Hispanic	34	−0.0349	0.0155
Native American	15	−0.0433	0.0114

ANOVA table

SOURCE	SS	df	MS	F
Between	0.0028	2	.00141	8.511
Within	0.0149	90	.00017	
Total	0.0178	92	P-value $< .001$	

Fisher's least significant difference test
Significance level = 5%
Critical value = 2.000
LSD Groupings:
{Native American} < {Hispanic} < {Anglo}

(e) The rank and power transformation methods have sensitized the analysis in the sense of breaking the grouping of Anglos and Hispanics made by the classical method.

12.31 The 10% trimmed mean t value is $t = -.872$, with P-value > 0.25. Consequently, there is no evidence of a difference in average prices for the two years.

Chapter 13

13.1 (a)

Goodness of fit table

DAY OF ACCIDENT	OBS. FREQ.	HYP. PROP.	EXP. FREQ.	z	SYMBOL
Mon.	37	0.200	22.8	2.974	@
Tues.	14	0.200	22.8	−1.843	o
Wed.	18	0.200	22.8	−1.005	.
Thurs.	25	0.200	22.8	0.461	.
Fri.	20	0.200	22.8	−0.586	.

Sample size = 114
Degrees of freedom = 4
$X^2 = 13.807$

(b) $P < .01$. At the 5% significance level, we conclude that accidents are not equally likely for the five work days of the week.

(c) The indication is of a higher proportion of accidents on Monday. This is most likely a "good weekend" effect.

(d) 95% confidence limits for the true proportion of Monday accidents are $L =$ 23.9% and $U = 41.1\%$. With 95% confidence, the true proportion exceeds the hypothetical value of 20% by an amount between 3.9% and 21.1%.

13.3

Goodness of fit table

SUICIDE AGE	OBS. FREQ.	HYP. PROP.	EXP. FREQ.	z	SYMBOL
20–29	34	0.253	31.9	0.376	.
30–39	26	0.201	25.3	0.134	.
40–49	23	0.148	18.6	1.008	.
50–59	22	0.155	19.5	0.559	.
over 59	21	0.243	30.6	−1.738	o

Sample size = 126
Degrees of freedom = 4
$X^2 = 4.509$

The P-value for the test satisfies the inequality (for 4 degrees of freedom): $.25 < P < .50$. There is no evidence that the suicide rates in the different age groups differ from the corresponding population proportions.

13.5 (a)

DIABETIC STATUS		BIRTH DEFECT STATUS		
		1+	NONE	ROW SUMS
Nondiabetic	Obs.	31	754	785
	Exp.	34.5	750.5	
	z	−0.6	0.1	
	Sym.	.	.	
Prediabetic	Obs.	13	362	375
	Exp.	16.5	358.5	
	z	−0.9	0.2	
	Sym.	.	.	
Diabetic	Obs.	9	38	47
	Exp.	2.1	44.9	
	z	4.8	−1.0	
	Sym.	@	.	
Column sums		53	1154	1207

Degrees of freedom = 2
$X^2 = 25.511$
P-value < .001

The P-value is less than .001. We conclude at the 5% significance level that there is an association between diabetic status of the mother and the occurrence of birth defects in the child.

(b) The indication is of a significantly higher than expected frequency of children with birth defects among diabetic mothers. This suggests (but does not prove) that maternal diabetes is responsible for an increased incidence of birth defects in children.

13.7 (a)

AGE		ETHNICITY			
		ANGLO	HISPANIC	NATIVE AMERICAN	ROW SUM
20–29	Obs.	34	25	11	70
	Exp.	46.4	18.4	5.2	
	z	−1.8	1.5	2.6	
	Sym.	o	.	0	
30–39	Obs.	26	9	2	37
	Exp.	24.5	9.7	2.7	
	z	0.3	−0.2	−0.4	
	Sym.	.	.	.	
40–49	Obs.	23	6	1	30
	Exp.	19.9	7.9	2.2	
	z	0.7	−0.7	−0.8	
	Sym.	.	.	.	
50–59	Obs.	22	6	0	28
	Exp.	18.6	7.4	2.1	
	z	0.8	−0.5	−1.4	
	Sym.	.	.	.	
60+	Obs.	21	4	0	25
	Exp.	16.6	6.6	1.8	
	z	1.1	−1.0	−1.4	
	Sym.	.	.	.	
Column sums		126	50	14	190

Degrees of freedom = 8
$X^2 = 21.213$
P-value < .01

The P-value is smaller than .01, so it would be concluded at the 5% level that age and ethnicity are associated.

(b) The nature of the association is that there is a higher than expected incidence of suicide among young Native Americans (20–29 year age group).

13.10 (a) The P-value for the two-sided test is smaller than 0.000001. (The value for the high-school-educated minus college-educated test statistic is $z = -5.196$.) At the 5% significance level, the proportion of college-educated individuals who express a low need for further education exceeds that for high-school-educated individuals.

(b) With 95% confidence, the amount by which the proportion of college-educated who express a low need exceeds the proportion of high-school-educated expressing low need is between 9.2% and 18.1%.

13.13 The confidence coefficient is set at 98% for each interval in order to achieve joint confidence 95%. The confidence limits for the proportion of boys minus the proportion of girls counseled and released are $L = -.398$ and $U = -.144$. The confidence limits for the difference in proportions sent to court are $L = 0.0681$ and $U = 0.237$. Thus, with 95% confidence, between 15.9% and 39.8% more girls than boys are counseled and released, while between 6.8% and 23.7% more boys are referred to court.

13.15

CAUSE OF DEATH		ALTITUDE			
		LOW	MEDIUM	HIGH	ROW SUMS
SIDS	Obs.	27	38	12	77
	Exp.	25.9	41.2	9.9	
	z	0.2	−0.5	0.7	
	Sym.	.	.	.	
Other	Obs.	6601	10512	2511	19624
	Exp.	6602.1	10508.8	2513.1	
	z	−0.0	0.0	−0.0	
	Sym.	.	.	.	
Column sums		6628	10550	2523	19701

Degrees of freedom = 2
$X^2 = 0.767$
P-value $> .50$

The P-value for the test of no difference in SIDS death rates for the three altitude levels exceeds .5. At the 10% significance level, there is no evidence of a difference in SIDS death rates for the three altitudes.

13.17 (a)

		NEED LEVEL		
EDUCATIONAL LEVEL		LOW	HIGH	ROW SUMS
Less than high school	Obs.	198	192	390
	Exp.	204.1	185.9	
	z	−0.4	0.5	
	Sym.	.	.	
High-school diploma	Obs.	386	481	867
	Exp.	453.8	413.2	
	z	−3.2	3.3	
	Sym.	@	@	
Some college	Obs.	803	924	1727
	Exp.	904.0	823.0	
	z	−3.4	3.5	
	Sym.	@	@	
Bachelor's degree	Obs.	453	301	754
	Exp.	394.7	359.3	
	z	2.9	−3.1	
	Sym.	@	@	
Graduate degree	Obs.	383	126	509
	Exp.	266.4	242.6	
	z	7.1	−7.5	
	Sym.	@	@	
Column sums		2223	2024	4247

Degrees of freedom = 4
X^2 = 170.433
P-value < .001

The test of no difference in need level for the various education levels is highly significant (P < .001). In particular a difference in education need distributions would be concluded at the 5% significance level.

(b) The trend is that the high need (low need) proportions decrease (increase) with education level.

13.19

Goodness of fit table

PEA VARIETY	OBS. FREQ.	HYP. PROP.	EXP. FREQ.	z	SYMBOL
Round-yellow	315	0.563	312.7	0.127	.
Wrinkled-yellow	101	0.188	104.2	−0.318	.
Round-green	108	0.188	104.2	0.367	.
Wrinkled-green	32	0.063	34.7	−0.467	.

Sample size = 556
Degrees of freedom = 3
X^2 = 0.470
P-value > .50

The P-value for the test exceeds .5, so there is no evidence to support a difference between actual and theoretical proportions. In fact, the value of chi-square is so unusually small that Mendel's helpers have been suspected of doctoring the data in favor of his theory. Fortunately, the theory has been amply confirmed in more recent experiments.

13.21

CONDITION		TREATMENT		ROW SUMS
		NOXOLINE	SALINE	
Dead	Obs.	2	5	7
	Exp.	2.9	4.1	
	z	−0.5	0.4	
	Sym.	.	.	
Normal	Obs.	3	1	4
	Exp.	1.6	2.4	
	z	1.1	−0.9	
	Sym.	.	.	
Mildly spastic	Obs.	4	0	4
	Exp.	1.6	2.4	
	z	1.8	−1.5	
	Sym.	o	.	
Very spastic	Obs.	0	7	7
	Exp.	2.9	4.1	
	z	−1.7	1.4	
	Sym.	o	.	
Column sums		9	13	22

Degrees of freedom = 3
X^2 = 12.988
P-value < .005

The P-value for the test is smaller than .005, so the hypothesis of no differences in the two treatments would be rejected at the 5% level. The data suggest that noxoline helps control the degree to which the cats who survive, but do not recover, suffer from the injury. Among those getting noxoline, a higher proportion than expected are mildly spastic, and a lower than expected proportion are severely spastic.

13.23

POT RATING		SITE TYPE		ROW SUMS
		HIGH DENSITY	LOW DENSITY	
Low	Obs.	5	8	13
	Exp.	6.9	6.1	
	z	−0.7	0.8	
	Sym.	.	.	
Medium	Obs.	12	8	20
	Exp.	10.7	9.3	
	z	0.4	−0.4	
	Sym.	.	.	
High	Obs.	7	5	12
	Exp.	6.4	5.6	
	z	0.2	−0.3	
	Sym.	.	.	
Column sums		24	21	45

Degrees of freedom = 2
X^2 = 1.633
P-value > .25

The test of no difference in the pot rating distributions for different site types is insignificant at the 10% significance level (P > .25).

13.25

POSSESS B12		CLASSIFICATION			
		ALCOHOLIC-NONCIRRHOTIC	ALCOHOLIC-CIRRHOTIC	CONTROL	ROW SUMS
Yes	Obs.	3	14	15	32
	Exp.	8.3	9.7	14.0	
	z	−1.8	1.4	0.3	
	Sym.	o	.	.	
No	Obs.	20	13	24	57
	Exp.	14.7	17.3	25.0	
	z	1.4	−1.0	−0.2	
	Sym.	.	.	.	
Column sums		23	27	39	89

Degrees of freedom = 2
$X^2 = 8.313$
P-value < 0.25

The P-value is smaller than .025, so the hypothesis of no difference in B12 distributions for the three categories of individuals would be rejected at the 5% level. There is a mild indication that among alcoholic Anglos, fewer than expected men without cirrhosis of the liver possess B12. This leads to the speculation that B12 may be a marker for cirrhosis-prone Anglo alcoholics.

13.27

BRAND TYPE		NUMBER OF BRANDS			
		ONE	TWO	THREE+	ROW SUMS
Brand name	Obs.	6	17	4	27
	Exp.	5.5	13.2	8.3	
	z	0.2	1.1	−1.5	
	Sym.	.	.	.	
Generic	Obs.	2	2	8	12
	Exp.	2.5	5.8	3.7	
	z	−0.3	−1.6	2.2	
	Sym.	.	.	0	
Column sums		8	19	12	39

Degrees of freedom = 2
$X^2 = 11.039$
P-value < .005

The conclusion is that there is a difference in the type of drug furnished for different numbers of brands on hand ($P < .01$). The data suggest that when three or more brands are on hand, a generic is furnished more often than if fewer brands are available.

13.29

AGE		SPEED (MPH)			
		< 11	11–15	> 15	ROW SUMS
≤17	Obs.	14	9	0	23
	Exp.	10.6	8.8	3.6	
	z	1.0	0.1	−1.9	
	Sym.	·	·	o	
18–29	Obs.	11	11	5	27
	Exp.	12.4	10.3	4.3	
	z	−0.4	0.2	0.4	
	Sym.	·	·	·	
≥30	Obs.	10	9	7	26
	Exp.	12.0	9.9	4.1	
	z	−0.6	−0.3	1.4	
	Sym.	·	·	·	
Column sums		35	29	12	76

Degrees of freedom = 4
$X^2 = 7.526$
P-value $> .10$

The P-value exceeds .10, suggesting that there is insufficient evidence to conclude a difference in speeds for the three age groups.

13.31

BLOOD GROUP		CAUSE OF DEATH		
		SIDS	CONTROL	ROW SUMS
O	Obs.	69	55	124
	Exp.	62.0	62.0	
	z	0.9	−0.9	
	Sym.	·	·	
A	Obs.	40	71	111
	Exp.	55.5	55.5	
	z	−2.1	2.1	
	Sym.	0	0	
B	Obs.	17	7	24
	Exp.	12.0	12.0	
	z	1.4	−1.4	
	Sym.	·	·	
AB	Obs.	9	2	11
	Exp.	5.5	5.5	
	z	1.5	−1.5	
	Sym.	·	·	
Column sums		135	135	270

Degrees of freedom = 3
$X^2 = 18.860$
P-value $< .001$

There is strong evidence ($P < .001$) of a difference in blood group distributions for children dying of SIDS and those dying of other causes. Fewer SIDS victims than expected have type A blood. The AB categories approach significance, yielding more SIDS victims than expected. However, this category is sufficiently rare in the population that this information would not be very useful as a predictor of which babies are at high risk for SIDS.

Chapter 14

14.1

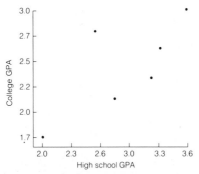

The linearity assumption appears to have some support. One out-of-line observation is seen in the diagram, and a suggestion of curvature can be seen in the regression for the remaining points.

14.2 A summary of the linear regression calculations follows:

$n = 6$ $S_{xx} = 1.648$ $\hat{\beta}_0 = 0.632$
$\bar{x} = 2.917$ $S_{xy} = 1.008$ $\hat{\beta}_1 = 0.612$
$\bar{y} = 2.417$ $S_{yy} = 1.148$ $s = 0.365$

The average increase in college GPA for each unit increase in high school GPA is .612 units.

14.5 $r^2 = 0.031$. Only about 3.1% of the variation in the y variable can be explained by its linear regression on the x variable. A look at the following scatter diagram shows why.

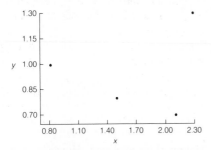

One "wild point" appears to "confuse" the nearly linear relationship suggested by the other three points.

14.7 The coefficient of variation is $r^2 = .537$. This suggests a moderate linear relationship between the two variables. The residuals and residual plot are as follows.

X	OBS. Y	PRED. Y	RESIDUAL
2.000	1.700	1.856	−0.156
2.600	2.800	2.223	0.577
2.800	2.100	2.345	−0.245
3.200	2.300	2.590	−0.290
3.300	2.600	2.651	−0.051
3.600	3.000	2.835	0.165

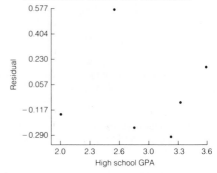

A single observation (observation 2) is seen to dominate the residual plot and is the major cause of the poor linear fit.

14.11 **(a)** The basic statistics for the sulfur versus precipitation least-squares line are as follows:

$n = 12$ $\bar{x} = 42.083$
$\Sigma x = 505$ $\bar{y} = 52.583$
$\Sigma x^2 = 26,425$ $S_{xx} = 5,172.917$
$\Sigma y = 631$ $S_{xy} = 6,714.417$
$\Sigma y^2 = 45,383$ $S_{yy} = 12,202.917$
$\Sigma xy = 33,269$

$$\hat{\beta}_0 = -2.041$$
$$\hat{\beta}_1 = 1.298$$
$$s = 18.675$$
$$r^2 = .714$$

Both the scatter diagram and the value $r^2 = .714$ indicate a reasonable degree of linearity for the sulfur versus precipitation data. The deduction that sulfur is precipitated by rainfall comes from a knowledge of the physical system involved. However, this deduction could well have been made after first noting the association between the amount of sulfur found in collectors and the amount of precipitation. The association itself does not prove the deduction to be true.

MONTH	PRECIPITATION	SULFUR	PREDICTED	RESIDUAL
1	35	55	43.4	11.6
2	25	30	30.4	-.4
3	12	25	13.5	11.5
4	36	43	44.7	-1.7
5	81	135	103.1	31.9
6	19	38	22.6	15.4
7	55	63	69.3	-6.3
8	63	93	79.7	13.3
9	69	64	87.5	-23.5
10	23	17	27.8	-10.8
11	52	34	65.5	-31.5
12	35	34	43.4	-9.4

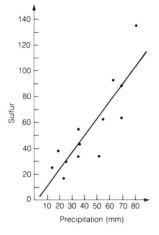

(b) The table of residuals from the least-squares lines is shown above. The resid-

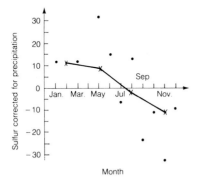

uals in the last column are plotted against month (1 = Jan., 2 = Feb., and so forth) on the following graph. The result of an exploratory regression is also shown.

The pattern of high sulfur levels in winter and spring, with decreasing levels over the summer, follows closely the needs for heat and light in Sweden. This result suggests that additional coal is being burned to furnish these commodities during the cold months of the year. It is possible that coal is being burned in homes as well. It is also possible that prevailing wind patterns change with the seasons, causing variations in the amount of sulfur seen by the recording station.

14.13 (a) With X = number of cigarettes smoked by the mother and Y = number smoked by the father, the summary data are as follows:

$$n = 14 \qquad \bar{x} = 12.570$$
$$\Sigma x = 176 \qquad \bar{y} = 17.500$$
$$\Sigma x^2 = 3168 \qquad S_{xx} = 955.429$$
$$\Sigma xy = 3384 \qquad S_{xy} = 304.000$$
$$\Sigma y = 245 \qquad S_{yy} = 539.500$$
$$\Sigma y^2 = 4827$$

$$\hat{\beta}_0 = 13.500$$
$$\hat{\beta}_1 = .318$$
$$s = 6.074$$
$$r^2 = .179$$

The scatter diagram and least-squares line are shown next. Note that duplicated points are indicated by 2. The scatter diagram suggests a curvilinear relationship. In any event, the linear fit is rather poor.

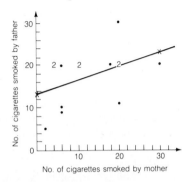

(b) The linear association, as measured by $r^2 = .179$, is weak.

14.15 (a) The computations for the least-squares lines, based on the given summary statistics, are as follows:

For females ($n = 15$):

$\bar{x} = 2.293$ $\hat{\beta}_0 = 2.578$
$\bar{y} = 8.760$ $\hat{\beta}_1 = 2.696$
$S_{xx} = 1.189$ $s = 1.139$
$S_{xy} = 3.206$ $r^2 = .339$
$S_{yy} = 25.496$

For males ($n = 16$):

$\bar{x} = 2.788$ $\hat{\beta}_0 = -1.155$
$\bar{y} = 10.831$ $\hat{\beta}_1 = 4.299$
$S_{xx} = 1.518$ $s = 1.588$
$S_{xy} = 6.526$ $r^2 = .443$
$S_{yy} = 63.374$

The combined scatter diagrams and least-squares lines are given in the graph that follows.

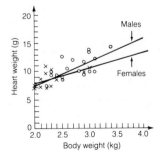

(b) The scatter diagrams show a moderate degree of linearity. Because of the somewhat wider range of body weights with roughly the same dispersion in the y coordinate, the male data show somewhat more linearity than do the female data. This result is supported by the values of r^2. We might expect better prediction for males on this basis. However, we will need the material in chapter 15 to properly evaluate the quality of linear predictability.

(c) The predicted heart weights for 2.3-kilogram cats are 8.78 grams for females and 8.73 grams for males.

14.17 (a) The summary data for the least-squares line are as follows:

$\bar{x} = 46.109$ $\hat{\beta}_0 = 231.227$
$\bar{y} = 123.566$ $\hat{\beta}_1 = -2.3395$
$S_{xx} = 13,494.98$ $s = 38.707$
$S_{xy} = -31,571.57$ $r^2 = .4915$
$S_{yy} = 150,271.02$

(b) The least-squares line and exploratory regression curve are plotted in the following graph.

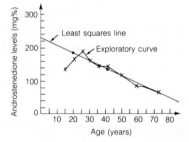

It appears that the least-squares line will provide good estimates of the true regression for women 30 years of age and older.

(c) The bias will be greatest in the age range from 15 to 30 years of age. Androstenedione levels will be overestimated (positive bias) from about 15 to 23 years of age and underestimated (negative bias) from about 24 to 29 years of age.

14.19 The accompanying scatter diagram shows virtually no linear relationship (or any other kind) between heating and cooling. The following summary supports this conclusion. An outlier in the heating variable is seen in the scatter diagram. Note that the regression line is very close to this point. This is no accident, as is seen in chapter 15.

$$n = 10 \qquad \bar{x} = 3.030$$
$$\Sigma x = 30.300 \qquad \bar{y} = 8.452$$
$$\Sigma x^2 = 110.443 \qquad S_{xx} = 18.634$$
$$\Sigma xy = 255.212 \qquad S_{xy} = -.884$$
$$\Sigma y = 84.520 \qquad S_{yy} = 53.349$$
$$\Sigma y^2 = 767.712$$

$$\hat{\beta}_0 = 8.596$$
$$\hat{\beta}_1 = -.047$$
$$s = 2.581$$
$$r^2 = .0008$$

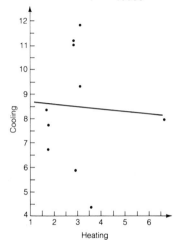

Chapter 15

15.1 With 95% confidence, the true slope is between −0.177 and 1.400.

15.3 Individual and joint 95% confidence limits at the given x values are given in the following table.

X	PREDICTED Y	INDIVIDUAL L	CONFIDENCE LIMITS U	JOINT CONFIDENCE LIMITS L	U
2.000	1.856	1.023	2.688	0.739	2.973
2.600	2.223	1.740	2.706	1.575	2.871
2.800	2.345	1.922	2.769	1.777	2.913
3.200	2.590	2.120	3.060	1.960	3.220
3.300	2.651	2.139	3.163	1.964	3.338
3.600	2.835	2.156	3.514	1.923	3.746

The graph of these intervals is as follows.

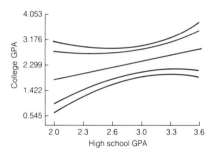

15.5 The 95% prediction interval for college GPA given a high school GPA of 2.3 has limits 0.843 and 3.236.

15.9 **(a)** With 95% confidence, the mean N level for stimulated runners with an E level of 200 is between 2884 and 3211.

(b) The limits for a 95% prediction interval for N levels at an E level of 200 are 2516 and 3579.

(c)

PREDICTION LINE AND LOWER AND UPPER 95% PREDICTION LIMITS FOR N-LEVELS VERSUS E-LEVELS

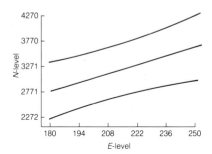

(Note: A residual plot shows one rather badly out-of-line observation, and the inference results would be viewed with some suspicion.)

15.11 **(a)** The value of the t statistic for testing zero slope is $t = 1.616$, yielding a P-value > .10.

(b) 95% confidence limits for the slope are −.1100 and .746.

15.13 **(a)** With 95% confidence, a 2.5-kg female cat would have heart weight between 6.734 and 11.901 grams. The limits for a 2.5-kg male cat are 5.994 and 13.195 grams.

(b) The joint 95% confidence limits are given in the following table.

X (BODY WEIGHT)	PREDICTED Y (HEART WEIGHT)	JOINT 95% CONFIDENCE LIMITS L	U
2.000	7.444	4.464	10.425
2.500	9.595	8.110	11.080
3.000	11.745	10.426	13.064

15.15 (a) The ordered-height stem and leaf plot with subdivisions indicated is as follows:

```
5*  4
5   7
6*  02333|444
6   555666|77789
7*  0
```

Height medians:

$$\text{group 1: } x_1 = 62$$
$$\text{group 3: } x_2 = 67.5$$

The weight stem and leaf plots for the first and third groups are as follows:

HEIGHTS 54–63 (7)	**HEIGHTS 67–70 (6)**
8 | 2	12 | 50
9 |	13 | 33
10 | 5	14 |
11 | 00	15 |
12 | 250	16 |
	17 | 0
	18 | 0

Weight medians:

$$\text{group 1: } y_1 = 110$$
$$\text{group 3: } y_2 = 133$$

Then

$$\beta_0^* = \frac{67.5(110) - 62(133)}{67.5 - 62} = -149.273$$

$$\beta_1^* = \frac{133 - 110}{67.5 - 62} = 4.182$$

(b) The eye-fit line is $y = -97.67 + 3.37x$ and the median lines is $y = -149.27 + 4.18x$. The agreement between the two lines seems to be reasonably good over the range of the data that excludes the two height outliers (54 and 57 inches). The median line appears to be a slightly better fit in that it passes more accurately through the central scatter of points. The maximum difference in weights predicted by the two lines over

the range 55 to 70 inches occurs at 55 inches and is $87.7 - 80.7 = 7.0$ pounds.

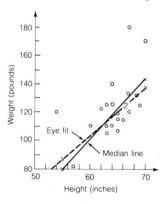

15.17 (a) The scatter diagram suggests the presence of an outlier in the distance coordinate. This suggestion is verified by the box plot.

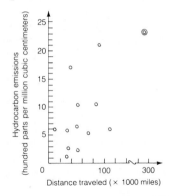

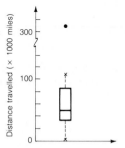

(b) For the full data set, $r^2 = .437$ and $t_0 = 2.922$. With $v = 11$, the P-value bounds are $.01 < P < .02$.

For the data set with the point $(312, 2352)$ deleted, $r^2 = .113$ and $t_0 = 1.130$. Now $v = 10$ and $.2 < P < .5$.

(c) Apparently the indication of linearity

shown by the entire data set was due almost entirely to the high leverage point (312, 2352). When this point is removed, the indication of linearity is insignificant. Moreover, the scatter diagram does not appear to indicate any form of nonlinear regression. Apparently factors such as automobile type, quality of maintenance, and so forth are more important than distance driven insofar as hydrocarbon emissions are concerned.

15.19 (a) The limits of a 95% confidence interval for the slope are .6447 and 1.0983.

(b) The limits for a 95% confidence interval for the intercept are −.3041 and .3632.

(c) Joint 95% confidence limits (using individual confidence coefficient 98%) are

$$.5951 \le \beta_1 \le 1.1479$$

and

$$-.3771 \le \beta_0 \le .4362$$

(d) We would accept the hypothesis that $\beta_0 = 0$ and $\beta_1 = 1$ at the 5% significance level, since these parameter values are contained in the respective intervals. The conclusion would be that the measure is suitable for measuring electrical consumption.

Chapter 16

16.1 (a) The ANOVA table is as follows.

ANOVA

SOURCE	SS	df	MS	F
Model	12.669	3	4.223	
Error	5.161	8	0.645	6.546
Total	17.830	11		
Model R-square = .711			$.01 < P < .05$	

(continued top of next column)

The *P*-value is between .01 and .05, indicating a significant relationship (at the 5% level) between dependent and independent variables.

(b) Only the auditory skill predictor variable is significant at the 5% level. 95% confidence limits are $L = 0.112$ and $U = 2.299$.

16.4 (a) Box plots of the variables show a minor outlier in the verbal skills variable, with modest left-skewness in this and the visual skills variable. A power transformation with index 1.5 corrects the skewness, but has no essential influence on the regression.

(b) No systematic structure is seen in the residual plot.

(c) The residual plot shows two out-of-line points corresponding to observations 1 and 11, and the partial regression plot for the verbal skills variable indicates that observation 7 is a high-influence point for this variable. The removal of observation 7 changes the slope but not the insignificance of verbal skill. Removing observations 1 and 11 as well decreases the overall significance of the model as well as the significance of the auditory skill variable. However, since these observations are not high-influence points for this variable, the effect is primarily due to the small sample size.

16.7 (a) The F value for the screening test is $F = .428$, which is insignificant at any interesting level. The conjecture that collinearity may be playing a role is supported by the following variable summary (below left).

Variable summary

VARIABLE	REG. COEF.	SE	t	R-SQUARE	INFL. FAC.
Intercept	109.76307				
Age	−7.61443	19.37977	−.393	.999773	66.372
Age^2	.25290	.60702	.417	.999945	135.333
Age^3	−.00278	.00616	−.451	.999796	70.066

(b) and (c) Recentering the age variable at its median of 30 years (and removing observations 7 and 27, which show up as outliers in the residual plot), though

it improves the collinearity problem, does not significantly improve the fit of the cubic polynomial. The resulting variable summary is the following.

VARIABLE	REG. COEF.	SE	t	R-SQUARE	INFL. FAC.
Intercept	34.281282				
Age $-$ 30	$-.64040$	.79942	$-.801$	.845982	2.548
$(\text{Age} - 30)^2$	$-.00610$	.06556	$-.093$	.512427	1.432
$(\text{Age} - 30)^3$	$-.00623$	.00885	.704	.887379	2.980

It follows that the basic problem is not collinearity but simply the poor fit of the cubic polynomial model.

16.8 The 25 observations for the two starch types are arranged in a single variable, the 12 observations for rice starch coming first, followed by the 13 canna starch observations. Then an indicator variable with 1's in the first 12 positions is constructed and used as the predictor variable in a regression analysis. The following ANOVA is obtained.

ANOVA table

SOURCE	SS	df	MS	F
Model	409293	1	409293	
Error	263007	23	11435.1	35.793
Total	672300	24		
Model R-square = .609				

The two-sample t statistic has a value of 5.983. The square of this number is seen to be the value of F in the table. The critical value (at any significance level) of F is the square of the two-sided critical value for the t test. Thus, the F test is equivalent to the two-sided t test.

16.11

ANOVA table

SOURCE	SS	df	MS	F
Model	2214.62	3	738.208	
Error	5940.81	10	594.081	1.243
Total	8155.43	13		
Model R-square = .27155195			$P > .25$	

The P-value for the screening test exceeds .25.

16.13 (a) Except for a minor outlier in the cholesterol variable, the box plots show the shapes of the distributions of all variables to be rather unremarkable.

(b) The accompanying residual plot suggests that observations 4, 8, and 12 have high influence on the regression analysis.

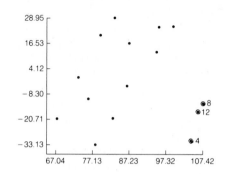

RESIDUALS VERSUS PREDICTED VALUES

(c) After removing the influence points found in the residual plot, a new analysis yields a screening test with P-value less than .05, confirming that the outliers have desensitized the analysis. Observation 13 appears out of line in the new analysis, and its removal further improves the model fit. This suggests that the model should be tried on a much larger data collection before being abandoned. It is possible that the poor fit is due to artifacts in the data whose influence might not be so great in a larger data set. This is the kind of data set for which a robust regression method would be especially useful.

Chapter 17

17.1

Effects table

	TREATMENT		
BLOCK	1	2	
1	-4.417	4.417	-2.917
2	5.583	-5.583	4.083
3	-1.917	1.917	$-.417$
4	-2.417	2.417	2.083
5	2.083	-2.083	-6.417
6	1.083	-1.083	3.583
	.417	$-.417$	11.917

ANOVA table

SOURCE	SS	df	MS	F
Treatment	2.0833	1	2.0833	.079
Blocks	167.4167	5	33.4833	
Error	131.4167	5	26.2833	
Total	300.9167	11		

Treatment effects are insignificant at the 5% level (P > .25).

17.3 The critical value for 95% confidence is $t = 2.571$. With 95% confidence, the difference in treatment effects (treatment 1 minus treatment 2) is between -6.777 and 8.443.

17.5 (a)

Effects table

FACTOR 1	FACTOR 2			FACTOR 1 MAIN
	1	2	3	
1	.952	2.974	-3.926	-.719
2	-.104	-3.681	3.785	-1.263
3	-.848	.707	.141	1.981
Factor 2 Main	1.559	1.437	-2.996	12.141

(b)

ANOVA table

SOURCE	SS	df	MS	F
Factor 2	121.2674	2	60.6337	44.852
Factor 1	54.3385	2	27.1693	20.098
Interaction	162.8860	4	40.7215	30.123
Error	24.3333	18	1.3519	
Total	362.8252	26		

(c)

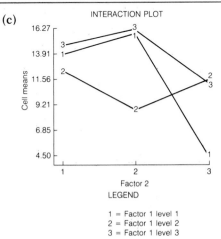

INTERACTION PLOT

LEGEND

1 = Factor 1 level 1
2 = Factor 1 level 2
3 = Factor 1 level 3

The trend for factor 1 levels 1 and 3 is an increase from level 1 to level 2 of factor 2, then a drop to level 3, but the reverse is observed for level 2 (first a decrease, then an increase). This difference, relative to the small dispersion of values within cells as measured by the MSE, accounts for the strong interaction effect. An overall decrease in the averages over factor 1 levels at each level of factor 2 accounts for the strong factor 2 main effect, while the differences in average levels for the three curves accounts for the strong factor 1 main effect.

17.7 (a) 95% confidence limits for $\mu_{11} - \mu_{12}$ are $L = 3.111$ and $U = 14.889$.

  (b) 95% confidence limits for $.5\mu_{11} + .5\mu_{12} - .5\mu_{21} - .5\mu_{22}$ are $L = -30.672$ and $U = -23.328$. With 95% confidence, the main effect for level 2 of factor 1 exceeds that for level 1 by an amount between 23.328 and 30.672.

17.9 (a)

ANOVA table

SOURCE	SS	df	MS	F
Test	0.0942	3	0.0314	3.948
Judge	0.0879	5	0.0176	
Error	0.1193	15	0.0080	
Total	0.3014	23		.01 < P-value < .05

(b)

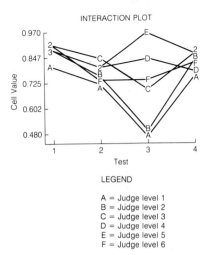

INTERACTION PLOT

LEGEND

A = Judge level 1
B = Judge level 2
C = Judge level 3
D = Judge level 4
E = Judge level 5
F = Judge level 6

It appears that test 1 has the highest average reliability, closely followed by tests 2 and 4. The variability among judges is small for these tests. While test 3 has the highest reliability for one judge, the variation among judges would make it undesirable to use.

17.11 (c) The P-values for all factors are less than .001.

(d) The interaction is due primarily to varieties 1, 3, and 5. Varieties 1 and 3 have relatively low means for stations 1, 2, and 4, but have the highest means for station 3. Variety 5 has a mean much higher than those of the other varieties for station 1, but then is only second or third best for the other stations.

No variety performs best at all stations. However, the best overall performance (ignoring the high value for variety 5 at station 1) appears to be given by variety 2.

(e) The variety with the largest mean is #5 (due primarily to the excessively large value at station 1.) Varieties 3 and 4 appear to be tied for worst performance. The best average performance was at station 4 and the worst at station 1. The overall average volume for a loaf of bread was 646.4 cubic centimeters.

17.13 (a)

Cell sample sizes, means, and standard deviations

		NUMBER OF BRANDS		
BRAND TYPE		1	2	3 OR MORE
Brand name	n	6	15	4
	$\bar{x}$	614.000	551.267	520.000
	s	191.478	160.633	75.609
Generic	n	2	2	8
	$\bar{x}$	368.50	377.50	356.375
	s	37.477	24.749	88.342

(b)

Effects table

	NUMBER OF BRANDS			
BRAND TYPE	1	2	3 OR MORE	BRAND TYPE MAIN
Brand name	25.601	−10.265	−15.336	97.149
Generic	−25.601	10.265	15.336	−97.149
Number main	26.643	−.224	−26.419	464.607

(c) $n_H = 3.73$.

(d)

ANOVA table

SOURCE	SS	df	MS	F
Number	10504.4610	2	5252.2305	.263
Brand type	211251.4600	1	211251.4600	10.591
Interaction	7431.3205	2	3715.6603	.186
Error	618357.8200	31	19947.0260	
Total	847545.0600	36		

Only the brand type main effect is significant ($P < .01$). The effects table suggests that the average price for generic drug prescriptions is smaller than that for brand-name prescriptions.

(e)

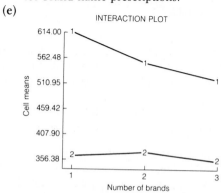

INTERACTION PLOT

LEGEND: 1 = Brand name
          2 = Generic

The large difference in average levels for the two curves represents the brand-type main effect. The "flatness" of the two curves (relative to the within-cell variation) leads to the insignificant interaction and number-of-brands main effect.

17.15 The brand type (brand-name minus generic) main effect difference is $.333\mu_{11} + .333\mu_{12} + .333\mu_{13} - .333\mu_{21} - .333\mu_{22} - .333\mu_{23}$. The 95% confidence limits for this contrast are $L = 72.309$ and $U = 315.897$. Thus, with 95% confidence we conclude that the brand-name prescriptions cost between $0.79 and $3.16 more than prescriptions filled with generic hydrocortisone, on the average.

Index